4th International Meeting for Researchers in Materials and Plasma Technology 2017

AA006395

Journal of Physics: Conference Series Volume 935

Santa Marta, Colombia
23 - 26 May 2017

Editor:

Ely Dannier V. Nino

ISBN: 978-1-5108-5516-8
ISSN: 1742-6588

Printed from e-media with permission by:

Curran Associates, Inc.
57 Morehouse Lane
Red Hook, NY 12571

Some format issues inherent in the e-media version may also appear in this print version.

Printed by Curran Associates, Inc. (2018)

For permission requests, please contact the Institute of Physics
at the address below.

Institute of Physics
Dirac House, Temple Back
Bristol BS1 6BE UK

Phone: 44 1 17 929 7481
Fax: 44 1 17 920 0979

techtracking@iop.org

Additional copies of this publication are available from:

Curran Associates, Inc.
57 Morehouse Lane
Red Hook, NY 12571 USA
Phone: 845-758-0400
Fax: 845-758-2633
Email: curran@proceedings.com
Web: www.proceedings.com

Table of contents

Volume 935

4th International Meeting for Researchers in Materials and Plasma Technology (4th IMRMPT)

23–26 May 2017, Santa Marta, Colombia

Accepted papers received: 11 December 2017
Published online: 29 December 2017

Preface

Papers

IOP Conf. Series: Journal of Physics: Conf. Series **935** (2017) 011001 doi:10.1088/1742-6596/935/1/011001

Preface

These proceedings present the written contributions of the participants of the 4th International Meeting for Researchers in Materials and Plasma Technology (4th IMRMPT) which was held from 23 to May 26, 2017, at the Santa Marta city, Colombia, organized by Foundation Of Researchers In Science and Technology Of Materials (FORISTOM). This was the fourth version of biennial meetings that began in 2011.

The proceedings consist of 75 contributions that were presented as plenary talks at the conference. The abstracts of all participants contributions were published in the Abstract Book with ISSN 2422-3824. The website of the conference is available at http://foristom.org:8080/4imrmpt.

The four-day scientific program of the 4th IMRMPT consisted of 7 Magisterial Conferences, 64 Oral Presentations, 112 Poster Presentations, 3 Courses and 1 Discussion Panel with the participation of undergraduate and graduate students, professors, researchers and entrepreneurs from Colombia, Russia, Spain, Unite Kingdom, Mexico, Uruguay, Brazil, Venezuela, among others. Moreover, the objective of IMRMPT was to bring together national and international researchers in order to establish a network of scientific cooperation with a global impact in the area of the science and the technology of materials; to promote the exchange of creative ideas and the effective transfer of scientific knowledge, from fundamental research to innovation applied to industrial solutions and to advance in the development of new research allowing to increase the lifetime of the materials used in the industry by means of efficient transference of the knowledge between sectors academia and industry.

All papers in these Proceedings refer to one of the following topics covered in the 4th IMRMPT: Additive Manufacturing, Biomedical Coatings, Characterization of Materials, Modelling, Simulation and Diagnostics, New Materials, Plasma Applications and Technologies, Renewable Energy, Surface Hardening Processes, Structural Integrity, Surface Modification (Ion Implantation, Ion Nitriding, PVD, CVD), Thin Films and Nanomaterials and Wear and Corrosion/Oxidation. The editor hopes that those interested in the area of the science and technology of materials and plasma, can to enjoy this reading, that reflects a wide variety of current issues.

On behalf of the organizing committee of the 4th IMRMPT, we are extremely thankful to all authors and participants for providing their valuable contributions for these Proceedings as well as the reviewers for their constructive recommendations and criticism aiding to improve the presented articles. Besides, especially we appreciate the great support provided by the sponsors and partners, for making possible this new version international meeting of researchers, which allow continuing with the dream of strengthen and contributing to Research, Innovation, and technological development in the present and future of Colombia.

Finally, it very important noted that the event organized by FORISTOM foundation, is a very significant contribution to the national and international scientific community, achieving the universal interaction of the different research groups of the sector academia and business sector of Colombia with other countries.

The Editor

Ely Dannier V. Niño

IOP Conf. Series: Journal of Physics: Conf. Series **935** (2017) 011001 doi:10.1088/1742-6596/935/1/011001

Organizing Committee
Harold Paredes Gutiérrez (Chairman)
Alejandro David Martínez Amariz
Héctor Jaime Dulce Moreno
José José Barba Ortega
Alfonso Santos Jaimes
Ely Dannier V. Niño

National Scientific Committee
Dr. Héctor Jaime Dulce Moreno
Dr. José José Barba Ortega
Dra. Miryam Rincón Joya
Dr. Hugo Armando Estupiñan Durán
Dr. Alejandro David Martínez Amariz
Dr. Fredy Antonio Rodríguez Prada
Dr. Henry Riascos Landázuri
Dra. Diana Padilla Rueda
Dr. Harold Paredes Gutiérrez
Dr. Darío Yesid Peña Ballesteros
Dr. Valeriy D. Dugar-Zhabon
Dr. Eduardo Alberto Orozco Ospino
Dr. William Gutiérrez Niño
Dr. Carlos Leonardo Beltrán Ríos
Dr. David Alejandro Miranda Mercado
Dr. Rafael Cabanzo Hernández
Dra. Claudia Paulina González Cuervo
Dr. (c) Jesús David González Acosta
Dr. (c) Fredy Humberto Vera Rivera
Dr. (c) Boris Raniero Pérez Gutiérrez
Dr. (c) Fredy Fabián Parada Becerra
MSc. Hernán Alfonso Garnica
MSc. Fernando Durán Flórez
MSc. Alfonso Santos Jaimes

International Scientific Committee
Dr. Petr Tsygankov - Russia
Dr. Andrés Díaz Lantada – Spain
Dr. Juan Manuel Muñoz Guijosa – Spain
Dr. Iván Fernández Martínez - Spain
Dr. Eduardo Méndez Morales - Uruguay
Dr. Mariana Pereyra - Uruguay
Dr. Elvis Oswaldo López Meza - Brazil
Dr. Humberto José Jiménez - Venezuela
Dra. Linda Elcida Gil de Fuentes - Venezuela
Dr. (c) Jaime Andrés Pérez Taborda - Spain
Dr. (c) Manuel José Andrade Romero - Brazil
Dra. (c) Angela Lorena Cardenas Manosalva - Brazil
Dra. (c) Mónica Arenas - Brazil
Dra. (c) Sugheis Liscano Durán – Venezuela

IOP Conf. Series: Journal of Physics: Conf. Series **935** (2017) 011001 doi:10.1088/1742-6596/935/1/011001

Sponsor and Partners

- ✓ Foundation of Researcher in Science and Technology of Materials
- ✓ Grupo de Investigación y Desarrollo de Ingeniería del Software
- ✓ Física y Tecnología del Plasma y Corrosión
- ✓ Física Computacional en Materia Condensada
- ✓ Grupo de Investigación en Materiales
- ✓ Universidad Nacional de Colombia
- ✓ Centro Colombiano de Tecnología
- ✓ Universidad Pontificia Bolivariana
- ✓ Grupo de Física Mesoscópica
- ✓ Universidad de Santander
- ✓ Canaán Multiservicios
- ✓ S & S Ingnieria S.A.S
- ✓ Nanovea
- ✓ Avianca
- ✓ Icetex

IOP Conf. Series: Journal of Physics: Conf. Series **935** (2017) 011001 doi:10.1088/1742-6596/935/1/011001

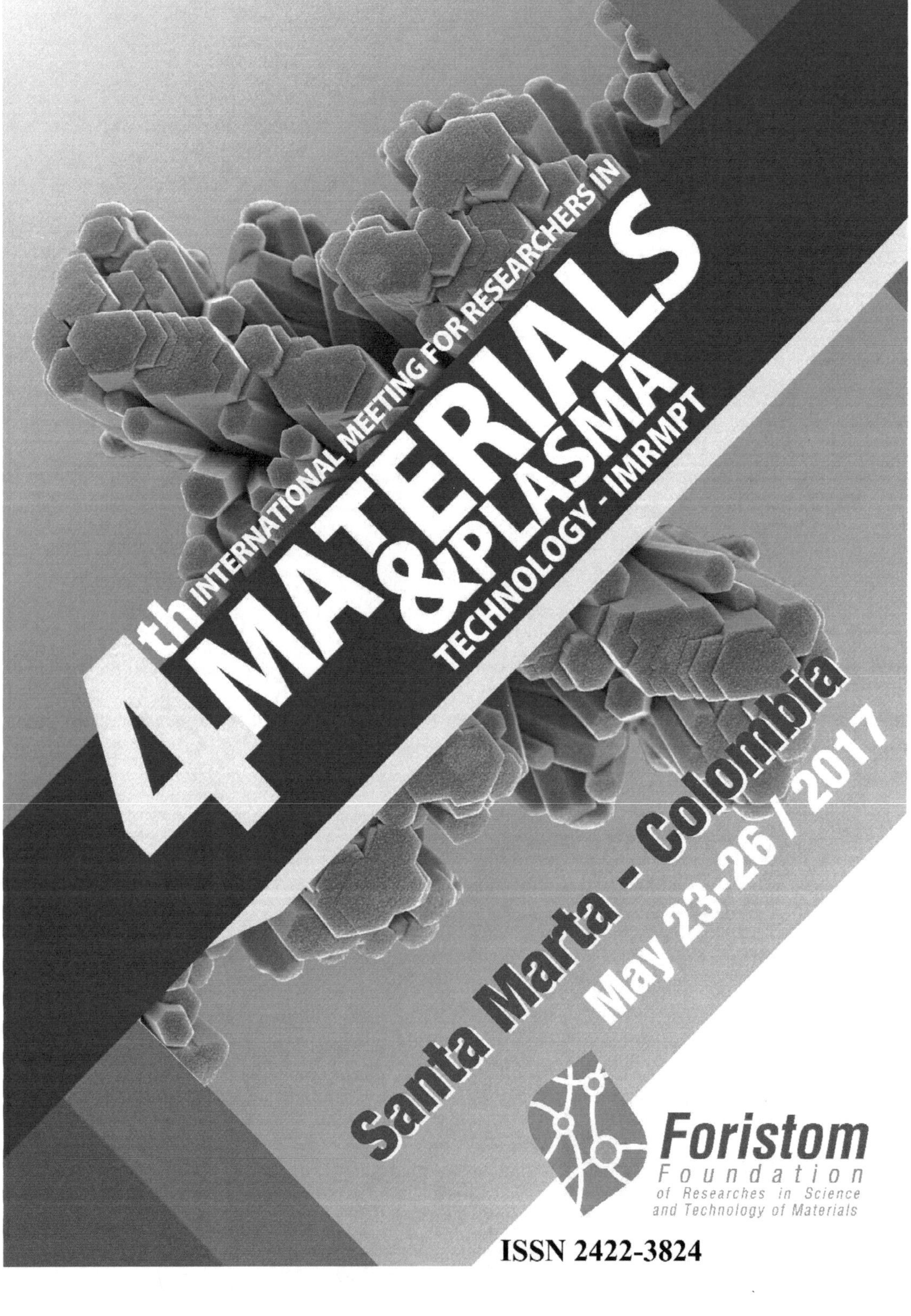

ISSN 2422-3824

IOP Conf. Series: Journal of Physics: Conf. Series **935** (2017) 011002 doi:10.1088/1742-6596/935/1/011002

Peer review statement

All papers published in this volume of *Journal of Physics: Conference Series* have been peer reviewed through processes administered by the proceedings Editors. Reviews were conducted by expert referees to the professional and scientific standards expected of a proceedings journal published by IOP Publishing.

IOP Conf. Series: Journal of Physics: Conf. Series **935** (2017) 012001 doi:10.1088/1742-6596/935/1/012001

A theoretical calculation of the electronic and magnetic behaviour of C-doped zincblende AlN

Miguel J Espitia R[1], J F Murillo G[2] and C Ortega López[2]
[1] Universidad Distrital Francisco José de Caldas, Bogotá, Colombia
[2] Universidad de Córdoba, Montería, Colombia

E-mail: mespitiar@udistrital.edu.co

Abstract. Based on structural spin-polarized density functional theory, the electronic and magnetic properties of carbon-doped zincblende AlN were investigated. The calculations were carried out by means of the pseudopotential method employed exactly as implemented in Quantum Espresso code. For 0.625% C-doped AlN, we found a ferromagnetic and half-metallic behaviour. Per supercell, the $AlC_{0.0625}N_{0.9375}$ has a net magnetic moment of $1.0\mu_{\beta}$. The main contribution to the magnetic moment comes from the 2p-C orbitals, with a lesser contribution of the 2p-N and 2p-Al orbitals. Our results suggest that C-doped AlN has potential applications in semiconductor spintronic devices.

1. Introduction

Aluminium nitride (AlN) is one of the compounds of group-III nitrides characterized by a wide band gap, low dielectric constant, high thermal conductivity, and large bulk modulus [1–4]. Due to these properties, currently there has been more interest in AlN, owing to its wide application in opto-electronics areas such as optical detectors, semiconductor lasers, and short-wavelength light-emitting diodes, as well as in high-temperature, high-power, and high-frequency devices [5–8]. AlN normally crystallizes in the wurtzite structure [9]. However, experimental studies of AlN epitaxially grown on most substrates yield the stable zincblende phase. This shows that AlN can also crystallize in the zincblende structure [10,11]. On the other hand, many theoretical and experimental studies have shown that AlN-doped transition metals are good candidates for use as Diluted Magnetic Semiconductors (DMS), with potentials applications in the rising field of spintronics [12–19]. Nevertheless, to make these applications a reality the compounds must exhibit ferromagnetism at room temperature. Recently, many researchers have reported high-temperature ferromagnetism in several types of Transition-Metal (TM)-doped semiconducting oxides and nitrides [20–22]. However, several investigators have proven that magnetic transition metal dopants-doped DMS segregate to form ferromagnetic clusters, secondary phases, or precipitates [23–25]. To avoid this problem, many researchers are investigating the effect of nonmagnetic ion doping on semiconductors such as C and N in order to obtain high-temperature ferromagnetic semiconductors [26–28].

In the present paper, we carry out first-principles total energy calculations in order to investigate the effects of the incorporation de C in the zincblende-AlN volume on the structural, electronic, and magnetic properties of the $AlC_{0.0625}N_{0.9375}$ compound.

2. Computational method

We used the pseudopotential method [29,30] within the framework of density functional theory [31] as implemented in the Quantum ESPRESSO computational code [31]. To take account of the correlation and exchange effects of the electrons, we used the generalized gradient approximation (GGA) of Perdew, Burke, and Ernzerhof (PBE) [32]. A kinetic energy cutoff of 40Ry was used for the expansion of the wave function in the plane wave. Brillouin zone integrations were performed with the special k-point method over a 6×6×4 Monkhorst-Pack mesh [33] for each unit cell. All calculations were carried out with spin polarization. To calculate the structural, electronic, and magnetic properties of pure AlN, we employed a supercell with 32 atoms, which corresponds to a 2a×2b×1c zincblende supercell (see Figure 1). For the doping of $AlC_{0.0625}N_{0.9375}$, only one Al atom was substituted for a C atom within the supercell. All atomic positions of pure c-AlN and $AlC_{0.0625}N_{0.9375}$ compounds were relaxed until all forces became smaller than 10^{-4}eV.

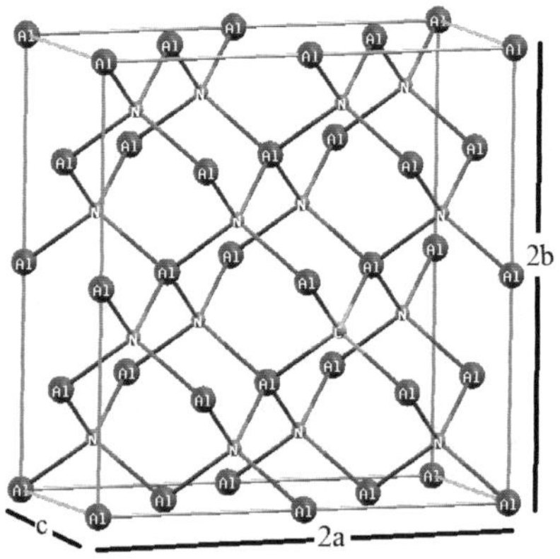

Figure 1. Unit cell of $AlC_{0.0625}N_{0.9375}$ compound in the zincblende structure.

3. Results and discussions

3.1. Structural properties

The main structural parameters, such as the lattice constant (a_0), the equilibrium volume (V_0), and the bulk modulus (B_0) of AlN and the ternary compounds $AlC_{0.0625}N_{0.975}$ calculated in this paper are listed in Table 1, along with the available experimental and theoretical values. One can see that for zincblende AlN, our calculated lattice constant of 4.3815Å agrees well with values reported theoretically, 4.380Å [34] and 4.376Å [35], and experimentally, 4.370Å [36], since it differs by less than one percent. Additionally, regarding the bulk modulus of 206.52GPa calculated in this paper, this value agrees well with values reported theoretically (212.4GPa [34] and 210GPa [35]), in comparison with the experimental value of 202GPa [36]. The error in the calculations is small, with a value of 2.23%. Since the errors in the lattice constant and the bulk modulus are small, this shows that the reported results are acceptable.

Table 1. Structural parameters: lattice constant (a_0), equilibrium volume (V_0), and bulk modulus (B_0) of the compounds.

Compound	a_0 (Å)	V_0 (Å^3)	B_0 (GPa)
AlN	4.3815	84.545	206.52
	4.3800[a]	-	212.4[a]
	4.3760[b]	-	210.0[b]
	4.3700[c]	-	202.0[c]
$AlC_{0.0625}N_{0.975}$	4.3800	84.528	202.0

[a,b] Theoretical reference [34,35], [c] Experimental reference [36].

On the other hand, when one N atom in the 2a×2b×1c zincblende supercell is replaced with a C atom, the lattice constant of the $AlC_{0.0625}N_{0.975}$ compound changes only slightly with respect to pure zincblende AlN. This may be because the radius of the C atom (0.914Å) is very close to the atomic radius of N (0.92Å).

3.2. Electronic properties

The lattice constants calculated for pure-AlN and $AlC_{0.0625}N_{0.975}$ compounds in the zincblende phase shown in the Table 1 were used to calculate the spin-polarized Density Of States (DOS) along the high-symmetry paths in the first Brillouin zone.

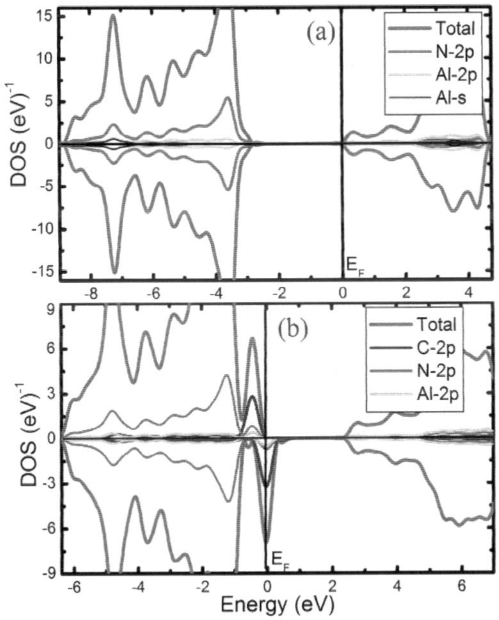

Figure 2. Total and partial density of state of (a) pure-AlN and (b) $AlC_{0.0625}N_{0.975}$ in the zincblende phase.

Figure 2(a) shows the total and partial density of states of pure-AlN. This compound exhibits a semiconductor behaviour with a forbidden energy gap of 3.41eV. The magnitude of this gap calculated in this paper agrees well with values reported in other theoretical papers, for example 3.38eV [37] and 3.40eV [38]. On the other hand, the spin-polarized DOS of the ternary compound $AlC_{0.0625}N_{0.975}$ is shown in Figure 2(b). We note that when a carbon atom occupies the position of a nitrogen atom in the cubic AlN, it loses its semiconductor behaviour. Thus, the $AlC_{0.0625}N_{0.975}$ compound acquires a half-metallic character, because the up spins are semiconductors and the down spins are metallic. The ternary compound has magnetic properties, with a total magnetic moment of $1.0\mu_\beta$/cell. These magnetic properties are determined by the hybridization between the 2p-C, 2p-N, and 2p-Al states.

4. Conclusions

We carried out a study of pure-AlN and the $AlC_{0.0625}N_{0.975}$ compound. The structural, electronic, and magnetic properties were studied using DFT calculations. The calculated lattice constants of pure AlN and the experimental values are in good agreement. Analysis of the density of states reveals that the $AlC_{0.0625}N_{0.975}$ compound exhibits a half-metallic behaviour, with a total magnetic moment of $1.0\mu_\beta$ per cell, which originates mainly from the 2p-C states and in lesser proportion from the 2p-N and 2p-Al states. These compounds are good candidates for potential applications in spintronics.

Acknowledgments

The authors wish to thank the Research Division of the "Universidad Distrital" CIUD for its financial support.

IOP Conf. Series: Journal of Physics: Conf. Series **935** (2017) 012001 doi:10.1088/1742-6596/935/1/012001

References

[1] Murilo J F, Ortega C and Espitia M J 2016 *Journal of Physics: Conference Series* **687** 012114
[2] Arbouche O, Belgoumène B, Soudini B and Driz M 2009 *Computational Materials Science* **47** 432
[3] Nakamura S 1997 *Solid State Commun.* **102** 237
[4] Espitia M J, Moreno M G, Rodríguez J A and Takeuchi N 2016 *J. of Phy.: Conference Series* **687** 012048
[5] Porowski S and Grzegory I 1994 *Properties of Group III Nitrides (EMIS Datareviews)* ed Edgar J H, (London: Inst of Engineering & Technology)
[6] Petrov I, Mojab E, Powell R C, Greene J E, Hultman L and Sundgren J-E 1992 *Appl. Phys. Lett. 60* **1491**
[7] Davis R F 1993 *Physica B* **185**
[8] Murillo J F, Ortga-López C and Espitia M J 2015 Results in Physics **5** 281
[9] Nakamura S, Senoh M, Nagahama S, Iwasa N, Yamada T, Matsushita T, Sugimoto Y and Kikoyu H 1997 *Jpn. J. Appl. Phys.* **36** L1059
[10] Paisley M J, Sitar Z, Posthill J B and Davis R F 1989 *J. Vac. Sci. Technol. A* **7** 701
[11] Xiao X I, Gao F, Zu X T and Weber W J 2010 *Journal Alloys Compound* **490** 537
[12] Wistrela E, Bittner A, Schneider M, Reissner M and Schmid U 2017 *J. of Applied Physic* **121** 115302
[13] Espitia-Rico M, Rodríguez-Martínez J, Moreno-Armenta M G and Takeuchi N 2015 *App. Surf. Sci.* **326** 7
[14] Pan D, Jian J K, Ablat A, Li J, Sun Y F and Wu J 2012 *Journal of Applied Physic* **112** 053911
[15] Espitia M J, Díaz J H and Castillo LE 2016 *International Journal of Physical Sciences* **11(1)** 11
[16] Luo J T, Li Y Z, Kang X Y, Zeng F, Pan F, Fan P, Jiang Z and Wang Y 2014 *J. Alloys Comp.* **586** 469
[17] Shah A, Mahmood A, Ali Z, Ashraf T, Ahmed I, Mehmood M, Rashid R and Shakir I 2015 *Journal of Magnetism and Magnetic Materials* **379** 202
[18] Espitia M J, Díaz J H and Ortega C 2015 *International Journal of Physical Sciences* **10(17)** 520
[19] Frazier R M, Thaler G T, Leifer J Y, Hite J K, Gila B P, Abernathy C R and Pearton S J 2005 *Applied Physic Letter* **86** 052101
[20] Espitia M J, Díaz J H and Rodriguez-Martínez J A 2016 *Solid State Communications* **244** 23
[21] Vargas-Hernández C, Espitia M, Báez R E 2015 *Computational Condensed Matter* **4** 1
[22] Espitia-Rico M J, Díaz-Forero J H and Castillo-Méndez L E 2015 *Revista Facultad de Ingeniería* **76** 143
[23] Park J H, Kim M G, Jang H M, Ryu S and Kim Y M 2004 *Applied Physics Letter* **84** 1338
[24] Kaspar T C, Droubay T, Heald S M, Engelhard M H, Nachimuthu P and Chambers S A 2008 *Physic Review B* **77** 201303
[25] Casiano-Jiménez G, Ortega-López C and Espitia-Rico M J 2015 *J. of Magn. and Magn. Mater.* **402** 156
[26] Herng T S, Lau S P, Wei C S, Wang L, Zhao B C, Tanemura M and Akaike Y 2009 *Applied Physic Letter* **95** 133103
[27] Cruz M M, Da Silva R C, Franco N and Godinho M 2009 *Condensed Matter* **21** 206002
[28] Yu L, Wang Z Y, Guo M, Liu D, Dai Y and Huang B 2010 *Chemical Physics Letter* **487** 251
[29] Laasonen K, Pasquarello A, Car R, Lee C and Vanderbilt D 1993 *Phys. Rev. B* **47** 10142
[30] Vanderbilt D 1990 *Phys. Rev. B* **41** 7892
[31] Giannozzi P, Baroni S and Bonin N 2009 *J. Phys. Condens. Matter* **21** 395502 (http://www.quantum-espresso.org)
[32] Perdew J, Burke K and Ernzerhof M 1996 *Physical Review Letter* **77** 3865
[33] Monkhorst H and Pack J 1976 *Phys. Rev. B* **13** 5188
[34] Verma U P and Bisht P S 2010 *Solid State Sciences* **12** 665
[35] Ramos L E, Teles L K, Scolfaro L W R, Castineira J L P, Rosa A L and Leite J R 2001 *Physics Review B* **63** 165210
[36] Stampfl C and Van de Walle C G 1999 *Physics Review B* **59** 5521
[37] Zhao Y J, Shu-Hong M and Ji-Fei Y 2011 *Solid State Sciences* **13** 331
[38] Christensen N E and Gorczyca I 1994 *Physics Review B* ***50*** 4397

IOP Conf. Series: Journal of Physics: Conf. Series **935** (2017) 012002 doi:10.1088/1742-6596/935/1/012002

Structural properties of $La_{3.5-x}Ba_{4.5+x-y}Ca_yCu_8O_{18}$ system (0.2<x<0.4 and 0.3<y<0.5)

L A Pérez Díaz[1], **A F Cruz Pacheco**[1], **J A Gómez Cuaspud**[1], **C A Parra Vargas**[1] **and C A Ortiz Otalora**[1]
[1] Universidad Pedagógica y Tecnológica de Colombia, Tunja, Colombia

E-mail: linda.perez@uptc.edu.co

Abstract. This work describes the synthesis and the structural characterization of the $La_{3.5-x}Ba_{4.5+x-y}Ca_yCu_8O_{18}$ (0.2<x<0.4 and 0.3<y<0.5) systems, obtained by the conventional solid-state reaction method. The samples were characterized by X-ray diffraction comparing the main signals of the experimental and theoretical diffractograms with the PCW software, revealed the obtaining of an orthorhombic structure with space group *Pmm2* (25) and unit cell volume V=465.5370Å^3. The analysis by Raman spectroscopy allows to determine the main modes of vibration of the different bonds that are formed between the atoms in the synthesized perovskite structure.

1. Introduction

In the last decades have been appearing important discoveries in the field of materials, the most remarkable are in ceramic materials. Those discoveries have required a rethinking in concepts, methodologies and the use of sophisticated new experimental procedures. Reciprocally, the progressive technologic advance has increase interest in searching new components with proprieties potentially relevant in industry in order to improve processes, reduce costs and preserve the environment. These factors create the impetuous necessity to search on some continuous basis new synthesis techniques, thermic treatment techniques or apply techniques that already exist to make compounds with a collection of proprieties to meet the currently needs [1]. In this context, the combination of specific proprieties in a compound permit to have varying multifunction degrees which is a challenge in field physical-chemical in the solid state. Different physical proprieties of materials of type perovskite are linked to variables like chemical composition, purity of its phases, superficial defects, crystalline structure, morphology and sinter conditions, the control of these parameters is critic inside the production process in this type of compounds [2]. One of the most relevant applications of perovskite type materials is superconductors, where the $YBa_2Cu_3O_7$ (YBCO) system is the most important because it has a high critical temperature.

Structurally the $CaLaBa_2Cu_3O_7$ (CLBCO) presents similarities to the YBCO, becoming an important system to carry out the study of superconducting properties and the effect caused by the superconducting fluctuations. The CaLaBaCuO was discovered in 1988 by W. T. Fu and D. M. De Leeuw [3], with studies on substitution and doping with elements of Rare Earth (RE). They observed that the CLBCO has an orthorhombic isomorph structure type YBaCuO, with Ca and La located on Y and Ba sites, it presents a critical temperature approximately 80K. The orthorhombic phase represents symmetric of space group *Pmm2* (25). This type of materials is commonly used in in fields as various as in medicine for the vast majority of image scanners which use Nuclear Magnetic Resonance where those base on

IOP Conf. Series: Journal of Physics: Conf. Series **935** (2017) 012002 doi:10.1088/1742-6596/935/1/012002

superconducting magnets [3] and in train systems which works with magnet levitation. Taking into account the properties of the CLBCO 123 systems, it is intended to verify that the CLBCO system 358 presents similar proprieties to the aforementioned system further to meet with phase diagram proposed by West and collaborators for the system $La_{3.5-x}Ba_{4.5+x-y}Ca_yCu_8O_{18}$ in regular conditions of temperature and pressure with 0<X<0.5 and 0<Y<0.5. In this context, this research studies the structural properties of the $La_{3.5-x}Ba_{4.5+x-y}Ca_yCu_8O_{18}$ system by means of X-ray diffraction and Raman spectroscopy obtained by the solid-state reaction method, in order to obtain a new perovskite type material with a single phase.

2. Experimental

The $La_{3,5-x}Ba_{4.5+x-y}Ca_yCu_8O_{18}$ (0.2<x<0.4 and 0.3<y<0.5) systems, were produced by the traditional Solid-State Reaction (SSR) method from corresponding precursor oxides of La_2O_3 (99.99%), CuO (99.999%), $BaCO_3$ (99.999%) and $CaCO_3$ (99.999%) carbonates, all from Sigma-Aldrich. Initially, the precursors were heat treated to remove residual moisture for 12 hours at 200°C. The stoichiometric amounts were weighed and milled in agate mortar for 4 hours in order to obtain a more homogeneous powder. Thermal treatments initially involved a first stage at 750°C for 24 to volatilize carbonate species. Subsequently, a second heat treatment was carried out for 24 hours at 780°C in order to consolidate the crystalline phase sought. The resulting material was ground and compressed uniaxially at 5.0MPa and subjected to a sintering process at 850°C for 24 hours to achieve the final structure of the proposed system. The structural characterization of the solid was performed by X-ray diffraction using PANalytical X'Pert High Score equipment using Co Kα rad radiation (1.789Å) between 20° and 90°. The vibrational characterization of systems synthetized, were made using the Raman spectroscopy technique.

3. Results and discussion

The crystallography analysis of the $La_{3,1}Ba_{4.5}Ca_{0.4}Cu_8O_{18}$ and $La_{3,1}Ba_{4.4}Ca_{0.5}Cu_8O_{18}$ systems were based on a comparison between the experimental diffractograms obtained and the diffractograms proposed by S. Gholipour *et.al.* [4]. These results are show in the Figure 1(a), and allow to establish that the synthesized samples have a crystalline structure orthorhombic with a space group *Pmm2* (25) and lattice parameters a=3.8450Å, b=3.8920Å and c=31.0930Å and α=β=γ=90°, with a cell volume of V=465.538Å^3.

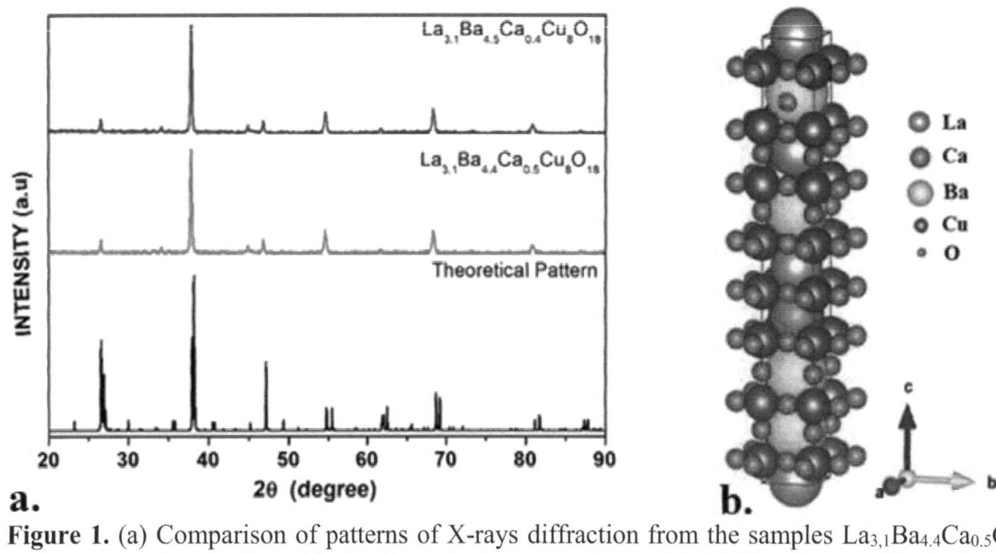

Figure 1. (a) Comparison of patterns of X-rays diffraction from the samples $La_{3,1}Ba_{4.4}Ca_{0.5}Cu_8O_{18}$ and $La_{3,1}Ba_{4.5}Ca_{0.4}Cu_8O_{18}$ with the theoretical pattern obtained in PCW software. (b) Ideal structure of system $La_{3,5-x}Ba_{4.5+x-y}Ca_yCu_8O_{18}$ obtained with Vesta software.

According to the comparison of the experimental diffractograms with the theoretical diffractogram, it is possible to affirm that the $La_{3,1}Ba_{4.5}Ca_{0.4}Cu_8O_{18}$ and $La_{3,1}Ba_{4.4}Ca_{0.5}Cu_8O_{18}$ samples present a single phase 358, since all the signals of main intensity are adjusted to the theoretical diffraction pattern. Figure 1(b) shows the ideal structure for the $La_{3,5-x}Ba_{4.5+x-y}Ca_yCu_8O_{18}$ system, obtained with the vesta program. The size of the crystalline domains was calculated with the Deybe–Scherrer equation taking in account the signal of major intensity in diffractogram and in average was 35nm for both samples [5].

The Figure 2 show us the Raman spectrum relevant from system $La_{3,5-x}Ba_{4.5+x-y}Ca_yCu_8O_{18}$ which exhibits consisting broadband with the literature [6], that are link with the vibrational proprieties of the atoms in the unit cell from CLBCO-358. The spectrum was measured at 785nm which allowed us to obtain the qualitative characteristics of this spectrum that are representative of the samples of the system currently studied [7]. A bottom was presented without specific signals indicating presence of fluorescence and dispersion effects which are typical when carrying out these measurements for the aforementioned system despite the care taken during the measurements made this comparing it with previously performed work, where it is observed that there is presence of fluorescence and dispersion. The signals showed in Raman are allocated in the following way [8]. The 5 signals belong to the link oscillations Cu-O (291.5cm^{-1}) and Ba-O (403.5cm^{-1}) making an axial stretching of the same [9], the signal of major intensity (534.7 cm^{-1}) correspond to axial movement which origins flexion between the links La-O likewise [10], there are flexion vibrations between links Cu-O (1113cm^{-1}) and axial stretching of Cu-O (1235cm^{-1}) [11]. These signals permit determine the symmetrical structure of CaLaBaCuO-358. The data obtained with Raman for the $La_{3,5-x}Ba_{4.5+x-y}Ca_yCu_8O_{18}$ system and in comparison, with the results obtained by C. Thomsen *et. al.* [12].

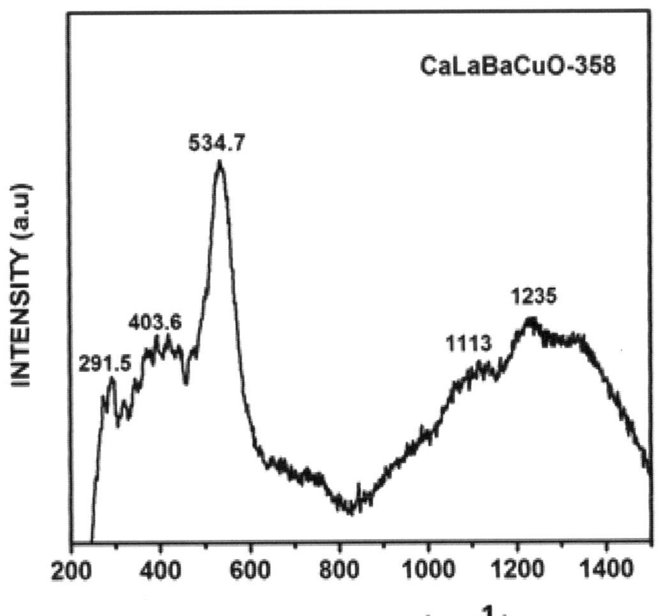

Figure 2. Raman spectrum relevant from system $La_{3,5-x}Ba_{4.5+x-y}Ca_yCu_8O_{18}$.

Some characteristics present in the spectra as mentioned correspond to the vibrations of oxygen and lanthanum atoms that are present in both the orthorhombic and tetragonal phases [13]. This the mode in 291.5cm^{-1}, which is absent in the tetragonal phase, has a greater intensity of both this signal and all the peaks in the orthorhombic phase of the CaLaBaCuO-358 system compared to the results obtained for the CLBCO-123 system [14]. The highest intensity signal appearing at 534.7cm^{-1} indicates the presence of Ca doped La_2CuO_4 (not shown). Own modes of all the vibrations in La_2CuO_4 are shown in Figure 2. In Raman, some peaks or signals that are induced by defects are formed when the oxygen bonds with the La or Ba are not completely occupied. These peaks, where the largest is at $534.7cm^{-1}$, were first identified as defects by de Liu *et. al.* [15]. The Raman spectrum is dominated by the vibrations of oxygen and copper bonds [16] so we can say that the results are similar to the results confirmed in DRX where the symmetry is orthorhombic.

4. Conclusions

The solid-state reaction method allowed to produce the samples of the $La_{3,5-x}Ba_{4.5+x-y}Ca_yCu_8O_{18}$ system, demonstrating the effectiveness of synthesis method to obtain a pure phase of this material in agreement with the modelling process predicted for this sample. The analysis of Raman spectroscopy allowed us to see the main vibration types present in the CaLaBaCuO-358 system, which allows corroborating the crystal structure obtained by the X-Ray Diffraction (XRD) characterization.

References

[1] Boudali A, Khodja M D, Amrani B, Bourbie D, Amara K and Abada A 2009 *Phys. Lett. A* **373** 879
[2] Canaría-Camargo C C, Sarmiento-Santos A, and Parra-Vargas C A 2016 *Dyna* **83** 108
[3] Rillo C 1991 *Rev. B* **30** 407
[4] Gholipour S, Daadmehr V, Rezakhani A T, Khosroabadi H, Shahbaz Tehrani F and Hosseini Akbarnejad R 2012 *J. Supercond. Nov. Magn.* **25** 2253
[5] Slimani Y, Hannachi E, Ben Salem M K, Hamrita A, Varilci A, Dachraoui W, Ben Salem M and Ben Azzouz F 2014 *Phys. B Condens. Matter.* **25** 2253
[6] Richter A, Irmer G, Kebler G, Panzner M and Herzog K 1992 *J. Alloys Compd.* **187** 59
[7] Wang H-Y, Ding F-Z, Gu H-W and Zhang T 2015 *Chinese Phys. B* **24** 97401
[8] Widder K, Münzel J, Göppert M, Lüerßen D, Becker R, Dinger A, Geserich H P, Klingshirn C, Kläser M, Müller-Vogt G, Geerk J and Burlakov V M 1998 *Phys. C Supercond.* **300** 115
[9] Sanchez H, Sanchez H, Bustamante A, Osorio A, Santos L D los, González J C and Barnes C H W 2015 *Rev. Peru. Química e Ing. Química* **17** 15
[10] González J, González J C, Osorio A, Marín R, Castro A, Sandivar J, Rodriguez A, Flores R, Salas N, Elescano G, Villegas I, Loyaga H, Santos L and Díaz A 2014 *Rev. Peru. Química e Ing. Química* **8** 12
[11] Bustamante A, Bustamante A, Osorio A, González J C, Carhuancho M, Salas N, Santos L D L, Cruz N D L and Díaz A 2014 *Rev. Peru. Química e Ing. Química* **7** 3
[12] Wang H-Y, Ding F-Z, Gu H-W and Zhang T 2015 Strongly enhanced flux pinning in the $YBa_2 Cu_3 O_{7-x}$ films with the co-doping of $BaTiO_3$ nanorod and $Y_2 O_3$ nanoparticles at 65 K *Chinese Phys. B* **24** 97401
[13] Boekholt M, Erle A, Splittgerber-Hünnekes P C and Güntherodt G 1990 *Solid State Commun.* **74** 1107
[14] Erle A and Güntherodt G 1990 *Phys. C Supercond.* **171** 216
[15] Liu R., Thomsen C, Kress W, Cardona M, Gegenheimer B, de Wette F W, Prade J, Kulkarni A D and Schroder U 1988 *Phys. Rev. B.* **37** 7971
[16] Krol D M, Stavola M, Weber W, Schneemayer L F, Waszczak J V, Zahurak S M and Konsinski S G 1977 *Phys. Rev. B.* **36** 8325

Blast furnace dust and phosphorous slag, new materials for use in road engineering

R Ochoa Díaz[1]
[1] Universidad Pedagógica y Tecnológica de Colombia, Tunja, Colombia

E-mail: ricardo.ochoa@uptc.edu.co

Abstract. This article proposes an alternative to the use of phosphorus slag and blast furnace dust, by-products of the steel industry, due to the negative environmental impact caused by its accumulation. Taking into account the above, the pertinence of the use of these by-products in asphalt mixtures for the construction of roads is studied. In this way, the origin and its properties are presented, as well as their physical and chemical characteristics. Once the tests have been carried out, it is determined that these by-products have adequate characteristics for their use since they do not present toxicity problems. Following this, the design of the mixtures is carried out to determine the mechanical and dynamic properties and thus determine the proportion to be replaced with the conventional materials. Taking into account the results it is concluded that its use is feasible since the mixture with these by-products presents acceptable resilient modulus parameters and improvement in some verification parameters.

1. Introduction

When you look at the current state of highways in the country, where large resources are invested and pavement structures do not last for the time they were designed, it is necessary to look for new construction methods and new materials that meet the specifications of construction [1]. Taking into account the above, and also, the environmental impact generated by the exploitation of natural materials and the accumulation of steel residues, it could be proposed to use such steel residues within the whole of the road as substitutes for natural aggregates. Thus, it is proposed that the steel residues be incorporated into embankments, granular layers or forming asphalt mixtures, either in the form of coarse aggregate, fine aggregate or mineral fillers [2].

Once the situation is understood, this work is oriented towards the sustainable use of steel residues, especially for phosphorus slag and blast furnace dust in parallel. These residual materials could be incorporated in the design of new asphalt mixtures, a previous step of vital importance for the construction of roads [3]. The residues used in this study are produced by the company Acerías Paz del Río S.A. The dust is produced in the blast furnace and the phosphorus slag is generated in converters [4]. It should be added that these steel residues should comply with the general specifications for road construction materials of INVIAS [5].

2. Materials and methodology

The dust is generated during the operation of the blast furnace for the production of pig iron, using as raw material iron ore, limestone, and coke. In the combustion process, gas is generated, which is composed of small particles that are trapped and evacuated by the pipe to hoppers. The phosphorus slag is generated in the converters during the transformation of pig iron into steel. In Figure 1, a photo of the dust is shown. In Figure 2, a photo of the phosphorus slag is shown.

The methodology used in this research is divided into three stages: the chemical and mechanical characterization of the materials, the comparison of the results with the required specifications, and the realization of tests to obtain the working formula using the RAMCODES methodology [6,7].

Figure 1. Dust of blast furnace.

Figure 2. Phosphorous slag.

The chemical evaluation was performed by means of a Scanning Electron Microscope (SEM) to know the texture of the residues used [8]. Figure 4 shows the SEM image of blast furnace dust at 100μm scale, observing a porous structure that potentially promotes asphalt absorption. Figure 5 shows a SEM image of the phosphorous slag at 10μm scale, which has a rough texture.

The study of the elemental chemical composition was performed by means of an X-ray fluorescence spectrometer in order to establish the components of each of the residues. In order to determine the optimum percentage of asphalt in the mixture, the criteria established in article 450-13 of the road construction specifications of INVÍAS for a transit level of NT3 design were followed: Marshall Stability, flow, and air voids. After verifying the compliance with these parameters, the design was verified by the following tests: adhesion to evaluate the effect of water on the indirect tensile strength of mixtures, resistance to plastic deformation or rutting, indirect stress test to determine the resilient modulus in the NAT (Nottingham Asphalt Tester) equipment, and determine the fatigue laws of asphalt mixtures subjected to dynamic bending.

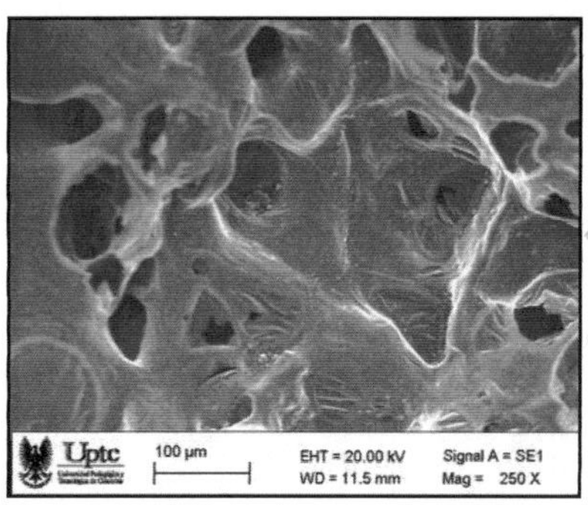

Figure 3. SEM micrographs of dust of blast furnace.

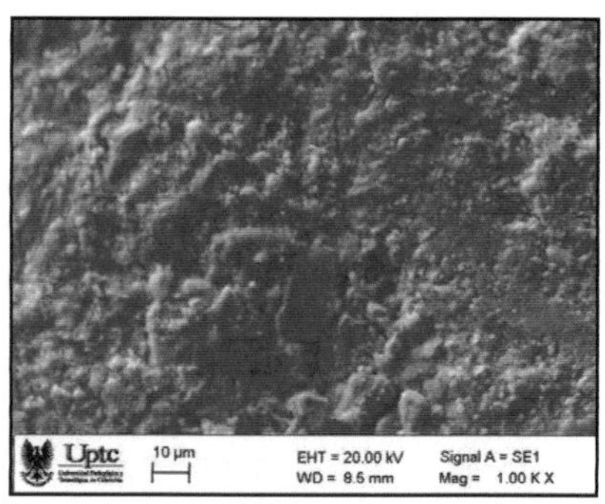

Figure 4. SEM micrographs of phosphorous slag.

3. Experimental design

The mixture chosen for the development of the research is a hot-dense mixture type 19 (MDC-19), article 450-13 INVIAS [5]. In order to carry out an analysis of the behaviour of the waste, three blends were designed as follows: mixture 1 (M1) with conventional materials (gravel and sand); Mixture 2 (M2) replacing 50% of gravel per slag and replacing 50% of sand with dust; Mixture 3 (M3) completely replacing gravel and sand for slag and dust respectively. Asphalt 60-70 was used for all mixtures.

IOP Conf. Series: Journal of Physics: Conf. Series **935** (2017) 012003 doi:10.1088/1742-6596/935/1/012003

4. Results

4.1. Characterization of waste

The chemical composition of the wastes is important characteristics that must be considered before inclusion in the asphalt mixtures. Table 1 shows the elemental composition obtained by X-ray fluorescence spectrometry of the dust and slag. The predominant component in the dust is Fe with 77.5% and in the slag, is Ca with 46.8%.

Table 1. Elemental chemical composition of dust and phosphorous slag.

Component	Mg	Al	Si	P	S	K	Ca	Ti	Mn	Fe	Zn
Dust. Cont. Unit. (%)	1	3.6	5.5	0.2	1.8	0.51	4.95	0.11	3.32	77.5	1.5
Slag. Cont. Unit. (%)	2.7	4.9	10.8	2.1	0.58	0.52	46.8	0.1	2.0	28.8	0.53

The physical and mechanical properties of dust and slag are presented in Table 2. These tests were carried out according to the INVIAS road materials test standards [9].

Table 2. Physical and mechanical properties of dust and phosphorous slag.

Feature	Unit	Requirement	Result	Standard
"Los Ángeles" wear (E) max.	%	25	20	INV E-218/ASTM C-131
Mechanical resistance (E) min.	kN	110	112	INV E-224/SABS Metth 842
Flat and elongated particles (E) max.	%	10	1	INV E-240/ASTM D-4791
Loss of solidity (P) max.	%	18	3	INV E-220/ASTM C-88-05
Plasticity index (P) max.	%	NP	NP	INV E-126/ASTM D-4318
Sand equivalent (P) min.	%	50	94	INV E-133/ASTM D-2419

E: Phosphorous slag.
P: Dust

4.2. Characteristics of the mixtures

The physical and mechanical characteristics of the designed mixtures are presented in Table 3.

Table 3. Mechanical and dynamic properties of the mixtures.

Feature	Unit	M1	M2	M3	Standard
Asphalt content	%	4.8	6.5	5.2	-
Marshall Stability	N	11967	10561	10896	INV E-748
Flow	mm	3.4	3.5	3.4	INV E-748
Air vents	%	4.6	6.0	5.4	INV E-736

As regards stability, it is observed that mixtures containing phosphorous slag behave very efficiently, and if the coarse aggregate is also related to the stability of the mixture, it is clear that all mixtures meet the minimum value for NT3 which is 9000N. Similarly, the flow meets the required parameters, since it is within the range of 2.0 to 3.5. It is also observed that the requirement of voids with air is fulfilled in the three mixtures. The results of the verification tests of the design of the asphalt mixtures are shown in Table 4.

Table 4. Verification parameters of the mixtures design.

Feature	Unit	M1	M2	M3	Standard
Adhesion RRT	%	87	57	58	INV E-725
Rutting	mm	3.3	1.8	1.5	INV E-756
Resilient module at 13°C	MPa	7200	6600	8100	INV E-749
Fatigue-Radial strain	Def.	0.912	0.054	0.002	INV E-784

With regard to the rutting test, the specifications require that the deformation rate of the specimen during a range of 105 to 120 minutes should not exceed 15μm/min [10]. The tests performed on the different mixtures show that for this variable the mixtures are satisfactory.

Observing the resilient modulus [11] obtained for the conditions of a rolling layer in a city with a temperature of 13°C, an improvement of the module in the mixture M3 was obtained, and in the case of the mixture M2 the module decreased.

As for the radial deformation, the equations were determined based on a transit of 5 million axes. It was observed that the values meet the criterion required since the asphalt binder is subjected to a number of load repetitions, a situation that must allow a smaller radial deformation, in order to avoid damages to the structure during the service [12].

5. Conclusions

The use of non-conventional materials such as phosphorus slag and blast furnace dust, waste from the steel industry, have a high level of functionality when used as aggregate in asphalt mixtures.

Blends containing blast furnace dust present a higher asphalt content, due to the high absorption capacity and texture of the dust, a situation that directly affects the economic terms.

In general terms, mixtures of dust and slag proportions are technically feasible in design. Although these mixtures are susceptible to moisture and saturation conditions, this drawback can be solved with the application of specialized additives for the improvement of the adhesion of the asphalt mixtures.

The phosphorus slag and the dust meet the requirements and specifications required for the materials to be used in asphalt mixtures, which is why its use is feasible for the manufacture of asphalt mixtures in road construction. In addition to the above, the use of these industrial wastes contributes to the preservation of the environment by preventing their accumulation, and also, reduces the exploitation of natural materials that could be replaced at least in part by such industrial by-products.

References

[1] Jun X, Juyong Ch, Shaopeng W, Juntao L, and Eei W 2013 Performance characteristics of asphalt mixture whit basic oxygen furnace slag *Construction and Building Materials* **38** 796

[2] Francisco S 2002 Utilización de residuos en la construcción de capas de firmes de carreteras *Ingeniería Civil* **128** 71

[3] Wang G, Wang Y and Gao Z 2010 *Use of steel slag as a granular material* Volume expansion prediction and usability criteria. Journal of Hazardous Materials **184** 555

[4] Gómez A 2002 *Proceso siderúrgico planta Belencito* (Colombia: Acerías Paz del Río S.A.)

[5] Instituto Nacional de Vías 2013 *Especificaciones generales de construcción de carreteras* (Bogotá: INVIAS)

[6] Sánchez F 2007 Gradation chart for asphalt mixes: Development *Journal of Materials in Civil Engineering* **19** 185

[7] Sánchez F, Garnica P, Larreal M and López D 2011 Polyvoids: Analytical tool for superpave HMA design *Journal of Materials in Civil Engineering* **23** 1129

[8] Yanping S, Ben Z, Yu Y, Huaxin Ch, Rui X and Jiuguang G 2017 Effect of phosphorus slag powder and polyester fiber on performance characteristics of asphalt binders and resultant mixtures *Construction and Building Materials* **141** 292

[9] Instituto Nacional de Vías 2013 *Normas de ensayos de materiales para carreteras* (Bogotá: INVIAS)

[10] Barugahare J, Liantong M, Fang H, Benan S and Shaopeng W 2017 Multi-stress loading effect on rutting performance of asphalt mixtures based on wheel tracking testing *Construc. and Building Materials* **148** 6

[11] Muhammad K, Hamid N, Surya S and Laksmi I 2017 Laboratory experiment on resilient modulos of BRA modified asphalt mixtures *International Journal of Pavement Research and Technology* **10** 5

[12] Gourab S and Krishna P 2017 Stato-dynamic response analyses through semi-circular bending test: fatigue life prediction of asphalt mixtures *Construction and Building Materials* **150** 668

Hydrothermal synthesis of a photovoltaic material based on $CuIn_{0.5}Ga_{0.5}Se_2$

Y T Castellanos Báez[1]**, D A Fuquen Peña**[1]**, J A Gómez-Cuaspud**[1]**, E Vera-López**[1] **and Y Pineda-Triana**[1]

[1] Universidad Pedagógica y Tecnológica de Colombia, Tunja, Colombia

E-mail: yeimy.castellanos01@uptc.edu.co

Abstract. The present work report, the synthesis and characterization of the $CuIn_{0.5}Ga_{0.5}Se_2$ system (abbreviated CIGS), by the implementation of a hydrothermal route, in order to obtain a solid with appropriate properties in terms of surface, morphological and texture properties for potential applications in the design of photovoltaic cells. The synthesis was carried out using the corresponding stoichiometric quantities (Cu:In:Ga:Se 1:0.5:0.5:2), which were mixed in a Teflon vessel under stirring conditions. The homogeneous solution was treated in a steel autoclave at 300°C for 72 hours at the end of which the resulting material was characterized by X-Ray Diffraction (XRD) and Rietveld refinement. The results of the structural characterization allowed to confirm the obtaining of a chalcopyrite type structure, with a *I-42 d* (122) structure and cell parameters a=0.570, b=0.570, c=1.140nm, α=90, β=90, γ=90° oriented along (1 0 4) facet, detecting the presence of a secondary phases, related with CuInSe and CuIn metallic selenides, derived from synthesis process. The structural refinement allowing to validate the obtaining of a nanometric crystalline material (10-20nm) for potential applications in field of photovoltaic technology.

1. Introduction

The current photovoltaic industry has found in ternary and quaternary materials derived from $CuInSe_2$ (CIS) and $CuIn_{1-x}Ga_xSe_2$ (CIGS) chalcopyrite compounds, promising materials for the effective conversion of solar energy, due to its semiconductor behaviour and band-gap value of approximately 1.45eV, with conversion levels around 19.2% [1-4]. These materials can be synthesized by a series of physical routes such as Physical Vapor Deposition (PVD), Chemical Vapor Deposition (CVD), Pulsed Laser Deposition (PLD) and other techniques related with the use of high vacuum conditions in which the stoichiometric control of elements is not the most effective or simple, reason why the obtaining of these compounds only is available under extreme reaction conditions [6]. Although the efficiency values in materials obtained by physical synthesis methods is around 21.7-22.6%, nowadays the use of chemical synthesis methods based on liquid solution reactions, get allow solids with a high degree of purity and provide an excellent composition control up to atomic or molecular level as in the case of sol-gel, controlled precipitation or solvothermal methodologies among others [5].

In the specific case of hydrothermal synthesis method, several advantages over other synthesis routes are possible, since is possible to generate a better dissolution of the used precursors under controlled conditions of reaction, which only depend of time and synthesis temperature as the main thermodynamic parameters with an optimal stoichiometric control [6,7]. As result of previous advantages related with the use of hydrothermal methodology for the production of advanced materials, the objective of current work is focused on the obtaining of a $CuIn_{0.5}Ga_{0.5}Se_2$ photovoltaic material, under 300°C and 72 reaction hours, with the aim to provide a simple methodology to development of photovoltaic materials ant micro or nanometric scale for potential application in solar technology.

2. Experimental

The procedure used in this work to perform the synthesis of CIGS material start with the corresponding dissolution of the metal precursors of copper acetate $Cu(CH_3COO)_2$, indium nitrate $In(NO_3)_3$, gallium acetylacetonate $Ga(C_5H_8O_2)_3$ and metallic selenium Se, all from Sigma-Aldrich. The reagent addition order was established in terms of the corresponding dissolution constants in a ratio (1:0.5:0.5:2). The adjust of the pH was performed with concentrated nitric acid around 1.0 until a homogeneous mixture was obtained. The solution was deposited in a Teflon vessel under stirring conditions (250rpm) and placed in a steel autoclave in which the system was sealed and held under a temperature of 300°C for 72 hours. At the end of the process, the system was cool to room temperature, then a green solid precipitate was recollected and washed with absolute ethanol, in order to remove the remnant impurities. The solid was dried at 90°C for 6 hours in an electric oven and grounded in an agate mortar to characterization process. The structural characterization of material was performed by X-ray diffraction in a PANalytical X´Pert PRO-MPD equipment, using a Pixel detector between 10 and 90°, with a Cu Kα source of 1.54Å. The corresponding refinement was performed by the Rietveld routine and the GSAS software in order to analyse the crystalline structure of the material and its main crystalline parameters

3. Results and discussion

The X-ray diffraction pattern of CIGS obtained material is shown in Figure 1, in which is possible to observe the main signal indexed along (1 0 4) facet.

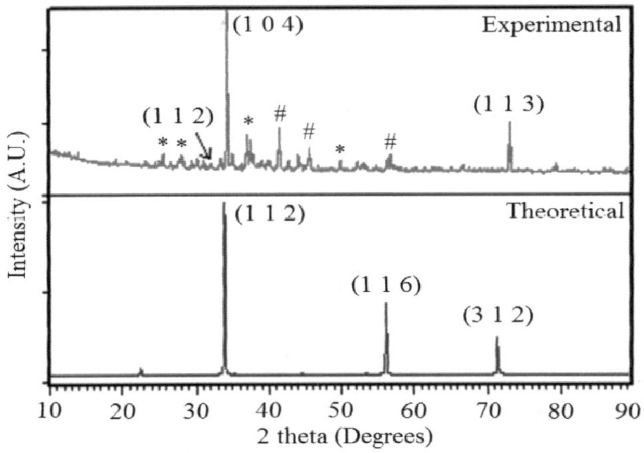

Figure 1. XRD experimental and theoretical pattern of the $CuIn_{0.5}Ga_{0.5}Se_2$ material, obtained at 300°C for 72 hours.

The detailed analysis of XRD results, confirm the presence of secondary phases of CuInSe (*) and CuSe (#), due to reaction conditions. The crystalline structure related with the pure phase of $CuIn_{0.5}Ga_{0.5}Se_2$ is around 85.7%, while the 14.3% is related with the presence of secondary phases. According to E. Romero *et. al.* [10], the effect of an acid medium could propitiate the formation of CuInSe phase due to stability related with these elements contrary to gadolinium effect, that preserves as isolated oxide species in form of Ga_2O_3 [11]. The main diffraction signal of experimental CIGS material is located along (1 0 4) facet, indicating a structural variation with respect to theoretical value of (1 1 2), this effect is provoked by the gallium cation and its composition level in obtained material, taking into account that the refined $CuIn_{0.5}Ga_{0.5}Se_2$ structure only could be compared with the

$CuIn_{0.7}Ga_{0.3}Se_2$ reference structure, generating a shift of the signal towards the left and promoted by the change in the atomic radius of the Gallium in the structure in accordance with previous reports [1,15].

The analysis of the 14.3% of impurities shown in Table 1, by means the Rietveld refinement permit to corroborate the presence of another compounds bonded in obtained CIGS material as follows:

Table 1. Obtained percentages from the secondary phases in the Rietveld refinement process.

Compound	Percentage
$CuIn_{0.5}Ga_{0.5}Se_2$	85.7%
$NaNO_3$	6.6%
Cu_2O	1.1%
$Cu_6Ga_{17}O_{68}$	1.2%
Na_2SeO_3	3.1%
InO_3	0.8%
$CuInSe_2$	1.4%

The above results observed in the material synthesis affect the performance of the reaction since the undesirable products can affect the electrical performance of material, although some authors [12,14], has been established that the necessity of a second thermal stage could reduce the presence of this secondary phases to increase the presence of a pure crystalline phase of CIGS, its employ in current work was not considered due to necessity of know the synthesis methodology to future works related with the obtaining of selenides. In this context, must to be considered that the hydrothermal synthesis pathway need to be tested in alkaline medium because the effects of mass transfer can improve getting a more effective crystalline growth of the desired material in accordance with previous works [7,13].

The analysed structure using the X´Pert High Score software in the ICDD databases and based on the 85.7% of the obtained material, permit to confirm the presence of a chalcopyrite-type structure, corresponding with a *I-42 d* (1 2 2) structure with cell parameters a=0.570, b=0.570, c=1.140nm, α=90, β=90, γ=90° oriented along (1 0 4) facet. With this data was possible to obtain the corresponding unit cell and expected X-ray diffraction pattern based on experimental data as shown in Figure 2 [9].

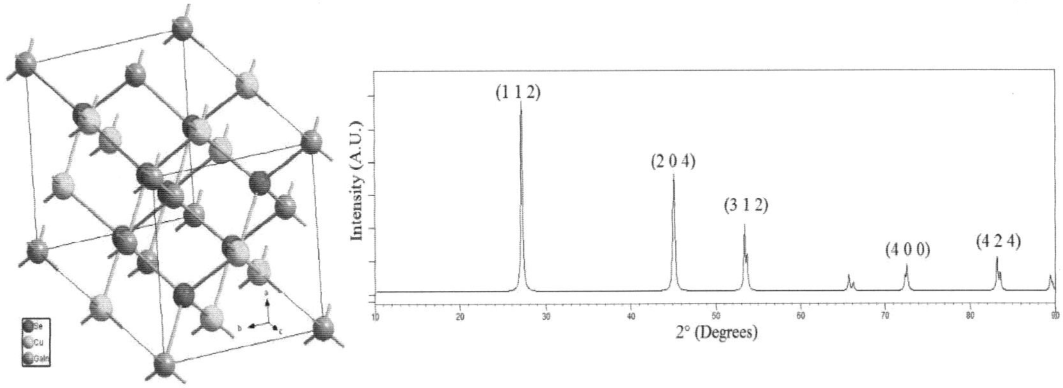

Figure 2. Unit cell and expected X-ray diffraction pattern of $CuIn_{0.5}Ga_{0.5}Se_2$ material.

The structural data obtained by ELMIX software and Rietveld refinement is shown in Table 2, using the GSAS software, permit to validate the obtaining of a prevalent crystalline phase of $CuIn_{0.5}Ga_{0.5}Se_2$ material with its main expected diffraction signals, in accordance with experimental pattern for potential applications in field of photovoltaic cells.

IOP Conf. Series: Journal of Physics: Conf. Series **935** (2017) 012004 doi:10.1088/1742-6596/935/1/012004

Table 2. Main structural parameters obtained for $CuIn_{0.5}Ga_{0.5}Se_2$ material by ELMIX and GSAS software.

$CuIn_{0.5}Ga_{0.5}Se_2$					
Crystal system	tetragonal	Space group	I -42d (1 2 2)	Cell Volume (Å)	370.39
Lattice parameters					
a (Å)	5.7000		α		90.0000
b (Å)	5.7000		β		90.0000
c (Å)	11.4000		γ		90.0000
Atomic positions					
	Cu	In	Ga		Se
X	0.00000	0.00000	0.00000		0.24260
Y	0.00000	0.00000	0.00000		0.25000
Z	0.50000	0.00000	0.00000		0.12500

4. Conclusions

By means of the hydrothermal synthesis route obtains of a $CuIn_{0.5}Ga_{0.5}Se_2$ material is performed, with a purity level around 85.7% with presence of a secondary phases, which could be reduced by the implementation of a second thermal stage or development of a longer reaction time. The X-ray diffraction data and the Rietveld refinement allows to confirm the obtaining of a chalcogenide phase for use as photovoltaic material in solar cell technology with the use of a scalable and easy procedure without necessity of extreme vacuum conditions or sophisticated equipment.

References

[1] L. Zhang, W. Zhang, J. Liu, et al 2014 *Materials Chemistry and Physics* **147** 390-394

[2] D. Portela 2015 *Síntesis y caracterización de materiales fotovoltaicos para paneles solares de bajo coste y alta eficiencia* (España: Universidad da Coruña)

[3] L. M. Hernandez 2007 *Digital universitaria, UNAM* **8(12)** 4-13

[4] L. J. Chen, J. D. Liao, Y.J Chuang 2011 *Thin Solid Films* **519** 3658-3692

[5] S. H. Mousavi, T. S. Müller. R.Karos, P. W. Oliveira 2016 *Journal of Alloys and Compounds* **659** 178-183

[6] R. Martí, V 2016 *Obtención de estructuras calcopirita (CIGS) y kesterita (CZTS) como absorbentes para dispositivos fotovoltaicos de capa fina mediante métodos de síntesis de bajo coste* (España: Universitat Jaume I de Castellón)

[7] L. Olmedo 2013 Síntesis y caracterización de catalizadores de Ni-CeO_2 por el método hidrotermal (México: Universidad Tecnológica del Valle de Toluca)

[8] P. Luo, R. Zuo, L. Chen 2010 *Solar Energy Materials & Solar Cells* **94** 1146-1151

[9] F. Chen, J. Ma, J. Sung, C. Lu 2014 *Solar Energy Materials & Solar Cells* **124** 166-171

[10] E. Romero, C. Calderón, P. Bartolo-Pérez, F Mesa and G. Gordillo 2006 *Brazilian Journal of Physics* **36** 1050-1053

[11] I. Lopez Garcia 2014 *Propiedades físicas y dopado de nanoestructuras de óxido de galio* (España: Universidad Complutense de Madrid)

[12] K. Liao, C. Su, et al 2013 *Applied Surface Science* **270** 139-144

[13] M. Castro, S. Mesa, et al 2013 *Rev Colombiana de Materiales* **5** 158-164

[14] J S. Ma, C Y Yang, J Ch Sung, et al 2015 *Ceramics International* **41** 10980-10987

[15] K. Chen, C. Chiang, et al 2013 *Materials letters* **95** 172-174

IOP Conf. Series: Journal of Physics: Conf. Series **935** (2017) 012005 doi:10.1088/1742-6596/935/1/012005

Structural and magnetic analysis of the $Pr_{1.5}Eu_{1.5}Ba_5Cu_8O_{18}$ system

J A Parra-Borda[1], F G Rojas-Cruz[1], A F Cruz-Pacheco[2], S Segura-Peña[1,2] and C A Parra Vargas[2]
[1] Universidad Santo Tomás, Tunja, Colombia
[2] Universidad Pedagógica y Tecnológica de Colombia, Tunja, Colombia

E-mail: sully.segura01@usantoto.edu.co

Abstract. This research describes the synthesis and characterization of the $Pr_{1.5}Eu_{1.5}Ba_5Cu_8O_{18}$ system, obtained by the solid-state reaction method. For the synthesis of this material, oxides of Pr_6O_{11}, Eu_2O_3, BaO and CuO were used. The sintering temperature was 900°C. The structural analysis of the synthesized systems was carried out using X-Ray Diffraction (XRD), with a comparison with theoretical patterns and validated with Rietveld refinement analysis, obtaining a type P2mm crystallographic orthorhombic system with a=3.892454Å, b=3.894411Å and c=31.155851Å. The characterization by means of Scanning Electron Microscopy (SEM) let us conclude that the material presents homogeneous granular morphology. Finally, the magnetic response of this material is determined by the magnetization curves as a function of the applied magnetic field 50K and as a function of temperature 2000Oe.

1. Introduction

Crystallographically, the perovskite type materials are represented by the general expression ABO_3, where A is a representative element and B is a transition metal which has a smaller ionic radius than the cation at position A [1-3]. In recent years attention has been paid to the incorporation of earth elements into the position of cation A, since these atoms create structural defects that allow the achievement of certain characteristics and properties, in particular ferromagnetic and superconducting ones, the most of the rare earths in their valence state 3+. And in coordination site VIII they have a similar radius and therefore can be replaced in different systems forming the same crystalline structure.

Currently, much interest has arisen about oxide composites with perovskite type materials due to their diversity of crystalline structures and physical properties; these compounds are known for having unique properties such as ferroelectricity, electrical conductivity, superconductivity and magnetoresistance that can be used in different technological and industrial applications [4]. Some of the most relevant applications of this type of materials are the superconducting magnets [5], electric motors [6] and magnetic shields [7].

One of the most important perovskite type materials for superconductive applications are compounds such as $YBa_2Cu_3O_7$, Bi-2223 and $REBa_2Cu_3O_7$, where RE refers to a rare earth element such as Nb, Nd, Sm, Gd, Eu, among others. The most used material for research in terms of superconducting properties is $YBa_2Cu_3O_7$ since this one presents a great facility in its synthesis process and it has excellent superconducting properties [8] due to its characteristics, thus, allowing the

IOP Conf. Series: Journal of Physics: Conf. Series **935** (2017) 012005 doi:10.1088/1742-6596/935/1/012005

modelling of complex systems to explain their electrical and magnetic properties. The development of these models has contributed to the creation of more complex systems, among them Re124, Re358 [9], which have ostensibly improved its characteristics. Many compounds with different stoichiometric weights have been used structurally varying the primitive cell of the perovskite, which can present a simple, or rather complex structure depending on the amount of rare earth atoms that are arranged in its stoichiometry. It is necessary to highlight that one of the first compounds is from the Y123 family [10].

The properties of these compounds and its use in various technologies are affected by the synthesis method applied, in which, parameters such as purity of the precursors and temperature, are indispensable for the improvement of the magnetic and morphological structural properties. For this reason, this work pretends to enhance the above described characteristics for the Re358 type compound through the new $Pr_{1.5}Eu_{1.5}Ba_5Cu_8O_{18}$ system, synthesized by means of the solid state reaction method, using high purity precursor oxides.

2. Experimental

The $Pr_{1.5}Eu_{1.5}Ba_5Cu_8O_{18}$ system was synthesized by using the standard method of solid-state reaction (SSR) [11]. Stoichiometric amounts of Pr_6O_{11} (99.9%), Eu_2O_3 (99.9%), BaO (99.99%) and CuO (99.999%) oxides were used. The oxides were mixed and macerated for 4 hours, it was compressed at a pressure of 5MPa and were calcined at a temperature of 850°C for 34 hours. The synthesis process was made at 900°C for 34 hours. After completing the previously described stages of the process, the synthesized and oxygenated correspondent final stage was carried out. The sample was set at a synthesis temperature of 900°C, with a ramp for this process that lasted a 116-hour period, and when 44 hours had passed, the oxygenation process started, and coming to an end when it arrived at a temperature of 450°C. The sample oxygenated process lasted approximately 51.5 hours. An oxygen flux of 100mL/min followed by slow cooling at 300°C was used and lasted 9 hours until achieving room temperature.

The structural characterization of the sample was analysed by XRD Method, in a PAN'alitycs PRO "MPD" diffractor, equipped with an accelerator detector X 'Ultra-Fast' and Bragg-Brentano configuration, using Cu Kα (λ=1.54186Å) between 20° y 90°. The measurements were performed with a tension of 40kV and a current of 20mA [2]. For the Rietveld refinement analysis PCW and GSAS software was used. The refinement was made with the purpose of establishing the network parameters that make up this new material.

3. Results and discussion

3.1. Structural analysis

For the elaboration of a theoretical Rietveld plot and the refinement of the system $Pr_{1.5}Eu_{1.5}Ba_5Cu_8O_{18}$ the superconducting material Y358 [12] was taken as a reference point. Figure 1 shows the analysis of Rietveld refinement, where you obtained a cell with a crystallographic space group P2mm and the following parameters: a=3.892454Å, b=3.894411Å and c=31.155851Å; this structure has been reported as perovskite-orthorhombic [13]. Figure 2 shows the structure of the system, and the crystallographic cell in the Table 1 shows the refined atomic positions and occupations of the Pr358. A comparison between theoretical Rietveld plot and the experiment shows evidence of the emergence of two impurities, $PrBa_2Cu_3O_7$ and BaO which can be removed if the sintering temperature is increased. With the refinement we found a χ^2=4.6 and R(F*2)=0.09% 0.0742 error.

IOP Conf. Series: Journal of Physics: Conf. Series **935** (2017) 012005 doi:10.1088/1742-6596/935/1/012005

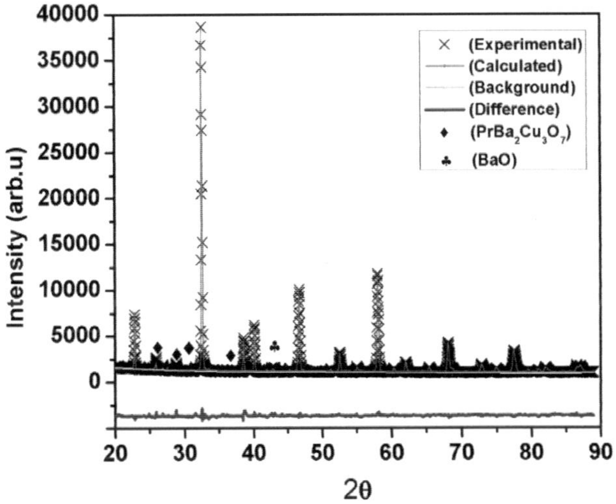

Figure 1. Rietveld refining of the sample $Pr_{1.5}Eu_{1.5}Ba_5Cu_8O_{18}$ through the general structure analysis of the program (GSAS).

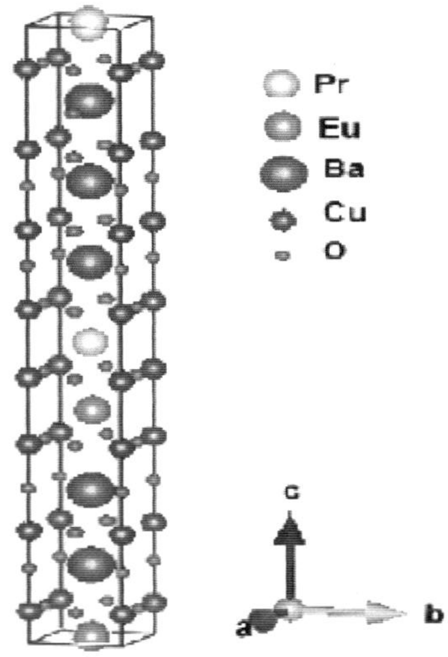

Figure 2. Crystal structure of $Pr_{1.5}Eu_{1.5}Ba_5Cu_8O_{18}$ obtained with the Vesta software.

Table 1. Refining of the atomic positions and atoms occupancy in the sample $Pr_{1.5}Eu_{1.5}Ba_5Cu_8O_{18}$.

Atoms	x	Y	z	Occupancy
Pr(1)	0.5263	0.6552	0.0000	0.3871
Eu(1)	0.4968	1.5716	0.0000	0.4141
Pr(2)	0.5465	0.6093	0.3547	0.5307
Eu(2)	0.4163	1.4368	0.3628	0.4561
Pr(3)	0.4174	0.5000	0.5156	0.4775
Eu(3)	0.5381	0.5000	0.5117	0.5861
Pr(4)	0.5706	0.5000	0.0334	0.4862
Eu(4)	0.3783	0.5000	0.0128	0.4854
Ba(1)	0.4754	0.5256	0.1191	0.9553
Ba(2)	0.5371	0.3520	0.2212	0.9021
Ba(3)	0.5032	0.5000	0.3871	0.9923
Ba(4)	0.4777	-0.4677	0.2512	0.8563
Ba(5)	0.5144	0.5000	0.8571	0.8672
Cu(1)	1.0256	0.0000	0.9420	1.0000
Cu(2)	1.0019	0.0000	0.8129	0.8507
Cu(3)	1.0239	0.0000	0.6865	1.0000
Cu(4)	0.9892	0.0000	0.5627	0.9655
Cu(5)	0.9965	0.0000	0.4379	0.9685
Cu(6)	1.0248	0.0000	0.3176	1.0000
Cu(7)	1.0021	0.0000	0.1940	0.8626
Cu(8)	1.0073	0.0000	0.0690	1.0000
O(1)	0.4388	0.0000	0.9466	0.9202

O(2)	1.0320	0.5000	0.9747	1.0000
O(3)	1.0624	0.0000	0.8999	1.0000
O(4)	0.0472	0.1267	0.8204	1.0000
O(5)	1.0820	0.0000	0.7706	1.0000
O(6)	0.5333	0.0000	0.6974	1.0000
O(7)	1.0843	0.5000	0.6829	0.4031
O(8)	0.5220	0.0000	0.5648	0.9203
O(9)	1.0050	0.5000	0.5747	0.9971
O(10)	0.5677	0.0000	0.4261	0.6453
O(11)	0.0221	0.4459	0.4407	1.0000
O(12)	1.0104	0.0000	0.3557	1.0000
O(13)	0.0367	0.6075	0.3167	1.0000
O(14)	1.0535	0.0000	0.2328	0.7458
O(15)	-0.0460	1.1201	0.2095	1.0000
O(16)	0.0095	-3.4793	0.1777	1.0000
O(17)	0.4384	0.0000	0.0533	0.9088
O(18)	1.0430	0.5000	0.0382	0.9699

3.2. Morphological analysis

Figure 3 shows the results obtained through SEM for different magnifications x (3000, 5000 and 10000) of a cross-section taken for the sample $Pr_{1.5}Eu_{1.5}Ba_5Cu_8O_{18}$, in which there is a clear presence of porosity, and the grain size is not uniform, indicating that the sintering process which was carried out for this sample is not satisfactory [14].

Figure 3. SEM images for the cross section of $Pr_{1.5}Eu_{1.5}Ba_5Cu_8O_{18}$ with different magnifications, increases (A) 3000, (B) 5000, (C) 10000.

Figure 4 shows the results obtained through SEM for three magnifications x (3000, 5000 and 10000) that correspond to the superficial part of the material. It is evident the presence of porosity in different parts of the material, and coplanar defects at high and low angles but presents domain in the crystallographic size between 10 and 20nm, which is in agreement with the results obtained by XRD. As reported in the references [12] and [15], the grain size of the sample is a function of the oxygen content applied to it, therefore, the samples of $Pr_{1.5}Eu_{1.5}Ba_5Cu_8O_{18}$ must have a process of oxygenation.

IOP Conf. Series: Journal of Physics: Conf. Series **935** (2017) 012005 doi:10.1088/1742-6596/935/1/012005

Figure 4. Images of SEM, surface part of samples Pr1.5Eu1.5Ba5Cu8O18 at different magnifications, increases (D) 3000, (E) 5000, (F) 10000.

3.3. Magnetic properties

This test was carried out in a commercial magnetometer, with a SQUID (Superconducting Quantum Interference Device) sensor, Quantum Design (MPMS-5S) device. Figure 5 shows the graph of magnetization as a function of the magnetic field at a constant temperature of 50K. The curve shows the typical behaviour of a paramagnetic material, as that can be seen in Figure 6, a graphic of magnetic susceptibility as a function of the temperature, at a constant magnetic field of 200Oe, which follows the law of Curie. At the same time at right top of the graph, the inverse of the magnetic susceptibility was plotted as a function of temperature. The slope of the graph determines the effective magnetic moment of the sample, from which it was obtained a value for $Pr_{1.5}Eu_{1.5}Ba_5Cu_8O_{18}$ of 5.74μ_B and a theoretical effective magnetic moment of 5.79μ_B [16].

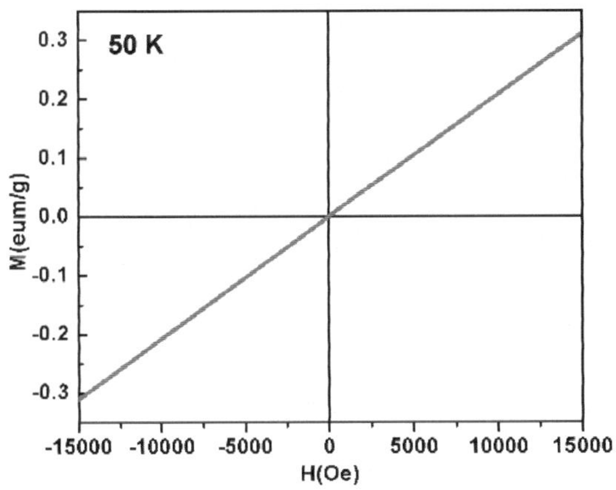

Figure 5. M vs H curves of $Pr_{1.5}Eu_{1.5}Ba_5Cu_8O_{18}$ samples at 50 K.

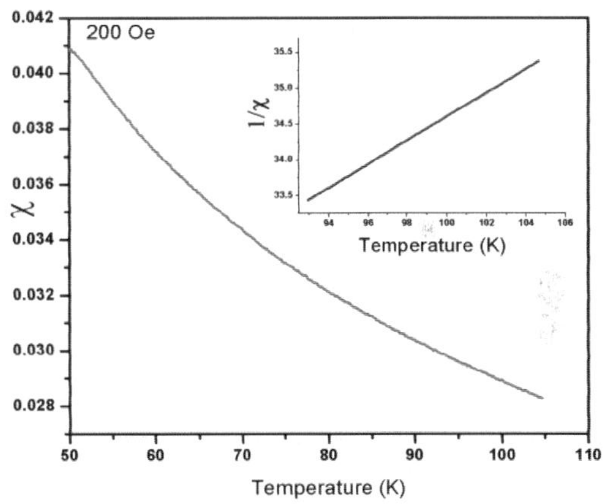

Figure 6. Susceptibility curve and its inverse susceptibility curve for sample $Pr_{1.5}Eu_{1.5}Ba_5Cu_8O_{18}$.

4. Conclusion

Using the method of solid-state reaction for the sintering of the new material $Pr_{1.5}Eu_{1.5}Ba_5Cu_8O_{18}$ with orthorhombic perovskite type structure and space group 25 with network parameters a=3.8960Å, b=3.8890Å and c=31.1500Å. The structural characterization of the sample was made with X-ray diffraction finding the main phase to be Pr358. It also showed the evidence of secondary phases. The secondary phase, which manifested itself in a greater proportion, was $PrBa_2Cu_3O_7$. The effective magnetic moment of the samples was obtained and it is expected the thermal and oxygenated processes to be performed in a slower way, in order to reach the superconducting state.

References

[1] Alvarado Flores J J 2016 Análisis de la estructura perovskita LaxSr1-xCryMn1-yO3-δ con potencial aplicación como ánodo para celdas de combustible de óxido sólido *Boletín la Soc. Española Cerámica y Vidr.* **56** 73–82

[2] Abdel-Aal S K 2017 Synthesis, characterization, thermal, and electrical properties of new diammonium hybrid perovskite [NH3-(CH2)7-NH3]CaCl2Br2 *Solid State Ionics* **303** 29–36

[3] Nishiyama A, Doi Y and Hinatsu Y 2017 Magnetic interactions in rhenium-containing rare earth double perovskites Sr2LnReO6 (Ln=rare earths) *J. Solid State Chem.* **248** 134–141

[4] Fujimoto H 2003 Flexural strength of melt-processed Y–Ba–Cu–O bulk superconductors with Ag addition measured at 77 K *Supercond. Sci. Technol.* **16** 1115–1119

[5] Jirsa M, Rameš M, Ďuran I, Melíšek T, Kováč P and Viererbl L 2017 Electric currents in REBaCuO superconducting tapes *Supercond. Sci. Technol.* **30** 45010

[6] Kawamura M and Jones J A 2017 Superconducting Super Motor and Generator *IEEE Trans. Appl. Supercond.* **27** 1–5

[7] Terao Y, Sekino M, Ohsaki H, Teshima H and Morita M 2011 Magnetic Shielding Characteristics of Multiple Bulk Superconductors for Higher Field Applications *IEEE Trans. Appl. Supercond.* **21** 1584–7

[8] D.A.LandinezTellez, H.Tovar O O-D J R 2007 Estructura cristalina del nuevo oxido tipo perovskita compleja Ba ′ 2NdZrO5,5 *Rev. Mex. Fis.* **S53** 299–302

[9] F P-C, J A-S and H T-H 2013 Contribución de los orbitales f en la densidad de estados de los superconductores NdBa2Cu3O7 y SmBa2Cu3O7 *Superf. y Vacío* **26** 84–89

[10] Gutiérrez S S, G I S, Roa-Rojas J and Vargas C A P 2016 Production a characterization of new system superconductor $TR_3 Ba_8 Cu_{11} O_\delta$ *J. Phys. Conf. Ser.* **687** 12089

[11] Vargas C A P, Jr. J L P, Pureur P, Téllez D A L and Roa-Rojas J 2012 Behavior of the irreversibility line in the new superconductor La1.5+xBa1.5+x−yCayCu3Oz *Phys. B Condens. Matter* **407** 3128–3130

[12] Gholipour S, Daadmehr V, Rezakhani A T, Khosroabadi H, Shahbaz Tehrani F and Hosseini Akbarnejad R 2012 Structural Phase of Y358 Superconductor Comparison with Y123 *J. Supercond. Nov. Magn.* **25** 2253–2258

[13] Pacheco A F C, Cuaspud J A G and Vargas C A P 2017 Synthesis, characterization and magnetic evaluation of praseodymium modified cerium oxide *J. Phys. Conf. Ser.* **786** 12023

[14] Guerrero U F, Parra Vargas C A and Santos A S 2011 Procesamiento en descarga luminiscente de la perovskita YBa 2 Cu 3 O 7-d **78** 201–205

[15] Hurtado P C P, Téllez D A L, Vásquez J A C, Vargas C A P and Roa-Rojas J 2015 Synthesis, structural characterization, electric and magnetic behaviour of a Sr_2DyNbO_6 double Perovskite *J. Phys. Conf. Ser.* **614** 12002

[16] Gómez J A M, García G I S, Palacio C A and Vargas C A P 2015 Production and structural and magnetic characterization of a $Bi_{1-x} Y_x FeO_3$ (x = 0, 0.25 and 0.30) system *J. Phys. Conf. Ser.* **614** 12003

IOP Conf. Series: Journal of Physics: Conf. Series **935** (2017) 012006 doi:10.1088/1742-6596/935/1/012006

Use of steel slag as a new material for roads

R Ochoa Díaz[1], M Romero Farfán[1], J Cardenas[1] and J Forero[1]
[1] Universidad Pedagógica y Tecnológica de Colombia, Tunja, Colombia

E-mail: ricardo.ochoa@uptc.edu.co , miguel.romero@uptc.edu.co

Abstract. This research paper aims to analyse the behaviour of MDC-19 hot dense asphalt mixtures with steel slag as coarse aggregate, by using asphalt 80-100, in order to verify if this residue has suitable characteristics that allow its use. The physical and mechanical characterization was accomplished using phosphorous slag from the company Acerías Paz del Río S.A. The working formula was then determined for each mixture using the RAMCODES methodology, the briquettes were produced in the laboratory and then, the design verification was performed. Taking into account the results obtained, it is concluded that the use of phosphorous slag as coarse aggregate in asphalt mixtures is workable, since acceptable design parameters and verification are obtained that meet the specifications for use as a rolling layer.

1. Introduction

Engineering is constantly faced with problems such as the demand for materials that meet the previously established technical requirements and ensuring their continuous supply. The design of asphalt mixtures with waste can be the solution to reuse waste from the steel industry and mitigate the environmental effects produced by the indiscriminate exploitation of natural resources [1]. Furthermore, this would enable offering pavement structures that meet all INVIAS specifications [2]. In addition, residues that offer geotechnical and hydraulic properties equivalent or superior to that of conventional granular materials can be found [3].

The phosphorous slag used in this study is a residue from the company *Acerías Paz del Río S.A.*, which is generated in converters for the transformation of pig iron into steel [4]. With the results obtained it will be possible to implement the steel waste in the elaboration of pavement structures for different levels of traffic, which would be of great benefit economically and environmentally, in order to reduce the exploitation of natural resources, for use of aggregate in Roads, and environmental pollution caused by waste in the steel industry [5].

2. Materials and methodology

The steel residue used in this study is phosphorous slag, as shown in Figure 1, which is produced in integrated steelworks, in the converters [6]. This process is carried out for the transformation of pig iron from the blast furnace into steel. Figure 2 shows a SEM image at 10μm scale, in which the rough surface texture of the slag can be observed [7].

The stone materials used in this work are from "La Roca" Shredder, located in the municipality of Moniquirá (Boyacá) and the sand is supplied by the "Solarte and Solarte" consortium in the city of Tunja (Boyacá). Asphalt type 80-100 was the binder used.

The methodology used in this research is based on three stages: physical, chemical and mechanical characterization of the aggregates to be used; preparation of the laboratory mixtures for the working formula using the RAMCODES methodology [8,9] and, verification of the design to make the comparison of the most representative parameters of the mixtures and thus, to be able to define and establish the use of this residue. All the tests were carried out in the pavement laboratory of the "Universidad Pedagógica y Tecnológica de Colombia".

Figure 1. Phosphorous slag

Figure 2. SEM micrographs of phosphorous slag.

3. Experimental design and results

Three blends were made for the development of the research: mixture A, with conventional stoneware (sand and gravel); mixture B, substituting 50% of the coarse aggregate for slag and C mixture, replacing 100% of the coarse aggregate with phosphorous slag. All mixtures were adjusted to the granulometry of a MDC-19 blend.

3.1. Characterization of aggregates and slag

The characterization of the stone aggregates and phosphorous slag must show an excellent behaviour when subjected to conditions of work in the field. The aggregates are submitted to laboratory processes in accordance with the requirements of the INVIAS general road rules and specifications [10] and thus to evaluate the characteristics of the asphaltic mixtures. The results of the characterization of the stone aggregates are shown in Table 1.

Table 1. Properties of stone aggregates (gravel and sand).

Feature	Unit	Requirement	Result	Standard
"Los Ángeles" wear max.	%	25	20	INV E-218
Loss of solidity max.	%	18	3.2	INV E-220
Fractured faces min.	%	85	94	INV E-227
Impurities max.	%	0.5	0.45	INV E-237
Sand equivalent min.	%	50	68.5	INV E-133
Plasticity index max.	%	NP	NP	INV E-126

As shown in Table 2, phosphorus slag used as coarse aggregate meets the requirements stipulated by the standards for coarse aggregates to be used in asphalt mixtures.

Table 2. Properties of phosphorous slag.

Feature	Unit	Requirement	Result	Standard
"Los ángeles" wear max.	%	25	20	INV E-218
Loss of solidity max.	%	18	1.7	INV E-220
Fractured faces min.	%	85	91	INV E-227
Impurities max.	%	0.5	0.2	INV E-237

IOP Conf. Series: Journal of Physics: Conf. Series **935** (2017) 012006 doi:10.1088/1742-6596/935/1/012006

3.2. Features and comparison of mixtures

Observing Figure 3, it can be established that the three designed mixtures exceed the minimum stability of 9000N, required by INVIAS specifications [1]. The stability results indicate that the coarse material (gravel and phosphorous slag) have good friction and bonding between particles, which makes the mixtures not vulnerable to high rutting problems during their useful life, maintaining their shape when under compression from loads produced by traffic.

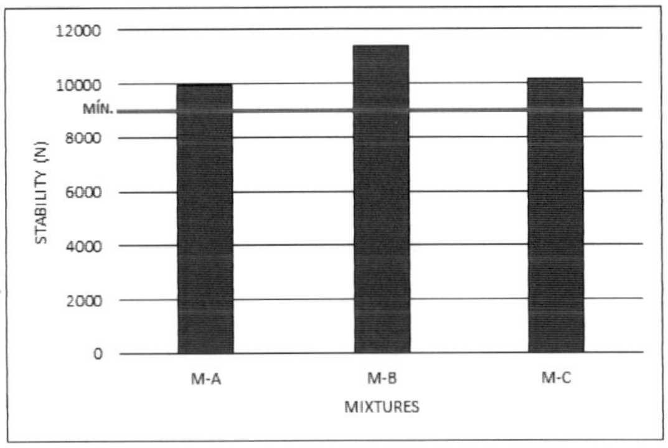

Figure 3. Stability of the mixtures.

As for the resilient modules, it was calculated for three specimens made from the working formula for each type of mixture. The Bulk gravity and the module by the principle of indirect voltage at a temperature of 25°C were determined [11]. The principle states that by applying a compressive load across the diameter of a cylindrical sample a voltage is produced on an orthogonal diameter to which the load is applied. The resilient modules were made at a frequency of 10Hz and at temperatures of 5.25 and 40°C. Each temperature was tested twice: first in its initial position and second with a specimen turn of 90°, with respect to the starting point.

The plastic deformation (rutting) was determined in the laboratory in a test consisting of subjecting a briquette of each mixture to the passage of a wheel [12], for which the depth of the produced deformation was periodically measured at 60°C.

The results of the plastic deformation test and other verification tests of the design of the mixtures are shown in Table 3. In this it is observed that the mixtures M-A and M-B comply with the adhesion parameter (minimum 80%), while the mixture M-C does not comply. The rutting presented in the three blends is very similar, fulfilling the established requirements.

Table 3. Features and properties of the mixtures.

Feature	Unit	M-A	M-B	M-C	Standard
Adhesion RRT	%	87	80	50	INV E-725
Rutting	mm	1.85	2.20	2.30	INV E-756
Resilient module at 13°C	MPa	5100	3970	1830	INV E-749
Fatigue-radial strain	Def.	0.26	0.24	0.22	INV E-784

4. Conclusions

The design of type MDC-19 hot dense asphalt mixtures, with phosphoric slag as coarse aggregate was carried out using the RAMCODES methodology. With the fulfillment of the void percentages. The optimum asphalt content of 4.7% was obtained for the M-A mixture, 6.4% for the M-B mixture and a percentage of 9.3% for the M-C mixture (for steel residues). The increase in the percentage of asphalt in the modified mixtures is due to the fact that the slag is a material with great porosity, resulting in increased costs for the design of mixtures with these materials.

IOP Conf. Series: Journal of Physics: Conf. Series **935** (2017) 012006 doi:10.1088/1742-6596/935/1/012006

The phosphorous slag used as coarse aggregate meets the requirements stipulated by the standards, especially the wear on "Los ángeles" machine, a fundamental parameter in the behaviour of the asphalt mixture.

When determining the resilient module for the temperature of Tunja (13°C) for each of the mixtures, it was possible to establish that the value of the modified mixture M-C (1810MPa) is low in relation to the mixtures M-B (3970MPa) and mixture M-A (5100MPa). If actual operating conditions are used in which repeated loads of traffic and temperature increase are generated, there will be a shortening of the life of the tread layer, due to the appearance of exudation problems and generation of repetitive deformations.

References

[1] Francisco S 2002 Utilización de residuos en la construcción de capas de firmes de carreteras *Ingeniería Civil* **128** 72

[2] Instituto Nacional de Vías 2013 *Especificaciones generales de construcción de carreteras* (Bogotá: INVIAS)

[3] Rahman A, Imteaz M, Arulrajah A, Piratheepan J and Disfani M 2015 Recycled construction and demolition materials in permeable pavement systems: geotechnical and hydraulic characteristics *Journal of Cleaner Production* **90** 184

[4] Gómez A 2002. *Proceso siderúrgico planta Belencito* (Colombia: Acerías Paz del Río S.A.)

[5] Gulnur M, Filiz B, Tanju K and Ulku Y 2016 The environmental impacts of iron and steel industry: a life cycle assessment study *Journal of Cleaner Production* **130** 195

[6] Wu W, Meng H, and Liu L 2013 Melting characteristics of recycling slag in descarburization converter and its application effects *Journal of Iron and Research, International* **20** 7

[7] Chao L, Zongwu Ch, Shaopeng W, Bo L, Jun X and Yue X 2017 Effects of steel slag fillers on the rheological properties of asphalt mastic *Construction and Building Materials* **145** 384

[8] Sánchez F, Garnica P, Larreal M and López D 2011 Polyvoids: Analytical tool for superpave HMA design *Journal of Materials in Civil Engineering* **23** 1129-1137

[9] Sánchez F, Garnica P, Gómez J and Pérez N 2002 Ramcodes: Metodología racional para el análisi de densificación y Resistencia de geomateriales compactados *Instituto Mexicano del Transporte* **200** 29

[10] Instituto Nacional de Vías 2013 *Normas de ensayos de materiales para carreteras* (Bogotá: INVIAS)

[11] Muhammad K, Hamid N, Surya S and Laksmi I 2017 Laboratory experiment on resilient modulos of BRA modified asphalt mixtures *International journal of Pavement Research and Technology* **10** 5

[12] Preeda Ch and Hussain U 2017 Mechanisms of asphalt mixture rutting in the dry Hamburg Wheel Tracking test and the potencial to be alternative test in measuring rutting resistance *Construction and building Materials* **146** 176

IOP Conf. Series: Journal of Physics: Conf. Series **935** (2017) 012007 doi:10.1088/1742-6596/935/1/012007

Structural and magnetic analysis of the $Bi_{x-1}Sm_{0x}FeO_3$ (x=0.04 and 0.07) system

J D Suárez Londoño[1], S Segura Peña[1,2], E H Sáchica[1,2], A F Cruz Pacheco[2], A Sarmiento Santos[2] and C A Parra Vargas[2]
[1] Universidad Santo Tomas, Tunja, Colombia
[2] Universidad Pedagógica y Tecnológica de Colombia, Tunja, Colombia

E-mail: sully.segura01@usantoto.edu.co

Abstract. This work reports on the synthesis and characterization of the $Bi_{x-1}Sm_xFeO_3$ (x=0.04 and 0.07) systems, produced by the solid-state reaction method. The samples obtained were characterized by X-Ray Diffraction (XRD), Rietveld refinement and magnetization measurements as a function of the magnetic field applied and temperature. The structural characterization was performed by Rietveld refinement and allowed the identification of a perovskite type structure on all the materials with rhombohedral crystalline system and $R3_C$ space group. The characterization by Scanning Electron Microscope (SEM) allowed to determine that the sintered solids present a homogenous microstructure. The magnetization curves as a function of magnetic field applied at 50 and 300K and as a function of the temperature at 5000Oe of the sintered solids exhibited a ferromagnetic behavior, with a rise of the permanent magnetization (Mr), the coercive field (Hc) and the magnetic susceptibility when the samarium concentration rose up. The structural and magnetic characteristics that are generated in the obtained solids make these materials of great interest as multiferroic components.

1. Introduction

Electronic devices play an important role in people's daily life. This is why there is a high demand for the evolution of these systems, producing everyday smaller, more efficient and affordable devices. However, the advance of this field is threatened by an invisible wall that limits the miniaturization of components for these elements.

During the last years, there has been a notable increase in the interest for components based on bismuth ferrite $BiFeO_3$ [1], several studies have been done in the area of multiferroics based on their potential applications in multiple state memory elements, sensors and spintronic devices. The advantages of $BiFeO_3$ over other materials are high ferroelectric transition temperature (Curie Temperature) which is around T_C=1083*K* and its high antiferromagnetic temperature (Neel temperature) which is around T_N=643 *K*, which places it really above its RT where it possesses a space group R3c, making it viable for practical applications [2].

This material is interesting not just for their multiferroic properties, there is not lead in their structure and therefore is a candidate to replace the traditional PZT in free lead ceramics. The $BiFeO_3$ has three big problems when it is obtained, as it is not pure due to the secondary phases presented, such as: $Bi_2Fe_4O_9$, $Bi_{25}FeO_{40}$ and $Bi_{36}Fe_{24}O_{57}$, because of metastable state of BFO, phases that depend of the synthesis technique used, the other problem found is a decrease of the density at the moment of obtaining pastilles [3–5] , and last, the big leakage current and the difficulties to achieve the net magnetization since the magnetic order of these materials is the antiferromagnetic type [6], thus limiting the applications of this material in various devices. However, in order to improve the magnetic properties, various studies suggest the application of a big magnetic field or the ionic doping of rare earths, occupying the place of the Bi compound, favoring the magnetism and the stability of distorted structures ferroelectrically [6]. Others studies show an improvement in the introduction of lanthanides such as La^{3+}, Nd^{3+}, Sm^{3+}, Eu^{3+}, Gd^{3+}, etc. Currently, a lot of studies with respect to the magnetic properties $BiFeO_3$ doping with Sm over 10% of the doping concentration have been reported. But, there is a lot of work to be done below this level [7–10]. The present research is based on the synthesis and characterization of a ceramic oxide, perovskite type, based on the $Bi_{1-X}Sm_XFeO_3$ system with a level of Samarium cation (x=0.04 and 0.07).

2. Experimental

The Spuds software is used to predict the tolerance factor of formation from structures of $Bi_{0.9}Sm_{0.1}FeO_3$ system with R3c space group as the most viable. For the synthesis of $Bi_{0.9}Sm_{0.1}FeO_3$ system, solid state method (RES) was used, starting with Bi_2O_3 (99,999%), Sm_2O_3 (99,9%) and Fe_2O_3 (99,995%) oxides, mixing and macerating them in a mortar of agate for 4 hours until the oxides get completely homogenized. The samples are compressed at 4MPa and calcined at 973K for 24 hours. The sintering process is performed at 1103K for 36 hours.

The crystallographic structure was determined with XRD using a X´Pert PRO PANalitycal diffractometer with Cu-K radiation (□=1.540598Å). The Rietveld refinement analysis was carried out using PCW and GSAS software. The images of SEM were obtained in a microscope JEOL JSM 6490-LV brand. Finally, the M-H and M-T curves were made in a Versalab magnetometer with magnetic fields between -3T and 3T and temperatures between 50-320K.

3. Results and discussions

The Figure 1 shows the Rietveld refinement for $Bi_{0.96}Sm_{0.04}FeO_3$, it has not manifested secondary phases. The Rietveld refinement analysis for $Bi_{0.93}Sm_{0.07}FeO_3$ system, (Figure 2), show secondary phases of $Bi_2Fe_4O_9$, which are traces of around 4% and a principal phase of $Bi_{0.93}Sm_{0.07}FeO_3$ with 96%.

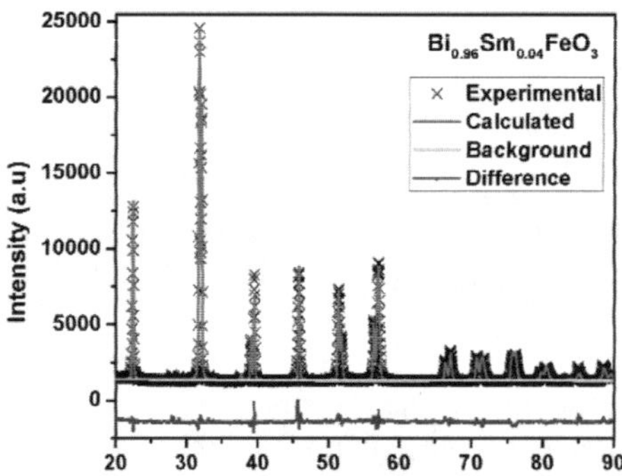

Figure 1. Rietveld refinement for $Bi_{0.93}Sm_{0.07}FeO_3$.

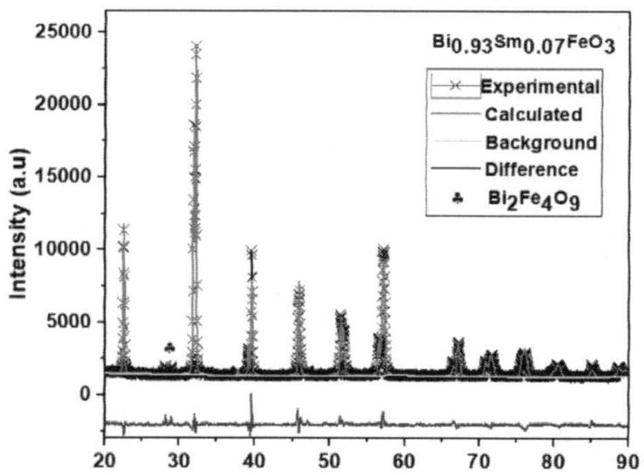

Figure 2. Rietveld refinement for $Bi_{0.96}Sm_{0.04}FeO_3$.

In Table 1, are shown the lattice parameters and atomic positions obtained by Rietveld refinement, with symmetry of space group *R*3c or rhomboidally. The a, b and c values become higher with an increasing samarium concentration. The statistical parameters of the refinements were: For $Bi_{0.96}Sm_{0.04}FeO_3$ system χ^2=4.941 and R(F*2)=0.0742, and for $Bi_{0.93}Sm_{0.07}FeO_3$ system χ^2=5.124 and R(F*2)=0.0565.

Table 1. Structural parameters for $Bi_{1-x}Sm_xFeO_3$ (X=0.04 and 0.07) obtained with Rietveld refinement.

Lattice parameters				$Bi_{0.96}Sm_{0.04}FeO_3$			$Bi_{0.93}Sm_{0.07}FeO_3$	
a (Å)				5.5805			5.5819	
b (Å)				5.5805			5.5819	
c (Å)				13.8550			13.8440	
Atomic positions								
Atoms	x	y	z	OCC	x	y	z	OCC
Bi	0.0000	0.0000	0.7767	0.9333	0.0000	0.0000	0.7756	0.8688
Sm	0.0000	0.0000	0.2281	0.0519	0.0000	0.0000	0.7575	0.0119
Fe	0.0000	0.0000	0.9993	1.0132	0.0000	0.0000	0.9969	0.9828
O	0.8976	0.6534	0.4965	1.0045	0.8918	0.6596	0.4941	1.1345

Figure 3 (a), (b) and (c) shows the SEM micrographs for the $Bi_{0.96}Sm_{0.04}FeO_3$ system with a magnification degree of X3000, X5000 and X10000 respectively, and Figure 4(a), (b) and (c) shows the $Bi_{0.93}Sm_{0.07}FeO_3$ system with a magnification degree of X3000, X5000 and X10000 respectively.

(a) (b) (c)

Figure 3. SEM micrographs of the $Bi_{0.96}Sm_{0.04}FeO_3$ system at different magnifications, (c) X3000, (b) X5000 and (c) X10000.

(a) (b) (c)

Figure 4. SEM micrographs of the $Bi_{0.93}Sm_{0.07}FeO_3$ system at different magnifications, (a) X3000, (b) X5000 and (c) X10000.

Both present grains with rounded shape and high interconnection between the grains and low porosity, in other words, a big homogeneity. Moreover, x=0.04 and x=0.07 have grain sizes around 5μm, but only x=0.04 has a samarium grain size less than it. A visual inspection reveals the presence of a single second phase for x=0.07 like the XRD measurements. For more concentrations of samarium the grains are a little larger. On other hand, the x=0.04 micrographs do not have high concentrations of samarium grains.

Both magnetization hysteresis (M-H) loops for Bi1-XSmXFeO3 Perovskite systems (x=0.04 and 0.07) taken at room temperature and 50K, show a very thin hysteretic feature but non-zero remnant magnetization as shown in Figures 5 and 6. The systems with the samarium samples exhibit higher values of remnant magnetization and low coercive field, so, it is possible to infer that the substitution of samarium in the A-site influences the magnetic properties of BFO system.

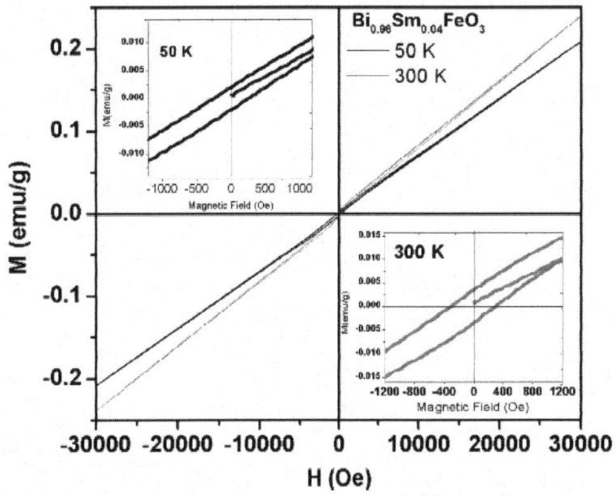

Figure 5. M-H curves hysteretic feature of $Bi_{0.96}Sm_{0.04}FeO_3$ system at 50K (red line) and 300 K (black line).

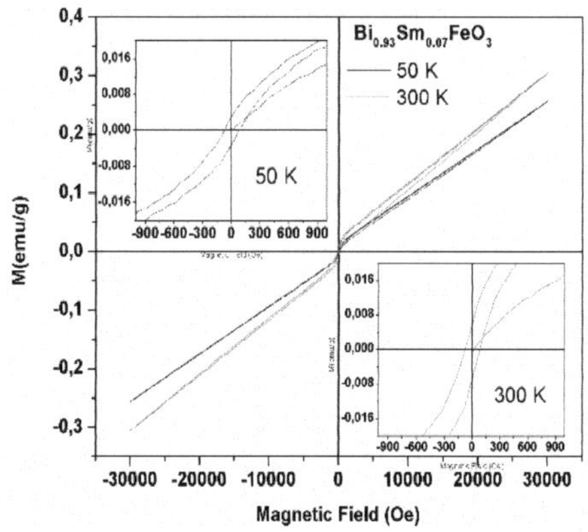

Figure 6. M-H curves hysteretic feature of $Bi_{0.93}Sm_{0.07}FeO_3$ perovskite at 50K (red line) and 300 K (black line).

As it can be seen in Table 2, the remnant magnetic fields are higher respect to the Sm concentration, but the coercive field decreases with that concentration. The ionic radium of Sm^{+3} is lower compared with the one for Bi^{3+}. Doping with this rare earth element affects the links between Bi-O and Fe-O, according with the reports by Gómez and collaborators [11].

Table 2. Magnetic values obtained for the $Bi_{x-1}Sm_{0x}FeO_3$.

	T=50K		T=300K	
X	H_C (Oe)	Mr (emu/g)	H_C (Oe)	Mr (emu/g)
0.04	262	0.001919	296	0.0035
0.07	79	0.003015	80	0.0059

Figure 7 and 8 show the graphics of magnetization as a function of the temperature for the sample 0.04 and 0.07 respectively, with a constant field of 5000Oe. If we compare both samples it can be observed that the sample 0.04 is ferromagnetic, but the sample 0.07 shows a weak ferromagnetic state that transforms into anti-ferromagnetic at 130K; this behavior has been reported for Zatsiupa A. A. *et.al.* [12].

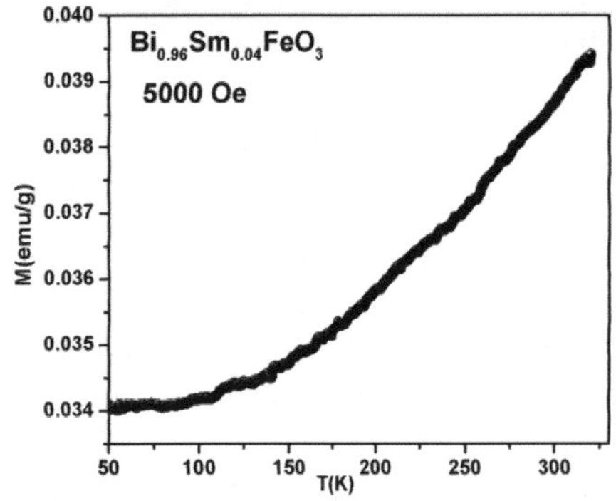

Figure 7. ZFC curves for $Bi_{0.96}Sm_{0.04}FeO_3$ samples at 5000Oe.

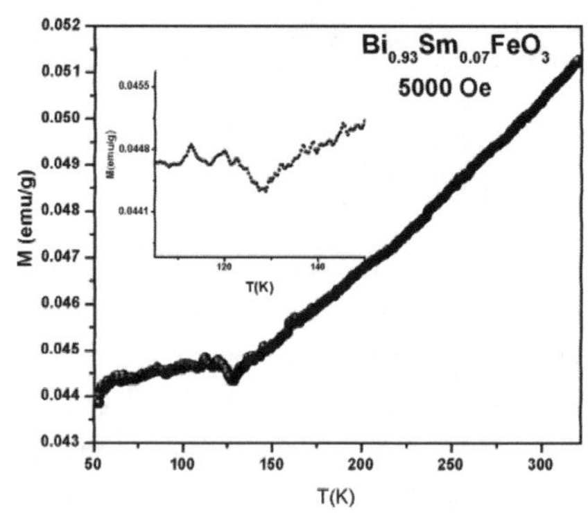

Figure 8. ZFC curves for $Bi_{0.93}Sm_{0.07}FeO_3$ samples at 5000Oe.

IOP Conf. Series: Journal of Physics: Conf. Series **935** (2017) 012007 doi:10.1088/1742-6596/935/1/012007

4. Conclusions

The crystallographic structure was obtained for both samples 0.04 and 0.07. The 0.04 one presented a single crystalline phase, while the second one presented a 96% pure phase and 4% Bi2Fe4O9. At the same time was observed that the behaviour of the grain is the same for the produced samples. The magnetic answer corresponds to the typical behaviour of the BiFeO3 materials. It was found that the remnant magnetic field increases with depending on the Sm concentration, but the coercive field decreases with the Sm concentration.

References

[1] Mejía Gómez J A, Canaria C, Ochoa Burgos R, Ortiz C A, Supelano G I and Parra Vargas C A 2016 Structural study of yttrium substituted $BiFeO_3$ *J. Phys. Conf. Ser.* **687** 12091

[2] Catalan G and Scott J F 2009 Physics and applications of bismuth ferrite *Adv. Mater.* **21** 2463–2485

[3] Mazumder R, Sujatha Devi P, Bhattacharya D, Choudhury P, Sen A and Raja M 2007 Ferromagnetism in nanoscale BiFeO3 *Appl. Phys. Lett.* **91** 62510

[4] Carvalho T T and Tavares P B 2008 Synthesis and thermodynamic stability of multiferroic BiFeO3 *Mater. Lett.* **62** 3984–3986

[5] Zhang X, Qi X, Yang Z and Zhong R 2014 Comparative studies of multiferroic BiFeO3 powders prepared by hydrothermal method versus sol-gel process *Applied Mechanics and Materials* **670-671** 113-116

[6] Godara P, Agarwal A, Ahlawat N, Sanghi S and Dahiya R 2014 Crystal structure transformation, dielectric and magnetic properties of Ba and Co modified BiFeO3 multiferroic *J. Alloys Compd.* **594** 175–181

[7] Lebeugle D, Colson D, Forget A, Viret M, Bataille A M and Gukasov A 2008 Electric-Field-Induced Spin Flop in BiFeO 3 Single Crystals at Room Temperature *Phys. Rev. Lett.* **100** 227602

[8] Khomchenko V A, Kiselev D A, Vieira J M, Jian L, Kholkin A L, Lopes A M L, Pogorelov Y G, Araujo J P and Maglione M 2008 Effect of diamagnetic Ca, Sr, Pb, and Ba substitution on the crystal structure and multiferroic properties of the BiFeO3 perovskite *J. Appl. Phys.* **103** 24105

[9] Yuan G L, Or S W, Liu J M and Liu Z G 2006 Structural transformation and ferroelectromagnetic behavior in single-phase Bi1−xNdxFeO3 multiferroic ceramics *Appl. Phys. Lett.* **89** 52905

[10] Chakrabarti K, Das K, Sarkar B and De S K 2011 Magnetic and dielectric properties of Eu-doped $BiFeO_3$ nanoparticles by acetic acid-assisted sol-gel method *J. Appl. Phys.* **110** 103905

[11] Gómez J A M, García G I S, Palacio C A and Vargas C A P 2015 Production and structural and magnetic characterization of a $Bi_{1-x}Y_xFeO_3$ (x=0, 0.25 and 0.30) system *J. Phys. Conf. Ser.* **614** 12003

[12] Zatsiupa A A, Bashkirov L A, Troyanchuk I O, Petrov G S, Galyas A I, Lobanovskii L S, Trukhanov S V. and Sirota I M 2013 Magnetization, magnetic susceptibility, and effective magnetic moment of the Fe3+ ions in Bi2Fe4O9 *Inorg. Mater.* **49** 616–620

IOP Conf. Series: Journal of Physics: Conf. Series **935** (2017) 012008 doi:10.1088/1742-6596/935/1/012008

Characterization and obtainment of phosphate rock concentrates of Turmequé, Boyacá

S Quijano Zanguña[1], L F Lozano Gómez[1] and Y Pineda Triana[1]
[1] Universidad Pedagógica y Tecnológica de Colombia, Tunja, Colombia

E- mail: sonia.quijano@uptc.edu.co

Abstract. The work focuses on the use and exploitation of the mineral concentrates from phosphate rock (PR) coming from mines with a low percentage of phosphorus. The procedure was based on the collection of a source of phosphate rock from the department of Boyacá (municipality of Turmequé), using a randomized design with three replications. The samples were initially milled and sifted using meshes between 140 and 200 US standard, homogenizing them and improving the process of solubility of the phosphorus in the soil. We conduced Z-potential tests, which show that by performing a prior wash on the mineral and maintaining certain concentrations and pH defined, better results are achieved in terms of the buoyancy of the particles in the flotation process. The results obtained from the microflotation tests; both direct and inverse, and the results of chemical composition, with X-Ray Fluorescence (XRF) and X-Ray Diffraction (XRD), before and after the microflotation process, were carried out to obtain of commercial laws grade phosphate rock concentrate, confirm that the protocol used increases by 9% the value of total phosphorus in the collected sample. These concentrates from phosphate rock, could be used in the future for the attainment of simple superphosphate (SSP), with the help of sulphuric acid and ammonium thiosulphate mixtures.

1. Introduction

Fertility problems limit the production of crops in many acid soils around the world, especially in the tropics. These soils generally have a low content of phosphorus (P), available to the plants [1]. Phosphorus used in agriculture comes from non-renewable sources, mainly from phosphate rock [2], which is insoluble in water and cannot be used directly in agriculture, in neutral or basic soils. In acid soils, finely ground phosphate rock is often applied, phosphorus pentoxide (P_2O_5) slowly enters the solution, throughout months or years, according to the pH of the soil [3]. However, for the most part, before phosphate rock can be used as a fertilizer, it is essential to treat it in some way in order to turn phosphorus into a soluble form [4]. The classic method used is acidulation, in order to produce normal superphosphate.

Colombia which has a significant agricultural potential of around four million arable hectares, imports virtually all of its phosphorus needs, despite having significant deposits of phosphate rock (PR) in the departments of Boyacá, Huila and Santander [5]. Some of these deposits were partially explored by "Ingeominas" in the early 1960s. According to this entity, the thickness of the mineralized levels varies between 0.5 to 5.4 meters and the tenors of P_2O_5 range between 10 and 37% [6]. In Boyacá, mines with a high percentage of phosphorus are being exploited, ignoring those with a low tenor. This study intends to harness and leverage the areas that have a low percentage of this element, as is the case of the ores located in the municipality of "Turmequé, Boyacá", exploiting the mineral and turning it into raw material for the attainment of phosphate rock concentrates and their subsequent use in the production of simple superphosphate type (SSP) type fertilizers [7], with the help of sulphuric acid and ammonium thiosulphate mixtures [8].

2. Experimental methodology

The samples of PR were homogenized, cracked and sieved successively until smaller samples were obtained, a part of them was wet grinded and dried in the oven at 60°C. for 24h. and sieved. For each treatment, approximately 3g was weighed and the respective flotation in the hallimond tube was performed, and then reproduced on a larger scale to obtain the phosphate rock concentrate, which will be the raw material for the production of simple superphosphates (SSP) [9].

2.1. Materials and methods

2.1.1. Sources of phosphorus. Phosphoric rock from the municipality of "Turmequé", Rosales sidewalk, was used; with a pH of 5.6 and a total phosphorus content of 16%.

2.1.2. Sample size. After carrying out the granulometric classification of the sample and were realized several tests of micro-buoyancy of the ore, it was found that the ideal size to find a greater performance in the phosphorus concentration was between 140 and 200 US standard mesh.

2.1.3. Experimental procedure. The phosphoric rock samples were homogenized, cracked and sieved successively until granulometric smaller samples were obtained, some of them were wet grinded and dried in the oven at 60°C. for 24h, then sifted. For each treatment, 3g of sample was weighed and the respective microflotation process in hallimond tube; to obtain the phosphate rock concentrate, for further analysis.

2.1.4. Analytical determinations. The total and assimilable phosphorus in ammonium citrate, are determined for the washed "Turmequé" phosphoric rock sample. (See Table 1), with the spectrophotometric method of the phosphor-molybdovanadate under the Colombian Technical Norm NTC 234.

Table 1. Phosphorus results of the initial sample.

Sample	P total (%)	P assimilable (%)
Turmequé	16	2.15

3. Results and discussion

3.1. Characterization

3.1.1. XRF analysis. The sieve analysis of washed and unwashed samples was carried out in order to know the elemental semiquantitative percentage retained in each mesh, besides selecting the appropriate size for the microflotation process. (See Figure 1 and Table 2). This analysis was determined by x-ray fluorescence, In the equipment Scattered Energy Fluorescence Spectrometer (Brand: PANalytical. Model: MiniPal 2). Located in the laboratory of the Institute for Research and Innovation in Materials Science and Technology (INCITEMA) of the "Universidad Pedagógica y Tecnológica de Colombia (UPTC)."

The previous Figure shown signals correspond to elements found in the phosphoric rock after having been subjected to previous washing in a ball mill.

IOP Conf. Series: Journal of Physics: Conf. Series **935** (2017) 012008 doi:10.1088/1742-6596/935/1/012008

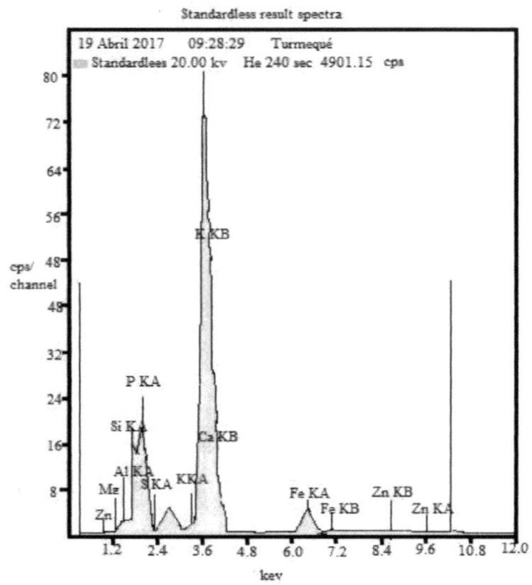

Figure 1. Spectrum of the sample.

Table 2. XRF Results sample retained in each mesh.

Sample	Mesh (mm)	Semiquantitative composition (%)									
		Na	Mg	Al	Si	P	S	K	Ca	Fe	Zn
Washed in the mill	Representative sample		2.4	4.3	23.3	21.6	0.16	0.53	46.4	1.20	0.09
Unwashed	Representative sample	9.5	2.0	5.9	26	14.8	0.07	0.82	39.2	1.69	0.11
Washed in tray	0.6		2.1	5.0	27.4	18.0	0.1	0.76	45.2	1.27	0.12
	0.250		2.4	5.1	25.6	18.9	0.1	0.74	45.6	1.31	0.13
	0.150	11	2.3	4.8	21.5	18.1	0.10	0.62	40.5	0.883	0.11
	0.106	10	2.3	4.9	21.7	18.6	0.1	0.63	40.6	0.985	0.10
	0.075		2.6	4.7	22.9	20.3	0.1	0.69	47.1	1.47	0.12
	0.063	10	2.5	5.0	22.6	18.2	0.09	0.64	39.2	1.49	
	Fine	9.8	2.2	4.9	23.4	18.2	0.07	0.68	38.8	1.83	

The advantage of washing the samples is the reduction of the clay content, which will lead to an increase in the recovery of the ore of interest in the microflotation process. This is observed in the reduction of silica and aluminium content [10]. After performing the grading of the samples and conducting several tests of mineral buoyancy, it was found that the ideal size to find a higher concentration of the sample was between 140 and 200 US standard mesh.

3.1.2. XRD analysis. These analyses were performed on X-ray diffraction equipment. (Brand: PANalytical. Model: XPERT-PRO), Located in the laboratory of the INCITEMA, UPTC, with the objective of knowing the mineralogical phases present in the sample. (See Figure 2 and Table 3).

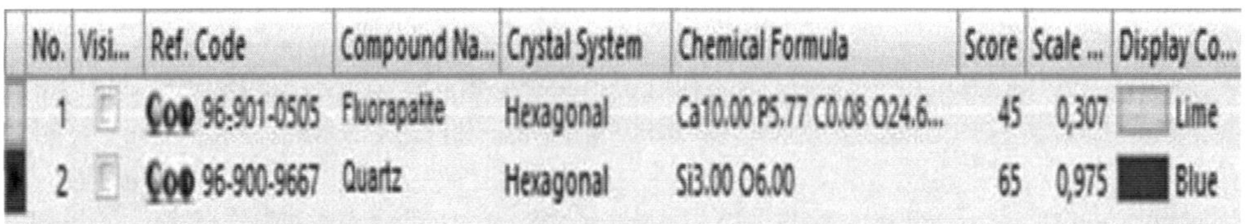

No.	Visi...	Ref. Code	Compound Na...	Crystal System	Chemical Formula	Score	Scale ...	Display Co...
1		COD 96-901-0505	Fluorapatite	Hexagonal	Ca10.00 P5.77 C0.08 O24.6...	45	0,307	Lime
2		COD 96-900-9667	Quartz	Hexagonal	Si3.00 O6.00	65	0,975	Blue

Figure 2. Mineralogical phases.

Table 3. Mineralogical species of interest.

Sample	Fluorapatite (%)	Quartz (%)
Washed in the mill	55.4	44.6
Washed in tray	52.5	57.5

3.1.3. Potencial zeta. These tests were carried out on the potential equipment Z, Zetasizer, located in the Institute of minerals CIMEX, belonging to the faculty of mines of the "Universidad Nacional de Colombia, Medellín" (See Table 4.).

Table 4. Potential results Z.

Reference of the analysis sample	PZ (mV)	Average
	-44.3	
Turmequé, washed. pH 10	-45.7	-45.13
	-45.4	
	-19.8	
Turmequé, without washing. pH 5,8	-20.3	-20.26
	-20.7	

The efficiency of this process depends on the degree of adsorption between the collector and the mineral [11]. When applying the collectors, it is observed that the zeta potential of the unwashed ore tends to values close to zero, which causes the mineral a to maintain an area of instability and enable depression in the process of rock flotation. Moreover, there is no selective flotation. As a general rule, if there is to be repulsion between particles, the zeta potential value should be greater than +30mV or below -30mV [12], which suggests that washing is necessary when the flotation reagents are applied.

3.1.4. Direct floating. The following results were found for the Turmequé sample, after the direct flotation in hallimond tube (See Figure 3 and Table 5).

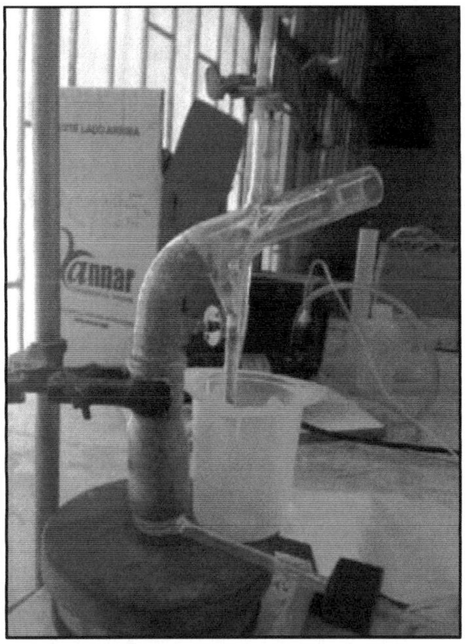

Figure 3. Flotation in hallimond tube.

IOP Conf. Series: Journal of Physics: Conf. Series **935** (2017) 012008 doi:10.1088/1742-6596/935/1/012008

Table 5. Results obtained with the spectrophotometer.

Fractions	P_2O_5 Total (ppm)	P_2O_5 Total (%)
Concentrates	13.31	25.0
Tailings	7.69	15.3

The data obtained with the direct microflotation process indicate that a good part of phosphorus is recovered from the PR, making it viable for later production on a larger scale in denver cells. It is expected to continue with reverse flotation, to clear the excess silica from the concentrate obtained.

4. Conclusions

The ideal size to find a greater performance in the phosphorus concentration was between 140 and 200 US standard mesh. Furthermore, if the samples are washed previously, either in a ball mill or in tray, the silica and aluminium content is reduced, thus contributing to a selective flotation of the mineral when applying the appropriate reagents in the process of mineral concentration.

After conducting flotation processes on phosphate rock, the total phosphorus was increased by up to 9%, from 16% to 25%. These results are attractive for future exploitation of mines with a low concentration of phosphorus, as in the case of "Turmequé, Boyacá".

References

[1] Chien N, Prochnow L, and Mikkelsen R 1992 Uso agronómico de la roca fosfórica para aplicación directa, *Inf. Agronómicas* **1** 13–16

[2] Calle Castañeda S 2016 *Evaluación de la acidulación de roca fosfórica empleando la bacteria acidófila acidithiobacillus thiooxidans* (Medellín: Universidad Nacional de Colombia)

[3] Miranda C S 2002 *Curso de Química Industrial II* (Bogotá: Universidad de los Andes)

[4] Soto Bubert A 2000 *Obtención de fertilizantes fosfatados (S.F.T.) a partir de roca fosfórica importada y ácido sulfúrico nacional, anteproyecto de una factibilidad técnico económica* (Chile: Universidad de Chile)

[5] Cantera V, Mateus E, Muñoz F, Montoya C, Herrera J and Loy J 2008 *Selección de la mejor alternativa para la industrialización de la roca fosfórica en el Departamento de Huila* (Colombia: ZEOLITAS AFL Ltda.)

[6] Unión Temporal GI Georecursos 2005 *Análisis de la estructura productiva y mercados de la roca fosfórica* (Bogotá; Unión Temporal GI Georecursos)

[7] Morillo A, SequeraO, and R. Ramirez 2007 Roca fosfórica acidulada como fuente de fósforo en un suelo ácido con o sin encalado *Bioagro* **19(3)** 12

[8] Sequera O and Ramírez R 2013 Roca fosfórica acidulada con ácido sulfúrico y tiosulfato de amonio como fuente de fósforo para frijol en dos tipos de suelo *Bioagro* **25(1)** 39–46

[9] Zapata F and Roy R N 2007 *Utilización de las rocas fosfóricas para una agricultura sostenible* (Roma: Organización de las Naciones Unidas para la Agricultura y la Alimentación y del Organismo Internacional de Energía Atómica)

[10] Mantilla C, Pedraza J and Laverde D 2008 Utilización de estudios de potencial zeta en el desarrollo de un proceso alternativo de flotación de mineral feldespático *Dyna* **75(154)** 65-71

[11] Dos Santos M A, Santana R C, Capponi F, Ataíde C H, and Barrozo M A 2010 Effect of ionic species on the performance of apatite flotation *Sep. Purif. Technol.* **76(1)** 15–20

[12] QuimiNet 2007¿Cómo ayuda el potencial zeta en las formulaciones de suspensiones y emulsiones? (https://www.quiminet.com/articulos/como-ayuda-el-potencial-zeta-en-las-formulaciones-de-suspensiones-y-emulsiones-21731.htm)

IOP Conf. Series: Journal of Physics: Conf. Series **935** (2017) 012009 doi:10.1088/1742-6596/935/1/012009

Analysis of synthetic profile of CZTS as photovoltaic material obtained with variations of titanium and TiN

M Muñoz[1], E Vera[1], J Gómez[1] and Y Pineda[1]
[1] Universidad Pedagógica y Tecnológica de Colombia, Tunja, Colombia

E-mail: helver.munoz@uptc.edu.co

Abstract. Semiconductor type Cu_2ZnTiS_4 (CZTiS) and Cu_2ZnSnS_4 (CZTS),were synthetized starting from a hydrothermal route from precursor powders such as copper, zinc, tin, titanium isopropoxide and tiocarbammide metal nitrates dissolved in deionized water in concentrations of $1molL_{-1}$. Dosed and placed in a steel autoclave equipped with a Teflon jacket under magnetic stirring (150rpm) and at a temperature of 300°C for 24 hours in order to promote the formation of the respective ceramic phases. Segregates have been repeatedly washed with ethanol at all times until obtaining crystalline-looking solids. Subsequently, in order to promote the production of pure crystalline phases, the materials were subjected to a second reaction stage in a tubular furnace at 400°C in flow (50mLmin-1) for the purpose of Reduce the concentration of secondary phases of sulphides. The characterization of the CZTiS and CZTS materials was performed by X-ray Diffraction (XRD) and Raman spectroscopy where the presence of Kesterite type crystalline structures was confirmed in the two materials revealing that the effect of titanium with a higher ionic radius than tin produces a distortion in the cell of the CZTiS material compared to the report for the CZTS system. The results of Scanning Electron Microscopy (SEM), confirm the regular aggregates obtained with composition consistent with the proposal theoretically and validated by Energy-Dispersion X-ray Spectroscopy (EDX) techniques and comparison between secondary emission spectra and Retro-dispersed.

1. Introduction

In the growing demand for energy consumption that currently lives in the world, the production of energy by means of photovoltaic modules has been one of the most attractive alternatives to solve this problem, currently the photovoltaic modules an expensive silicon base, which opens The field new research of new materials that replace silicon and its more economic view in its manufacture, materials with the structure CZTS and CZTiS result from the most attractive options [1-2].

Both Cu_2ZnTiS_4 (CZTiS) and Cu_2ZnSnS_4 (CZTS), are quaternary material with direct band-like semiconductor properties with a theoretical energy of 1.5eV and a light absorption coefficient of (>104cm^{-1}) in its visible range, theoretically Predict the absorption efficiencies at 32% depending on the dope of the unit cell [3,4].For the synthesis of the materials it was decided to use the routes of chemistry by means of hydrothermal routes (polymerization-combustion, hydrothermal), a technique highly used in the production of materials in glass for the simplicity in its processes and its low cost In order to carry out the process, it was necessary to use precursor solutions in the form of nitrates, chlorides, isopropoxide and acetates, dosed in a sulphur rich solution in order to promote the formation of the desired materials. In order to verify the formation of the kesterite phase, the aggregates obtained by XRD, Raman spectroscopy and SEM are characterized, using a two-material research methodology one with an interaction between them [5].

IOP Conf. Series: Journal of Physics: Conf. Series **935** (2017) 012009 doi:10.1088/1742-6596/935/1/012009

2. Materials and methods

To obtain kesterite (CZTS), copper nitrate Cu $(NO_3)_2$, zinc acetate Zn $(CH_3COO)_2$ and anhydrous tin chloride ($SNCL_2$) were each used in concentrations of $1molL^{-1}$ dissolved in a solution of tiocarbammide (H_2NCSNH_2) diluted in deionized water $3.3H_2O$, for CZTiS-type kesterite, anhydrous tin chloride ($SNCL_2$) was replaced by titanium isopropoxide ($C_{12}h_{28}0_4Ti$) in equal proportions, each of the solutions was subsequently stirred for 5min to obtain a Single solution [8,3].

For the synthesis of each material the respective precursors were dosed in a steel autoclave equipped with a Teflon jacket under magnetic stirring (150rpm) and at a temperature of 300°C for 24 hours in order to promote the formation of the respective compound; The materials obtained at the end of the preliminary stage were washed with absolute ethanol repeatedly until crystalline appearance solids were obtained. Subsequently, in order to promote the preparation of pure crystalline phases, the materials were subjected to a second reaction stage in a furnace at 400°C in He flow ($50mLmin^{-1}$), for the purpose of reduce the concentration of secondary phases of sulphides, as shown in Figure 1 [8,6].

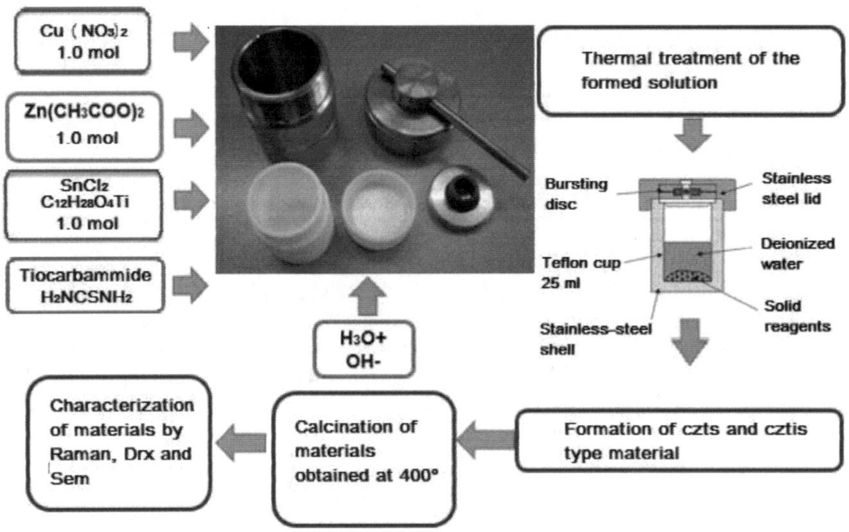

Figure 1. Process of synthesis and characterization of CZTS and CZTiS.

3. Results and discussion

To verify the obtaining of the crystallographic structure, x-ray diffraction tests were performed for each of the aggregates obtained in the synthesis processes of materials CZTS and CZTiS, the samples was identified by X-ray diffraction (XRD, PANanalytical XPert Pro Multi-Purpose Diffractometer, Co Kα, λ=0.1768nm) taking as reference data collection angles ranging from 20° to 80° 2 θ (theta), a sufficient range to verify the characteristic peaks of the kesterite phase, in order to make the comparison of theoretical patterns with the experimental ones We used the "Cod" Crystallography Open Database.

The results of the obtained diffraction patterns (Figure 2 and 3) revealed the presence of kesterite mainly in three characteristic peaks Attributed to the diffraction planes (1 1 2), (0 2 4) and (1 3 2), together with secondary phases of covellite mainly oriented in the diffraction planes (0 1 3), (1 1 0), (0 1 2) and sulphur present in each of the samples with an orthorhombic and monoclinic crystalline system respectively for CZTS and ZCTiS, two very common structures presented in the formation of sulphur rich compounds, a polymorphic characteristic Attributed to the allotropic nature of the sulphur where small changes of temperature or pressure can cause the transition of structure, where in either case the sulphur is forming ring-shaped S_8 molecules (Table 1 and 2) [7].

The formation of the secondary phases is mainly attributed to the electronegativity of the elements since in the period of composition of the acetate of copper and zinc acetate carried out at temperatures higher than 200°C, the sulphur tends to react more Easy with elements such as copper or tin due to its high electronegativity, in this way the formation of secondary phases in the compound is highly probable; The effect of annealing in an inert atmosphere at 400°C shows positive results for obtaining kesterite as the only compound present, but previous studies report that to obtain a single phase of CZTS we will need a temperature higher than 500°C in an atmosphere of He+H_2S (5%) favouring not only the consolidation of the phase but also it could lead to increase in its crystallinity and grain size [8,12].

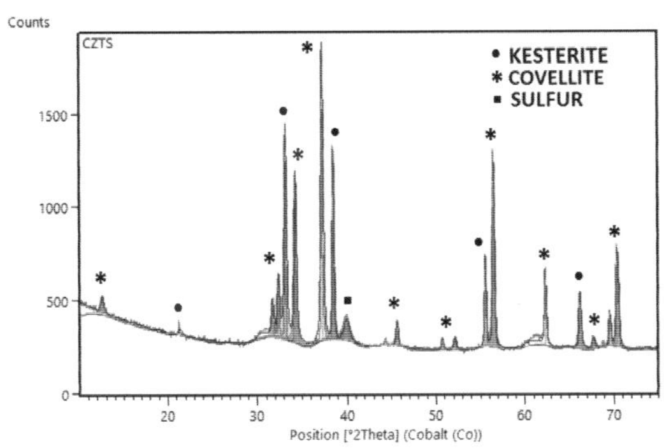

Figure 2. X-ray diffraction pattern obtained synthesis of CZTS.

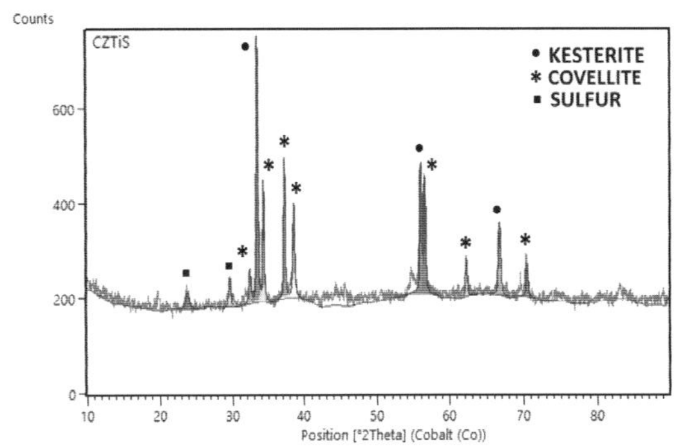

Figure 3. X-ray diffraction pattern obtained synthesis of CZTiS.

Table 1. Species found in CZTS diffraction patterns.

Compound name	Ref cod.	Crystal system	Chemical formula
Kesterite	96-900-4751	Tetragonal	$Cu_{2.0}Zn_{1.0}Sn_{1.0}S_{4.0}$
Covellite	96-900-8390	hexagonal	$Cu_{1.0}S_{1.0}$
Sulfur	96-901-1363	Orthorhombic	$S_{1.0}$

Table 2. Species found in CZTiS diffraction patterns.

Compound name	Ref cod.	Crystal system	Chemical formula
Kesterite	96-900-0010	Tetragonal	$Cu_{2.0}Zn_{1.0}Sn_{1.0}S_{4.0}$
Covellite	96-900-8390	hexagonal	$Cu_{1.0}S_{1.0}$
Sulfur	96-901-1363	Orthorhombic	$S_{1.0}$

In the two x-ray diffraction spectra, there are notable differences, where the most important is the displacement of the CZTiS spectrum with respect to the CZTS of 0.321 degrees at the 2θ (theta) position which can be observed in Figure 4, This behaviour in the patterns is due to the tin atom was replaced by the titanium Atom in the unit cell the interplanar distances thereof are modified [9], resulting in different diffraction planes for the two materials.

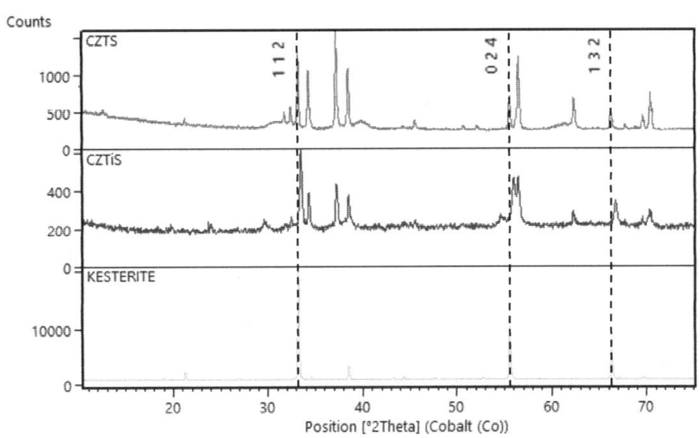

Figure 4. Comparison patterns obtained with theoretical patterns of each compound present in the sample.

IOP Conf. Series: Journal of Physics: Conf. Series **935** (2017) 012009 doi:10.1088/1742-6596/935/1/012009

Even though the titanium atom has a higher atomic radius than the tin atom so it would expect its crystalline cell to increase in size, the shift of the right diffraction peak reflects that the cell size for the CZTiS type kesterite is lower than that of the kesterite type CZTS [4].

Raman spectroscopy was performed in order to further verify the presence of kesterite phase in the two samples obtained, the results show in Figure 5 that both spectra have significant peaks centred in 287cm^{-1} [10], this characteristic is attributed to a As a result of the vibration of the sulphur atoms present in the kesterite, the position of the most intense resonance peak present in the spectrum of CZTS at 338cm^{-1} is consistent with other investigations carried out for results obtained in synthesis of CZTS type materials; For the main signal present in the spectrum related to the CZTiS appears a resonance peak in the position 440cm^{-1} a very specific characteristic of the material since it presents a considerable deviation in the position of the primordial line of Raman with respect to the main peak Raman Presented in traditional CZTS type materials, which could be due to a difference in the interactions of resonant photons and to the repulsion of the states of photonic interference that the structure presents[11].

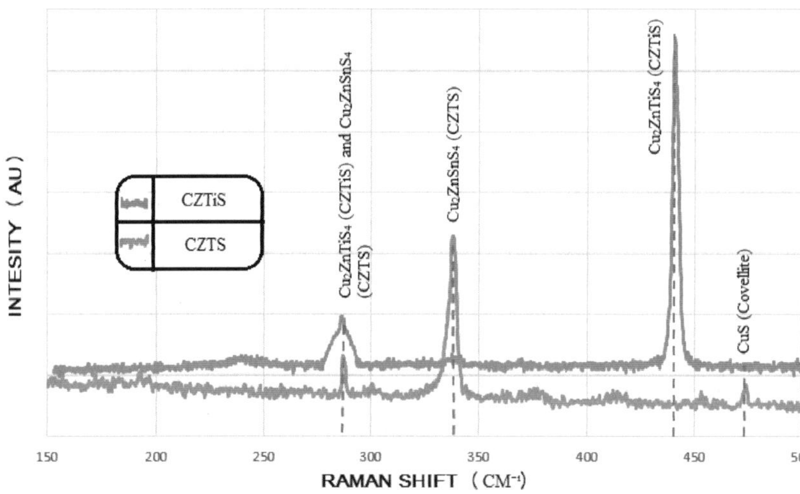

Figure 5. Raman spectra of materials obtained in CZTS and CZTiS synthesis, showing the main vibrational modes of each sample.

For each of the tests was taken micrographs at 5000X using the scanning electron microscope, the photos obtained are presented in Figures 6 and 7.

Figure 6. CZTS sample morphology set at 400°C.

Figure 7. CZTiS sample morphology set at 400°C.

The results of the two materials show a morphology of irregular agglomerates with sizes varying from 10 to 120nm, it is observed that the reaction in two stages homogenizes the material maintaining similar sizes in the two cases of synthesis, small formations of Crystals in circular form, generally observed in materials synthesized with high concentrations of tiocarbammide, the solvent can generate colloids which generates an effect of inhibitor in the growth of aggregates [8,7].

The analyses carried out by EDX made to CZTS and CZTiS material demonstrate an elemental composition homogeneous to the general stoichiometry of the kesterite where the levels of tin prove to be a little lower thanks to the formation of secondary phases such as copper sulphides type covellite confirmed in the spectrum by x-ray diffraction (see Figures 8 and 9).

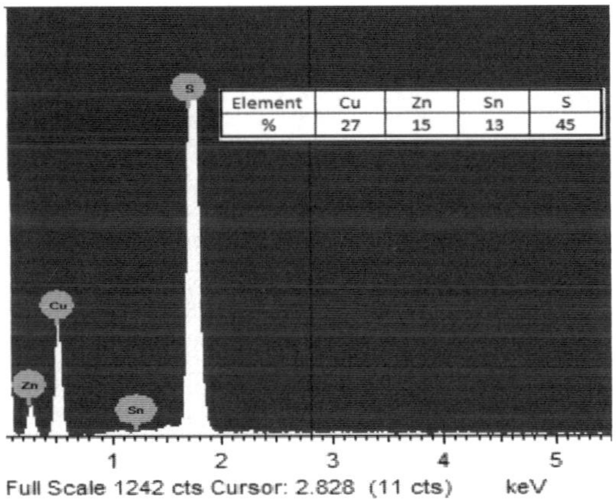

Figure 8. Chemical composition by EDX of CZTS.

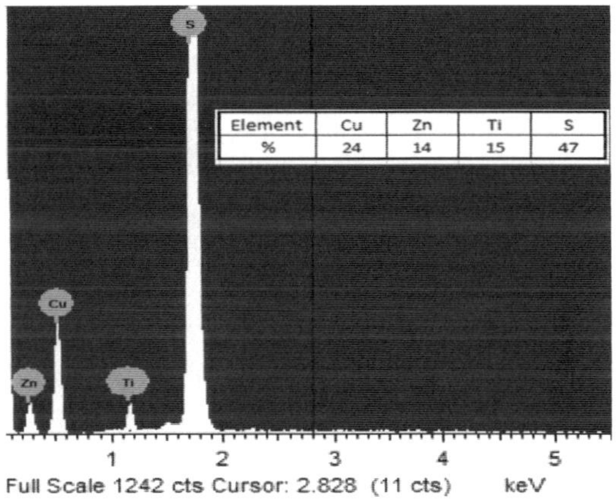

Figure 9. Chemical composition by EDX of CZTiS.

4. Conclusions

The synthesis of the crystallises kesterite phase type is possible to carry out by means of hydrothermal routes but taking into account that one of the crucial parts to decrease the secondary phases depend directly of the second phase of heating at higher temperatures than the first one.

The synthesis of the CZTiS ceramic phase demonstrates significant changes in its crystallinity since the shift of the x-ray diffraction peaks shows a smaller size in its unit cell.

Although the atomic volume size of titanium is greater than that of tin, the unit cell of the CZTiS type kesterite is less than the CZTS type kesterite, which means that the distances between atoms are smaller, a characteristic that may favour the band of conduction of electrons in the material

References

[1] García Verduch A 1994 *Cerámica tradicional y no tradicional* **33** 309-313
[2] Sánchez Muñoz L 2011 *Cerámica y vidrio* **50** 15-21
[3] Xia Y N, Yang P D, Sun Y G, Wu Y Y, Mayers B, Gates B, Yin Y D, Kim F and Yan H Q 2003 *Advanced Materials* **15** 353-389.
[4] Jianning D, Huafei G, Ning Y, Ningyi Y, Kezhi Z and Yan L 2016 *CN105908132 A* **10** 1-8
[5] Zhu Y F, Fan D H and Shen W Z 2008 *Langmuir* **24** 11131-11136
[6] Tiong V T, Bell J and Wang H 2014 *beilstein* **5** 438-446
[7] Mau T T, Ersan Y M and Kim K H 2015 *Journal of the Korean Physical Society* **67** 1028-1832
[8] Camara S M, L. Wang and X Zhang 2013 *Nanotechnology* **10** 89-94
[9] Adiguzel S, Kaya D, Cenisel M F, Celik O, Tombak A, Ocak Y S and Turan R 2017 *Journal of Electronic Materials* **46** 3976-3981
[10] Valakh M Y, Litvinchuk A P, Dzhagan V M and Yukhymchuk V O 2016 *RSC Adv.* **6** 67756-67763
[11] Dimitrievska M, Pistor P, Saucedo E and Roca V I 2016 *Solar Ener. Mater. and Solar Cells* **157** 462-467
[12] Maheshwari B U and Kumar V S 2014 *Inter. J. of Eng. Sci. Tech. and Res.* **16** 1458-1461

IOP Conf. Series: Journal of Physics: Conf. Series **935** (2017) 012010 doi:10.1088/1742-6596/935/1/012010

Preparation of concrete mixtures with electric arc furnace slag and recycled ground glass

Y Pérez Rojas[1], E Vera López[1], M López Rodríguez[1] and J Díaz Pita[1]
[1] Universidad Pedagógica y Tecnológica de Colombia, Tunja, Colombia

E-mail: yasmin.perez@uptc.edu.co

Abstract. The present work includes the first advances in the development of investigations that seek to include Ground Grinding Glass (GRR) and the Electric Arc Furnace Slag (EAFS) in the production of mixtures of hydraulic concrete mixing them simultaneously, so that it satisfies the specifications techniques to be used in the construction of rigid pavements. Firstly, we cite the tests carried out on the different materials to obtain their physical, chemical and mechanical characterization and determine their compliance, as well as the measurement of certain characteristics that may be somewhat empirical to standardize their control. Technique such as X-Ray Diffraction (XRD), X-ray Fluorescence Spectrometry (XFR) and Scanning Electron Microscopy (SEM) have been used. Once the results of the characterization tests and their correspondence with the Colombian technical standards have been obtained, it has become possible to select the use of the Transparent Recycled Ground Glass (TRGG) as the most suitable for the replacement of the sand in the dosage of new mixtures modified concrete.

1. Introduction

The present paper contains the advances in the development of the doctoral thesis denominated "Replacement performance in concrete aggregates, by EAFS and RGG, and corrosion analysis." in which it is tried to demonstrate the feasibility of the use of industrial residues in the construction of rigid pavements. The waste to be used is the Electric Arc Furnace Slag (EAFS) as a substitute for the coarse aggregate and the Recycled Ground Glass (RGG) as a substitute for the fine aggregate. The research is in an early development, approximately 25% so the data and observations recorded in this document are limited to the characterization of the new materials to be incorporated into the mixtures of the hydraulic concrete, especially to the characteristics related with RGG, which have not been easily obtained within the current literature, especially geographically demarcated for the department of Boyacá (Colombia). In addition, it presents an updated bibliographic revision of the use of EAFS and RGG as substitutes in the mixtures of hydraulic concrete, without being used simultaneously, as is the fundamental of this research, being a novel advance of great impact when implementing its use in the construction of the road infrastructure.

The research is experimental and seeks to evaluate the behaviour of reinforced hydraulic concrete, useful in the construction of infrastructure works, modified in its dosage gravel by EAFS and RGG in replacement of the sand. The properties to be evaluated are the simple compression strength, corrosion resistance of reinforcing steel, density, voids and absorption. Electronic Scanning (SEM/EDX) and X-Ray Diffraction (XRD) will be employed.

2. Industrial waste in Colombia

Industrial waste from steel has ceased to be by-products to be part of cement and construction raw materials [1]; however, there is a need to expand knowledge, techniques, methods and regulations that guarantee the efficiency of its use. Slag production in Colombia is between 0.1 and 0.3 tons per ton of steel produced. According to Diaco SA internal records, the amount of black slag generated for 2009 was 53.122 tons from 327.250 tons of scrap for a production of 280.354 tons of steel, i.e. 16.2% of the scrap is converted into Slag during the process [2]. In its use as an aggregate for concrete, the crystallized slag offers properties of excellent adherence with cement, stability of volume and durability, and fire resistance superior to that obtained with other aggregates.

Glass in turn and due to the population increase in Colombia and its settlement in urban centres has become one of the main post-consumer waste, especially its presentation in bottles that saturate at a great pace the dumps and landfills, and its improper disposal reach water sources [3].

Table 1 shows the volumes of glass (bottles) that are generated in the sanitary landfills of the main cities of Colombia.

Table 1. Glass volumes generated in landfills.

Department	Waste	Tonne/day	Landfills
Bogotá	Solids	6300	Doña Juana
	Bottles (4.6%)	290	
Antioquia	Solids	3260	Pradera y Guacal
	Bottles (3%)	98	
Valle del Cauca	Solids	3072	Yotoco
	Bottles (2.56%)	79	
Atlántico	Solids	1907	Los Pocitos
	Bottles (1.58%)	30	

3. Materials and testing

The research is carried out with natural silica sand with modulus of 2.41 fineness and specific gravity of 2.42 and limestone of 19mm of maximum nominal size as aggregates for concrete. The recycled ground glass is used with particles smaller than 4.75mm and specific gravity of 2.49. The EAFS has a maximum nominal size of 19mm and a specific weight of 3.28. The cement to be used is ordinary Portland cement in accordance with ASTM C150 and INV-307-13 with a specific gravity of 2.855 according to the test performed. The recycled glass was classified by colour by selecting two types, Green Recrystallized Ground Glass (GRGG) and Transparent Recycled Ground Glass (TRGG).

3.1. Aggregates

The main physical and mechanical characteristics of the aggregates are presented in Table 2.

Table 2. Physical and mechanical characteristics of gravel and sand.

Material	Materials finer 75μm	Specific gravity	Absorption	UWla	UWrb	Angeles machine	Micro-Deval
Gravel	0.05	2.420	1.746	1.326	1.442	36.08	23.74
Sand	-	2.600	1.208	-	-	-	-
EAFS	0.01	3.277	1.877	-	-	19.67	6.96
RGG	-	2.494	0.806				

3.2. Electric arc furnace slag

The oxidizing slag of the electric arc furnace is a by-product of the steel industry, generated after the melting and the refined acid preliminary liquid steel. It is a stone material that is easy to crush for use as an aggregate in concrete mixtures [4].

Slags have been used as an addition or in proportional combination with Portland cement for the manufacture of concrete or mortars. The use of slag in construction has recently increased [5]. In the United States and Europe, slags have shown good performance over time, indicating that they can be effective in reducing pore size, thus taking less waterproofing and longer durability of concrete mixtures. Concrete combined with slag has many advantages: increased durability, moderate heat of hydration, higher compressive strength, and lower permeability, superior resistance to alkali silica reaction due to penetration of chloride ions and sulphate ions, workability, environmental and economic value [6].

Slags are composed of lighter compounds than steel, which rise to the surface of the metal bath, forming a layer of variable thickness. These compounds are mostly oxides, silicates, sulphides, aluminates and phosphates, which form as a consequence of the technological process of obtaining steel [7].

In order to perform the EAFS characterization in the INCITEMA laboratory of the "Universidad Pedagógica y Tecnológica de Colombia", the X XFR test was carried out for the analysis of the chemical composition of the EAFS, observed in Table 3.

Table 3. Chemical composition of the EAFS (%)

Compound	Mg	Al	Si	S	K	Ca	Ti	Cr	Mn	Fe	Zn
Concentration	4.0	7.0	14	0.33	0.20	26.7	0.58	1.2	6.68	38.8	0.46

3.3. Recycled ground glass

Glass is an amorphous material produced by melting silica, sodium carbonate and calcium carbonate ($CaCO_3$), followed by cooling, during which solidification can occur without crystallization [8].

This material comes from the collection at the source of waste such as glass bottles and transparent green containers, which are washed in a soapy solution, to remove labels and residues and then disinfected with 0.1% chlorine bleach solution concentration. They are then dried to constant mass and broken, to be brought to the Los Angeles machine for 10 minutes at 500rpm. Once ground, it is passed through #4 sieve. The retained material is re-ground for the same period of time. The material to be incorporated into the new concrete mixtures is the one that passes through the #4 sieve.

The recycled glass used as an aggregate in the concrete causes surface cracks in the concrete due to the expansion generated by the ASR (alkali - silica) reaction between the amorphous silica present in the glass and the cement paste of alkaline nature [9]. For this reason, the expansion in both types of glasses was studied, so that it was possible to establish the parameter to be controlled and the convenience in the selection of the type of glass to be used. Is of major importance the study of the alkali-silica reaction due to the large amount of amorphous silica in the glass [10].

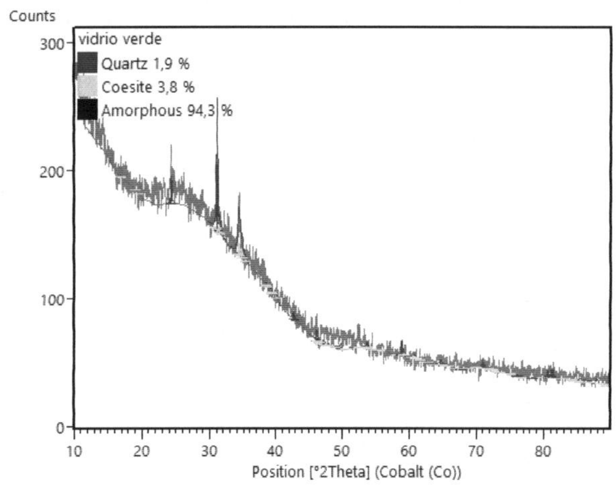

Figure 1. XRD in GRGG.

IOP Conf. Series: Journal of Physics: Conf. Series **935** (2017) 012010 doi:10.1088/1742-6596/935/1/012010

Within the glass characterization process, the XRD was used to define the chemical composition and the morphology and microstructure was studied by SEM. Green and clear recycled ground glass samples were obtained from material passing the #100 mesh (150μm). XRD tests on both green glass and clear glass showed that it was an amorphous material as seen in Figure 1. The green glass showed a diffraction peak which, when analysed by the test of Rutile, corresponds to quartz and coesite in a percentage of 1.9% and 3.8% respectively.

The green and transparent glass was analysed by scanning electron microscopy (SEM) to determine its morphology and microstructure. In Figures 2 and 3, the values or particle size and twins present in the milling process is observed.

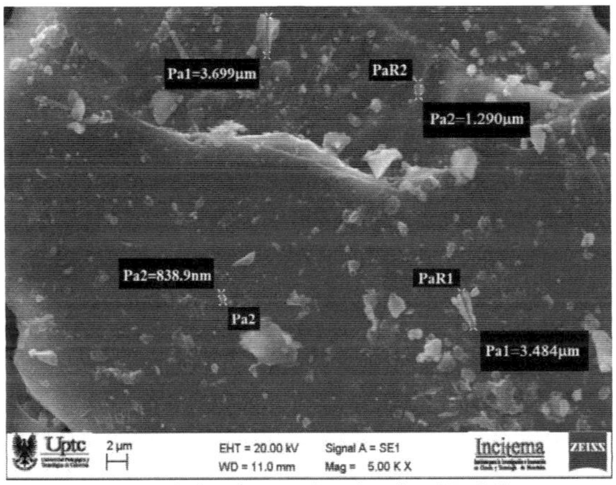

Figure 2. Particle size and twins in GRGG.

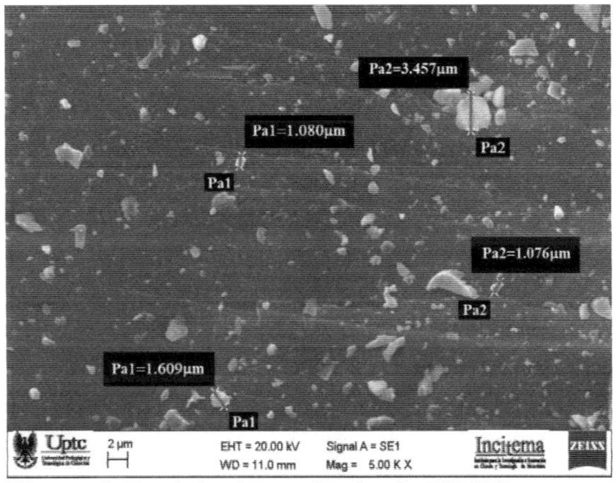

Figure 3. Particle size and twins in TRGG.

4. Discussion

According to current Colombian regulations for the production of concrete aggregates, the aggregates proposed in the research comply with the specifications. The coarse aggregate by its particle size distribution corresponds to an aggregate AG-2. For the gradation of the master mix the adjustment of the theoretical curves will be done according to Fuller [11].

The granulometry of the aggregates associated with the compactness of the concrete, as well as its quality, influences the workability of the fresh mixture, the mechanical resistance, the resistance to the chemical agents and in the economy in the production of hydraulic concretes [12]. For this reason, actors like Fuller and Bolomey developed ideal grain sizes to achieve maximum compactness and thus guarantee the quality of the concrete.

The design mix will be a blend of plastic consistency, 50 mm settling, 50% flowability, and ideal blend for reinforced retaining walls, foundations, normally compacted paving, slabs, beams and poorly, reinforced columns.

The method of design of the mixtures will follow the methodology ACI 211, the variation of the percentage of slag in 25%, 50%, 75% and 100% will be carried out in replacement of the coarse aggregate content in the standard sample and of 20% 30% and 40% of the content of fine aggregate.

In view of the results of the expansion test on samples of recycled ground glass, it is concluded that when the same expansion is present and because the generation of volumes of transparent glass is greater, experimentation will be carried out with the ground glass recycled transparent for the elaboration of the modified concrete mixtures as a replacement of the fine aggregate.

5. Conclusions

This study is in an initial stage of development and seeks to incorporate industrial waste in the manufacture of concrete mixes for pavements.

In the expansion test it was verified that there is no significant difference in the expansion presented by transparent glass and green glass.

IOP Conf. Series: Journal of Physics: Conf. Series **935** (2017) 012010 doi:10.1088/1742-6596/935/1/012010

The replacement materials comply with the Colombian technical specifications in terms of granulometry.

The artificial aggregates (glass and EAFS) have the same fineness modulus of sand and gravel respectively.

To find the percentages of the aggregates fitted the ideal curve of Fuller, the fineness modulus method was used. The method considers the modulus of fineness of the aggregates to be used and the fineness modules of the Fuller curve and a system of 2 equations was proposed with 2 unknowns that allow the determination of the proportions of the aggregate in the mixture.

References

[1] Lothenbach B, Scrivener K and Hooton R 2011 Supplementary cementitious materials *Cement and Concrete Research* **41(12)** 1244-1256

[2] Parra Araque L and Sánchez García D 2010 *Análisis de la valorización de escorias negras como material agregado para concreto en el marco de la gestión ambiental de la siderúrgica Diaco Municipio Tuta Boyacá* (Bogotá: Universidad de la Salle) p 48

[3] Ling T, Poon C and Wong H 2013 Management and recycling of waste glass in concrete products: Current situations in Hong Kong *Resources, Conservation and Recycling* **70** 25–31

[4] Arribas I, Santamaría A, Ruiz E, Ortega-López V and Manso J 2015 Electric arc furnace slag and its use in hydraulic concrete *Construction and Building Materials* **90** 68–79

[5] Aprianti E 2017 A huge number of artificial waste material can be supplementary cementitious material (SCM) for concrete production – a review part II *Journal of Cleaner Production* **142(4)** 4178-4194

[6] Muhmood L, Vitta S and Venkatrswaran D 2009 Cementitious and pozzolanic behavior or electric arc furnace steel slag *Cement and Concrete Research* **39** 102-109

[7] Valdés R, Bogachkov A, Gómez A and Robau D 2001 Utilización de las escorias de los hornos de arco eléctrico y de cuchara como materiales de construcción *Ingenieria Mecanica* **1** 19–26

[8] Omran A and Tagnit-Hamou A 2016 Performance of glass-powder concrete in field applications *Construction and Building Materials* **109** 84–95

[9] Peñafiel Carillo D A 2016 *Análisis de la resistencia a la compresión del hormigón al emplear vidrio reciclado molido en reemplazo parcial del agregado fino* (Ecuador: Universidad Técnica de Ambato)

[10] Catalan Arteaga C J 2013 Estudio de la influencia del vidrio molido en hormigones grado H15, H20, y H30 (Chile: Universidad Austral de Chile) p 81

[11] INVIAS 2013 *Especificaciones generales de construcción de carreteras, pavimentos de concreto* (Colombia: Instituto Nacional de Vías)

[12] Afshinnia K and Prasada R 2016 Impact of combined use of ground glass powder and crushed glass aggregate on selected properties of Portland cement concrete *Construction and Building Materials* **117** pp 263-272

IOP Conf. Series: Journal of Physics: Conf. Series **935** (2017) 012011 doi:10.1088/1742-6596/935/1/012011

Dyes removal from water using low cost absorbents

S Giraldo[1], A P Ramirez[1], M Ulloa[1], E Flórez[1] and N Y Acelas[1]
[1] Universidad de Medellín, Medellín, Colombia

E-mail: nyacelas@udem.edu.co

Abstract. In this study, the removal capacity of low cost adsorbents during the adsorption of Methylene Blue (MB) and Congo Red (CR) at different concentrations (50 and 100mg·L^{-1}) was evaluated. These adsorbents were produced from wood wastes (cedar and teak) by chemical activation ($ZnCl_2$). Both studied materials, Activated Cedar (AC) and activated teak (AT) showed a good fit of their experimental data to the pseudo second order kinetic model and Langmuir isotherms. The maximum adsorption capacities for AC were 2000.0 and 444.4mg·g^{-1} for MB and CR, respectively, while for AT, maximum adsorption capacities of 1052.6 and 86.4mg·g^{-1} were found for MB and CR, respectively.

1. Introduction

Dyes discharges to water bodies from different industrial sectors has led to the detriment of aquatic ecosystems by micro-toxicity. Due to this, the regulation related to the amount of dyes in water is every day more strict [1]. With the above mentioned, research related to new and efficient procedures to dyes removal from water is increasing. Biological and chemical methods are the most common to solve this problem. However, they generate high amounts of waste, have low selectivity and have high costs. Therefore, adsorption methods arise as suitable technologies because they are cost-effective and easy to operate. The most important challenge here is to find a selective, renewable and economical adsorbent material [2].

Agroindustry wastes become in a suitable source for the development of activated carbon to dyes removal. Some authors such as, Duman, D. *et. al.* [1] and Hameed, B. *et. al.*[3] produced activated carbon from wood and they found good porosity development and high adsorption capacities during dyes removal, with values of 370.37mg·g^{-1} and 294,14mg·g^{-1},respectively.

In this paper, we studied the removal of two typical dyes present in industrial water (MB and CR), using activated materials developed in our laboratory from wood wastes of cedar and teak [4]. To evaluate the removal rate of dyes on activated carbon, different kinetic models were used. Langmuir and Freundlich isotherms were evaluated to determine the interaction between the contaminants and the adsorbents.

2. Methodology

2.1. Adsorption experiments

To obtain the active carbon a methodology used and reported in our group was used [5] and the adsorption tests were carried out using low cost adsorbents produced from waste wood (cedar and teak) [4] . Adsorption experiments were carried out by batch methods using two different types of dyes (MB and CR), which included two different concentration of each 50 and 100mg·L^{-1}. By kinetic studies, the optimum contact time was determined. And equilibrium studies allowed to find the adsorption capacity of each material. To determine the final dyes concentration, calibration curves were performed. The absorbance to determine the concentration was measured on a spectrophotometer (VIS-DR 3900) at a wavelength of 665nm and 490nm for MB and CR, respectively.

2.2. Kinetic experiments

Kinetic experiments were conducted at room temperature (22ºC) by adding a fixed amount of activated carbon (0.05g) into a 250mL erlenmeyer flask that contained 50mL of the MB and CR solutions. The mixture was agitated at 150rpm, and samples were collected at different times as follow: between 5 and 240min for MB and between 5 and 1440min for CR. The final concentration of each dye was measured by using their absorbance. Adsorption kinetics experiments are important because they can predict the removal rate of a contaminant from aqueous solutions and provides data that facilitate the understanding of the adsorption mechanism [6]. To analyze the adsorption rates of dyes onto the activated materials and the amount of each dye adsorbed at time (t) a mass balance between initial concentration and concentration at time (t) should be done by using the follow equation (1).

$$q_t = \frac{(C_i - C_t)}{w} \times V \tag{1}$$

Where q_t is the adsorption capacity of the activated carbon (mg·g^{-1}), C_i and C_t are the dyes concentrations in solution at initial and time t (mg·L^{-1}), V is the solution volume (L), and w is the mass of the activated carbon (g). Through this experiments the optimal contact time can be determinate [6]. In this study, the kinetics models of pseudo first order, pseudo second order and the intra-particle diffusion were evaluated for the adsorption of MB and CR on low cost adsorbents. The kinetic model of pseudo first order assumes that the limiting step during the adsorption process is the physisorption. Dye mass transfer to the adsorbent surface is based on equation 12 in [7]. When plotting a $log\ (q_e - q_t)$ as a time function, q_e y k_1, can be determinate from the intercept and slope respectively [7]. The pseudo second order model is described by equation 14 in [7]. It assumes that chemisorption is the determining step of the adsorption process rather than mass transfer in solution. q_e and k_2 values can be determined from a plot of t/q_t Vs t using the slope and the intercept, respectively. Intra-particle diffusion model is useful to elucidate the diffusion mechanism, it assumes that the adsorption mechanism is developed by different steps: external diffusion, pore diffusion and adsorption at the pore surface, and it is described by equation 16 in [7]. The values of the intra-particular diffusion rate constant k_{di} y C_i can be calculated from the slope and intercept of q_t vs $\sqrt{t}$ [6].

2.3. Equilibrium experiments

Langmuir and Freundlich isotherm models can describe the interaction between the contaminant and the adsorbent material and measure the adsorption capacity of the activated carbon, at equilibrium. The Langmuir model assumes homogeneous surfaces with a single interaction energy [5]. The Langmuir equation 3 in [5] is widely used to quantify adsorption capacity. When plotting C_e/Q_e vs C_e, straight lines are obtained, where K_L and Q_{max} can be calculated using the intercept and slope, respectively. From equation 4 in [5], the dimensionless separation factor, R_L, can be calculated. It allows to indicate if the adsorption process is favorable or unfavorable [8]. Freundlich model is used for heterogeneous adsorption and it is described by equation 3 in [5]. K_f y n can be determined from the intercept and slope of the linear relationship of log Q_e vs log C_e, respectively.

3. Results and discussion

3.1. Effect of contact time and initial concentration on the adsorption process

Figure 1 shows the contact time effect and initial concentration (50 and 100mg·L^{-1}) of both dyes for the adsorption process on activated carbon (AC and AT). Results indicate fast adsorption at the beginning which decreases over time until reaching equilibrium. This tendency could be related to the existence of high number of active sites (porosity and functional groups) available for the adsorption process. Then, the process begins to stabilize due to the saturation of the active sites. This trend is similar to some literature reports [9]. For the adsorption of MB on AC and AT (Figure 1(A) and (B)), it is observed that the process is strongly influenced by the initial concentration, since the adsorption capacity increases with the concentration, which is in agreement with other studies [5]. Authors reported that for a concentration of 150mg·L^{-1} the adsorption capacity of MB removal was 884.5mg·g^{-1}, while for a lower concentration of 50mg·L^{-1} its adsorption capacity decreased to 671.1mg·g^{-1}. The opposite case occurs when CR is used as an adsorbate on AC and AT (Figure 1(C) and (D)), where the adsorption capacity decreases with the increase of concentration. This behavior can be explained because CR molecule is larger than MB molecule. High concentrations of CR make a rapidly saturation of the adsorbent material, decreasing the adsorption capacity, which is similar to reports done by Lorenc-Grabowska, *et. al*.[9,10]. y Tor, *et. al*.[9,10].

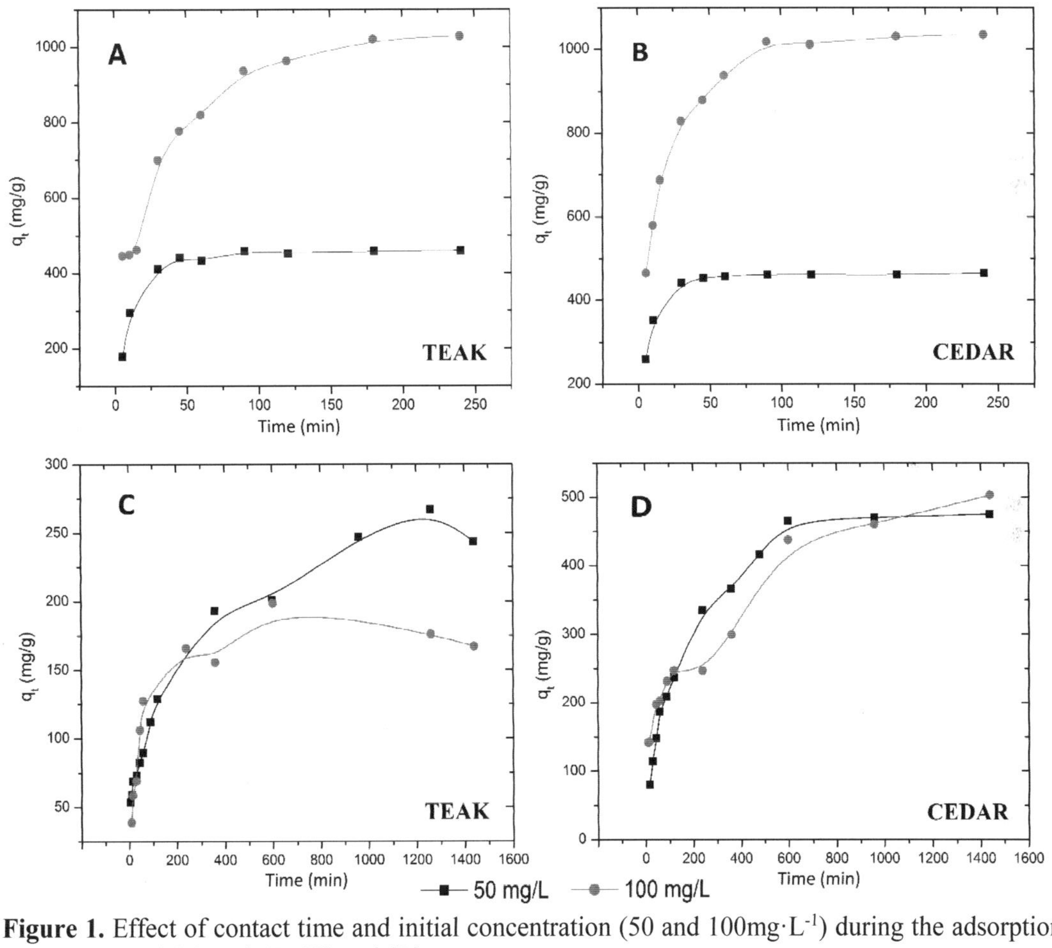

Figure 1. Effect of contact time and initial concentration (50 and 100mg·L^{-1}) during the adsorption on MB (A) and (B) and CR (C) and (D).

3.2. Kinetics experiments

Table 1 shows kinetic parameters for pseudo first order, pseudo second order and intra-particle diffusion models. For the two activated carbons (AC and AT) it was found that the optimal contact time for CR was 10h for both concentrations (50 and 100mg·L^{-1}), while for MB it was 1.5h for 50mg·L^{-1} and 2 h for 100mg·L^{-1}. Observing the correlation coefficients, $R^2 \geq 0.970$ (Table 1), for the adsorption of dyes (MB and CR) on each adsorbent material (AC and AT), these processes can be well represented by the pseudo-second-order kinetic model. This infers that chemisorption is the determining step of the adsorption process rather than mass transfer in solution. Similar kinetics results have also been reported by different authors using a variety of adsorbents [5,6].

In order to study a possible mechanism of MB and CR adsorption on the two activated materials, the intra-particle diffusion model was evaluated. A good adjustment of the experimental data was found. These plots present multilinearity, which indicates that two or more steps occur and that they are limiting the overall rate of adsorption. For the adsorption of MB and CR on AC, the value of k_{1d} is greater than the value of k_{2d}, indicating that surface adsorption is faster than adsorption into the pores, which is driven by the difference in the initial dyes concentration. Conversely, it occurs with AT, where the value of k_{1d} is lower than the value of k_{2d}. C_1 y C_2 values indicate the effects of the boundary layer. For AC, C_2 values are greater than the values of C_1, whereas for AT the opposite occurs. Larger values for C_i, indicate that surface adsorption contribution is greater to the rate-limiting step [6].

Finally, it can be said that adsorption of MB and CR on activated carbon is a process involving multiple steps (adsorption on the outer surface and diffusion within the pores). Similar results have been reported in the literature for the removal of MB and CR using different activated materials [5,9,11,12].

Table 1. Pseudo first order kinetic, Pseudo second order kinetic and intra-particle diffusion model constants for the adsorption of MB and CR by the activated carbon (AC and AT).

Material	Cedar				Teak			
Dye	MB		CR		MB		CR	
C_i (mg L^{-1})	50	100	50	100	50	100	50	100
$q_{e\ exp.}$ (mg g^{-1})	462.34	1010.88	466.00	437.77	458.32	963.2	200.89	198.58
			Pseudo first order					
q_e (mg g^{-1})	222.79	594.02	386.81	273.02	229.35	754.57	167.65	-
k_1 (min^{-1}) x 10^{-2}	6.91	3.5	-0.41	0.18	-3.48	3.39	0.83	-
R^2	0.972	0.987	0.989	0.858	0.976	0.957	0.988	-
			Pseudo second order					
q_e (mg g^{-1})	461.7	1075.3	512.82	523.56	471.7	1098.9	265.25	176.06
k_1 (min^{-1}) x 10^{-4}	4.49	0.86	3.8	3.65	4.49	0.83	14.2	32.26
R^2	0.999	0.999	0.995	0.970	0.999	0.998	0.988	0.995
			Intra-particle diffusion					
k_{d1} (mg g^{-1} $min^{1/2}$)	5.29	13.48	2.07	1.49	6.73	0.92	0.62	1.19
C_1 (mg g^{-1})	15.98	16.09	1.23	9.08	5.11	42.40	4.08	0.57
R^2	0.934	0.996	0.925	0.966	0.942	0.842	0.953	0.856
k_{d2} (mg g^{-1} $min^{1/2}$)	0.29	4.78	1.65	-	-	5.80	0.89	0.61
C_2 (mg g^{-1})	43.47	56.37	5.84	-	-	38.17	2.59	7.23
R^2	0.997	0.997	0.983	-	-	0.994	0.989	0.939

MB: methylene blue; CR: congo red; C_i: initial concentration; q_e: adsorption capacity at equilibrium; exp.: experimental; k_1: Adsorption rate constant; R^2: correlation coefficient; k_{di}: intra-particle diffusion rate constant; C_n: constant

3.3. Equilibrium experiments

Isotherms analysis allows to describe the interaction between the dye and the adsorbent material, and to know the adsorption capacity of the material in the equilibrium. Kinetics parameters are shown in Table 2. For AC, maximum adsorption capacities (q_m) of 2000.0 and 444.4mg·g^{-1} for the adsorption of 100mg·L^{-1} of MB and CC, respectively were found. For AT, maximum adsorption capacities (q_m) are 1052.6 and 86.4mg·g^{-1} para MB (100mg·L^{-1}) and CR (50mg·L^{-1}), respectively. Langmuir correlation coefficients are R2≥0.942, indicating that adsorption process for both AC and AT have homogeneous sites on the surface which can interact with dyes (MB and CR). For AC and AT, the R_L values for both dyes are in the range of $0<R_L<1$, indicating that the adsorption process is favorable. The adsorption of

MB on AC can be well represented by Freundlich model (Figure 2(A)), which indicates that in this process there are different types of interactions between adsorbent and adsorbate. That is, the MB can be adsorbed on AC by interaction with functional groups and by diffusion through the pores [4] , which could explain the high adsorption capacity of AC to remove MB.

Table 2. Calculated parameters for the two different adsorption models tested during this study.

Material	Cedar				Teak			
Dyes	MB		CR		MB		CR	
C_i (mg L^{-1})	50	100	50	100	50	100	50	100
Langmuir								
q_m (mg g^{-1})	1369.86	2000.00	381.68	444.44	416.67	1052.63	86.43	-
K_L (L g^{-1})	0.23	0.28	0.84	0.12	1.75	0.63	0.13	-
R_L	0.09	0.04	0.03	0.08	0.01	0.02	0.14	-
R^2	0.988	0.997	0.991	0.995	0.999	0.992	0.948	-
Freundlich								
K_F (mg g^{-1})	27.03	59.96	25.27	20.72	-	55.96	48.47	-
n (g L^{-1})	2.02	3.10	9.58	6.55	-	6.30	2.53	-
R^2	0.935	0.925	0.884	0.821	-	0.890	0.805	-

MB: methylene blue; CR: congo red; Ci: initial concentration; q_m: maximum adsorption capacity; K_L: Langmuir constant; R_L: dimensionless separation factor; R^2: correlation coefficient; K_F: Freundlich; constant; n: constant.

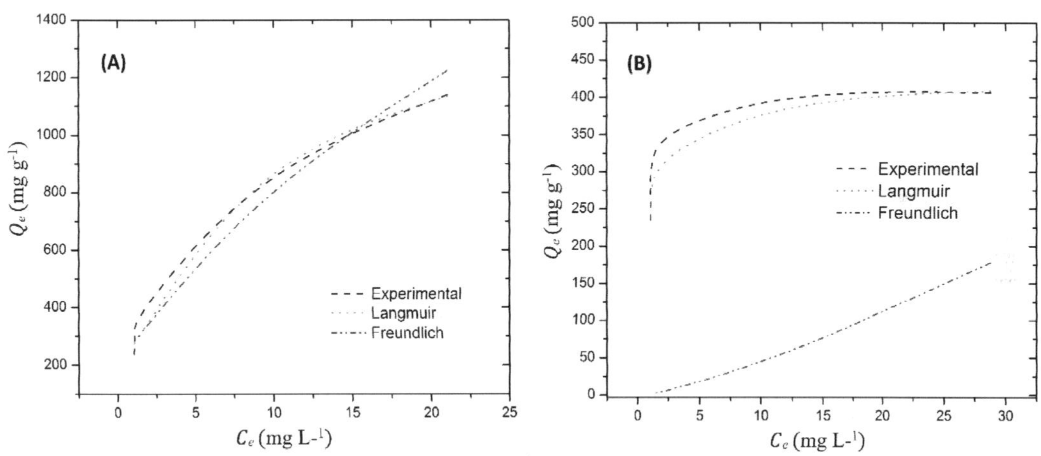

Figure 2. Langmuir and Freundlich Isotherms for MB adsorption with a concentration of 50mg·L^{-1} using: (A) Activated Cedar (AC) and (B) Activated Teak (AT).

3.4. Comparison with other adsorbents

To illustrate the potential in the use of these activated materials as MB and CR adsorbents a comparative evaluation of the adsorption capacities of various types of adsorbents is provided in Table 3. The high adsorption capacity of the materials evaluated in this study, compared with those reported by other authors suggests that our materials have great potential to be applied as adsorbents in the removal of dyes from aqueous solutions.

Table 3. Comparison of the activated materials (AC and AT) with other adsorbents for MB and CR.

Adsorbent	Type of dye	Q_m (mg·g^{-1})	Reference
AC	MB	2000.0	This study
AC	CR	444.4	
AT	MB	1052.6	
AT	CR	86.4	
Oil Palm Fiber	MB	763.4	[5]
Oil Palm Peel	MB	724.6	[5]
Potato wastes	MB	540.0	[13]
m-Cell/Fe_3O_4/ACCs	CR	66.1	[14]
BAF	CR	11.9	[15]
Sodium Bentonite	CR	19.9	[16]

4. Conclusions

Two activated carbon (AC and AT) were evaluated for the removal of different concentration of MB and CR. The kinetic and equilibrium studies showed that the pseudo second order kinetics model provided the best correlation with the experimental data ($R^2 \geq 0.970$), which indicates that the adsorption was controlled by a chemisorption process. Using the regression coefficient values, both the Langmuir and the Freundlich isotherms models fit MB and CR adsorption behavior on AC materials, from which the maximum adsorption capacities were calculated. For AC the capacities were 2000.0 and 444.4mg·g^{-1} for MB y CR, respectively. For AT the capacities were 1052.6 y 86.4mg·g^{-1} para MB and CR, respectively. This study demonstrate that the materials produced in our group are viable materials to remove dyes from natural aquatic environments.

Acknowledgments

The authors are grateful to Univ. de Medellín, for financing the project. N Y Acelas thanks "COLCIENCIAS" for the PhD scholar-ship.

References

[1] Duman G et al 2009 *Energy Fuels.* **23(4)** 2197-2204
[2] Pirsaheb M et al 2016 *Desalination Water Treat.* **57(13)** 5888-5902
[3] Hameed B, Ahmad A and Latiff K 2007 *Dyes Pigm.* **75(1)** 143-149
[4] Ramirez A et al 2017 *J. Phys.: Conf. Ser. (submitted)*
[5] Ramirez A et al 2017 *Rev. Colomb. Quim.* **46(1)** 33-41
[6] Acelas N et al 2015 *Chemosphere.* **119** 1353-1360
[7] Isah U et al 2015 *Int. Biodeterior. Biodegradation* **102** 265-73
[8] Mahapatra K, Ramteke D and Paliwal L 2012 *J. Anal. Appl. Pyrolysis* **95** 79-86
[9] Lorenc-Grabowska E and Gryglewicz G, 2007 *Dyes Pigm.* **74(1)** 34-40
[10] Tor A and Cengeloglu Y 2006 *J. Hazard. Mater.* **138(2)** 409-415
[11] Munagapati V and Kim D 2017 *Ecotoxicol. Environ. Saf.* **141** 226-234
[12] Aboua K et al 2015 *J. Environ. Manage.* **156** 10-14
[13] Zhang Z et al 2015 *J. Taiwan. Inst. Chem. Eng.* **49** 206-211
[14] Zhu H et al 2011 *Chem. Eng. J.* **173(2)** 494-502
[15] Mall I et al 2005 *Chemosphere* **61(4)** 492-501
[16] Vimonses V et al 2009 *Chem. Eng. J.* **148(2)** 354-364

IOP Conf. Series: Journal of Physics: Conf. Series **935** (2017) 012012 doi:10.1088/1742-6596/935/1/012012

Production and characterization of activated carbon from wood wastes

A P Ramirez[1], S Giraldo[1], M Ulloa[1], E Flórez[1] and N Y Acelas[1]
[1] Universidad de Medellín, Medellín, Colombia

Email: nyacelas@udem.edu.co

Resumen. Cedarwood *(Cedrela Angustifolia)* and teak *(Tectona Grandis)* woods are typically used for furniture manufacture because they have high durability, are light and easy to work. During these manufacturing process, large amount of these wastes is generated causing disposal environmental problems. In this paper, the residual wastes (sawdust) of Cedar (C) and Teak (T) are transformed into an activated material. The chemical composition of both biomass (C and T) was determinate by TGA (Thermogravimetric Analysis). Activated materials were characterized in surface area following the BET (Brunauer, Emmett and Teller) method, morphology using SEM (Scanning Electron Microscopy) and to know their functional groups a FTIR (Fourier Transform Infrared Spectroscopy) analysis was done. Their adsorption capacity was evaluated by removal of Methylene Blue (MB) and Congo Red (CR) from aqueous solutions.

1. Introduction

Activated commercial carbon is a very typical adsorbent to remove gaseous and aqueous contaminants. However, due to the high cost during its production and the environmental problems (it comes from fossil fuel) it is necessary to look for new sources to produce these materials. In a study published by British Petroleum, 2007 [1], coal reserves will only last 147 years (since 2006). Therefore, agroindustry wastes from oil palm [2], potato [3], wood [4, 5] become in good sources to produce activated carbon.

On the other hand, according to Food and Agriculture Organization of the United Nations (FAO) only about 28% of a tree is turned into sawn timber, the remaining 72% is waste that can be used as a fuel, in particles boards and others. These processes mentioned above generate about of 55% in wastes without any use [6], which makes the sawdust as a potential precursor for activated carbon.

Activated carbons produced from wood wastes are recognized for their good pore development, large surface area and high removal capacities. They are typical used as separators and purifiers in both gases and aqueous solutions, catalyst supports and, adsorbents to dyes removal or solvents recovery [7-10].

In this study, the production of activated carbon from two types of wood wastes, Cedar (C) and teak (T), was evaluated. Its chemical composition, morphology and textural properties were determined using different analysis techniques. And their application for dye removal from aqueous solutions was studied.

2. Methodology

2.1. Preparation and characterization of materials

C and T sawdust wastes were obtained from Sena-Calatrava, Itagüí, Antioquia. To get a homogeneous particle size, samples were passed through a sieve with size of <850μm. They were dried at 100°C for 24 hours. A chemical activation was performed with a ratio, biomass: $ZnCl_2$ 3M of 1:5 for C and 1:4 for

T (g:mL) and allowed to stir for 24 hours. After this time, both biomass was calcined at 550°C in a muffle for 60min with a heating rate of 15°C/min; then they were washed with a solution of HCl, 2M and kept under stirring at room temperature for three hours. After that, samples were passed through a filter paper (filter paper qualitative Advantec 2) and washed with distilled water, repeating the process until obtaining a pH between 6 and 7. Finally they were dried in an oven at 105°C for 12 h. Two activated carbons were obtained: Activated Cedar (AC) and Sctivated Teak (AT). To compare, both types of wood wastes were only subjected to calcination at 550°C for 30 minutes. They were called Calcined Cedar (CC) and Calcined Teak (CT).

Moisture, Volatile Material (VM), Fixed Carbon (FC) and ash of each biomass were determined by thermogravimetric analysis (TGA Q500, TA Instruments). The surface area of activated materials was determinate following the BET method (Brunauer, Emmett and Teller) [11], pore volume and size was found by nitrogen adsorption (ASAP 2020, Micrometrics) which is a standard procedure for the determination of adsorbents porosity [2]. Their morphology was determined by SEM analysis (Oxford Instruments). And FTIR (Shimadzu IRtracer 100) was used to know their functional groups.

2.2. Yield of dyes adsorption

Batch experiments were carried out in synthetic solutions of 50mg·L^{-1}, at 150rpm and 2 hours to evaluated the yields of C, T, CC, CT, AC and AT for the adsorption of MB and CR. To determine the final concentration of MB and CR, calibration curves were performed and the equation for each one was obtained with a R^2=0.9999 (equation (1)) for MB and R^2=0.9995 (equation (2)) for CR. The absorbance was measured by using a spectrophotometer (VIS-DR 3900) at a wavelength of 665nm and 490nm for MB and CR, respectively.

$$x = \frac{y + 0.5242}{0.5205} \tag{1}$$

$$x = \frac{y - 0.0354}{0.0859} \tag{2}$$

Where "*y*" is the absorbance and "*x*" is the dye concentration in mg·L^{-1}.

3. Results and discussion

3.1. Materials characterization

A typical chemical composition of different types of wood is: VM (78 and 84%), FC (6 and 16%) and ash (0.4 and 2.4%) [4,5]. As can be seen in Table 1, both biomass (T and C) are found within the previous ranges. With these analyses it is possible to say that during the calcination process, about 83% in mass was lost due to the elimination of VM, being available about 15% of FC to be activated. These results indicate that C and T can be used for the manufacture of activated carbon.

Table 1. Biomass properties.

Biomass	Moisture (%)	Proximate analysis (Wt %)[a]		
		VM	FC	Ash
C	6.02	83.76	15.41	0.83
T	5.02	83.92	15.67	0.41

C: cedar; T: teak; VM: volatile material; FC: fix carbon. [a]dry basis

Figure 1 shows SEM analysis for the two types of materials obtained here, calcined and activated (CC, CT, AC and AT). The micrographs show significant changes in the morphology of the studied biomass. Images for calcined materials, CC and CT, show a development of average porosity, which is in agreement with the determined values of surface areas 378.92m^{2}g^{-1} for CC and 311.51m^{2}g^{-1} for CT (Table 2); this porosity is generated by the release of volatile compounds during the calcination process

IOP Conf. Series: Journal of Physics: Conf. Series **935** (2017) 012012 doi:10.1088/1742-6596/935/1/012012

[3]. On the other hand, the micrographs for the activated materials (chemical activation), AC and AT, show a discontinuous surface caused by the dehydration action of the activating agent, in this case $ZnCl_2$ [12-14]. This dehydration leads to the good development of porosity and therefore to the increase of the surface area, with values of 1375.67m^2g^{-1} and 996.65m^2g^{-1} for AC and AT, respectively.

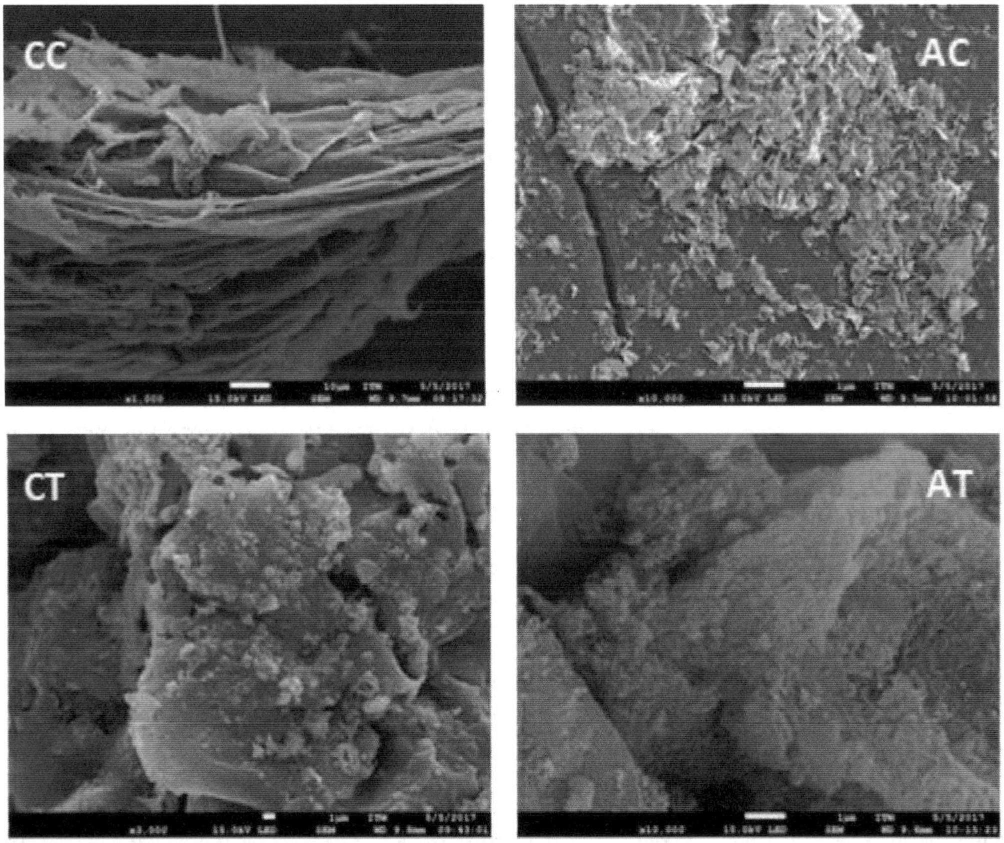

Figure 1. SEM micrographs for calcined and activated materials. CC: calcined cedar; AC: activated cedar; CT: calcined teak; AT: activated teak.

From the above, it can be said that the activation process is efficient since it manages to increase the surface area 3.6 times for AC and 3.2 times for AT respect to the calcination process. This parameter is one of the most important to take into account for activated carbons. It is widely known that in a greater surface area, there are more available active sites, which allows the material to be promising for using in adsorption processes [15].

The IUPAC (International Union of Pure and Applied Chemistry) classifies the micro-, meso- and macroporosity of the materials depending on the pore size: micropores ($dp<2nm$), mesopores ($2nm<dp<50nm$) and macropores ($>50nm$). Table 2 shows the pore size for calcined and activated materials. These values indicate that CC, CT, AC and AT have a mesoporous structure with tendency to microporous. In the same way, an increase in pore volume of AC and AT in comparison to CC and CT is observed. According to some authors [4,5,16] activation parameters such as, time, temperature and activation method strongly influence textural properties of the activated carbons. For example, they reported that as the temperature increases from 700 to 900°C, the yield of the material is reduced and the effect of the activation time is more pronounced (4 hours). The decrease in yield with increasing temperature is due to the removal of carbon atoms during the activation stage, which leads to the creation of a large number of new pores, enlargement of existing pores and opening of blocked pores [5,16].

Table 2. Properties of calcined and activated materials.

Materials	Surface area, BET (m^2 g^{-1})	d_p (nm)	Pore volumen (cm^3g^{-1})
CC	378.92	2.19	0.21
CT	311.51	2.19	0.17
AC	1375.67	2.73	0.94
AT	996.65	2.55	0.63

CC: calcined cedar; CT: calcined teak; AC: activated cedar; AT: activated teak; d_p: pore size

Other important parameters to consider for activated carbons are the functional groups on their surface which may influence the adsorption process [4,5,16]. The functional groups commonly found in activated carbon are: alcohols, aromatic rings, aromatic compounds and aliphatic chains [17].

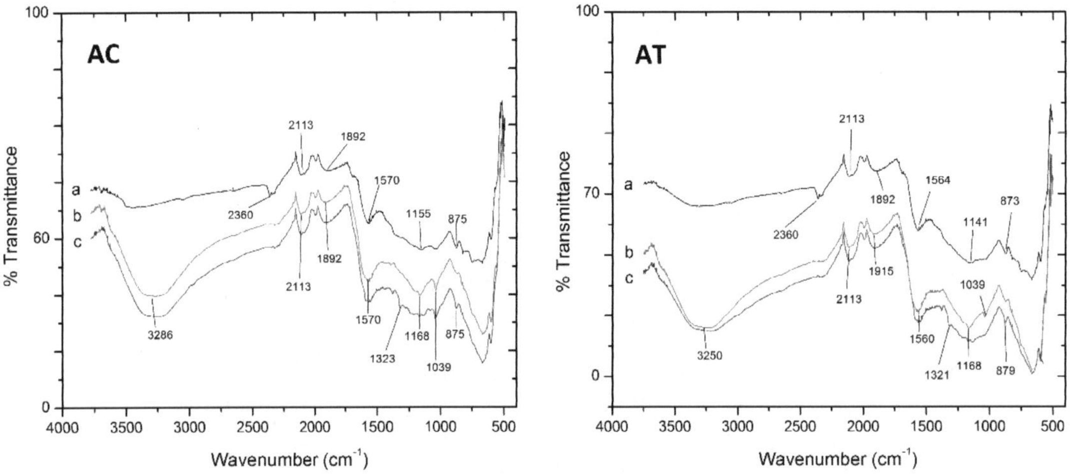

Figure 2. FTIR spectra for activated carbons. AC: activated cedar; AT: activated teak. a: before the adsorption process and b, c: after the adsorption process using CR and MB, respectively.

Figure 2 shows the FTIR spectrum for AC and AT before (a) and after the adsorption process with CR (b) and MB (c). It has been reported that the main interactions between the functional groups in activated carbon and dyes are: (i) hydrophobic interactions (Van der Waals - low energies), (ii) electron donor-acceptor interactions between the aromatic rings of activated carbon and -OH groups in dyes and, (iii) hydrogen bonds [17]. The adsorption process studied in this work may be governed by the interactions mentioned above. According to Figure 2, the main functional groups present in our materials are aromatic, aliphatic and alcohols, which according to their relative intensities can be explained as follows: interactions between 3280cm^{-1} and 3250cm^{-1} correspond to hydrophobic ones, between 2360cm^{-1} and 2113cm^{-1} are donor-acceptor electron interactions through C-H symmetrical stretches; signals at 1570cm^{-1} and 1560cm^{-1} correspond to donor-acceptor electron interactions through aromatic rings of activated materials and -OH group of dyes. Signals between 1323cm^{-1} and 1321cm^{-1} are donor-acceptor electron interactions by C-H inflections. And the ones between 1168cm^{-1} and 1039cm^{-1} are attributed to interaction donor-acceptor by stretching CO. These results are in very good agreement with the reported by Thue, *et. al.* (2006) [17].

3.2. Yield of the materials for dyes adsorption

Figure 3 shows the percentages in MB and CR removal using the materials before and after the transformation processes (C, T, CC, CT, AC, AT). It can be observed that for all cases, activated materials present a greater removal compared to calcined materials and biomass without any treatment. The percentages of removal for CR are given as: AC ~40%, CC ~15% and C ~6%.

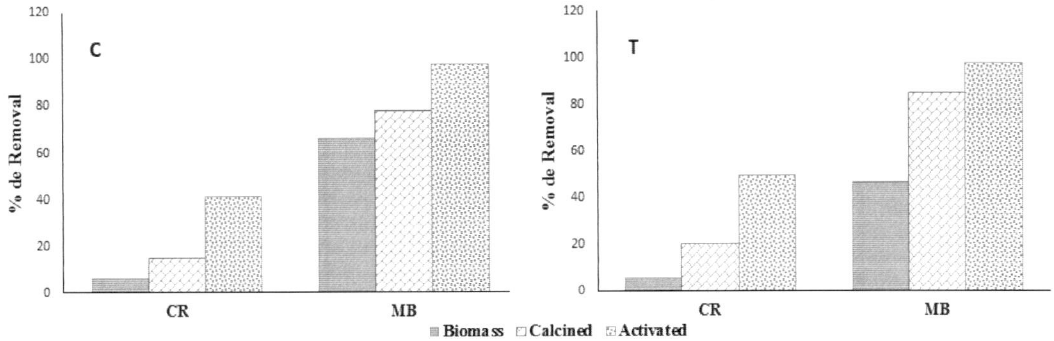

Figure 3. Dyes removal using materials before and after the transformation processes. C: cedar; T: teak; CR: Congo Red; MB: Methylene Blue. green: biomass; grey: calcined; orange: activated

The good performance of activated materials during dyes removal can be explained by adsorption inside the pores or by interaction of functional groups. In this case, it is suggested that the greater removal given by AC and AT for MB is related to dyes size. CR (Figure 4(b)) has a larger molecular size than MB. Therefore, for the adsorption of a CR molecule it is necessary more active sites (functional groups) than for a MB molecule. And, the pore size is a limiting factor for the adsorption of CR. Similar reports have been publish in the scientific literature [18,19].

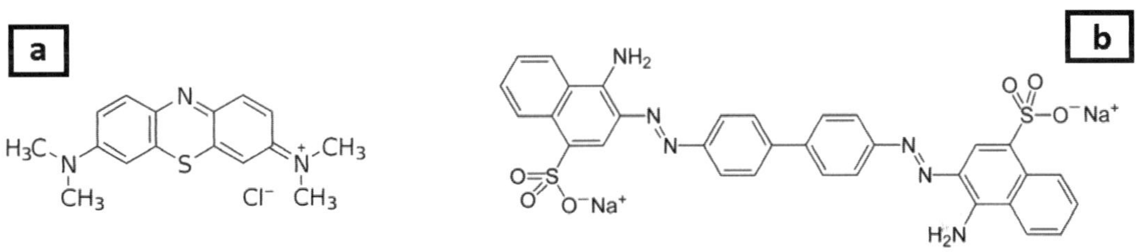

Figure 4. Chemical structure of (a) Methylene Blue (b) and Congo Red.

4. Conclusions

It was demonstrated that our activation method is suitable for the transformation of cedar and teak wastes into an activated carbon with a potential use for dyes removal from water. Activated materials presented surface areas of 1375.67m^2g^{-1} and 996.65m^2g^{-1} for AC and AT, respectively which are above the range of those that have been reported in the literature.

Acknowledgments

The authors are grateful to Universidad de Medellín, for financing the project. N Y Acelas thanks "COLCIENCIAS" for the PhD scholar-ship.

References

[1] Petroleum B 2007 *BP Statistical Review of World Energy* 66th edition (London: BP Energy)
[2] Ramirez A et al 2017 *Rev. Colomb. Quím.* **46(1)** 33-41
[3] Zhang Z et al 2015 *J. Taiwan. Inst. Chem. Eng.* **49** 206-211
[4] Botomé M et al 2017 *Chem. Eng. J.* **321** 614-621
[5] López F et al 2013 *J. Anal. Appl. Pyrolysis* **104** 551-558
[6] Forestry Department 1990 *Energy conservation in the mechanical forest industries* (Italy: Food and agriculture organization of the united nations) chapter 6.
[7] Li Y et al 2011 *Appl. Surf. Sci.* **257(24)** 10621-10627
[8] Nieto C, Terrones M and Rangel J 2011 *Biomass Bioenergy* **35(1)** 103-112
[9] Ahmadpour A and Do D 1996 *Carbon* **34(4)** 471-479

IOP Conf. Series: Journal of Physics: Conf. Series **935** (2017) 012012 doi:10.1088/1742-6596/935/1/012012

[10] Mudoga H, Yucel H, and Kincal N 2008 *Bioresour. Technol.* **99(9)** 3528-3533
[11] Gañán J et al 2006 *Appl. Surf. Sci.* **252(17)** 5976-5979
[12] Pezoti Jr O et al 2014 *Ind. Eng. Chem. Res.* **20(6)** 4401-4407
[13] Ahmed M and Dhedan S 2012 *Fluid Phase Equilib.* **317** 9-14
[14] Luna D et al 2007 *ContactoS* **64(10)** 39-48
[15] Islam M et al 2015 *J. Taiwan. Inst. Chem. Eng.* **52** 57-64
[16] Yorgun S andYıldız D 2015 *J. Taiwan. Inst. Chem. Eng.* **53** 122-131
[17] Thue P et al 2016 *J. Mol. Liq.* **223** 1067-1080
[18] Vimonses V et al 2009 *Chem. Eng. J.* **148(2)** 354-364
[19] Mall I et al 2005 *Chemosphere* **61(4)** 492-501

IOP Conf. Series: Journal of Physics: Conf. Series **935** (2017) 012013 doi:10.1088/1742-6596/935/1/012013

Synthesis by combustion in solution of Zn_2TiO_4+Ag for photocatalytic and photodynamic applications in the visible

A A Lopera[1], A M Velásquez[2], E A Chavarriaga[1], S Ocampo[1], M A Zaghete[2], M A Graminha[2] and C P Garcia[1]
[1] Universidad Nacional de Colombia, Bogotá, Colombia
[2] Universidade Estadual Paulista Júlio de Mesquita Filho, Araraquara, Brasil

E-mail: aalopera@unal.edu.co

Abstract. Zn_2TiO_4 + Ag compounds were synthesized by the solution combustion path seeking to enhance their photocatalytic and photodynamic response in the visible. X-ray diffraction tests confirmed the formation of the phase and the presence of metallic silver. Field emission electron microscopy evidenced the formation of aggregates formed by grains lower than 100nm. The diffuse reflectance tests allowed to detect compound absorption in the visible region and activation energy of 2.8eV. The evaluation of the photocatalytic properties was performed by the degradation of methylene blue while the photodynamic response in biological systems was performed by the antilesihmanicidal response of the compounds in promastigotes of Leishmania amazonensis. Indirect measurement of ROS species confirmed the formation of oxygen singlets and OH radicals of the compounds when subjected to the action of visible light.

1. Introduction

Semiconductor ceramic materials with photocatalytic properties are of great interest due to their multiple applications in different areas, from waste water treatment to applications such as photodynamic therapy. These applications are based on the principle of the production of ROS reactive species, which can interact with different types of inorganic and organic molecules until their breakdown. It allows applications in photodynamic systems for the treatment of skin diseases such as cancer or parasitic diseases such as cutaneous leishmaniasis. One of the main interests in photocatalytic and photodynamic systems is that their activation could do it in the visible spectrum and thus enhance their applications because approximately 40% of sunlight is within the range of 400-700nm while UV only covers about 4%.

The zinc titanate Zn_2TiO_4 is an inverse spinel often reported as a secondary or resulting phase of the TiO_2/ZnO system. It has an excellent thermal stability and is interesting in different applications such as pigments, catalysts, sensors, dielectric materials [1,2] adsorbent systems [3], and degradation of dyes by UV irradiation [4]. On the other hand, Ag is a material with excellent antimicrobial and antiparasitic properties. Silver nanoparticles have been used in multiple investigations against different strains of E.Coli, Stapilococcus aureous, and antileishmanial response [5,6]. Recently, a study of the antimicorbial and photocatalytic properties of the Ag/TiO_2/ZnO system under visible irradiation was carried out [7,8], reiterating their properties and potential use in water decontamination and disinfection. In the present work, we synthesized Zn_2TiO_4+Ag through combustion in solution and evaluated its potential photocatalytic response in the visible by the degradation of methylene blue and its photodynamic response in the visible with the effect aintilehismania in promastigotes L. amazonensis.

2. Experimental

2.1. Samples preparation and morphological - structural characterization

Zn_2TiO_4+Ag powders were synthesized by means of combustion in solution route. Titanium isopropoxide (Alfa Aesar) Zinc nitrate hexahydrate (Merck), commercial silver metal, nitric acid and glycine (Sigma Aldrich) were used as reagents. Titanium nitrate was initially obtained from titanium isopropoxide using the appropriate amount of water and nitric acid. Silver nitrate was obtained by reacting commercial silver and nitric acid. Zinc nitrate and silver nitrate were added to Titanyl nitrate solution. Glycine was used as fuel and the molar ratio between nitrates-fuel was adjusted to the previously reported parameter $\varphi=1$ [9]. The system was heated and magnetic stirred at 80°C and 300rpm respectively until combustion was reached. The ashes obtained were macerated in agate mortar and calcinated at 700°C for 2 hours. For crystalline phase identification, X-ray diffraction with a double circle multipurpose Xpert-Pro PANanalytical Radiation Cu-Ka (λ=0.15406nm) diffractometer in a range of diffraction of 0-80°(2θ) was used. The optical properties of powders were analysed by spectrophotometry measurements of diffuse reflectance UV-VIS using a Perkin Helmer equipment with a spectral range between 280-800nm. The powder morphology was analysed in a Scanning Electron Microscope JEOL JSM-7610F (FE-SEM).

2.2. Assessment of photocatalytic activity by indirect measurement of ROS species and degradation of methylene blue

For the photodynamic activation, a white led light source was used with an irradiation power of 100mW/cm^2. The detection of ROS species was performed indirectly by fluorescence of the probe 2',7'-dichlorofluorescein diacetate (H_2DCFDA) and Singlet Oxygen Sensor Green (SOSG) for hydroxyl radicals and singlet oxygen respectively. For the measurement of the hydroxyl radicals the compound was suspended in 200µl of deionized water and in 10µg/mL H_2DCFDA. For the detection of the oxygen singlets, 3µM of the SOSG probe dissolved in ethanol was used. The compounds were immediately exposed to visible irradiance by led light and the fluorescence was monitored for 60 minutes at excitation/emission in an equipment Infinite M200Pro, TECAN with wavelengths of 488/520nm and 505/525 nm respectively.

For the photocatalytic evaluation, a stock solution of methylene blue at 1mmol in water was prepared. 20mg of the compound were added. The system was sonicated for 20 minutes and subsequently was placed in a reactor under constant stirring and visible irradiation. The degradation of methylene blue was monitored by spectrophotometry at 670nm.

2.3. Assessment of the activity by measuring antipromastigotes of Leishmania amazonensis

Promastigotes of L. amazonensis MPRO/BR/1972/M1841-LV-79 were used for the photodynamic tests. The parasites were kept in culture medium at 28°C. 96-well plates and a concentration of parasites of 5×10^5 were used. A stock solution of Zn_2TiO_4+Ag with a concentration of 20mg/mL was prepared. This solution was added to the wells with Leishmania in the respective amounts starting from a concentration of 1000µg/mL to 25µg/mL. The plates with Leishmania+compounds were exposed to visible irradiation for 40 minutes (the tests were performed in triplicate) and incubated in a humid chamber at 27°C for 24 hours. Subsequently a cell viability test was performed [10].

3. Results and discussion

Figure 1(a) shows the diffractograms of the obtained powders subjected to 700°C for 2 hours. The predominant phase was the spinel Zn_2TiO_4 indexing with the pattern JDPCS 00-013-0536. It is possible to identify metallic silver indexed with the pattern JCPDS 04-0783 and anatase and rutile as the main titanium phases. Zn_2TiO_4 is usually formed under temperatures close to 1000°C when is obtained by solid state. Using the combustion route the temperature of the phase was 700°C. Other works report 900°C as the minimal temperature to obtain this phase by combustion route during 4 hours when urea was used as fuel [3]. In our case the fuel used was glycine, which could result in a more exothermic combustion reaction that makes easier the formation of the phase. Synthesis of the $Ag/TiO_2/ZnO$ system by a sol gel route was performed by Shalaby et al. finding that the Zn_2TiO_4 phase was not formed for temperatures lower than 800°C and only from this temperature appeared that phase associated to other secondary phases of ZnO, $ZnTiO_3$ and metallic silver [7]. UV-Vis spectrophotometry (Figure 1(b)) showed that the Zn_2TiO_4+Ag compounds showed visible absorption unlike the commercial TiO_2. The compounds Zn_2TiO_4+Ag presented an energy band gap of 2.8eV. The absorbance of the sample in the visible can be related to the formation of intermediate states of energy due to the possible doping of the structure with C and N atoms from the combustion reaction in solution [11]. This kind of reactions are characterized by releasing large amounts of CO_2 and NO_x. Similar behaviours had already been observed in synthesis of TiO_2 structures by combustion in solution [12].

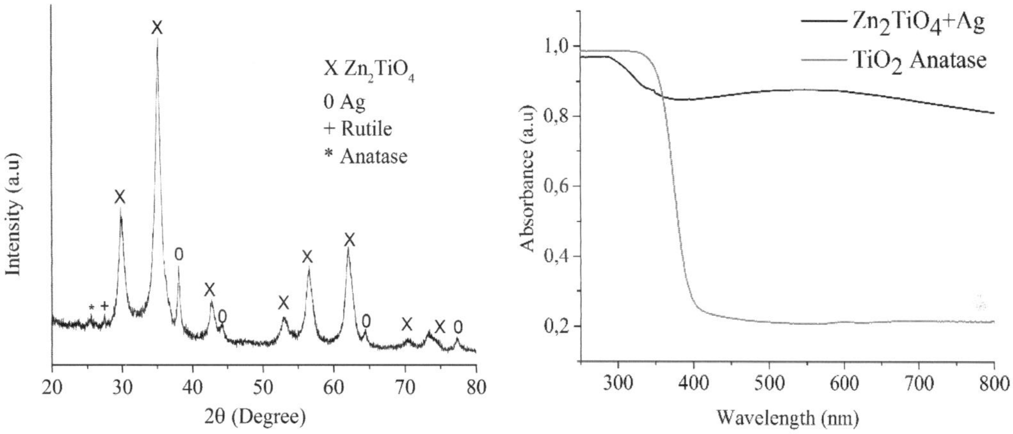

(a) (b)

Figure 1. (a) X-ray diffractograms for the sample treated at 700°C for two hours (b) diffuse reflectance of the calcined powders at 700°C.

Field emission microscopy showed micrometric aggregates with sizes ranging from 0.5 to 3μm (see Figure 2(a)). These aggregates are composed of small particles with geometries tending to be granular and lower than 100nm (Figure 2(b)). The combustion route is characterized by the facility to obtain nanometric systems with geometries tending to be spherical but with a considerable degree of aggregation [9]. In the combustion routes, the production of nanoparticles depends mainly on the local temperature and the distribution of it in the medium. Choosing an adequate fuel and an adequate ignition mechanism play a decisive role in the final product morphology [9].

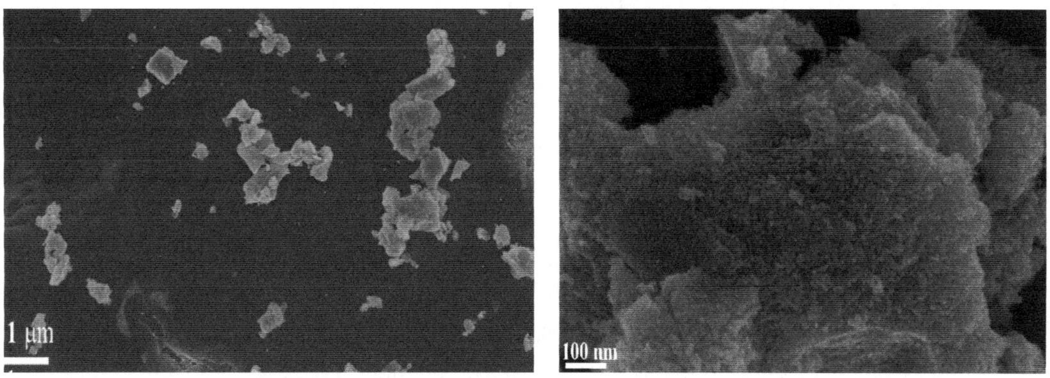

(a) (b)

Figure 2. Electron emission field photomicrographs for calcined powders.

Figure 3 (a) shows the fluorescence intensity of the test molecules over time, evidencing the formation of reactive oxygen species, both for the production of oxygen singlets and hydroxyl radicals. The fluorescence increases with the time of exposure to the light source corroborating the electronic excitation and the production of electron hole pairs in the structure of the Zn_2TiO_4.

Figure 3 (b) shows the absorbance measurements as a function of time for the solution of methylene blue exposed to visible irradiance by led light. The degradation measurements showed 50% of methylene blue decomposition at 70 minutes of irradiation and almost total discoloration at 180 minutes of irradiation. Figure 4 shows the percentage of cell viability of L amazonensis promastigotes incubated with different concentrations of the compound and irradiated with visible light during 40 minutes.

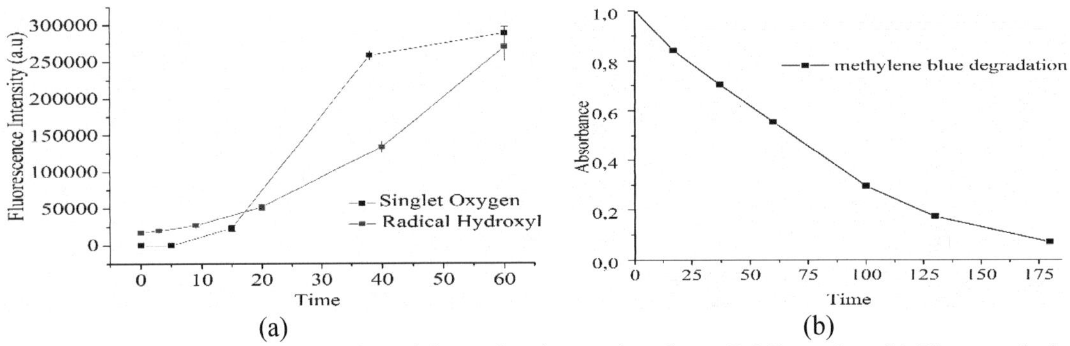

Figure 3. (a) Fluorescence intensity of the molecules used to detect ROS species. (b) Photocatalytic degradation of methylene blue undo visible radiation.

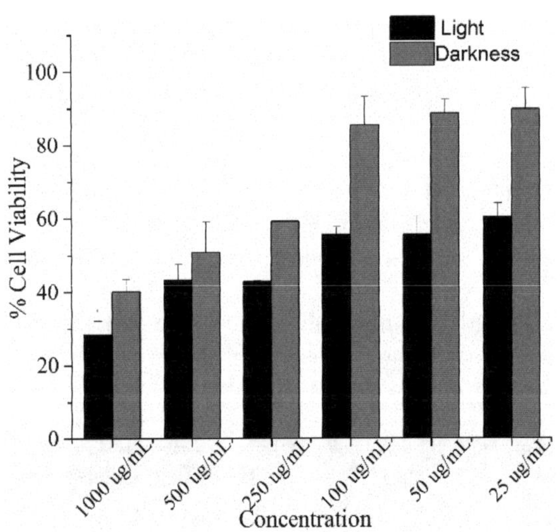

Figure 4. Cell viability of L. amazonensis promastigotes incubated with the compound at different concentrations and subjected to visible light.

A dose-dependent response was shown. Even for concentrations greater than 1000µg/mL cell viability was reduced by more than 50%. This behaviour could be associated with Ag properties acting as anti-parasitic agent. The anti-leishsmania capacity of the Ag nanoparticles and Ag/ZnO nanocomposites is well known [5]. The photodynamic effect of the compounds was higher with lower concentrations (25, 50, 100µg/mL). At these values the cell viability decreased between 25-30% with respect to controls when the systems were irradiated, whereas for the higher concentrations (250, 500, 1000µg/mL) the viability cell decreased between 5-10% with respect to the controls. This could be explained by assuming an antileishmanicide effect corresponding to the sum of the effect due to light irradiation (photodynamic response) and the independent effect of the silver particles. At low concentrations the photodynamic effect is greater than the cellular inhibition of Ag and in the case of high concentrations the antiparasitic response is due to a greater extent to the Ag present in the compounds. The antileishmania response of these compounds is related to the production of reactive species of oxygen both singlet oxygen and hydroxyl radicals which can interact with different biological molecules producing membrane peroxidation and stimulating the immune response of the cell. A. Nadhaman *et. al.* [13] reported that materials that can produce these reactive species are interesting as antilehismania compounds. Other studies as cytotoxicity, genotoxicity and hepatoxicity are required in order to support the biological safety of these compounds to be used as photodynamic systems.

4. Conclusions

Zn_2TiO_4+Ag compounds were synthesized by combustion in solution. The X-Ray diffractograms showed the presence of peaks associated with metallic silver. The compounds showed a visible absorption spectrum and a band gap of 2.8V. The compounds showed production of single ROS species of oxygen and hydroxyl radicals upon irradiation. These species were responsible for the degradation of methylene blue dye in 180 minutes and the photodynamic response antileishmaniasis. The presence of silver also plays an important role because of its antiparasitic capacity.

Acknowledgements

This work was partially funded by CNPq-TAWS of Brazil, Colciencias, Project "National Doctorates Call for 6172" and the National University of Colombia, Medellín.

References

[1] A C Chaves, S J Lima, R C Araújo, M A M Maurera, E Longo, P S Pizani, L G Simões, L E Soledade, A G Souza and I M G dos Santos 2006 *J. Solid State Chem.* **179** 985-992
[2] H T Kim, Y Kim, M Valant and D Suvorov 2001 *J. Am. Ceram. Soc.* **84** 1081-1086
[3] R Raveendra, P Prashanth and B Nagabhushana 2015 Synthesis, Characterization of Combustion Derived Zn 2 TiO 4 Nanocrystals and Its Application to Adsorption of Azo Dye *International Journal of Advanced Scientific and Technical Research* **5** 129-137
[4] P Zhang, C Shao, M Zhang, Z Guo, J Mu, Z Zhang, X Zhang, P Liang and Y Liu 2012 *J. Hazard Mater.* **229** 265-272
[5] P Baiocco, A Ilari, P Ceci, S Orsini, M Gramiccia, T Di Muccio and G Colotti 2011 *ACS. Med. Chem. Lett.* **2** 230-233
[6] I Sondi and B Salopek-Sondi 2004 *J. Colloid Interf. Sci.* **275** 177-182
[7] A Shalaby, A Bachvarova-Nedelcheva, R Iordanova, Y Dimitriev, A Stoyanova, H Hitkova, N Ivanova and M Sredkova 2015 *J. Optoelectron. Adv. M.* **17** 248-256
[8] S Manik and S Pradhan 2006 *Physica E.* **33** 69-76
[9] K C Patil, S T Aruna and T Mimani 2002 *Curr. Opin. Solid St. M.* **6** 507-512
[10] T Mosmann 1983 *J Immunol. Methods* **65** 55-63
[11] K Nagaveni, G Sivalingam, M Hegde and G Madras 2004 *Appl. Catal. B-Environ.* **48** 83-93
[12] A D Mani, V Laporte, P Ghosal and C Subrahmanyam 2012 *Mater. Res. Bull.* **47** 2415-2421
[13] A Nadhman, S Nazir, M I Khan, S Arooj, M Bakhtiar, G Shahnaz and M Yasinzai 2014 *Free Radical Bio. Med.* **77** 230-238

IOP Conf. Series: Journal of Physics: Conf. Series **935** (2017) 012014 doi:10.1088/1742-6596/935/1/012014

Influence of MWCNT/surfactant dispersions on the mechanical properties of Portland cement pastes

B Rodríguez[1], J H Quintero[1], Y P Arias[2], O A Mendoza-Reales[3], J C Ochoa-Botero[2] and R D Toledo-Filho[3]
[1] Universidad de Medellín, Medellín, Colombia
[2] Universidad Nacional de Colombia, Medellín, Colombia
[3] Universidade Federal do Rio de Janeiro, Rio de Janeiro, Brazil

E-mail: oscar@coc.ufrj.br

Abstract. This work studies the reinforcing effect of Multi Walled Carbon Nanotubes (MWCNT) on cement pastes. A 0.35% solid concentration of MWCNT in powder was dispersed in deionized water with sodium dodecyl sulfate (cationic surfactant), cetylpyridinium chloride (anionic surfactant) and triton X-100 (amphoteric surfactant) using an ultrasonic tip processor. Three concentrations of each surfactant (1mM, 10mM and 100mM) were tested, and all samples were sonicated until an adequate dispersion degree was obtained. Cement pastes with additions of carbon nanotubes of 0.15% by mass of cement were produced in two steps; first the dispersions of MWCNT were combined with the mixing water using an ultrasonic tip processor to guarantee homogeneity, and then cement was added and mixed until a homogeneous paste was obtained. Direct tensile strength, apparent density and open porosity of the pastes were measured after 7 days of curing. It was found that the MWCNT/surfactants dispersions decrease the mechanical properties of the cement based matrix due to an increased porosity caused by the presence of surfactants.

1. Introduction

Multi Walled Carbon Nanotubes (MWCNT) reinforce cement based matrices submitted to flexural and tensile stresses by working as "bridges" across micro cracks and pores, forming a network capable of enhancing the load transmission capacity through the matrix [1], depending on the length of the nanotubes [2]. It has been proposed that MWCNT have potential to control crack propagation phenomena, giving additional strain capacity to the matrix by dissipating energy through multiple pull out of nanotubes from the matrix before reaching a critical state [3].

Surfactants, along with other dispersing agents, have been identified as an effective aid to disperse MWCNT in water [4]. A good dispersion degree of MWCNT throughout the matrix is necessary to obtain crack propagation control [5]. However, it is not clear what Another section of your paper the maximum reinforcing capacity of MWCNT is, and how it is affected by the use of surfactants. Small additions of MWCNT have been found to increase the flexural strength [6-9], modulus of rupture [10], fracture energy and fracture toughness [11] of cement pastes and mortars. This work aims to quantify the reinforcing efficiency of MWCNT/surfactant dispersions in cement pastes submitted to direct tensile stress.

2. Materials and methods

The materials class G cement, industrial grade NC7000 multi walled carbon nanotubes (MWCNT) produced by Nanocyl, sodium dodecyl sulfate (SDS) of 96% purity produced by J.T. Baker, cetylpyridinium chloride (CPC) 100% pure produced by Sigma Aldrich, triton X 100 (TX 100) of 98% purity produced by Sigma Aldrich, and Rheomac UW 410 viscosity modifying agent (VMA) produced by Basf Chemicals were used. Cement and VMA were characterized by X-ray Diffraction (XRD) in a Bruker D8 diffractometer equipped with a Cu radiation source (Kα, λ=1.5418Å). Diffraction patterns were measured from 4 to 70° 2θ with a step size of 0.05 and an accumulation time of 30 seconds.

Aqueous dispersions containing 0.35% MWCNT were prepared using the three types of surfactants in 1mM, 10mM and 100mM concentrations. Dispersions were sonicated using an ultrasonic tip, which applied a total energy of 390J/g to guarantee an adequate dispersion degree. Sonication was carried out in a 500W ultrasonic processor set to 40 % amplitude in 20 seconds on/off. These dispersions were used to produce cement pastes in a 600mL Chandler constant speed. All liquids were stirred first at 500rpm for 1 minute; then, cement was added and speed was increased to 2070rpm during 1 minute and maintained constant for 5 more minutes. The proportions of the pastes studied in this work are presented in Table 1.

Table 1. Proportions of the pastes studied.

Paste Name		w/c	VMA (% bwoc)
Surfactant and concentration	MWCNT solid addition (% bwoc)		
Ref	-		
SDS 1mM			
SDS 10mM			
SDS 100mM			
CPC 1mM			
CPC 10mM	0.15% MWCNT	0.55	1.00
CPC 100mM			
TX 100 1mM			
TX 100 10mM			
TX 100 100mM			

After mixing, pastes were poured in 11.5×11.5×2.5cm molds and cured in a high humidity environment for 7 days. At testing age, each sample was cut in four 115×25×12.5mm slabs with two 5mm notches at midspan in the 12.5mm faces. Two metal plates were glued at the edge of one the notches to couple a clip gauge used to measure crack mouth opening displacement (CMOD). Samples were held in place using pneumatic claws, leaving a gauge length of 50% of the total length of the sample. The load application rate was varied to maintain a constant deformation rate at the CMOD of 0.03%/min. Testing was carried out in a Shimadzu AGX 100 universal testing machine equipped with a 100kN load cell. Immediately after testing, apparent density and open porosity were measured for each sample by water absorption following equations (1) and (2).

$$P = \frac{m_s - m_d}{m_s - m_h} x100 \quad (1)$$

$$\rho = \frac{m_d}{m_s - m_h} \rho_w \quad (2)$$

Where: P: Open porosity (%); ρ: apparent density; m_s: saturated mass (g); m_d: dry mass (g); m_h: hydrostatic mass (g) and ρ_w: density of water (g/cm^3).

IOP Conf. Series: Journal of Physics: Conf. Series **935** (2017) 012014 doi:10.1088/1742-6596/935/1/012014

3. Results and discussion

3.1. Raw materials characterization

XRD patterns obtained for cement (Left) and VMA (Right) are presented in Figure 1. Anhydrous calcium silicates, calcium carbonate and calcium sulphate were found for cement. It should be noticed that anhydrous calcium sulphate (anhydrite) was found to be predominant over hydrated calcium sulphate (gypsum), which indicates that clinker was grinded at a higher than expected temperature, causing gypsum de dehydrate [12]. VMA was found to be composed mainly of cellulose mixed with sodium chloride and sodium sulphide.

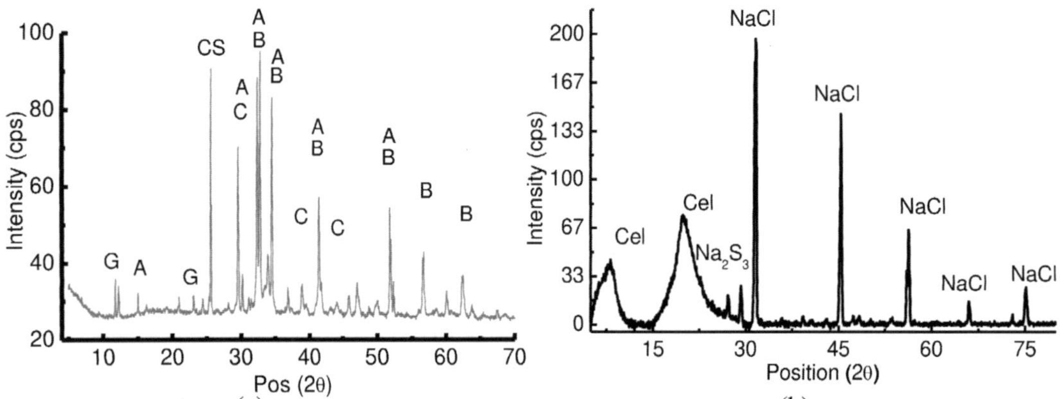

(a) (b)

Figure 1. (a) X-Ray diffraction pattern of cement (G: Gypsum (CaSO42H2O), CS: Anhydrite (CaSO4), A; Alite, B: Belite, C: Calcite, cps: counts per second) and (b) X-Ray diffraction pattern of VMA (Cel: Cellulose).

3.2. Density and porosity

Apparent density and open porosity results for all the pastes studied are presented in Figure 2. It was found that for a constant amount of MWCNT, both apparent density and open porosity of the pastes after 7 days of hydration decreased proportionally with the surfactant concentration. This can be related to a decrease in surface tension of water caused by surfactants. Air becomes entrapped in the paste in the mixing process, and this allows surfactant molecules to adsorb on the water-air boundaries, stabilizing the bubbles and maintaining them individual [12]. These individual bubbles will generate a non-connected porous structure, accounting for the fact that while apparent density decreases, open porosity also decreases rather than increasing.

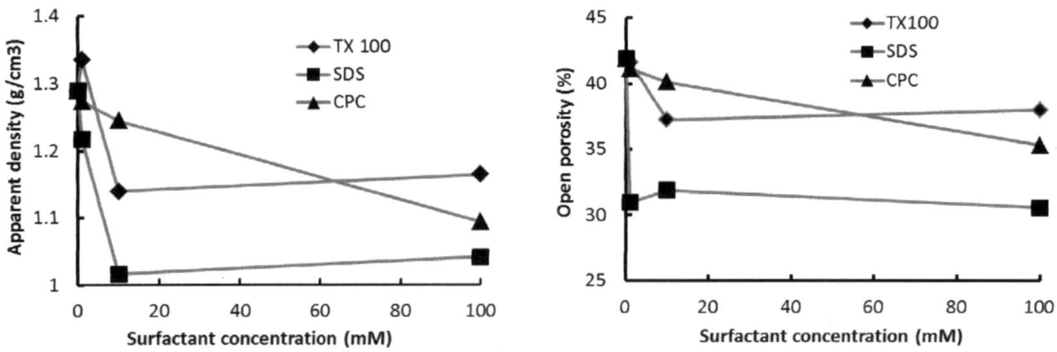

(a) (b)

Figure 2. (a) Apparent density of the paste cured 7 days versus surfactant concentration and (b) open porosity of the paste cured 7 days versus surfactant concentration.

3.3. Tensile strength of pastes

Tensile strength results of all the pastes studied after 7 days of curing is presented in Figure 3. It was found that the MWCNT/surfactant dispersions considerably decreased the tensile strength of the cement matrix when comparing to the reference sample. Additionally, it can be seen that all three surfactants have similar effects, regardless of their concentration. This result is the opposite of the expected, since a reinforcing effect of MWCNT in the cement matrix would increase its tensile strength. Similar results have been reported in the literature before but for compressive strength [13] and have been related to agglomeration of the nanotubes which create stress concentration spots.

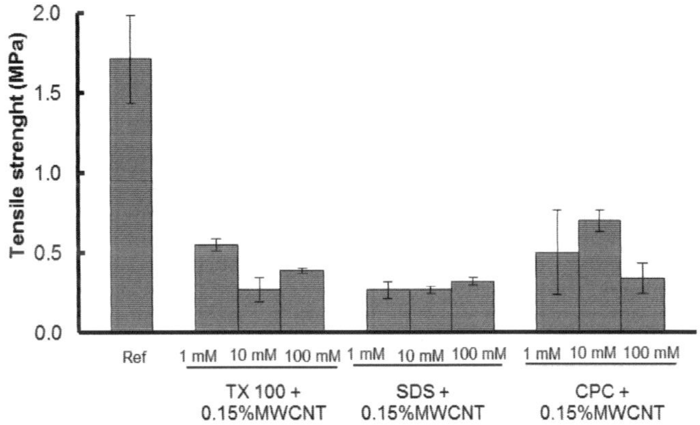

Figure 3. Tensile strength results of pastes after 7 days of curing.

4. Final discussion

A correlation between tensile strength and apparent density for all pastes studied is presented in Figure 4. It can be seen clearly that lower densities are associated with lower tensile strengths for all three types of surfactants studied. As it was previously shown, the decrease in density is directly related to the surfactant concentration used to produce the MWCNT/surfactant dispersion, rather than to the presence of MWCNT. This indicates that any reinforcing effect from MWCNT is much lower than the strength decrease induced by surfactants. This does not mean that the reinforcing effect of MWCNT does not exist, but rather that it is masked by the effect of surfactants. It has been reported that high amounts of surfactants are necessary to adequately disperse MWCNT in water [14], but these high amounts of surfactants seem to be incompatible with an application in the studied cement matrix.

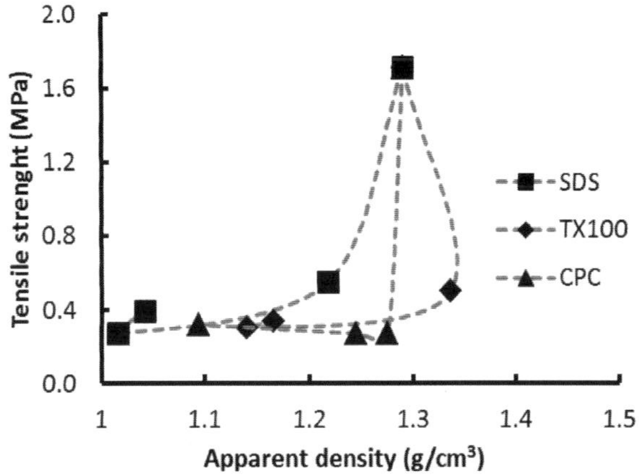

Figure 4. Tensile strength versus apparent density of the pastes studied.

IOP Conf. Series: Journal of Physics: Conf. Series **935** (2017) 012014 doi:10.1088/1742-6596/935/1/012014

5. Conclusions

The reinforcing effect of MWCNT/surfactants dispersions is masked by the effect of surfactants on the density and porosity of the cement matrix.

Decreases in density induced by surfactants are directly related to decreases in tensile strength of the cement matrix.

The amount of surfactant necessary to obtain an adequate dispersion of MWCNT in water is incompatible with a positive reinforcing effect of the MWCNT/surfactant dispersion in the studied cement matrix.

References

[1] B. Wang, Y. Han, and S. Liu 2013 *Constr. Build. Mater.* **46** 8
[2] R. K. Abu Al-Rub, A. I. Ashour, and B. M. Tyson 2012 *Constr. Build. Mater.* **35** 647
[3] L. Y. Chan and B. Andrawes 2010 *Comput. Mater. Sci.* **47N4** 994
[4] J. Rausch, R.-C. Zhuang, and E. Mäder 2010 *Compos. Part A Appl. Sci. Manuf.* **41N9** 1038
[5] O. Mendoza and R. Toledo 2016 *Carbon Nanomaterials Sourcebook: Nanoparticles, Nanocapsules, Nanofibers, Nanoporous Structures, and Nanocomposites, 1st edition* vol II, ed K. Sattler (New York: CRC Press) p 573
[6] S. Xu, J. Liu, and Q. Li 2015 *Constr. Build. Mater.* **76** 16
[7] M. S. Konsta-Gdoutos, Z. S. Metaxa, and S. P. Shah 2010 *Cem. Concr. Compos.* **32N2** 110
[8] M. S. Konsta-Gdoutos, Z. S. Metaxa, and S. P. Shah 2010 *Cem. Concr. Res.* **40N7** 1052
[9] F. Torabian Isfahani, W. Li, and E. Redaelli 2016 *Cem. Concr. Compos.* **74** 154
[10] S. Musso, J.-M. Tulliani, G. Ferro, and A. Tagliaferro 2009 *Compos. Sci. Technol.* **69** 1985
[11] Y. Hu, D. Luo, P. Li, Q. Li, and G. Sun 2014 *Constr. Build. Mater.* **70** 332
[12] P. Hewlett 2004 *Lea's Chemistry of Cement and Concrete, fourth edition* (Oxford: Elsevier)
[13] F. Collins, J. Lambert, and W. H. Duan 2012 *Cem. Concr. Compos.* **34N2** 201
[14] R. Siddique and A. Mehta 2014 *Constr. Build. Mater.* **50** 116

IOP Conf. Series: Journal of Physics: Conf. Series **935** (2017) 012015 doi:10.1088/1742-6596/935/1/012015

Structural and mechanical study of concrete made from cementitious materials of low environmental impact

A K González[1], A M Montaño[1], C P González[2] and A Santos[2]
[1] Universidad Industrial de Santander, Bucaramanga, Colombia
[2] Universidad Pontificia Bolivariana, Floridablanca, Colombia

E-mail: claudia.gonzalez@upb.edu.co

Abstract. This work shows the results obtained by replacing Type I Portland®, by cementitious geopolymers materials, derived from minerals, in concrete mixtures. Synthesis of both geopolymers through alkaline activation of two alluminosilicates: Bentonite and Pumice with sodium silicate (Na_2SiO_3). XRD, SEM and XRDE are used to structural study of new geopolymers. Concrete mixtures with replacement of Portland have 10% and 30% of geopolymer. Finally, concrete mortars formed were mechanically analysed according to ICONTEC 220 at 7, 14, 28, 41, 90 and 120 days of cure. Results shows that compressive strength of concrete from Bentonite and Pumice are almost the same for the standard concrete at 28 days of cure. At 90 days of cure, compression resistance of concrete from Pumice at 10% is even higher than those that standard concrete shows.

1. Introduction

Because of the socioeconomic development and the population increase, demand of Portland Cement (PC) as construction material has increased. Its manufacturing process is one of the most polluting practices, since 6 to 7% of the CO_2 emissions of the planet correspond to the cement industry. This industry together with research groups seek solutions for the conservation and protection of non-renewable resources through the development of new raw materials that improve the production processes and quality of cement achieving sustainability [1-3]. Alumni and silica oxides are geopolymerized for partial cementitious replacement reducing greenhouse gases producing in PC clinkerization. This study presents the results of the synthesis of IP from the alkaline activation of two clays: bentonite and pumice, to obtain a new material with properties similar to those of PC. X-ray diffraction (XRD), Scanning Electron Microscopy (SEM), Energy-Dispersive X-ray Microscopy (EDS) and Mechanical Resistance (MR) according to the ICONTEC 220 standard are employ to characterization of IP.

2. Experimental

Geopolymerization of pumice (PP) from Sabaneta, Antioquia, Colombia and bentonite (BT) (Quimicos Campota®) was carried out at 80±5°C for 31 hours in metallic containers with sodium silicate (Na_2SiO_3) (Aldrich®). Dried product reached at 4.77nm of particle size according to ASTM C 94/C 94 M-09) [5]. XRD (Bruker D8 ADVANCE, DaVinci geometry, CuKα1 radiation), SEM (FEI Quanta 650 FEG SEM, magnification=2000-5000X, WD=9.2-10.0mm, HV=5-20kV), EDS (HV:

20.5kV, Tilt: 0.00, Take-Off: 35.88, Amplitude: 14.4.) are used to structural analysis. It was performed mechanical strength for PP, BT and IP.

3. Result and discussion

3.1. Geopolymerization reaction

Dissolution, hydrolysis and polycondensation of raw materials it happens because of the sodium silicate media to obtain polymeric matrix for each aluminosilicate source (PP and BT) [7].

3.2. Structural analysis

3.2.1. X- Ray diffraction. Structural study by XRD shows for BT: albite, gypsum, calcite, anortoclase, kaolinite, montmorillonite and quartz; for the IP from BT and because of the alkaline activation, new phases (orthoclase at 28°, muscovite at 16°, 27° and 39°) appear instead of gypsum and anortoclase. On the same way, for PP there are present: quartz, illite, albite, anorthite and iron oxide. In IP from PP there are not presents anorthite, illite and iron oxide while moscovite (28°) and cristobalite (22°, 37° and 60°) appears; this cristobalite gives a greater crystallinity degree to IP [3,5].

It is remarkable that PP has a porous structure and it is composed of vitreous particles of various forms, mainly elongated and angular, and shows amorphous structure, which undergoes a structural rearrangement of tetrahedra of Si and Al inducing in the system a certain degree of crystallinity. This fact is proof in the appearance of new mineralogical phases and in the increase in the intensity of the initially existing phases [2,7-9], as in the case of quartz and albite with peaks at 27°, 28° and 36° 2θ respectively.

3.2.2. Analysis by scanning electron microscope. The SEM micrographs of the raw materials and the synthesized PIs are presented in Figure 1, where in *a* and *c* (starting materials) homogeneous morphology and a similar particle size are observed. In turn, in b and d (IP of PP and IP of BT, respectively), the observed conformation is different, not homogeneous, exhibiting remains of unpolymerized raw materials. Presumably, the above is due to albite does not react by a low concentration of activating solution and it does not geopolymerized all the raw material. Thus, reaction medium should have a higher Na_2SiO_3 concentration [8]. Thus, various morphologies can exist in the geopolymeric paste as unreacted particles, "semi-reacted" particles attacked by the activating solution and the reaction products, as evidenced by the respective micrographs [11].

Finally, the particle size is clearly smaller for the synthesized IP than for the BT and PP clays, due to the compaction of the materials in the cement mixture, which presents a better molecular organization, which generates the crystallinity in each IP formed.

IOP Conf. Series: Journal of Physics: Conf. Series **935** (2017) 012015 doi:10.1088/1742-6596/935/1/012015

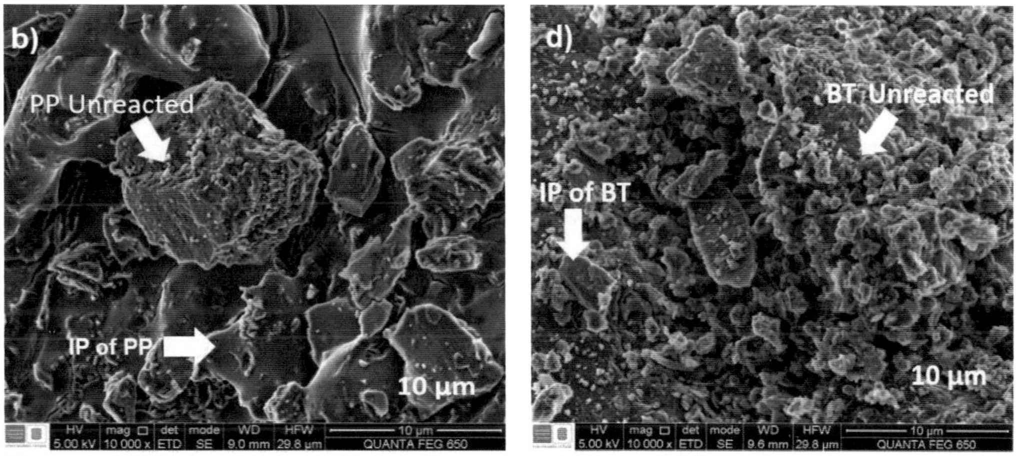

Figure 1. SEM micrographs of (a) PP, (b) IP of PP, (c) BT and (d) IP of BT.

3.2.3. Analysis by X-ray energy dispersive spectroscopy. Based on the results obtained in the EDS analysis (Figure 2), it is possible to determine that the geopolymers are mainly constituted by Na-Si-Al and some traces of Ca and K as is the case of PI of PP. These traces come from the impurities of the raw materials, being able to be present in a dispersed and random way in the microstructure, because they not are dissolved in the geopolymer matrix. [5,9]

The results showed that the Si, Al and Na contents of the IP differ from the initial values of the aluminosilicate sources (BT and PP), as a consequence of alkaline activation with Na_2SiO_3, finding that the IP of BT presents more content of Si and for IP of PP greater content of Al.

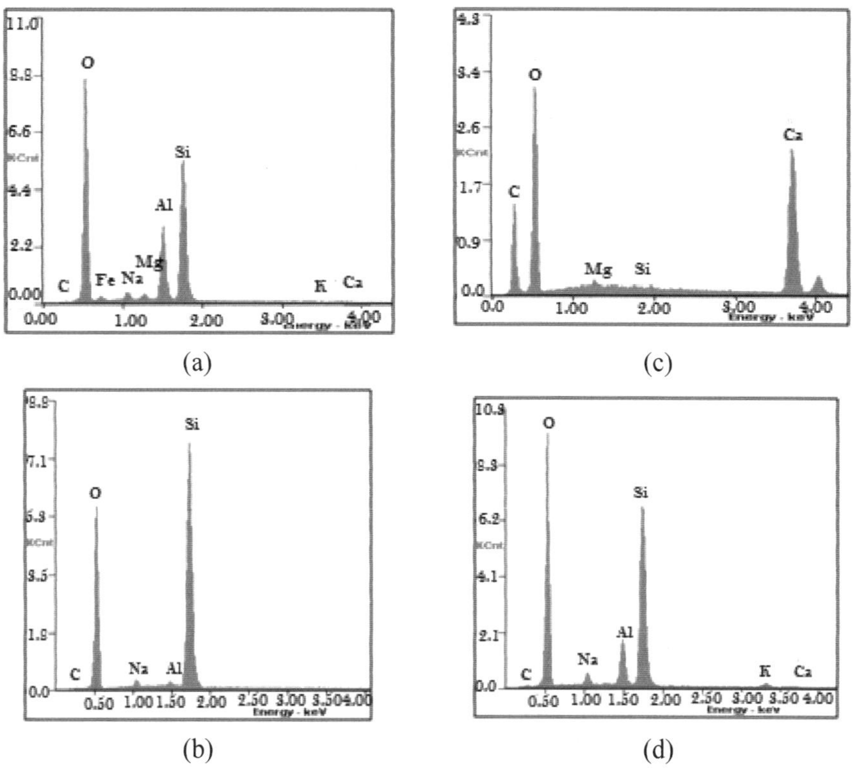

Figure 2. EDS of (a) BT, (b) IP of BT, (c) PP and (d) IP of PP.

3.3. Mechanical analysis

The compressive strength in MPa was determined according to the ICONTEC 220 standard at curing ages of 7, 14, 28, 41, 90 and 120 days, for each of the 10 and 30% substituted mortars in the PC for each IP (BT and PP). Results obtained appear in Figure 3, where it is evident that in the case of IP of 10% and 30% BT, the best values of mechanical resistance correspond to 28 days of curing in comparison with the other IP substitutions and the PC, where higher exposure time shows higher mechanical performance. Likewise, there is a significant difference in the values of MR for substitutions of 10% compared to those of 30% of each IP in the PC, where the lower percentage of substitution in the PC, generated greater mechanical resistance [6].

On the other hand, the results of MR are almost linear until the 28 days of curing for all the mortars where they reach their longest hydration time and by increasing the curing time this increases but steadily. This is because in concrete, hydration reactions occur at high speed, generating hydrated calcium silicates that contribute to increase the mechanical resistance and decrease the porosity of the same. However, it is noteworthy that at 90 days of curing the concrete resistance derived from IP of PP exceeds that of CP [4,10,12].

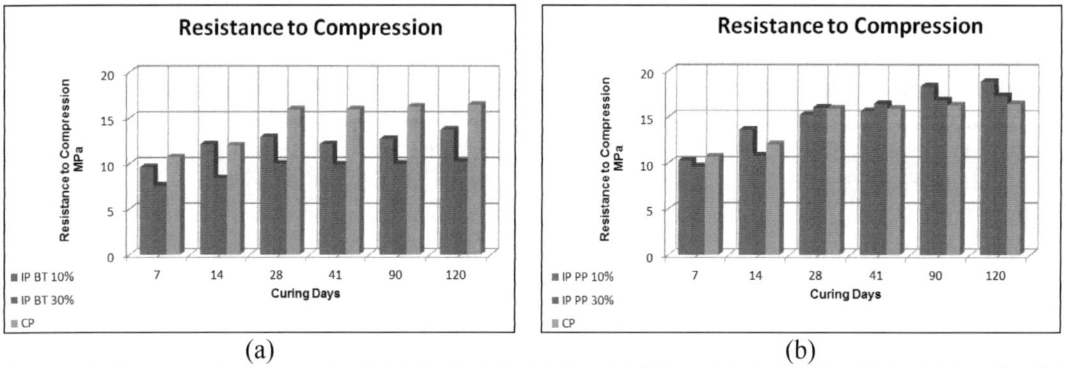

Figure 3. Compressive strength of (a) IP de BT al 30 and 10% and PC; (b) IP of PP al 30 and 10% and CP

4. Conclusions

The Physicochemical and mechanical properties of the synthesized IP depend directly on the mechanism of the geopolymerization reaction, in which the concentration of the raw material, concentration and type of the activating solution and temperature are parameters that allow the formation of the sialate monomer for subsequent condensation and formation of the polymer matrix.

Results of XRD, SEM and EDS, allow verifying that geopolymerization reaction occurs in the reaction medium, through an alkaline activation. The above implies a structural rearrangement in the aluminosilicates (BT and PP), due to formation of the Si-O-Al bond in each IP synthesized with the improvement in the degree of crystallinity.

Not all the raw material polymerized. More concentrate activating solution (Na_2SiO_3) would be all alluminosilicates reacts.

From MR test, it is evident that the mortars that presented the best mechanical performances of the synthesized IP are the 10% PP, which implies that it is as possible cementing agent.

Acknowledgements
The authors express their gratitude to the School of Chemistry, X-ray diffraction Laboratory and Microscopy Laboratory of the Universidad Industrial de Santander and the "Universidad Pontificia Bolivariana, seccional Bucaramanga", for the financial support of the project 016-0615-9560.

References

[1] Davidovits J. 1991 Geopolymers: inorganic polymeric new materials *J. Therm.* **37** 1633-1656

[2] Alomayri T 2017 Effect of glass microfibre addition on the mechanical performances of fly ash-based geopolymer composites *J. Asian Ceram. Soc* **5(3)** 334-340

[3] Rodríguez E, Mejía de Gutiérrez R, Bernal S, Gordillo M 2009 Efecto de los módulos SiO_2/Al_2O_3 y Na_2O/SiO_2 en las propiedades de sistemas geopoliméricos basados en un metacaolín *Rev. Fac. Ing. Univ. Antioquia* **49** 30-40

[4] ASTM C94/C94M–17a 2017 *Standard Specification for Ready-Mixed Concrete* (USA: American Society for Testing and Materials)

[5] González C P, Montaño A M, González A K, Ríos C A 2014 Synthesis and characterization of inorganic polymers from the alkali activation of an aluminosilicate *IOP Conf. Series: Materials Science and Engineering* **59** 1-5

[6] Favier A., Habert G., D'espinose de Lacaillerie J.B, Roussel N 2013 Mechanical properties and compositional heterogeneities of fresh geopolymer pastes *Cem. Concr R.* **48** 9–16

[7] Nikolić V, Komljenović M,Baščarević Z, Marjanović N, Petrović R 2015 The influence of fly ash characteristics and reaction conditions on strength and structure of geopolymers *Constr. Build. Mater.* **94** 361-370

[8] Hongling W, Haihong L, Fengyuan Y. 2005 Synthesis and mechanical properties of metakaolinite-based geopolymer *Coll. Surf. A: Physicochem. Eng. Asp.* **268** 1–6

[9] Wan Q, Rao F, Song S, Cholico-González D. F, Ortiz N. L 2017 Combination formation in the reinforcement of metakaolin geopolymers with quartz sand *Cem. Concr. Comp.* **80** 115-122

[10] Okoye F. N. Durgaprasad J, Singh N. B 2015 Mechanical properties of alkali activated fly ash/Kaolin based geopolymer concrete *Constr. Build. Mater.* **98** 685–691

[11] Puligilla S, Mondal P 2013 Role of slag in microstructural development and hardening of fly ash-slag geopolymer *Cem. Concr. R.* **43** 70–80

[12] Temuujin J, Williams R P, Van Riessen A 2009 Effect of mechanical activation of fly ash on the properties of geopolymer cured at ambient temperature *J. of Mater. Proce. Tech.* **209** 5276–5280.

IOP Conf. Series: Journal of Physics: Conf. Series **935** (2017) 012016 doi:10.1088/1742-6596/935/1/012016

Inclusion of geopolymers derivate from fly ash and pumice in reinforced concrete

A M Montaño[1], C P González[2], D Castro[3], G Gualdron[1] and R Atencio[4]
[1] Universidad Industrial de Santander, Bucaramanga, Colombia
[2] Universidad Pontificia Bolivariana, Floridablanca, Colombia
[3] Instituto Politécnico Nacional, Ciudad de México, México
[4] Instituto Venezolano de Investigaciones Científicas, Maracaibo, Venezuela

E-mail: claudia.gonzalez@upb.edu.co

Abstract. This paper presents results of a research project related to the development of alkali-activated geopolymers, synthesized from alumina-silicate minerals (fly ash and pumice) which are added to concrete. Alkali sources used in geopolymer synthesis were sodium hydroxide and sodium silicate solution. New materials were structurally characterized by Infra-Red spectroscopy (IR) and X-Ray Diffraction (XRD). Concretes obtained after geopolymers addition as Portland cement substitutes at 10%, 20% and 30%, were mechanically analysed by compression resistance at 7, 14, 28 and 90 drying days. Results were referred to standard (concrete of Portland cement) allows to know cementitious characteristics of geopolymers are lower than those for standard, but it keeps growing at longer drying time than Portland cement. By Electrochemical Impedance Spectroscopy (EIS) it is found that this new material shows high electrical resistance and have been proved as a protection agent against corrosion in reinforced concrete exhibiting anticorrosive properties higher than those showed by the conventional concrete mixture.

1. Introduction

Geopolymers (GPs) are produced from aluminosilicates, sources of silicon oxides and aluminium, which are activated by an alkaline agent. GP have characteristics similar to those of cement, which allows them to be used as a cementing agent. In previous works [1-5] it is reported that the substitution of GPs confers to the mixture mechanical properties and durability. In this paper, we report the results obtained by synthesizing geopolymers from the activation of fly ash (FA) and pumice (PP) with different concentrations of NaOH (3M, 7.5M and 12M) as alkaline agent. Obtained geopolymers were structurally characterized by experimental techniques such as: Infra-Red spectroscopy (IR) and X-Ray Diffraction (XRD). Subsequently, GPs were added to a concrete mixture, replacing the cement content in 10%, 20% and 30%. Obtained concretes were mechanically evaluated by compression resistance in specimens of 7, 14, 28 and 90 days of setting. Finally, the electrical response of the GPs with better mechanical behaviour was evaluated by Electrochemical Impedance Spectroscopy (EIS), placing the test specimens at accelerated corrosion conditions for 44 days.

2. Experimental

Synthesis of the GPs was initiated by adding the aluminosilicate to the solution of the alkaline agent to the base products: FA from thermoelectric plant TERMOZIPA, located in the municipality of "Tocancipa, Cundinamarca, Colombia" and PP extracted from "Sabaneta, Antioquia, Colombia". Alkaline agents used were a solution of NaOH and a solution of sodium silicate (Na_2O=8.6%, SiO_2=27.8% and H_2O=63.2%). Obtained mixture was then heated at a temperature between 60 and 100°C for 6-12h, obtaining a viscous paste, which was allowed to dry at room temperature for 24 hours, until curing. Thereafter, this slurry was ground to a particle size of 4.75mm (No. 4 sieve, according to ASTM E-11-87). NaOH concentrations of 3M, 7.5M and 12M were used. Structural characterization was performed using IR and RXD. Concretes were then made using the following aggregates: crushed stone with a maximum size of 19mm and fineness modulus sand of 1.97, both coming from the river "Pescadero, Santander, Colombia"; Portland cement type I, Diamante®. Cylinders with 75mm diameter and 150 mm high were prepared, with a setting time of 7, 14, 28 and 90 days. The designs of the concrete mix were made according to the process followed by Tecnoconcretes S.A. [6].

3. Result and discussion

3.1. Structural characterization

3.1.1. Infra-red spectroscopy IR. Figures 1 and 2 shows IR spectra of the geopolymers obtained from fly ash GF and those obtained from pumice (GP), respectively. It is observed the characteristic band of the vibrations produced by the Si and Al oxides, displaced to lower values (1030-1018cm^{-1}) for both geopolymers, with respect to the base products, with a variation also in the intensity.

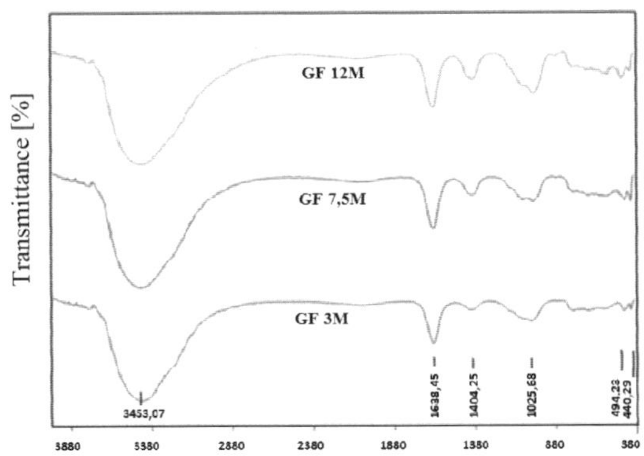

Figure 1. IR spectra of GF with different NaOH concentrations.

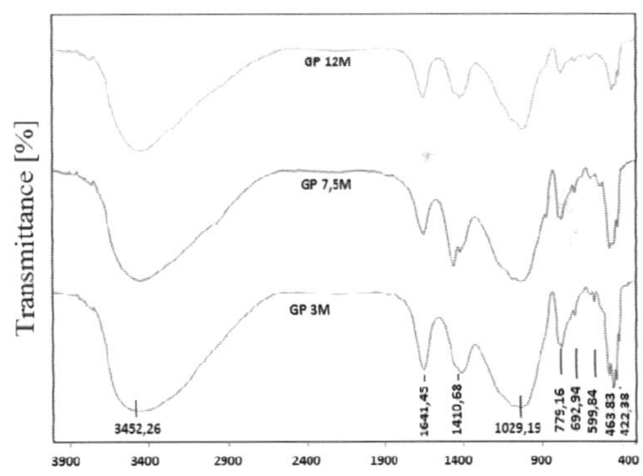

Figure 2. IR spectra of GP with different NaOH concentrations.

The behaviour of Figures 1 and 2, possibly is due to the transformation of the precursor aluminosilicate ($Al_2O_3.2SiO_2$) to the ion orthosialate, by the presence of the alkaline agent. This agent activates the bonds of the oxides, forming the two species, the aluminate ion ($Al(OH)_4^-$) and the monosylic acid ($OSi(OH)_3^-$), responsible for forming the monomer ($n(OH)_3$-Si-O-Al-$(OH)_3$) intermediate of the polycondensation. Bands located between 600cm^{-1} and 800cm^{-1} are related to the tension vibrations of the Al-O bond. The bands located between 400cm^{-1} and 600cm^{-1} are associated with vibrations by deformation of the Si-O-Si and Al-O-Si bond. A broadband from 2500 to 3700cm^{-1} approximately, indicates the presence of water in both minerals [7]. Peak located at 1641cm^{-1} for GP and at 1638cm^{-1} for GF, corresponds to the vibration by deformation of the H-OH bond, since the high alkali content in the polymer network decreases the pore size, thus avoiding the evaporation of water

[8]. The higher intensity of the band is associated to the greater number of water molecules present in GF 12M and GP 3M, because in these geopolymers the highest degree of geopolymerization is achieved. Broad and very intense band at 3452cm^{-1} for GP and at 3453cm^{-1} for GF, results from the tension vibrations of the group –OH of linking water, present in free molecular form, or adsorbed on the surface or going into pores of the geopolymer. Band located in the region of 1025-800cm^{-1} are due to Si-O bond flexion vibrations from the sodium silicate, this band is centered at 1029cm^{-1} for GP and at 1025cm^{-1}for GC. Signal at 1410cm^{-1} for the GP and at 1404cm^{-1} for the GC is attributed to CO tension vibrations of the carbonate present in the geopolymers. The presence of carbonates is due to the interaction of the material with the CO_2 of the environment, in this case to form sodium carbonates (Trona and Temonatrite).

3.1.2. X-Ray diffraction. Figure 3 shows the XRD patterns obtained for FA and GF, with the different concentrations of NaOH: 3M, 7.5M and 12M. In FA diffractogram mineral phases were identified: Quartz (SiO_2); Mullite [Al ($Al_{8.8}Si_{1.08}O_{4.85}$)]; Hematite ($Fe_2O_3$); Magnetite ($Fe_3O_4$); Berlinite ($AlPO_4$); Rutilo ($TiO_2$) and Anatasa ($TiO_2$) and the presence of an amorphous phase in between 20 to 30° 2θ, this phase is also found in the IR spectra [8]. This behaviour agrees with that reported in the literature for this mineral [9]. In diffractograms for GF, this amorphous phase is also observed, but with a lower intensity and with a shift of 0.5° in 2θ. All diffractograms of GF also present the formation of new crystalline phases that are not present in the ash; these phases are attributed to the dissolution of the structure of the aluminosilicate with the alkaline solution. Ions resulting from the reaction form phases such as: Termonatrite, Albite, Calcite, Muscovite, Clinochlorine, Microcline, Trona and Anortoclase, which generally consist of sodium, potassium, magnesium and aluminium aluminosilicates. Trona, Calcite and Temonatrite are carbonates, formed mainly by the reaction of the alkaline ions of the geopolymer network with the CO_2 present in the environment. In these materials two crystallographic phases of FA are conserved: Quartz and Mullite; The characteristic reflections of these phases have shifts of the order of 0.09° in 2θ. Phases mentioned above are found in the three GF, except trona, which is only found in GC 12M, mainly because this geopolymer is obtained with excess sodium. In GF 7.5M and GF 3M, anortoclase is present, which is not present in GF 12M, indicating that at higher concentration of NaOH, more carbonate phases appear, that is to 12M the highest degree of geopolymerization is achieved.

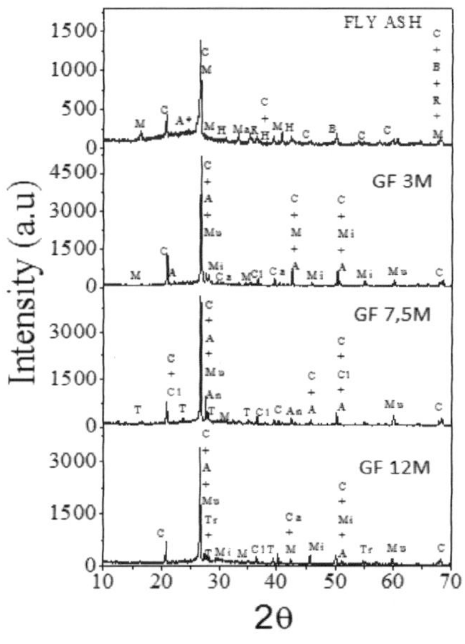

Figure 3. X Ray diffraction patterns of GF with different NaOH concentrations.

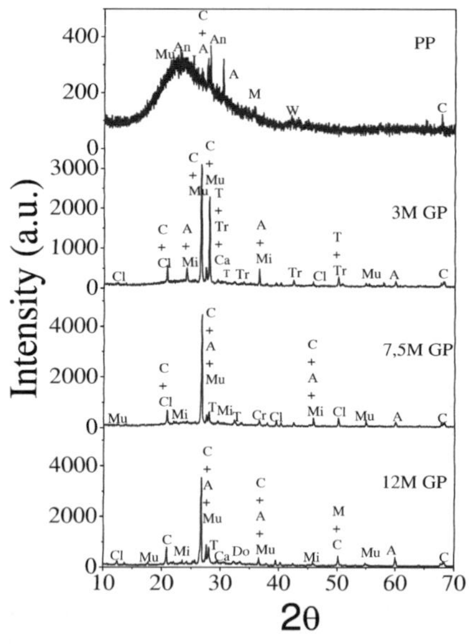

Figure 4. X Ray diffraction patterns of GP with different NaOH concentrations.

Figure 4 shows the XRD patterns obtained for PP and GP, with NaOH concentrations: 3M, 7.5M and 12M. In the PP diffractogram the mineral phases were identified: Anortita [$Al_2Ca\ (SiO_4)_2$]; Magnetite (Fe_3O_4); Anortoclase, syn ($Na_{0.71}K_{0.29}AlSi_3O_8$); Halite, potassium, syn ($K_{0.4}Na_{0.6}Cl$); Alpha quartz (SiO_2); Wuestita, syn ($Fe_{0.971}$); Muscovite [$(K, Na)\ (Al, Mg, Fe)_2\ (Si_{3.1}Al_{0.9})\ O_{10}(OH)_2$] and as in the case of FA, an amorphous phase in between 15°-35°2θ but of greater intensity. Diffractograms for the GP at different concentrations also present the amorphous phase existing in PP, but with less widening and intensity. This phase was also observed by IR spectroscopy. It is observed the formation of new crystallographic phases that are not present in PP, these phases result from the dissolution of the structure of the aluminosilicate in the alkaline medium. In these materials two crystallographic phases of PP: quartz and Muscovite are conserved, the characteristic reflections of these phases present shifts of 0.08° in 2θ. The phases named above are in the three GP, except Trona, which is only found in GP 3M; probably due to the fact that the SiO_2 and Al_2O_3 present in PP are activated at a lower concentration of NaOH, that is to say that with 3M has been achieved the highest degree of geopolymerization [10].

3.2. Mechanical analysis
In compressive strength analyzes performed on concrete made with substitutions of 10, 20 and 30% of the cement by GF and with GP for 7, 14, 28 and 90 days of setting [11], it was found that the highest values of resistance in the GC geopolymer are observed for a 12M NaOH concentration and 10% of substitution. On the other hand, GP exhibit the highest resistance for NaOH 3M and 10% of substitution. Concrete shows an increase in resistance as the drying time increases. Conventional concrete, has its maximum resistance to 28 days of drying, reaching a value of 300Kg/cm^2, whereas for concrete made with the geopolymers (GF and GP), this continues to increase even at 90 days, where the GF with 12M NaOH concentration and 10% substitution reached a value of 165kg/cm^2 and GPs with 3M NaOH and 10% of substitution reaches a value of 145kg/cm^2.

3.3. Electrochemical impedance spectroscopy
This test was performed for those specimens that exhibited the best mechanical resistance: GF 12M and GP 3M, for a total of 7 specimens, 6 with substitutions: 10%, 20% and 30% of each geopolymer (GF and GP), and one without replacement of the portland cement, used as a reference. In the electrochemical cell, the working electrode acts as reinforcing steel of the concrete structures, and the reference electrode allows to record the potential of the reinforcing steel. For electrochemical monitoring, the EIS in AC was used with an IM6 BAS-Zahner Impedance Unit, according to ASTM G3 [12] and it was supported with Resistance to Line Polarization (LPR) in DC. Nine monitoring were carried out at 0, 1, 2, 4, 8, 11, 15, 31 and 44 days of exposure to the corrosion medium. At the beginning of the immersion in the corrosive medium, the resistance increased in the geopolymer materials, evidencing the protection of the steel by the oxide layer formed in the surface of the same. As shown in Figures 4 and 5, the corrosion resistance of the concretes modified with geopolymer material increases as the time of exposure to the corrosive medium elapses, showing a greater anti-corrosion effect by replacing conventional cement by 20% of GF and GP. At the end of the process, the electrical resistance value is 40KΩ for the GF 20% and 18KΩ for the 20% GP, compared to only 4KΩ for the reference mixture. At the early stage of the corrosion process, around day eight, there was a decrease in the steel's resistance due to the amount of chloride around it, which causes localized breakage of the protective layer formed, resulting in material wear. Afterwards the resistance increases as the formation of the anticorrosive layer on the steel stabilizes, reducing the rate of corrosion. Increase in the electrical resistance indicates the low conduction of the electricity of GF and GP, responsible for a corrosion visibly lower than that occurred in the concrete made with portland cement. Stabilization of anticorrosive layer is caused by the decrease of the chlorides dissolved in the electrolyte. These anions react with the tricalcium aluminate ($Ca_3Al_2O_6$) present in the concrete mixture, forming the Friedel salt ($3CaO\ Al_2O_3\ CaCl_2\ 10H2O$). This reduces the conductivity and increases the strength of the concrete.

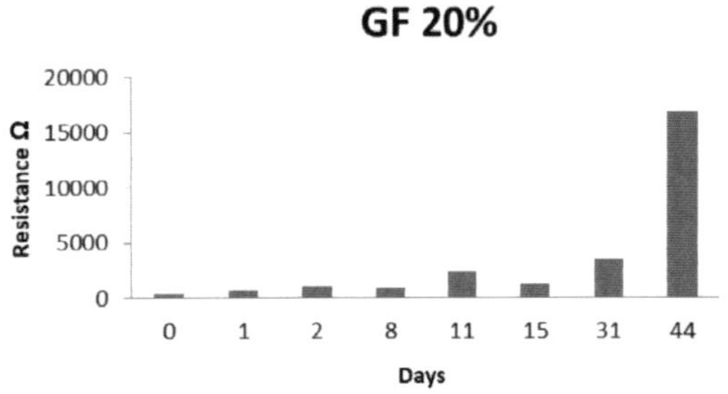

Figure 5. Electrical resistance over time of the modified concrete prepared with the substitution of 20% GF.

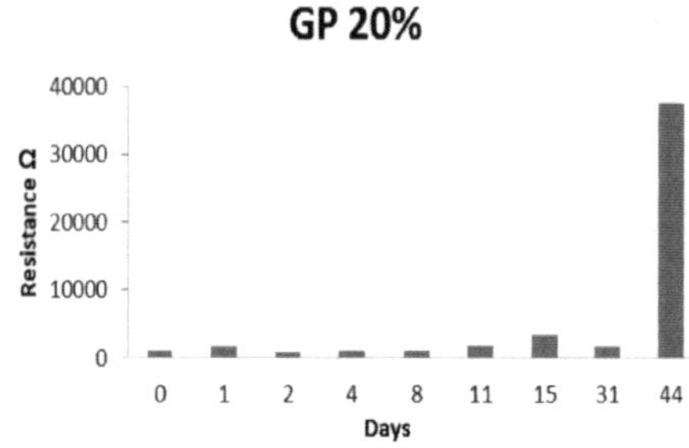

Figure 6. Electrical resistance over time of the modified concrete prepared with the substitution of 20% GP.

4. Conclusions

Sources of aluminosilicates used in this study have a high reactivity in the presence of a highly alkaline medium, where an optimum concentration of the activating solution allows to obtain geopolymeric materials with mechanical resistance up to 170kg/cm^2 at 90 days of setting.

Highest values of resistance in the GF geopolymer are observed for 12M and in the GP for 3M with 10% replacement of the concrete. Based on the persistent increase of the mechanical resistance of the GF and GP, it is very interesting to continue this research, allowing even greater setting times to 90 days.

Synthesized geopolymers have a marked tendency to protect the steel, forming an anticorrosive layer between this and the concrete. The increase in the strength of the reinforcing steel has its maximum value, in the study carried out, at 44 days of exposure to the medium of accelerated corrosion, this increase in the resistance indicates that the geopolymer material is bad conductor of the electricity, allowing to elaborate Concrete with a lower corrosion rate than conventional concrete.

References

[1] Xu H and Van Deventer JSJ 2000 The Geopolymerisation of alumino-silicate minerals *International Journal of Mineral Processing* **59** 247-266

[2] Sindhunata A, Provis JL, Lukey G, Xu H. and Van Deventer JSJ 2008 Structural evolution of fly ash based geopolymers in alkaline environments *Ind. Engi. Chem. Res.* **47(9)** 2991-2999

[3] Hardjito D and Rangan B 2005 Research Report GC 1 *Development and properties of low-calcium fly ash-based geopolymer concrete*. (Australia: Curtin University of Technology, Perth)

[4] Bernal S, Mejía de Gutiérrez R, Rodríguez E 2006 Nuevos concretos para el aprovechamiento de un sub-producto industrial *Proc. Congreso Iberoamericano de Metalurgia y Materiales (La Habana)* (Cuba: Iberomet IX) pp 793-802

[5] Komnitsas K and Zaharaki D 2007 Geopolymerisation: A review and prospects for the minerals industry *Minerals Engineering* **20** 1261

[6] NTC 3752 2002 *métodos de ensayo para determinar el contenido de agua en el concreto fresco* (Colombia: Instituto colombiano de Normas Técnicas y Certificación)

[7] Skoog D A and West D M 1984 *Análisis instrumental* (México: Editorial Interamericana)

[8] Lee W and Van Deventer JSJ. 2002. Physicochemical and Engineering Aspects. **211** 1 49

[9] Davidovits J. 1991 Geopolymers Inorganic polymeric new materials. Journal of Thermal Analysis. **37**, 1633

[10] Duxson P, Fernandez-Jimenez A, Provis JL, Lukey GC, Palomo A, Van Deventer JSS 2007 Geopolymer technology: the current state of the art *Jounal of Materials Science* **42(9)** 2917-2933

[11] NTC 673 2010 *Ingeniería civil y arquitectura. elaboración y curado de especímenes de concreto para ensayos de laboratorio* (Colombia: Instituto colombiano de Normas Técnicas y Certificación)

[12] ASTM G3-89 2004 *Standard practice for Convention Applicable to Electrochemical measurements in corrosion testing*. (USA: American Society for Testing and Materials)

IOP Conf. Series: Journal of Physics: Conf. Series **935** (2017) 012017 doi:10.1088/1742-6596/935/1/012017

Evaluation of the electrical conductivity and corrosion resistance for layers deposited via sputtering on stainless steel

J Blanco[1], Y Salas[2], C Jiménez[3], Y Pineda[1] and A Bustamante[2]
[1] Universidad INCCA, Bogotá, Colombia
[2] Universidad Pedagógica y Tecnológica de Colombia, Tunja, Colombia
[3] Pontificia Universidad Javeriana, Bogotá, Colombia

E-mail: juan.blanco@uptc.edu.co

Abstract. In some Engineering fields, we need that conductive materials have a mechanic performance and specific electrical for that they maintain conditions or corrosive attack if they are in the environment or if they are closed structure. The stainless steels have an inert film on their surface and it has the function to act in contrast to external agents who generates the corrosion, especially for stings, spoiling the film until to fail. We found a solution taking into account the electrical performance and the anticorrosive; into the process we put recovering of specific oxides on, stainless steel using the method of sputtering with Unbalanced Magnetron, (UBM) varying the oxygen in the reactive environment. The coating obtained had a thickness one micron approximately and we saw on serious structural uniformity [1]. The corrosion resistance was evaluated through the potentiodynamics polarization and electrochemical spectroscopy impedance in NACL according to the standard. The cathode protection is the most important method employed for the corrosion prevention of metallic structures in the soil or immersed on the water. The electrical resistivity was evaluated with the four points methods and it showed a behaviour of diode type in some substrates with a threshold potential in several volts. We noticed a simple resistance solution when it was analysed in the Nyquist graphics whit the Electrochemical Impedance Spectroscopy technique. With on equivalent circuit, for this reason we determinate a variation in the corrosion speed in almost two orders of magnitude when we analysed the potentiodynamics curve by Tafel approximation. The data obtained and analysed show that this type of surface modification maintains the conductivity condition at the interface, improving the resistance in relation whit the corrosion of these elements where the recovering allowed the ionic flow wished for overcoming threshold voltage, acting as an insulator in different cases.

1. Introduction

The growth of electrical systems, causes a greater number of failures and accidents of electrical origin, so that the issue of quality and safety in facilities should be a priority in the development of any territory, promoting the construction of facilities Safe and reliable. [2]

The development of the project seeks to evaluate the electrical performance and corrosion behaviour of stainless steel rods of a grounding system, seeking the protection of electromagnetic compatibility and the safety of people, as well as reducing the risks and problems caused by corrosion. [3]. Oxides are binary compounds in which oxygen has an oxidation state of -2. One of the techniques in which samples of stoichiometrically very precise oxides are obtained is the deposition of thin films by reactive sputtering. Controlled manipulation of the gases in the sputtering process can cause

IOP Conf. Series: Journal of Physics: Conf. Series **935** (2017) 012017 doi:10.1088/1742-6596/935/1/012017

variations in the crystal structure of the oxide. It has been noted that the relative size between the atomic motifs of the crystal lattice in the substrate and the molecules formed by oxidation of the target material in the deposition affects the adhesion of these oxides. It is also known the extraordinary property of titanium oxide when connecting with the metallic crystalline networks. This work explores the management of these oxides, creating vacancy defects or excess free electrons in this crystal lattice to provide semiconductor properties, taking advantage of the ability to generate both types of semiconductors, whether type P or type N. The manipulation is intended to provide desirable electrical properties at grounding electrodes, as this buried structure, in addition to the typical corrosion and corrosion performance properties, also requires an electrical performance determined by the function for which Design in a complex electrical system. Have been developed some series materials intended to mitigate the effect of corrosion, among which stand out the coatings, especially those deposited in the form of thin film. With the investigations that have been undertaken were discovered other properties of thin films. In the case of niobium, in addition to its superconducting properties, identified characteristics such as hardness, resistance to wear and corrosion [4].

Work on barriers against corrosion is generally oriented to provide the substrate coating system with mechanical properties, thermal and anti-degradation, but rarely considered elements such as grounding systems, which require a specific electrical performance [5]. The multicomponent coatings are developed as multi-tier heterostructures in order to improve the wear resistance and the oxidation of coated components.

The improvements occur in alternating deposition of two (or more) chemical and / or mechanically different layers, so that the stress concentration and conditions for the propagation of nano cracks can be controlled. Therefore, the multi-layered structure can act as an inhibitor of nano cracks and increase the resistance to fracture [6]. In addition to the characterization of the behaviour against corrosion, an evaluation of its performance against minimum working voltages of the coating scheme was done. The thermal performance produced by Joule effect was guaranteed when using oxides that supported high working temperatures.

2. Methodology technological product

Within the technological product development process, it must be taken into account that the coating to be deposited on the stainless steel substrate has a physical, chemical and ecological performance, being able to maintain electrical levels of transmission in case of failure, and Insulation otherwise; It must also be a corrosion protection agent. It must not harm the environment causing consumption of non-renewable materials and sterilizing the sectors where it is installed

The techniques to take into account to make the deposition of thin films and manufacture of nano structures Top Dow, are: Spray thermal, sol gel and plasma. For the work done, the Sputtering Plasma technique was used, which guarantees homogeneity in the coating and adhesion when exchanging electrons between the surface and the target (coating), being films of the order of nanometres become susceptible to abrasion.

The process of sputtering-Sputtering is mainly a process of ionic bombardment, which obtains the deposition in vapor phase, on a substrate of the bombarded material. In this technique, the ions formed in the plasma are accelerated towards the material to be deposited, through an electric field. The plasma is made up of process gases, in this case argon and oxygen, ionized by the strong electric field [7]. Under defined experimental parameters, the water spray technique can ensure a good anchoring surface between the coating and the substrate. This method extracts atoms on the surface of a negatively polarized called White electrode, the electrode eroded by the action of ions and atoms of high-energy contained in the Argon (Ar), from the formation of the plasma that bombards the atoms of the surface of the white. If it generates enough kinetic energy the atoms or molecules of the white they are torn [8].

In order to apply the sputtering technique, it is necessary to prepare the substrates for the most homogeneous surface polishing possible by removing scratches, holes or pores leaving a surface metallography to mirror brightness as seen in Figure 1, this will facilitate and optimize the deposition of the Thin films, maintaining a flat and homogeneous surface without concentrations where failures may occur.

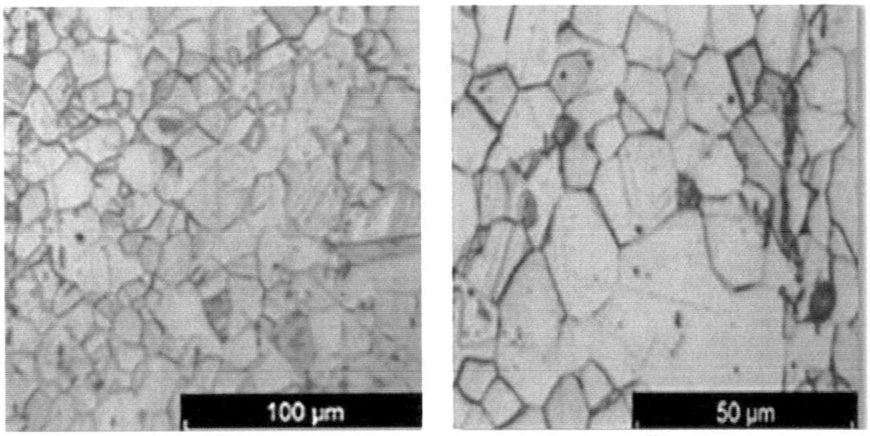

(a) (b)

Figure 1. Microstructure stainless steel at (a) 700 X y (b) 1400 X.

The targets to be considered are binary oxides that are stable in their free energies of Gibs which reacted with the use of titanium, zinc and copper targets, the results showed purity of 99.99999%.

The production of titanium, zinc and copper oxide coatings on stainless steel specimens is performed by dc reactive Sputtering in argon and magnetron assisted oxygen [9], whites are 3 inches in diameter in the configuration of parallel and coaxial flat electrodes, In a high vacuum chamber of approximately $0.023m^3$ and final vacuum 4mPa. Luminescent discharge is obtained with pressures ranging from 0.3 to 3Pa, voltages between 150 and 350V, voltage between 0.1 and 1A, distance between electrodes of 3 and 4.5cm, concentration of oxygen between 10 and 30%, substrate temperature and test specimens between 370 and 520K. With deposition times of 20 minutes, the yield on the substrate varies between 30 and 50nm/min. The possible species in the plasma are monitored through a NUV spectrometer VIS NIR with FO; the morphology, structure and stoichiometry of the coating, will be determined mainly by the stoichiometry of the surface of the white, the reactions of the material pulverized in its transit to the substrate and by the absorption and adsorption of oxygen of the coating of the substrate in the presence of the plasma in the luminescent zone. The absorption and adsorption of oxygen from the surface and the balance between the spray yields of the oxidized surface area and the non-oxidized zone of the target will determine the control of the oxidation state of this surface. If spraying is maintained under this yield competition regime, the type of pulverized species passing through the substrate will be controlled.

The morphology and structure of the coating is balanced between the amorphous and/or polycrystalline state of the oxides and it can be indirectly controlled and assisted with the balance between the anneals due to the substrate temperature and the temperature of the reactive plasma on the substrate [10]. The matrix can be amorphous with embedded polycrystalline, which determines the possible compromises between hardness and corrosion resistance. The substrate temperature is assisted and measured with termocoax and Pt100 and thermocouple. The monitoring and controlling of the growth of parameters were assisted with NI and LabVIEW cards. Post-deposition should be kept empty for 2 hours to avoid any thermal shock. Post-deposition should be kept empty for 2 hours to avoid any thermal shock.

3. Discussion and results

Oxides that occurred during the deposition reaction in the sputtering in its atmosphere with 16% oxygen and 84% argon were polar covalent bonds that generated elements as: TiO_2 , ZnO , TiN y CuO, With a percentage ionic character, according to the mathematical relationship of paulling of link with an ionic character of 25%, average for the 4 elements and in the same form of a 75% of polar covalent bond.

In Table 1 is reported the chemical composition obtain the characterization of stainless steel substrate before and after treating, where it shows the small proportions that reflect the additional compositions that generate the thin film, in the face of two techniques to quantitative and with other quantitative results may vary with respect to each other, for this reason, it makes for a single sample.

Table 1. Results chemical composition of stainless steel 304 (%) determined from the techniques of atomic emission spectroscopy without surface treatment and X-ray fluorescence technique with surface treatment.

The composition of the surface without treatment (%).									
C	Mn	P	S	Si	Cr	Mo	Ni	Al	Co
0.049	1.530	0.0249	~0.020	1.19	18.04	0.154	10.26	0.171	0.171
The composition of the surface with treatment (%).									
Mg	Al	Si	Ti	Cr	Mn	Fe	Co	Ni	Zn
3.8	0.6	0.2	4.01	11.1	0.76	46.6	0.1	6.01	26.8

The deposition on the substrates of 304 stainless steel with coatings ZnO and TiN and ZnO and uncle, evidence the homogeneity that generates the technique of sputtering, as seen in the images obtained by optical microscopy in Figure 2, By having an area of uniform appearance with small inclusions, at the rate of a pore or impurity in the preparation of the substrate. Usually when there is a presence of corrosive media, stainless steels and TiN ceramic layers formed on its surface a protective oxide layer. However, when the aggressive substance contains hard particles, these layers of oxides are seriously damaged due to the large number of shocks on its surface, a fact which implies a new formation of the oxide layer [11].

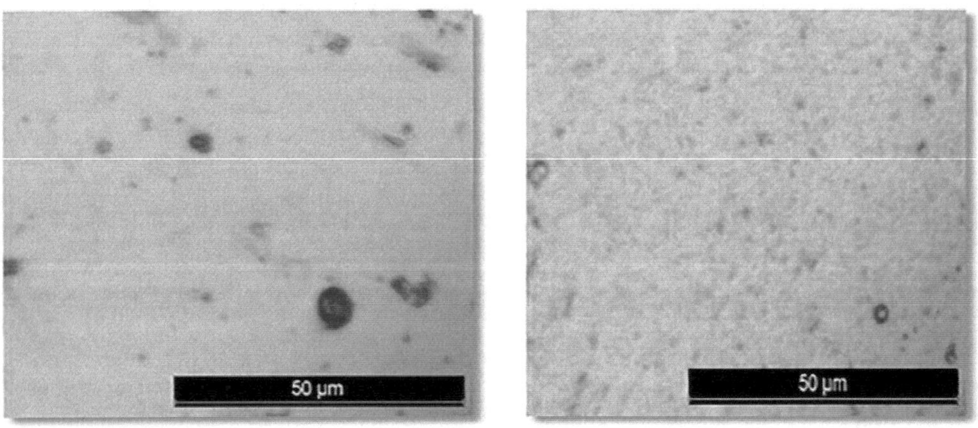

(a) (b)

Figure 2. Stainless steel with ZnO coatings (a) and (b) TiN and ZnO and TiO with an increase of 1400X.

A cross section of the substrate was cut to observe the interfaces of the coatings, determining that they have a uniform bilayer with values of approximately 200 and 450 nanometres, respectively, as can be seen in the images obtained by electron microscopy in the Figure 3.

Cutting the cross section to observe the interfaces should be very careful because the coatings can be detached as shown in Figure 4.

IOP Conf. Series: Journal of Physics: Conf. Series **935** (2017) 012017 doi:10.1088/1742-6596/935/1/012017

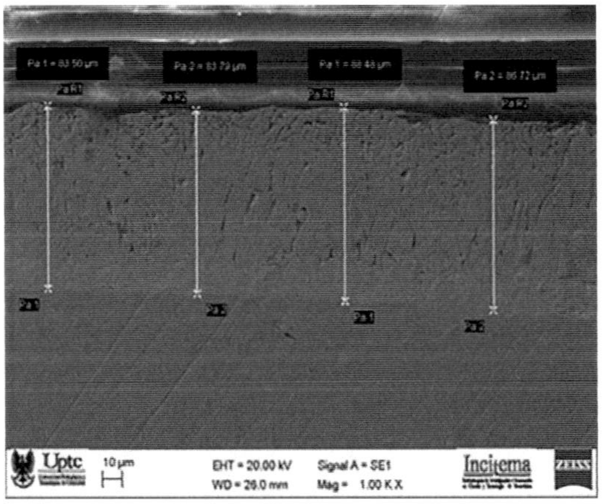

(a)

Figure 3. Stainless steel with ZnO and TiN and ZnO and TiO 500X

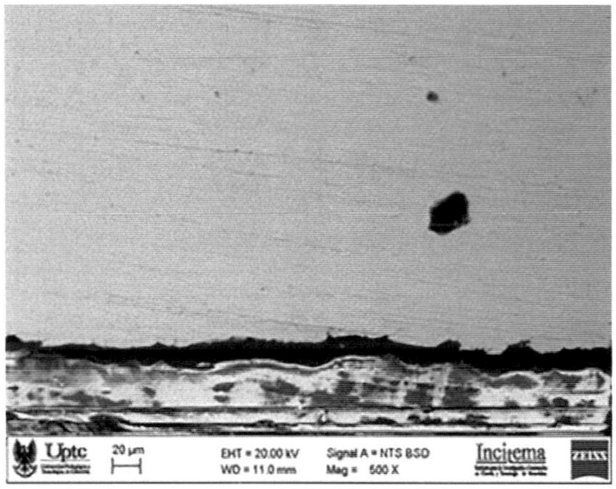

(b)

Figure 4. Stainless Steel with Coatings ZnO and TiN, ZnO and TiO 500X.

Surface properties, the microhardness of the surface treatments was determined and a comparison was made with a pattern that for this case would be the substrates without coating obtaining the following results (see Table 2).

Table 2. Microhardness of 304 stainless steel substrates with thin films and without them.

Measurement Register	TiN+ ZnO	ZNO + TiO2	Substrate
Measurement Force	100 gf	200 gf	1000 gf
Load Duration	25 seconds	25 seconds	25 seconds
Standard Deviation	209 HV	187 HV	199 HV

HV: Hardness Vickers Unit.

The standard deviation is within the range of the microhardness of the substrate, the characteristics that are allowable for the surface that was drawn up in order to be buried in its use the grounding system. With these tests elaborated topographic graphs observing the depth and the traces that were obtained with the test of microhardness; Taking into account the referent of the substrate, the actual notions of hardness in the coatings were obtained, as shown in Figure 5.

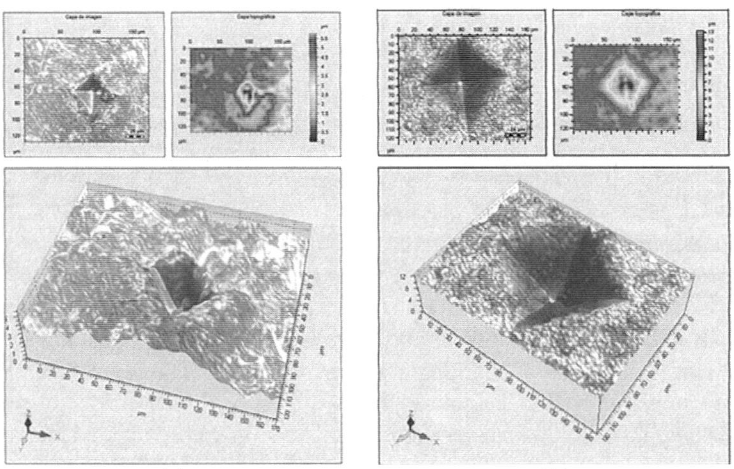

(a) (b)

Figure 5. Topography for the coatings (a) ZnO+TiO and (b) ZnO+TiN.

The electrical performance was evaluated by the 4-point method, which intends to see the curve of the diode in the coating in the form of bilayers simulating an active element, where one layer works as Type P and the other N, in the results obtained not all the configurations made with the coatings worked with the electrical purpose we intended as can be seen in Figure 6.

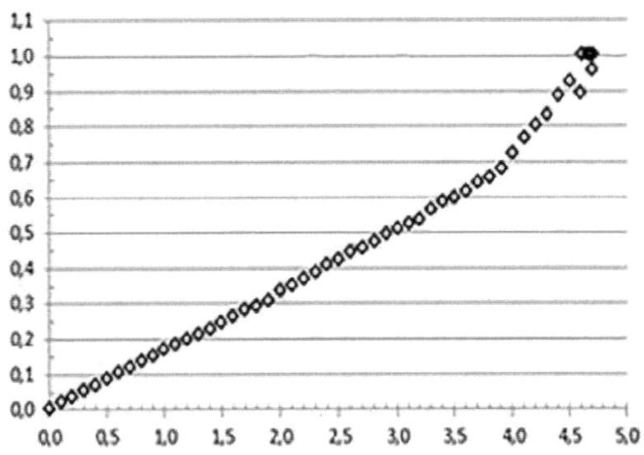

(a)

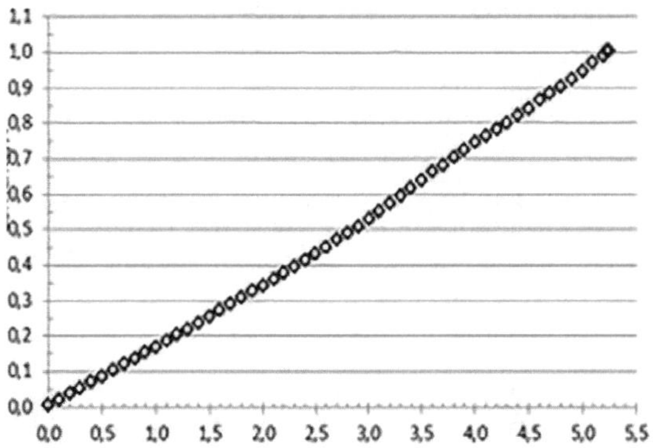

(b)

Figure 6. Diode curve where it behaves as a resistive element, where is observed the behaviour of current in function of tension. (a) ZnO+TiO and (b) ZnO+TiN.

The configuration that finally solved the problem and generated the expected electrical performance, consists of a layer of copper oxide that served as a P-type semiconductor; And then a slightly thicker layer of titanium oxide was deposited, which served as n-type material, resulting in the curve of the diode shown in Figure 7.

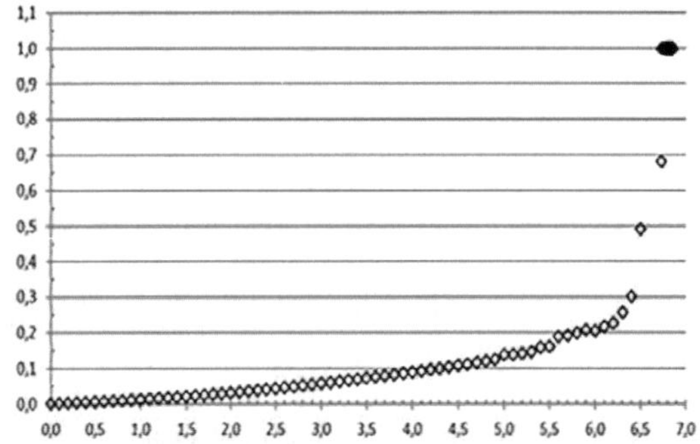

Figure 7. Diode curve where the device is activated at 7V.

Electrochemical impedance spectroscopy, the conditions of the scans in frequency for all experimental units were made with an initial frequency of 100KHz up to 0.2Hz; 10 points per decade were determined for an exposed area of 2.41cm^2 electrode, in aqueous NaCl solution. The experimental data obtained by means of IS-dynamic tests are see Table 3.

The corrosion rate and the polarization resistance, as a function of the number of bilayers, can clearly be seen that the lower value of the resistance to effect corrosive occurs in the substrate, which generates the highest speed value to corrosion, as the number of bilayers increases, an increase in polarization resistance is generated, indicating a significant decrease in the corrosion rate. See diagrams of Figures 8 and 9. This implies that the energy required for the solutions to migrate freely from the surface to the film / substrate interface, is greater with the increase in the number of bilayers. [12].

IOP Conf. Series: Journal of Physics: Conf. Series **935** (2017) 012017 doi:10.1088/1742-6596/935/1/012017

Table 3. Electrochemical impedance spectroscopy on the surface of 304 stainless steel substrates with thin films and without these.

Parameter	Value without treatment	Value whit treatment
lower Fit Limit	-87.18mV vs. Ref. (Pt#0)	-557.9mV vs. Ref. (Pt#0)
Upper Fit Limit	-67.22mV vs. Ref. (Pt#0)	-517.9mV vs. Ref. (Pt#0)
Beta A	455.8e-3V/decade	120.0e-3 V/decade
Beta C	116.8e-3V/decade	120.0e-3 V/decade
Icorr	285.3nA	5.168nA
Ecorr	-83.21mV	-537.9mV
Rp	141.5Kohms	5.041Mohms
Corrosion Rate	84.61e-3mpy	8.272e-3mpy

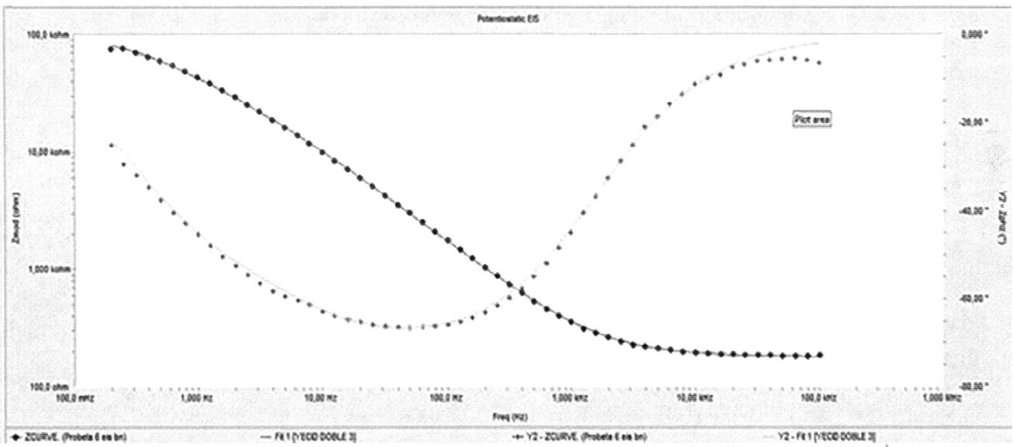

Figure 8. Bode diagram.

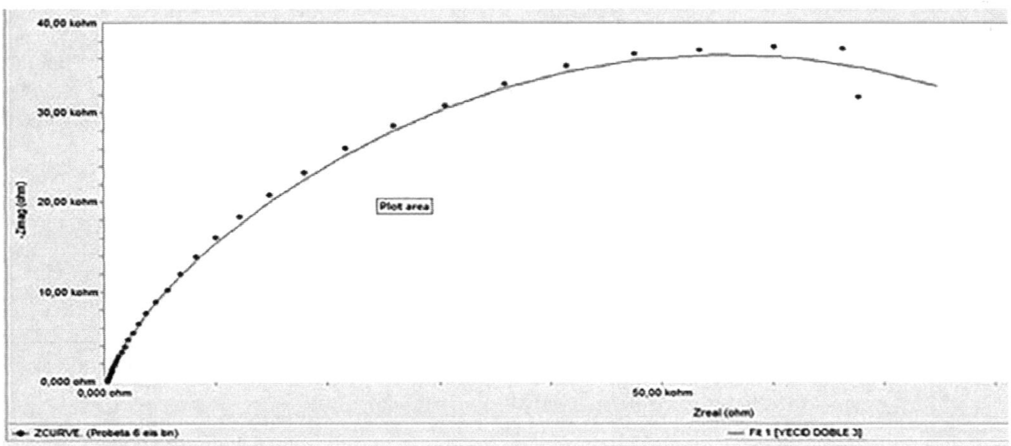

Figure 9. Nyquist diagram.

4. Conclusions

The behaviour of the semiconductor is acceptable for the purpose of earthing system with an excitation voltage of 7V.

The behaviour against corrosion for stainless steel with surface treatment improved by an order of magnitude, compared to the characterization tests performed in previous sequential projects that do not have surface treatment.

The preparation of the substrates for the application of the coatings may be different, depending on the characteristics of the coating to be deposited.

The electrical and corrosion-resistance characteristics of the coatings are defined by their deposition settings.

Surface treatments in the form of bilayer are a successful solution to maintain levels of conductivity and corrosion resistance.

References

[1] L. Ardila Rodríguez, J. Olaya Flórez y J. M. Arroyo Osorio 2014 Resistencia a la corrosión de recubrimientos de NbC sobre acero AISI 316L depositados por UMB *Ingeniare Revista Chilena de Ingeniería* **22(3)** 445-450

[2] J. A. Acosta Wihiler 2007 Estudio de diseño de las instalaciones eléctricas del hospital san rafael de leticia mediante la aplicación del retie (Bogotá: Universidad de la Salle)

[3] L Guerrero, Y Salas and J Blanco 2016 Corrosive effect of the type of soil in the systems of grounding more used (copper and stainless steel) for local soil samples from the city of Tunja (Colombia), by means of electrochemical techniques *Journal of Physics: Conference Series* **687** 012017

[4] W. Hernández Muñoz 2011 Caracterización de la microestructura y resistencia a la corrosión de las multicapas de nitruro de niobio y niobio depositadas con el sistema sputtering magnetrón desbalanceado (Bogotá: Universidad Nacional de Colombia)

[5] W. Aperado y E. Vera 2006 Obtención de electrorecubrimientos de cobre–níquel depositados sobre substratos de zamak por medio de la técnica de corriente pulsante inversa *Scientia et Technica* **XII(30)** 361-364

[6] W Aperador Chaparro, E Vera López and A Vargas Uscategui 2010 Estudio de la resistencia a la corrosión electroquímica de electro-recubrimientos níquel/cobre obtenidos por corriente pulsante Ingeniería y Desarrollo **27** 48-61

[7] J. M. S. Barricarte 2005 Deposición de Capas Constituyentes de Estructuras Multicapa con Funciones Controlables Eléctricamente (España: Universidad de Zaragoza)

[8] J Estrada Martínez 2008 Caracterización estructural y morfológica de películas delgadas de TiO2 depositadas por RF-sputtering sobre acero inoxidable 304 (México: Instituto Politécnico Naciona)

[9] E. Restrepo, A. Devia, R. Cogollo, A. Mariño y H. Sánchez 2002 Espectrocopía óptica de un sistema de magnetron sputtering *Revista Colombiana de Física* **34(2)** 488-492

[10] Ahumada Lazo R 2014 Crecimiento y caracterización de películas de óxido de ZINC por "sputtering" para su potencial aplicación en fotocatálisis (México: Universidad Autónoma de Nuevo León)

[11] H Payán, W Aperador, A Vargas 2008 Estudio de la sinergia corrosión-erosión de recubrimientos duros de TiN y CrN obtenidos sobre acero AISI 1045 *Scientia Et Technica* **XIV(38)** 177-180

[12] W Aperador, C Amaya, C España 2011 Resistencia a la corrosión de las multicapas de [TiN/AlTiN]n Depositadas sobre acero al carbono AISI 1045 *Dyna* **78(165)** 186-187

IOP Conf. Series: Journal of Physics: Conf. Series **935** (2017) 012018 doi:10.1088/1742-6596/935/1/012018

Evaluation of tribological wear and corrosion in coatings of diamalloy 4060NS deposited by thermal spray

S M Acuña R[1], C M Moreno T[1] and E J Espinosa C[1]
[1] Universidad Pedagógica y Tecnológica de Colombia, Tunja, Colombia

E-mail: sandramarcela.acuna@uptc.edu.co

Abstract. Surface engineering seeks the development of new techniques to improve the performance and life of components of machines or industrial facilities, always looking for low costs and the least possible environmental damage. Thermal projection is one of the techniques that is based on the projection of particles of compounds and alloys on properly prepared and heated substrates, these particles are driven by a stream of air passing through an oxyacetylene flame which gives the energy to the process; These coatings give the possibility to improve the properties of the materials or the maintenance of components to maximize the availability of service. In order to reduce the damage caused by wear and corrosion of a low carbon AISI 1020 steel, they were coated with a metal based alloy, studying the effect of the cobalt-chromium-silicon-tungsten carbide alloy coating (DIAMALLLOY 4060 NS). The coating was deposited with two different pressures in the gases supplied to the torch, obtaining two flames and working three thicknesses of coating that oscillate between 100-500µm, according to the number of deposited layers, making use of a projection gun Castolin Eutectic. Powder and substrate characterization was performed using X-Ray Diffraction (XRD) techniques, X-Ray Fluorescence (XRF), Scanning Electron Microscopy (SEM), spark emission spectroscopy and metallographic analysis. The results confirm the chemical nature and structure of the powder of the alloy and the substrate to be used, in addition, the thermal stability of the system was verified. The evaluation of the adhesion of the deposited layers was carried out by the implementation of pull-off tests according to ASTM D4541, in order to determine the type of failure that is presented. Mechanical wear was determined using a MT/60/NI microtest tribometer while electrochemical tests were performed using a suitable experimental unit for this purpose, confirming that the substrate exhibits lower wear levels when coated with Layers between 300-400µm, demonstrating that the implemented methodology allows to reduce the process of wear of mechanical parts subjected to aggressive operating conditions.

1. Introduction

Thermal spray coatings are widely used throughout the world; however, Colombian industry has not adopted this technique yet, as a common method to combat corrosion, wear and thermal fatigue. This is due to the lack of knowledge regarding the benefits of thermal projection. Another reason is that the medium and small metallurgical companies consider that this technique is expensive and opt for other alternatives less expensive, like paintings of high performance. Although the costs of these coatings are apparently high, the increase in efficiency is approximately 1.5 times higher than in high-performance paint systems [1].

In order to extend and improve the properties of an AISI 1020 steel, the projection parameters of the cobalt-chromium-silicon-tungsten carbide alloy (DIAMALLLOY 4060 NS) were determined and

IOP Conf. Series: Journal of Physics: Conf. Series **935** (2017) 012018 doi:10.1088/1742-6596/935/1/012018

deposited with two gas pressures supplied to the torch, obtaining two flames with three coating thicknesses ranging from 100-500μm, depending on the number of deposited layers (2.4 and 6 layers). This is about the way to determine the best performance regarding wearing and corrosion in a saturated salt medium.

2. Experimental procedure

The coatings were made in a chamber systematized with a Castolin Eutectic 8000 spray gun, where the parameters and variables with which the coatings were made were fixed (Figure 1).

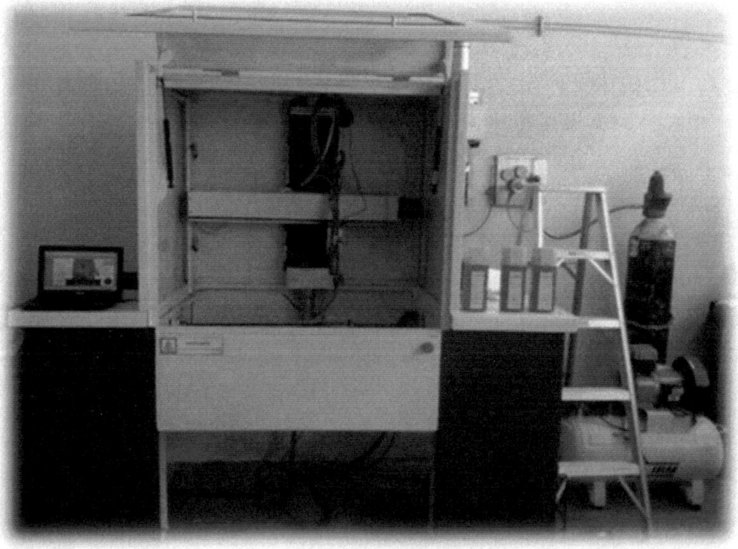

Figure 1. Thermal projection equipment. UPTC.

2.1. Material

The material used in the development of the tests corresponds to an AISI 1020 steel, 2cm in diameter and 1cm thick.

2.2. Terms

The samples used were sandblasting or sandblasting surface preparation achieving an average Ra roughness of 8.79μm in order to create a surface that allows a good adhesion between the coating and the substrate. The cleaning was done in a bath with isopropyl alcohol in an ultrasonic cleaner Brason-2510, during 20min.

The Diamalloy 4060NS powder was baked at 100°C to remove the existing moisture and then fed to the dispenser with a speed of 2.5rpm and a nitrogen pressure of 20psi and a flow rate of 5l/min.

The samples of AISI 1020 steel were heated before starting the projection with a temperature of approximately 160°C.

2.3. Settings

A projection distance of 150 mm between the substrate and the gun was used, a torch advance speed of 3.2mm/s, speed of rotation of the sample holder 63rpm, 4 heating passes, and two types of flame were obtained with oxygen pressures of 60psi and acetylene of 10psi, flame 1 with a flow of 4l/min of oxygen and 7.2l/min of acetylene and flame 2 with a flow of 4l/min of oxygen and 9.1l/min of acetylene. Obtaining a relation of the flame 1=1:3.27 and calls 2=1:5.75 respectively O2:C2H2. See Table 1.

Table 1. Projection parameters.

Thermal Spray Equipment Parameters	
Projection distance	150mm
Dosing plate speed	2.5rpm
Thickness per coat	40-60μm
Speed feed rate	3.2mm/s
Turning speed of the sample holder	63rpm
Preheating layers	4
Preheating temperature	140°C
Oxygen pressure	60psi
Flame 1 oxygen flow Flame 2 oxygen flow	4l/min
Acetylene pressure	10psi
Flow of acetylene flame 1	7.2l/min
Flow of acetylene flame 2	9.1l/min
Nitrogen Pressure	20psi
Flow of Nitrogen	5l/min

3. Results and discussion

3.1. Characterization of steel

The roughness Ra obtained for the flame 1 had an average of 12.0354μm and the flame 2=12.7642μm obtaining a rougher surface with respect to that of the bare steel. Figure 2 shows the microstructure of the steel, in which the phases of ferrite (light zone) and perlite (dark zone) are observed, which correspond to AISI/SAE 1020 steel.

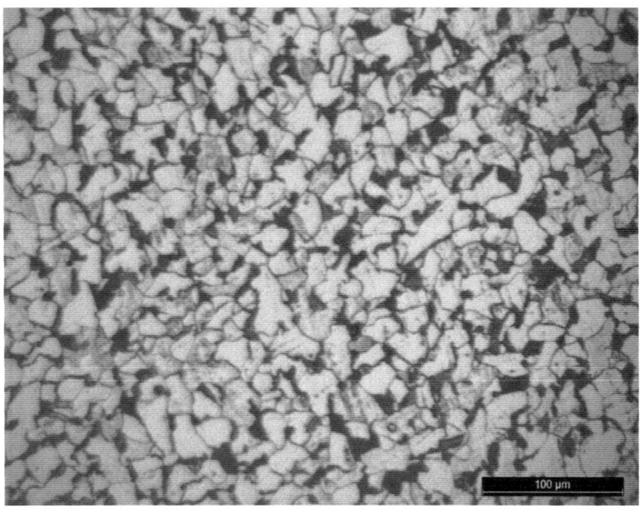

Figure 2. Microstructure Steel 1020 to 700X.

The chemical composition according to the results obtained in the optical emission spectroscopy with spark source corresponds to AISI/SAE 1020 Steel, with a difference in Mn of 0.16; With respect to the norm that is of 0.3-0.6% Mn (Table 2).

IOP Conf. Series: Journal of Physics: Conf. Series **935** (2017) 012018 doi:10.1088/1742-6596/935/1/012018

Table 2. Chemical composition of 1020 steel.

Element	Al	C	Cr	Cu	Fe	Mn	Ni	P	S	Si	Sn	Ti	W
%	1.01	0.15	0.05	0.13	98.1	0.76	0.04	0.02	0.05	0.16	0.03	0.01	0.02

3.2. Morphology and dust characterization Diamalloy 4060NS

3.2.1. Morphology obtained by SEM. Figure 3 shows the shape of the powder particles which are totally spherical due to their method of manufacture by atomized gas with average sizes of 11.53, 29.1, and 44μm.

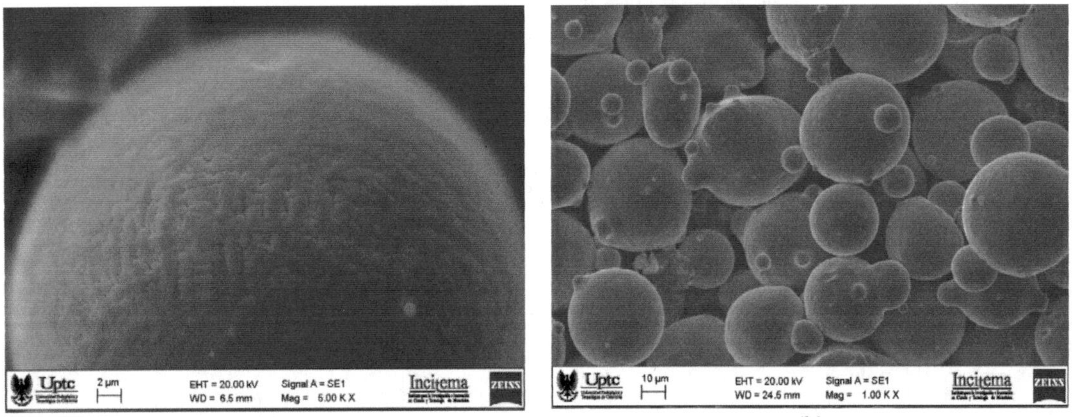

(a) (b)

Figure 3. (a) Morphology of the powder at 5000X. (b) Morphology of the powder at 700X.

3.2.2. Characterization made using DRX and FRX. Analyses were performed in order to corroborate the information obtained by the manufacturer of the Diamalloy 4060NS powder and confirm the chemical composition. Figure 4 and Table 3 shows the results obtained from the analysis of XRD and FRX to the powder, (a) the presence of Co, Cr and other elements in proportions very similar to those reported by the manufacturer is reported. (b) Presence of Co and Cr is confirmed as the elements in greater proportion with small traces of the other elements, Co and Cr peaks are present in the same crystallographic planes because they are in the same cubic crystalline system centred in The FCC faces.

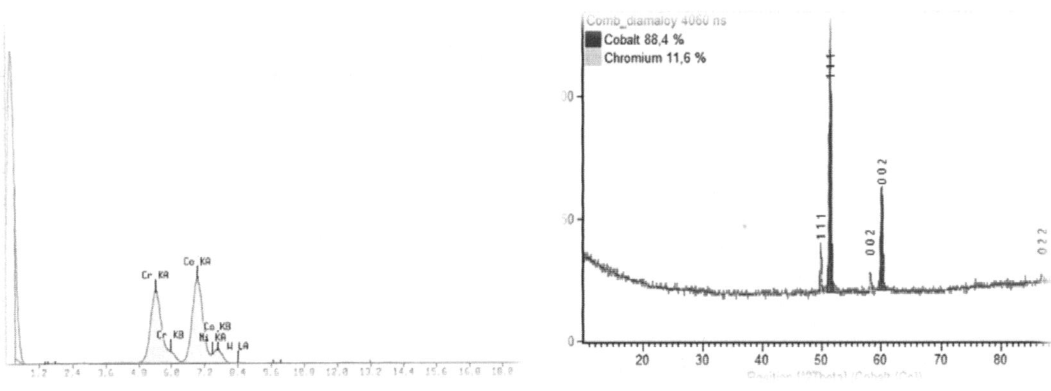

(a) (b)

Figure 4. Characterization of the Diamalloy 4060NS powder.

IOP Conf. Series: Journal of Physics: Conf. Series **935** (2017) 012018 doi:10.1088/1742-6596/935/1/012018

Table 3. Characterization of the Diamalloy 4060NS powder.

	Element	Co	Cr	Si	W
Casa Fabric	%	64.3	28.5	1.60	4.5
FRX	%	67.5	27.3	1.31	3.8

3.3. Characterization of coatings

3.3.1. Pull-off Test. According to the results we can say that the flame 2 has a better adhesion behaviour to the flame 1 with increases as the thickness of the coating grows. Table 4.

Table 4. Pull-off test results.

Flame	Type of fault	Adhesion value (psi)	N° Layers
Flame 1	adhesive	424	2
	glue	877	4
	cohesive	1557	6
Flame 2	cohesive	898	2
	adhesive	972	4
	glue	1124	6

3.3.2. Thickness using SEM. The coatings with the highest thickness reported were those that were worked with the flame 2, in all common characteristics were observed such as lamellae, non-melted particles, voids, starters, pores. Figure 5. (a) flame 1 and 2 layers: thickness of 146.5µm, (b) flame 2 and 2 layers: thickness of 186.3µm, (c) flame 1 with 4 layers: a detachment between coating and (D) flame 2 with 4 layers: thicknesses of 383.8µm, (e) flame1 with 6 layers: thicknesses of 379.9µm, (f) flame 2 with 6 layers. Layers: a slight detachment between the substrate and the coating with greater voids is observed unlike the others, thicknesses of 410.2µm.

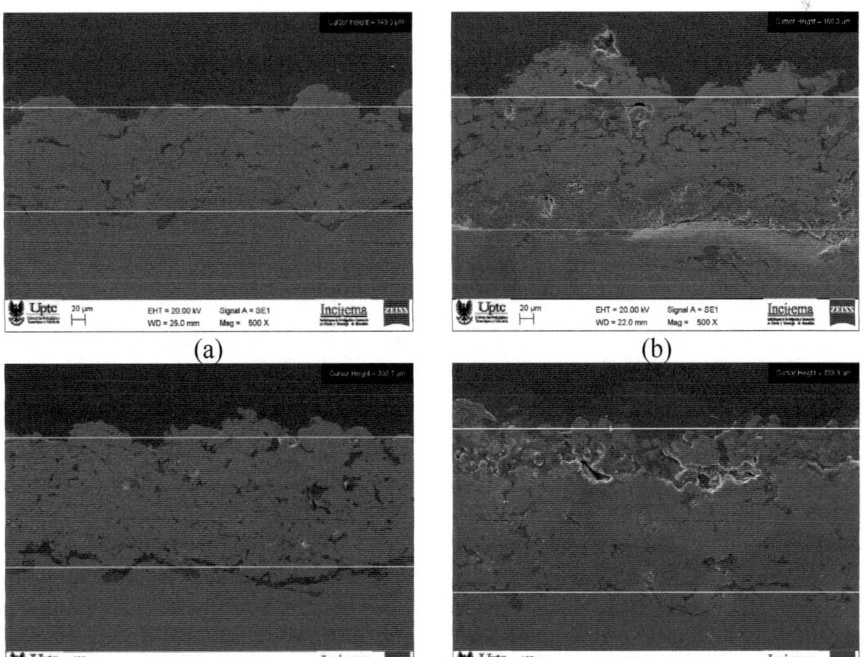

(c) (d)

IOP Conf. Series: Journal of Physics: Conf. Series **935** (2017) 012018 doi:10.1088/1742-6596/935/1/012018

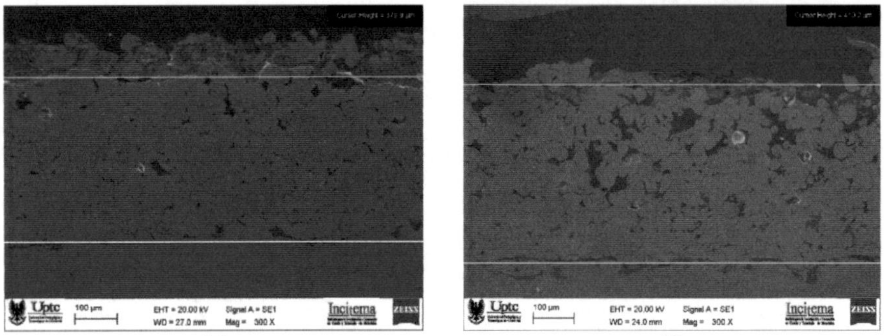

(e) (f)

Figure 5. Scanning electron microscopy.

3.3.3. Wear evaluation (Pin on disk). The wear coefficient was determined with a load of 5N, footprint of 6 mm of radius, distance of travel of 500m, speed of wear of 100rpm and an ambient temperature of 19 to 21°C. The coatings that presented the least wear were those of flame 1 with a thickness of approximately 300µm. Figure 6.

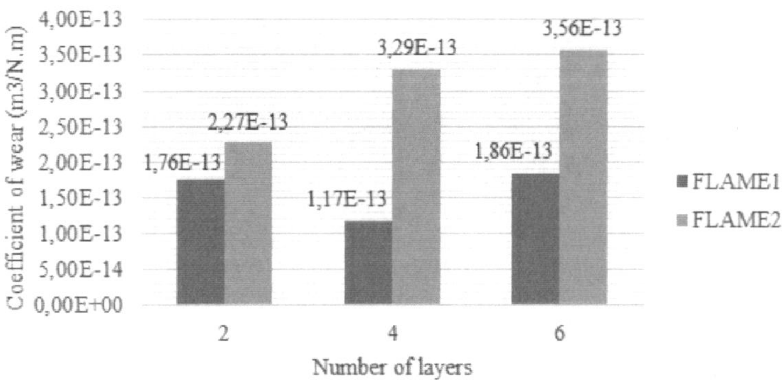

Figure 6. Wear evaluation.

3.3.4. Corrosion evaluation (electrochemical cell). The corrosion rate was determined by means of an electrochemical cell, using as reference electrode: Pt / ClPt and as counter electrode: platinum wire; In 1% NaCl salt saturated with CO2. The effectiveness of the flame 2 with respect to the flame 1 is evident, but considering that both protect the steel against the phenomenon of corrosion. For the flame 1 decreases as the thickness increases and in the flame 2 the opposite happens totally as the thickness increases the corrosion increases. Figure 7.

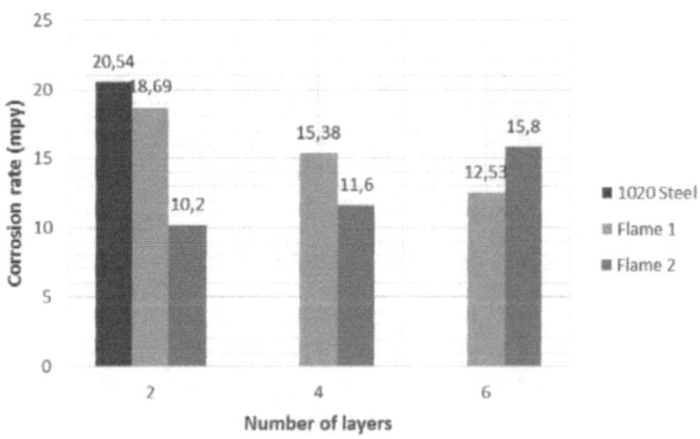

Figure 7. Evaluation of corrosion.

4. Conclusions

The tribological wear of the steel improves when a flame 1 is obtained with a flow of 4l/min O_2 and 7.2l/min C_2H_2 with thicknesses around 400μm obtained an approximate coefficient of wear of 1.17E-13m^3/Nm , with a flame ratio of 3.27:1. Unlike flame 2, as the thickness increases, the coefficient of wear increases proportionately; thus the flame 2 with a flow of 4l/min O_2 and 9.1l/min C_2H_2 reports an increase in the adhesion with respect to the flame 1, as its number of layers increased, the particles had greater adhesion between them and the substrate, compared with the electrochemical tests was determined that this is inversely proportional, that is to say; as the thickness of the coating increases, the wear increases but the corrosion decreases, which agrees since as the particles are better adhered and melted in the coating the better their resistance to corrosion and wear; since the particles are more united causing less abrasion and not allowing the passage of substance to the substrate.

In scanning electron microscopy both types of flames presented lamellae, unmelted particles, voids, pores and starts with a significant increase between layers, noting that flame 2 has an increase in thickness percentage with respect to flame 1, increase of thickness is due to the fact that the particles melted or semi-melted in a better way, being able to be located forming more bonded and resistant layers giving the coating better adhesion between layers and the substrate as less porosity.

References

[1] Cadenas P, Rodríguez M, and Staia M H 2007 Efecto del tratamiento térmico posterior sobre la resistencia al desgaste por deslizamiento de un recubrimiento base níquel depositado por HVOF *Revista de Metalurgia* **43(1)** 50-62

[2] J L Marulanda Arévalo, J L Tristancho Reyes, H Á Gonzáles B 2014 La tecnologia de recuperación y proteccion contra el desgaste esta en el rociado térmico *Prospectiva* **12(1)** 70-78

[3] Zhang H, Wang X, Zheng Z, and Jiang X 2001 Studies of splat morphology and rapid solidification during thermal spraying *Journal of heat and mass transfer* **44(24)** 4579-4592

[4] Pawlowski L 2008 *The science and engineering of thermal spray coatings* (England: John Wiley-Sons Ltd.)

[5] Rodríguez M de los A, Gil L, Pérez R and Ferrara S 2009 Desgaste de recubrimientos WC-Co-NiCrBSi tratados térmicamente *Revista Latinoamericana de Metalurgia y Materiales* **29(2)** 135-150

[6] Stoica, V, Ahmeda R, Itsukaichib, T and Tobe, S 2004 Sliding wear evaluation of hot isostatically pressed (HIPed) thermal spray cermet coatings *Wear* **257(11)** 1103-1124

[7] Guilemany JM, Miguel J, Vizcaino S, and Climent F 2001 Role of three-body abrasion wear in the sliding wear behaviour of WC-Co coatings obtained by thermal spraying *Surface & coatings Technology* 140(2) 141-146

[8] Picas JA, Forn A, Rilla R and Martin E 2005 Mejora de la resistencia al desgaste de aleaciones de aluminio mediante recubrimientos obtenidos por proyección térmica HVOF *Rev. de Metalurgia* **41(Extra)** 197-201

[9] Miguel JM, Vizcaino S, Dosta S, Cinca N, Lorenzana C and Guilemany JM 2011 Recubrimientos de materiales compuestos metal-cerámico obtenidos por nuevas tecnologías de proyección térmica: Proyección fría (CGS) y su resistencia al desgaste *Revista de Metalurgia* **47(5)** 390-401

[10] Morales JA, Olaya J and Rojas H 2012 Una aproximación a la tecnología de proyección térmica *Avances Investigación e Ingeniería* **9(2)** 60-71

[11] Santa Marin JF and Toro A 2015 Microestructura y resistencia adhesivo-cohesiva de recubrimientos de níquel, carburo de tungsteno y óxido de cromo aplicados por aspersión térmica *TecnoLógicas* **19(36)** 27-40

[12] Dimate LM, Olaya J and Orjuela JE. 2017. Corrosion resistance of Cr2O3 coating measured through EIS. *Ingeniería y Desarrollo* **35** 17-37

IOP Conf. Series: Journal of Physics: Conf. Series **935** (2017) 012019 doi:10.1088/1742-6596/935/1/012019

Sputtering of the 1020 AISI steel in abnormal glow discharge

J A García Zúñiga[1], A Sarmiento Santos[1] and B Álvarez Luna[1]
[1] Universidad Pedagógica y Tecnológica de Colombia, Tunja, Colombia

E-mail: asarmiento.santos@uptc.edu.co

Abstract. In all material treated in Sbnormal Glow Discharge (AGD) the phenomenon of sputtering occurs. In this work we study the sputtering suffered at different temperatures by AISI 1020 steel subjected to a DC discharge in two types of atmospheres. The steel samples were previously sanded until obtaining mirror brightness and subjected to the AGD plasma in the gaseous atmospheres of H_2 and Ar. The temperature for each sputtering process was set in the range of 420°C to 600°C. In these samples the mass variation was measured and the yield sputtering processes was determined. Next, the simulation of the sputtering process was performed in the SRIM/TRIM 2008 software, by adjusting sputtering yield computational computations to those experimentally measured, in order to determine the energy with which the responsible ions of the sputtering collide with studied target.

1. Introduction

Physical sputtering is the extraction of atoms from the surface of a solid due to the impact of the energetic particles [1]. Sputtering is present in any treatment performed with abnormal glow discharge. The process can be described as a moment transfer in a cascade of collisions initiated by the particle that impinges on the surface layer of the solid [2-5]. An atom is extracted from the surface if its binding energy to the surface is less than the energy of the incident ion. The sputtering yield is the parameter with which the process is quantified; it is a measure of the number of particles that are extracted from the surface by each incident ion. Then the physical sputtering depends on factors such as: Mass of interacting particles, its energy, angle of incidence, structural changes during the process [6] and roughness of the sample surface [7]. In the sputtering there is a transfer of mechanical energy through the exchange of the amount of movement between the ion, from the plasma of the glow discharge, and the atoms of the target [8]. The DC system is the easiest to perform sputtering. In it the particles ejected from the target (cathode) collide with the gas molecules and can be backscattering returning to the cathode or diffused in the opposite direction to the cathode depositing in the surfaces in front of the cathode. This since by the high gas pressure the mean free path of the ejected particles is less than the distance between the electrodes. It is estimated that about 90% or more of the sputtered material returns to the cathode by backscattering. The sputtering yield, i.e. the number of atoms removed from the target per incident ion can be obtained by the following empirical equation [9]

$$Y = \frac{6{,}02 * 10^{23} * W}{M * (Id/1{,}6 * 10^{-19}) * t} \qquad (1)$$

where Y is the yield sputtering in atoms/ion, W is the mass lost by the simple in grams, M is the molar mass of the target material, Id is the discharge current in amperes and t the time during which the target was subjected to the sputtering process in seconds. In this work, with equation (1), the sputtering yield of the 1020 steel was calculated in H_2 and Air glow discharges, and with this value, the energy of ions impinging the substrate was found, through the SRIM software [10].

2. Methodology

At the cathode of the abnormal glow discharge using confined geometry [11], AISI 1020 steel samples were subjected to the sputtering process. In order to compare the sputtering process in atmospheres of H_2 and Ar, discharges were generated in these two types of atmospheres under pressure of 2 torr and flow of 2scc/s. For this purpose, gases with 99.995% purities were used. The samples of 1020 steel were of cylindrical geometry with diameter of 10mm and height of 4mm and before subjecting them to the discharge were sanded and polished. A variable voltage pulsed DC source was used to initiate the discharge, the temperature of the cathode and the sample being adjusted to between 420°C and 600°C. The treatment time was 40 minutes. The samples were weighed before and after treatment, with a digital balance of 0.0001g resolution, and using equation (1) the rate of sputtering was determined. With the experimental value of the sputtering yield, using the SRIM 2008 software (TRIM), the energy of the ionic species responsible for sputtering was determined. The behaviour of the sputtering yield and its respective energy was observed as a function of the temperature for the two types of atmospheres. The samples surface after the sputtering process was also observed by optical microscopy.

3. Results

Figure 1 shows the behaviour of the current as a function of the voltage applied to abnormal glow discharge in atmospheres of H_2 and Ar to establish temperatures of 420°C, 500°C and 600°C. The increase of the current with the applied voltage, characteristic in the abnormal glow discharge [12], can be observed for the two atmospheres in Figure 1. For H_2 atmosphere is required higher voltages than for the Ar atmosphere, to reach the same temperature of the cathode. This because of the hydrogen atoms have a smaller mass, they must then have a higher velocity to impinging the cathode to transfer the necessary energy required to raise its temperature to the desired level. The temperature also increases with the voltage applied to the discharge as shown in Figure 2. This because of at higher cathode voltage more energetic particles impinges the cathode by raising its temperature. This allows correlate the behaviour of the temperature as a function of the voltage shown in Figure 2, for the two types of atmospheres, with that of the current versus voltage of Figure 1.

Figure 3 shows the behaviour of the mass variation as a function of temperature, with a higher loss of mass at higher temperatures for the same atmosphere due to the increase in the amount and energy of the particles impacting the cathode when the temperature is increased. This because of that the temperature is increased with voltage increases as mentioned above. Comparing the atmosphere type, there is a greater mass loss in the atmosphere with lower mass atomic species, similar behaviour observed for the sequence Xe, Kr, Ar where sputtering increases for less mass species [9].

The sputtering obtained by equation (1) is shown in Figure 4 as a function of temperature. A much higher sputtering yield is observed for the hydrogen atmosphere, atmosphere with smaller and energetic species as will be shown below.

Using the sputtering yield data of Figure 4, the energy for these values was adjusted using the SRIM 2008 software (TRIM), obtaining the energy of the particles responsible for sputtering as a function of the temperature shown in Figure 5. Thus, for the lightest species, energies of around 600eV were obtained, while for the heavier ones it was of the order of 30eV, differentiating by a factor comparable to the mass ratio (m_{Ar}/m_{H2}) of the two gaseous species. This higher ionic energy reached in the atmosphere of hydrogen is correlated with the greater voltage necessary to establish the respective temperatures as shown in Figure 2.

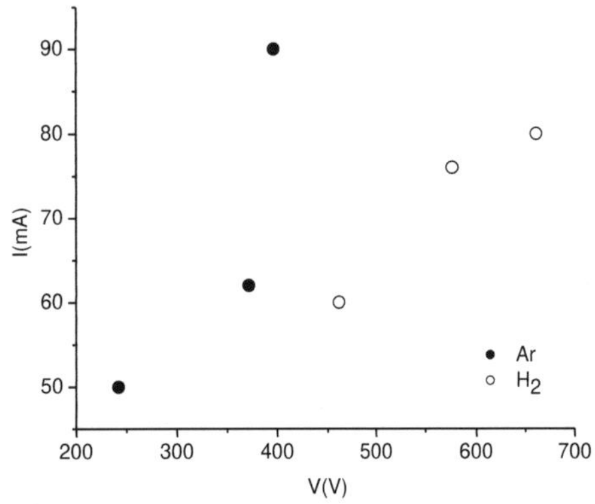

Figure 1. Current vs discharge voltage in hydrogen and argon atmospheres (Flow=2scc/s at 2torr).

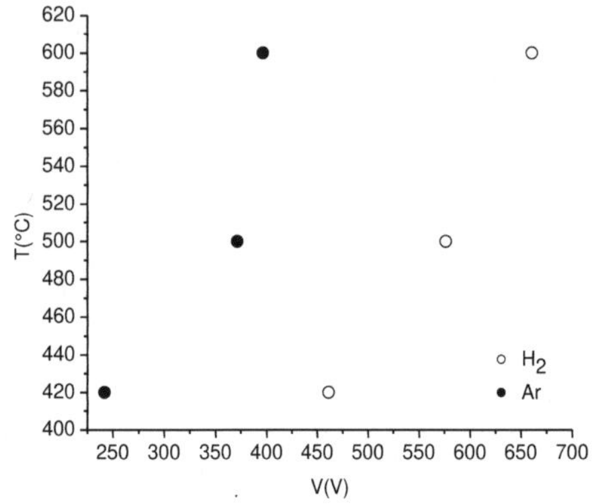

Figure 2. Temperature vs discharge voltage in hydrogen and argon atmospheres (Flow=2scc/s at 2torr).

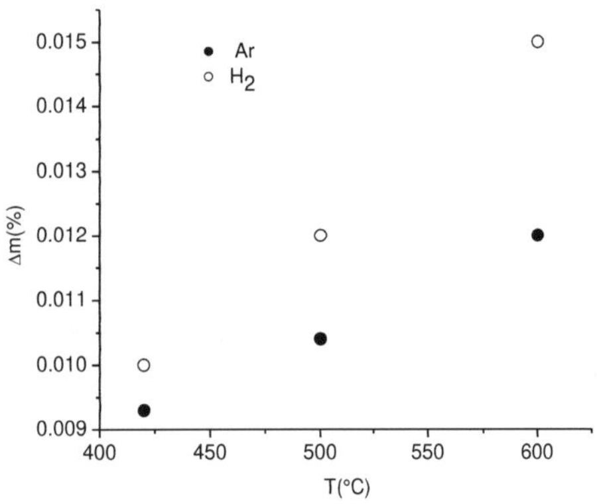

Figure 3. Sample lost mass vs temperature hydrogen and Argon (Flow=2scc/s at 2torr).

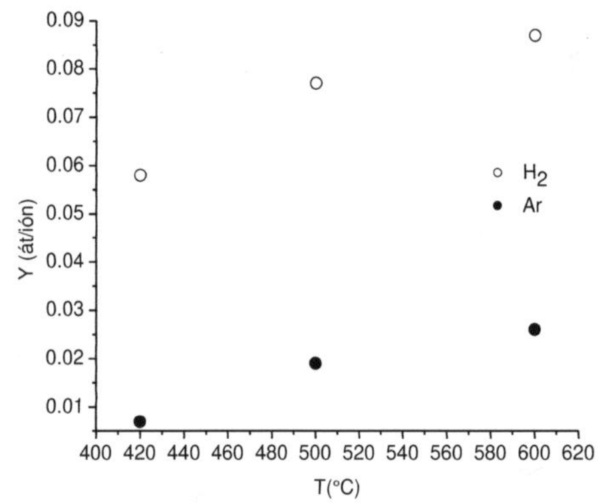

Figure 4. Sputtering yield vs sample temperature in hydrogen and argon atmospheres (Flow=2scc/s at 2torr).

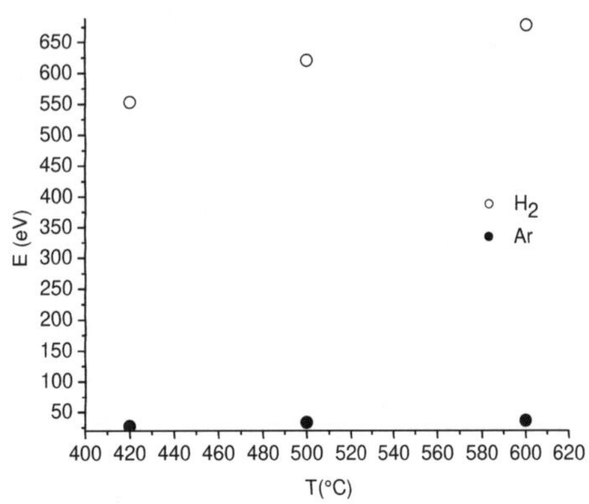

Figure 5. Ionic energy during the sputtering vs sample temperature in hydrogen and argon atmospheres (Flow=2scc/s at 2torr).

Figures 6 and 7 show the micrographs at 1000X of samples surface subjected to sputtering in H_2 and Ar atmospheres, respectively, at different temperatures used. In the samples subjected to a temperature of 420°C and 500°C in H_2 (Figures 6(a) and 6(b)) an incipient development of the microstructure of the material can be observed. These surface patterns disappear for the other surface (Figure 6(c)), possibly due to the greater energy of the incident particles that promote inner erosion of the sample. This can be sustained by the greater mass loss of this sample. In contrast, samples treated under argon atmosphere (Figure 7) reveal a surface pattern that could be a reveal of part of the microstructure of the material, due to the relatively low energy of the incident argon ions interacting with the surface region of the sample. The above suggests that these atmospheres can be controlled to promote the reveal of the microstructure of AISI 1020 steel.

(a) (b) (c)

Figure 6. Surface optical micrographies at 1000X of samples treated in the glow in H_2 at temperatures of a) 420°C, b) 500°C and c) 600°C (Flow=2scc/s at 2torr).

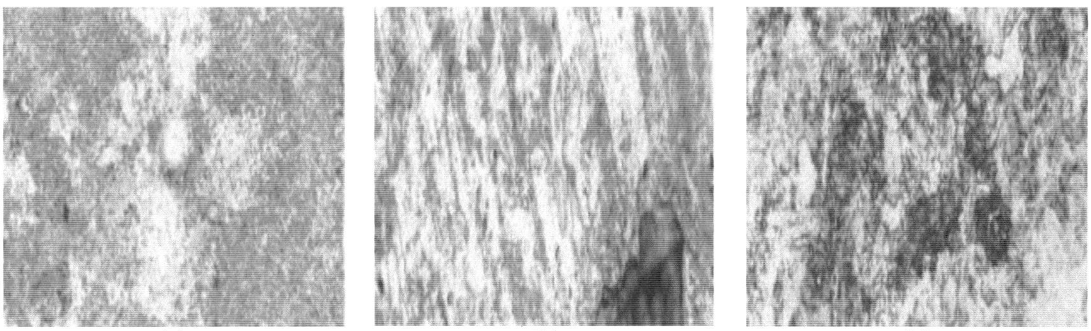

(a) (b) (c)

Figure 7. Surface optical micrographies at 1000X of samples treated in Ar at temperatures of a) 420°C, b) 500°C and c) 600°C (Flow=2scc/s at 2torr).

4. Conclusions

The sputtering process of the AISI 1020 steel was studied in two atmospheres used for the treatment of materials in the abnormal glow discharge as a function of temperature, with a higher cathodic sputtering yield in the H_2 atmosphere of lighter species. This is due to the greater energy that must be printed through the voltage to reach temperatures similar to those easily reached with Ar. This energy was determined using the SRIM/TRIM 2008 software. The cathodic sputtering yield also increases with the sample temperature due to the particle shock mechanism at the cathode used for its control. On the other hand, these atmospheres are potentially applicable for reveal of the microstructure of the AISI 1020 steel samples.

References

[1] R V Stuart 1969 *J. Appl.Phys.* **40** 2
[2] P Sigmund 1969 *Phys. Rev.* **184** 383
[3] J F Ziegler, J P Biersack and U Littmark 1985 *The Stopping and Range of Ions in Solids* (New York: Pergamon)

IOP Conf. Series: Journal of Physics: Conf. Series **935** (2017) 012019 doi:10.1088/1742-6596/935/1/012019

[4] P Sigmund 1914 *Particle Penetration and Radiation Effects* 2 (Switzerland: Springer International Publishing)
[5] E G Sheikin 2015 *Thin Solid Films* **574** 52
[6] A Chaoumead, Y-M Sung and D-J Kwak 2012 *Advances in Condensed Matter Physics* **2012** 651587
[7] A B Martin-Rojo, M Gonzalez and F L Tabares 2013 Informes Técnicos Ciemat 1273 *Glow Discharge Emission Spectrometry (GDOES): Theoretical Introduction, General Aspects, and its Applications within the Framework of the Technofusion Programs* (España: Centro de Investigaciones Energéticas, Medioambientales y Tecnológicas)
[8] E Galindo and M Albella 2008 *Técnicas de deposición física (PVD) mediante bombardeo catódico (sputtering)* (España: Instituto de Ciencia de Materiales de Madrid) p 8
[9] R S Manson and M Pichilingi 1994 *J. Phys. D: Appl. Phys.* **27** 2363
[10] http://www.srim.org
[11] A Cárdenas, Y Pineda, A Sarmiento-Santos and E Vera 2016 *J. of Phy.: Conference Series* **687** 012025
[12] B Brandt 1980 Rjnhuinzen Report 80-125 *Stable Operation of high current glow discharge* (Netherlands: Association Euratom-FOM)

Effect of surface roughness on nitriding processes by dielectric barrier discharges of carbon steel

Y P Quiñonez-Delgado[1], **H J Dulce-Moreno**[1], **G Peña-Rodriguez**[1] **and L G Arriaga-Hurtado**[2]
[1] Universidad Francisco de Paula Santander, San José de Cúcuta, Colombia
[2] Centro de Investigación y Desarrollo Tecnológico en Electroquímica, Querétaro, México

E-mail: yenypatriciaqd@ufps.edu.co , hectorjaimedm@ufps.edu.co

Abstract. Changes in the behaviour of steels to carbon, for three different finishes, surfaces were determined to be nitride by atmospheric pressure dielectric barrier discharge. The surfaces were characterized both morphologically and structurally before and after the treatment. The techniques used were: measurement of roughness and electrochemical testing (Tafel). Results show changes differential depending on the type of finish, but in any case, an improvement of corrosion resistance.

1. Introduction
Most of the industrial companies are affected by the increase of the stops of machinery and the costs of maintenance due to the change of the pieces with wear. It is claimed that surface treatments are one of the best techniques to increase the life cycles of a mechanical part [1] and to improve some properties such as corrosion resistance, fatigue resistance, even to increase hardness [2].

Nitriding is one of the most advanced surface treatment methods of research and development in recent years. Ionic nitriding is a process of thermochemical diffusion with which the formation of a hard surface layer is obtained by absorption and diffusion of nitrogen [3]. To improve the conditions of the surface layer, it is said that it depends on the application and use of the dielectric barrier discharge method for the formation of a uniform nitride layer [4]. This treatment allows to improve the properties of the steel so that it presents a better performance in different mechanical parts, such as axles, pinions, cutting tools or other industrial application [5].

Studies on nitriding processes and electrochemical tests are generally performed without regard to surface finishes [3].

The finishes of the pieces in the industry do not have surfaces with a type of mirror polished, then the techniques developed in prototypes cannot be implemented at a productive level, this is known from the perspective of fractal geometry, an object whose structure is irregular can have a behaviour in which the effective surface is greater than the geometric, thus varying the parameters considered in the estimations of surface densities [6], so that for this investigation the effects of the roughness of the material were determined, the nitriding was performed three types of surface finishes and electrochemical tests, such as polarization curves (Tafel) and corrosion rate.

2. Materials and methods

2.1. Preparation of samples
Samples were taken $2.54cm$ in diameter and $1cm$ in height. The cleaning of surface oxides and impurities of the specimens and the roughness profile were achieved by means of an abrasive blast

called sand blasting. This process was done in a private laboratory of blasting services. The anchor profile (roughness level) was given according to the European standard SIS 05 5900 for degree of rust A and finish SA 3 or SSPC SP5 according to the American standard. The roughness measurement was done by means of a digital surface roughness tester SRT 6200. The test pieces were cleaned with isopropyl alcohol by means of ultrasound to discard residues of other materials, before nitriding.

2.2. Nitriding of test pieces

The nitriding was done by means of a reactor composed of the electrodes (aluminium anode and as cathode each steel sample) with a tempered glass dielectric barrier of $3.0mm$ thickness under a nitrogen atmosphere of 99.9% purity. The assembly was made using Teflon and Mica accessories insulators. The dielectric barrier was maintained in contact with the anode at a distance of about $1.0mm$ from the cathode in a nitrogen atmosphere. Figure 1 shows the nitriding device used in this research.

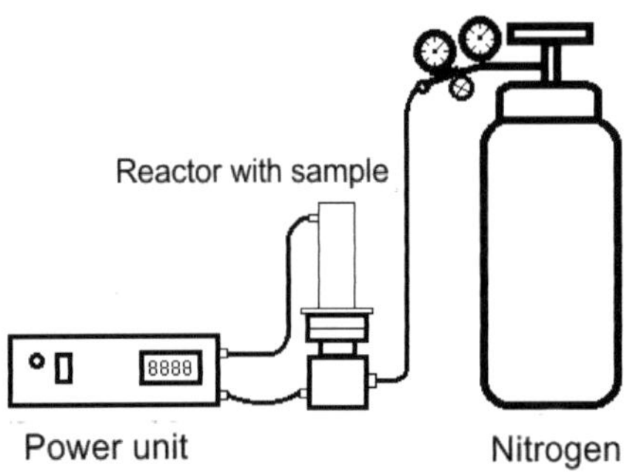

Figure 1. Dielectric barrier nitriding equipment.

The nitriding process of the samples had an average duration of 32 minutes. First, the nitrogen passage is opened to displace the reactor air for five minutes. Subsequently the nitriding discharge is done for 15 minutes with a gas pressure in the pipes of $12.5\,psi$. Then the unit is turned off and the gas flow is left open for five minutes. Finally, it takes seven minutes for the nitrogen tank gauges to reach zero. The flow measured by displacement with grade five high purity nitrogen is $3.6\,liters/hour$. The quality of nitriding depends on the flow rate of hydrogen, which has an optimum value [7].

2.3. Electrochemical tests

One of the most widely used electrochemical methods in scientific and practical characterization studies is the polarization curves method [8]. These tests are done to measure the corrosion rate of the samples, before and after the nitriding. An electrochemical cell was used, the GAMRY software version 6.30 of the Echem Analyst TM package was used for data acquisition and processing. The devices or electrochemical cells used in this type of test basically consist of five elements: container, probe holder, working electrode (test tube connected to potentiostat), reference electrode (standard to measure electrochemical effects) and auxiliary electrode. Figure 2 shows the assembly of the device which was started by placing the steel sample in the propeller holder which is secured to the acrylic structure (A). The sample was tightened with screws to press on a rubber (B). 3% sodium chloride solution is injected into the cell container, avoiding the presence of bubbles (C). After reaching the top of the solution, the reference electrode (silver electrode) with direct connection to the data processor (D) is inserted into one of the holes in the vessel. The resulting connectors are matched to the colour and the assembly is placed inside the protective plastic housing.

For the measurement of the corrosion rate, the Echem Analyst TM software reports the results on Tafel with its respective data Table. The polarization curves allow to find values of the anodic and cathodic slopes. The parameters found in the polarization curves enable corrosion current densities and corrosion rates to be found in milliseconds per year (mpy) [9].

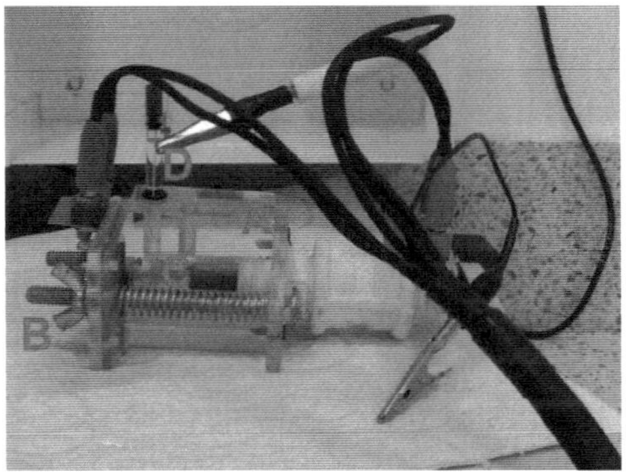

Figure 2. Assembly of the specimen. (A) Acrylic structure, (B) screws that press the sample. (c) Cell container and (D) reference electrode (silver electrode) with direct connection to the data processor.

3. Results

One of the important and relevant results is the change in the value of the roughness after the nitriding process by means of a dielectric barrier discharge, and it is presented by the roughness test 2 as shown in Table 1 with a value of $6.15\mu m$. The dielectric barrier discharges are characterized by the formation of micro arcs, which are directly linked to the type of topography of the surface of the electrodes and the surface of the dielectric material that separates them [6]. As a consequence of the above, the roughness specimen 2 presents changes in its surface structure as the higher peaks are devastated and allow a significant reduction in the value of its roughness.

Table 1. Comparison of roughness before and after nitriding.

Measurements	Without nitriding	With nitriding
Roughness machining	$0.58\mu m$	$1.32\mu m$
Roughness 1	$3.72\mu m$	$3.87\mu m$
Roughness 2	$6.38\mu m$	$6.15\mu m$

Table 2 shows the results of the corrosion rate in each of the test specimens, where the samples with roughness 1 present the best behaviour with a value of $3.238mpy$. And Figure 3 shows the comparative graph of each of the Tafel curves.

The results show that the method of nitriding by means of discharge of dielectric barrier at atmospheric pressure is an effective method to improve the mechanical properties and to obtain greater resistance against the phenomenon of corrosion [10], and experimental results indicate that the rapid nitriding process not only improves the nitriding rate remarkably but also maintains the high hardness of the nitride layer [11].

Table 2. Corrosion analysis of samples of Tafel curves.

Samples	Machined without treatment	Machined with treatment	R1 without treatment	R1 with treatment	R2 without treatment	R2 with treatment
E_{corr}(mV)	-547.0	-545.0	-612.0	-581.0	-570.0	-533.0
I_{corr}(μA)	20.40	18.20	14.50	7.09	20.20	13.70
Beta A (V/decade)	$82.80e^{-3}$	$238.0e^{-3}$	$101.2e^{-3}$	$78.4e^{-3}$	$84.8e^{-3}$	$109.1e^{-3}$
Beta C (V/decade)	$619.9e^{-3}$	$82.00e^{-3}$	$585.8e^{-3}$	$60.70e^{-3}$	$551.8e^{-3}$	$180.4e^{-3}$
Corrosion rate (mpy)	9.345	8.319	6.608	3.238	9.226	6.244
Chic square	341.0e-3	207.1e-3	467.7e-3	564.6e-3	286.0e-3	183.8e-3

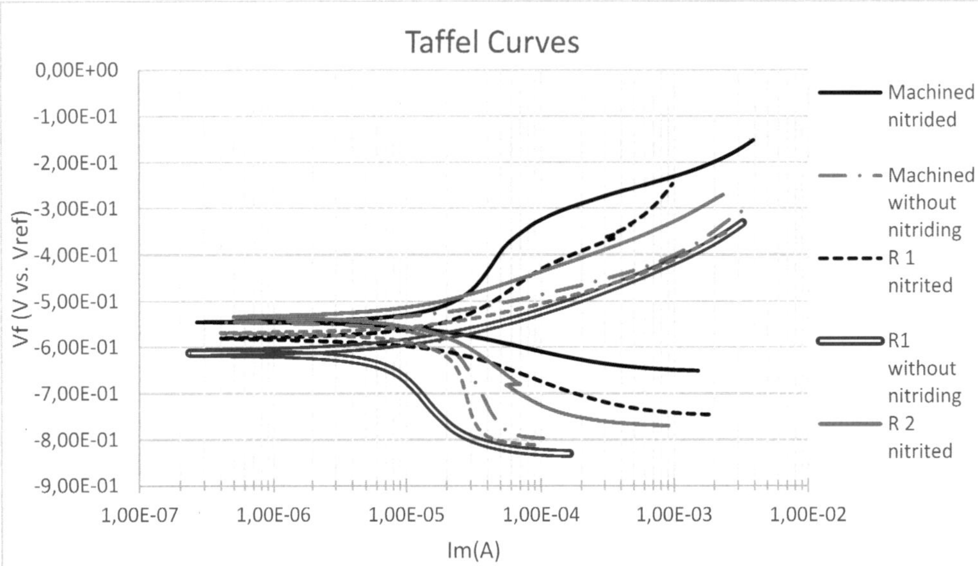

Figure 3. Comparative graph of Tafel curves in each of the samples.

4. Conclusions

From the investigation, it is concluded that the corrosion rate of the AISI/SAE 1020 steel is lower when a surface nitriding process has been done, but the corrosion rate is further reduced in the samples with a greater roughness, machined sample decreases to a very small amount because its surface layer does not exist some chemical elements that allow a greater reduction in its speed of corrosion. It should be noted that the test specimens of the blasting process reduce the corrosion rate without performing a nitriding process due to the content of some elements and the action they exert in the microstructure of the AISI/SAE 1020 steel, which will be used for future studies.

References

[1] Kalpakjian S and Schmid S R 2008 *Manufactura, Ingeniería y tecnología*. (México: Pearson Prentice Hall) p 10-110

[2] Belkin P N, Yerokhin A and Kusmanov S A 2016 Plasma electrolytic saturation of steels with nitrogen and carbon *Surface and Coatings Technology* **307** 1194-1218

[3] Dulce-Moreno H J & Contreras de la Ossa G J 2009 Comportamiento del acero AISI-SAE 1025 nitrurado mediante descarga de barrera dialéctica a presión atmosférica *Bistua: Revista de la Facultad de Ciencias Básicas* **7(2)** 65-69

[4] Miyamoto J, Inoue T, Tokuno K, Tsutamori H and Abraha P 2016 Surface modification of tool steel by atmospheric-pressure plasma nitriding using dielectric barrier discharge *Tribology* **11(3)** 460-465

[5] Apraiz B J 1984 *Tratamientos térmicos de los aceros* (Madrid: Nuevas gráficas S.A.) p 345

[6] Vargas W, Pineda L and Santaella L E 2006 Rugosidad y textura de superficies: experimento simulaciones *Ciencia e Ingeniería Neogranadina* **16(2)** 54-63

[7] Nagamatsu H, Ichiki R, Yasumatsu Y, Inoue T, Yoshida M, Akamine S and Kanazawa S 2013 Steel nitriding by atmospheric-pressure plasma jet using N_2/H_2 mixture gas *Surf. and Coat. Tech.* **225** 26-33

[8] Ugaz A y Díaz I. 1988 Principios de las técnicas electroquímicas en estudios de procesos de corrosión *Revista de Química* **11(1)** 23-31

[9] Aperador W, Vera L E and Vargas U A 2010 Electrochemical corrosion of electrocoating nickel/copper obtained by pulsating current *Engineering and Development Science Journal* **27** 58

[10] Borgioli F, Fossati A, Galvanetto E, Bacci T and Pradelli G 2006 Glow Discharge Nitriding of AISI 316L Austenitic Stainless Steel: Influence of Treatment Pressure *Surf. and Coat. Tech.* **200** 5505-5513

[11] Zhao C, Li C X, Dong H and Bell T 2006 Study on the active screen plasma nitriding and its nitriding mechanism *Surface and Coatings Technology* **201(6)** 2320-2325

IOP Conf. Series: Journal of Physics: Conf. Series **935** (2017) 012021 doi:10.1088/1742-6596/935/1/012021

Wear and corrosion behaviour of Al_2O_3-TiO_2 coatings produced by flame thermal projection

M Forero-Duran[1], **H J Dulce-Moreno**[1], **M Ferrer-Pacheco**[1] **and F Vargas-Galvis**[2]
[1] Universidad Francisco de Paula Santander, San José de Cúcuta, Colombia
[2] Universidad de Antioquia, Medellín, Colombia

E-mail: myriamfd@ufps.edu.co

Abstract. Evaluated the wear resistance and the coatings corrosion behaviour of Al_2O_3-TiO_2 prepared by thermal spraying by flame on AISI 1020 carbon steel substrates, previously coated with an alloy base Ni. For this purpose, were controlled parameters of thermal spraying and the use of powders of similar but different chemical composition is taken as a variable commercial reference for ceramic coating. SEM images allowed to know the morphology of the powders and coatings. Electrochemical techniques (Tafel) were applied to evaluate the protection against corrosion. Coatings were tested for wear with a tribometer configuration bola-disco. It was determined that the phases present in coatings are directly relate to the behaviour against corrosion and wear them. Keywords: wear, corrosion, thermal imaging.

1. Introduction

Ceramic coatings deposited are on the surface of a component in order to improve its properties [1], the inorganic coatings are deposited on the metallic surface without altering the original properties of resistance and hardness of the substrate [2]; there are a number of factors that need to be taken into account when making a composite coating such as: substrate compatibility, adhesion, porosity, wear resistance, corrosion resistance, resistance to sudden thermal changes and the possibility of their repair [3]. The properties of the coatings are evaluated by measuring the microhardness, ductility, wear resistance, coefficient of friction and corrosion resistance, which vary according to particle type, size, volume percentage and distribution in the tank [4,5], the thermal projection is a technique that consists of the deposition of coatings with torches that generate the necessary heat to melt the material and accelerate it towards the substrate by means of compressed gas and then the particles collide on the surface forming layers that adhere to irregularities of the surface and allows obtaining high-melting ceramic coatings [6]. Aluminides possess a sufficient concentration of aluminium to form a layer of alumina when exposed to atmospheres of air and oxygen, which makes them have excellent resistance to oxidation and corrosion in aggressive chemical environments [7]. TiO_2 is a hard material with high melting point, its coatings are used in applications such as photocatalysis, bone implants, electrical devices, and renewable energy and gas sensors [8].

This work uses commercial steel carbon AISI 1020, cutting and machining in specimens of $2.54cm$ in diameter and a cm thick, prepare surfaces using Sand-blasting, sand of zirconia, high speed abrasive sand particles impact clean, remove corrosion and give surface finishing (roughness) required for the coating to stick by mechanical anchor i.e. it serves as grip coating. Applies a layer basis with commercial nickel powder CPM 1205TM AISI 1020 steel substrates and coatings of alumina-titania using commercial powders of different chemical composition and different then House producer: Saint Gobain

108 (SG108), 25060 MetaCeram (MC25060), Metco™ 131VF (M131VF); prepared by thermal spraying by flame. The characteristics of the ceramic coatings are affect by projection parameters, by powders to projecting, chemical characteristics, morphology, and particle size distribution, among others [9]. All these characteristics influence the roughness, surface finish, porosity, hardness and tenacity of the coatings. At work are characterized Alumina-Titania ceramic powders used to produce coatings evaluates the chemical and morphological characteristics. Relate the morphology and particle size distribution of porosity and wear of coatings properties.

2. Experimental development

2.1. Characterization of starting powders

The anchor powders and alumina-titania ceramic materials used to apply coatings, chemical characteristics and particle size distribution are evaluated, each of which is assigned a code: CPM1205TM1 anchor layer sample 1=CAM1, Powder for coating SG108 sample 2=SGM2, Powder for coating M131VF sample 3=VFM3, Powder for coating MC25060 sample 4=MCM4; the results are shown in Tables 1 to 3. The data in Table 2 and 3 and Figure 1(b), (c), (d) were taken from the doctoral thesis, authorized by Fabio Vargas [10].

Table 1. Chemical composition of the dust of anchorage.

Material	Element % by weight				
	Ni	Si	Fe	Al	Cu
CAM1	97.62	1.16	0.34	0.18	0.07

Table 2. Chemical composition of powder coatings.

Ceramic powders	% by weight of compound		
	Al_2O_3	TiO_2	Others
SGM2	59.00	40.00	1
VFM3	60.00	40.00	
MCM4	59.64	39.12	1.24

Table 3. The powders grain size distribution.

Ceramic powder	$d_{10\%}$	$d_{50\%}$	$d_{90\%}$
SGM2	15.88	27.20	39.76
VFM3	2.17	11.12	25.02
MCM4	21.18	32.40	49.37

The Table reads $d_{10} = 15.88$ and $d_{90} = 39.76$, indicates that at least 10% by volume of the powder has size less than $15.88\mu m$ and 90% of the particles have one size $39.76\mu m$.

2.2. Morphology powder

Micrograph using SEM allowing to obtain images where the morphological characteristics of the powders can be seen. The microstructure of the anchor and to project the material of the coating materials is shown in Figure 1.

(a) (b) (c) (d)

Figure1. Powder morphology (a) CAM1, (b) SGM2, (c) VFM3, (d) MCM4.

3. Coatings

Deposited Areste 1, the GIPIMME group of the University of Antioquia, with camera armed with a torch Eutectic Castolin-Terodyn 2000, with an infrared pyrometer RAYTEK deposited CAM1, material layer that serves as anchor between the substrate and ceramic coating; and then apply the layer of commercial ceramic powder, SGM2, VFM3 and MCM4. For CAM1 layer, the projection distance remained constant at $15cm$ and an oxidizing flame is used ($59.46l/min$ of O_2 by $21.18l/min$ of C_2H_2), while for the ceramic layer 1:4 worked with $9cm$ as super oxidant projection, a burning distance in the relationship. For VFM3 powder, the flow was $9\ gr/min$ and 8 screening passes.

3.1. Characterization of coatings

The roughness of 1020 clean uncoated steel (AL) and coatings is determined with a model SRT-1000 roughness tester. The results are shown in Table 4.

Table 4. Roughness of coatings.

AL (μm)	AL-CAM1 (μm)	AL-CAM1-SGM2 (μm)	AL-CAM1-VFM3 (μm)	AL-CAM1-MCM4 (μm)
4.98 ± 0.06	8.02 ± 0.28	7.61 ± 0.09	6.98 ± 0.06	7.92 ± 0.21

The technique used for the analysis of surface coatings Energy-Dispersive X-rays Spectroscopy (EDS). The chemical composition of the coatings is presented in the Table 5.

Table 5. Chemical composition of coatings.

Material	Element % Wt				
	C	O	Al	Ti	Fe
AL-CAM1-SGM2	1.96	30.76	36.6	29.47	1.22
AL-CAM1-VFM3	-----	33.42	38.62	27.95	-----
AL-CAM1-MCM4	-----	29.46	6.53	64.01	-----

SEM Analysis of the ceramic coatings SEM shows the cross section of the coatings at 200X where the characteristics of the molten particles are observed. The microstructure of the coatings made by thermal projection is shown in Figure 2.

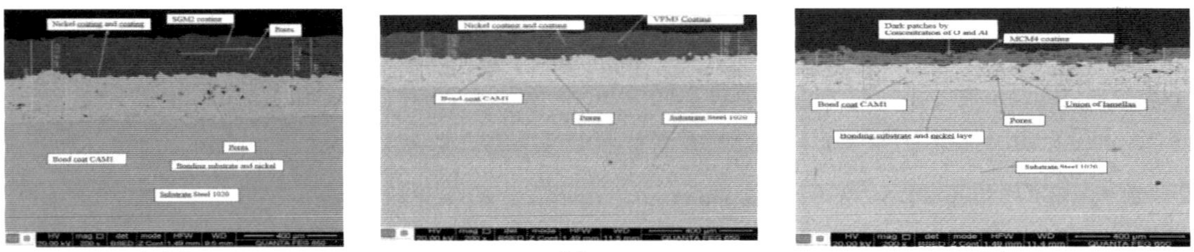

(a) (b) (c)

Figure 2. Coating microstructure (a)AL-CAM1-SGM2, (b)AL-CAM1-VFM3, (c) AL-CAM1-MCM4.

3.2. Electrochemical analysis

A solution of sodium chloride (NaCl) is used to 3.5% as an electrolyte, the reference electrode is Ag/AgCl and against electrode of coal. To perform the tests the electrochemical cell is connect to a team of electrochemical analysis using software and equipment of impedance GAMRY 1000. The results are presented in Table 6 and Figure 3. Tafel curves are shown of the coatings prepared.

In the potentiostatic tests, the uncoated steel has a corrosion potential of -547mV and a corrosion rate of 9.35mpy, the substrate with nickel layer has a corrosion potential of -299mV, quite high and a speed of corrosion of 4.98mpy, compared to the uncoated substrate the nickel-plated substrate is protected. The AL-CAM1-SGM2 has an Ecorr of -577mV lower than uncoated steel but presents a corrosion rate of 4.85mpy which indicates that it is self-protected by the effects of Ni and Ti having passivation properties. The AL-CAM1-VFM3 coating has an Ecorr of -548mV and a velocity of 7.07mpy and this

IOP Conf. Series: Journal of Physics: Conf. Series **935** (2017) 012021 doi:10.1088/1742-6596/935/1/012021

electrochemical behaviour must be the reason for new studies. The AL-CAM1-MCM4 coating has a corrosion potential of -512mV and a corrosion rate of 4.38mpy indicating that it is protected.

Table 6. Analysis of corrosion coatings Tafel curves.

Material	Beta A	Beta C	I_{corr} (μA)	E_{corr} (mV)	Corrosion rate (mpy)	Chi Squared
AL	82.8×10^{-3}	432.3×10^{-3}	20.50	-547.0	9.35	341.00×10^{-3}
AL-CAM1	172.1×10^{-3}	176.1×10^{-3}	10.90	-299.0	4.98	210.20×10^{-3}
AL-CAM1-SGM2	103.9×10^{-3}	73.6×10^{-3}	10.60	-577.0	4.85	30.55×10^{-12}
AL-CAM1-VFM3	125.7×10^{-3}	131.3×10^{-3}	15.50	-548.0	7.07	17.30×10^{-12}
AL-CAM1-MCM4	80.3×10^{-3}	122.2×10^{-3}	9.58	-512.0	4.38	28.50×10^{-12}

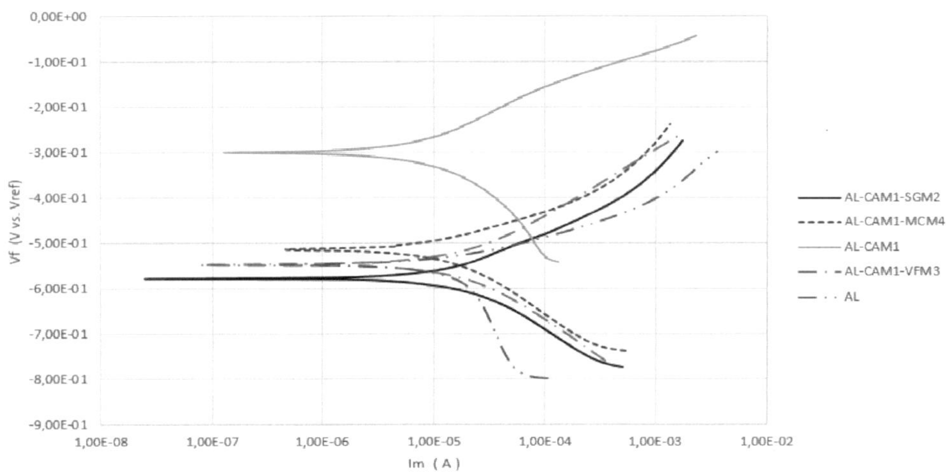

Figure 3. Comparative graph. Curves Tafel of the elaborated coatings.

The wear rate was determined by the average of five measurements of the transverse area in radial direction of traces of wear and tear, using a BRUKER Dektak XT profilometer with a force of $3mg$, a range of scanning of $524\mu m$ and a radius of $2\mu m$ probe. The coefficient of friction of the coatings tested was determined by tribometer Pin on disk the coefficient was obtained as a function of time. The results are presented in Table 7 and 8.

Table 7. Tribological analysis: wear.

Sample	Loss of volume (mm)	Wear rate samples ceramics (mm^3/Nm)	Mass loss Anti-body (g)
AL-CAM1-SGM2	1.03 ± 0.14	$5.13 \times 10^{-3} \pm 6.79 \times 10^{-4}$	1.5×10^{-6}
AL-CAM1-VFM3	0.93 ± 0.24	$4.67 \times 10^{-3} \pm 1.19 \times 10^{-3}$	8.2×10^{-6}
AL-CAM1-MCM4	1.30 ± 0.22	$5.48 \times 10^{-3} \pm 1.09 \times 10^{-3}$	6.9×10^{-6}

Table 8. Coefficient of friction coatings.

Type of coating	Friction coefficient
AL-CAM1-SGM2	0.70
AL-CAM1-VFM3	0.85
AL-CAM1-MCM4	0.80

4. Conclusions

Coatings made with Al_2O_3 and TiO_2 have good corrosion resistance compared to the uncoated substrate. Good corrosion resistance with different temperatures and applied techniques is reported in the investigations. Coatings made by APS with Al_2O_3 and nanostructured TiO_2 have smooth morphologies without obvious grooves that provide wear resistance compared to microstructure powders [11]. The chemical composition of powder coatings is very similar; the melting temperature of the alumina is higher than that of the nickel, it is for this reason that in the microstructure of the coatings patches of different shades are observed that are related to the size of the particle of the projected powder since the smaller particles are cast and the largest are semi funded.

Evaluated the corrosion resistance of metal pump pistons with coatings made with different percentages of Al_2O_3 and TiO_2 and report high corrosion resistance in acid and saline environments attributed to lack of interconnected pores [12]. The reduction of oxidation in the tests compared to the uncoated steel is attributed to the barrier forming the ceramic layer on the substrate and the ability of the nickel layer to prevent the passage of oxygen to the substrate.

The coating AL-CAM1-VFM3 presents an electrochemical behaviour that must be the reason of new studies, since the increase of the corrosion rate can be associated with more complex phenomena like tortuosity or electro-catalysis of some phases of the titanium oxide.

In the coatings, it is observed that the roughness of the coatings is due to the size and shape of the particles and the projection parameters. As the particle size is reduced, the roughness of the coating is reduced. The presence of cracks and pores in the coatings is influenced by the size of the particles and by the projection parameters, as they determine the heat at which the particle is melted, the velocity and the projection distance. The wear behaviour of the projected coatings improves as the particle size of the projected material decreases and is related to the roughness of the surface.

Acknowledgments

GIMACYR-GIPIMME Group of the "Universidad de Antioquia". To the Master's Program in Science and Technology of Materials of the "Universidad Francisco de Paula Santander".

References

[1] Pawloski, L 2008 *The science and engineering of thermal spray coatings, 2nd edition* (New York: Wiley)
[2] Espinosa A, Pacheco J, Ibarra L, Galván R & Espitia I 2011 Resistencia a la corrosión de películas cerámicas delgadas *Superficies y Vacío* **24(4)**, 121-125
[3] Garcia I, Fransaer J, Celis J P 2001 Electrodeposition and sliding wear resistance of nickel composite coatings containing micron and submicron SiC particles *Surface and Coatings Technology* **148** 171-178
[4] Orlovskaya L, Periene N, Kurtinaitiene M, Surviliene S 1999 Ni–SiC composite plated under a modulated current *Surface and Coatings Technology* **111** 234-239
[5] Benea l, Bonora P L, Borello A, Martelli S 2002 Wear corrosion properties of nano-structured SiC–nickel composite coatings obtained by electroplating *Wear* **249** 995-1003
[6] Coronado J 2006 Efecto de la preparación superficial y aplicación de recubrimientos en la resistencia a la fatiga del acero SAE 1045 *Scientia Et Technica* **13(34)** 213-218
[7] Liu C T, Stiegler J O 1990 Ordered Intermetallics *ASM Metals handbook: Properties and selection: nonferrous alloys and special purpose materials, 10th edition* vol 2 (USA: ASM International) pp 913-942
[8] Gardo M, Guilemany J 2014 Milestones in Functional Titanium Dioxide Thermal Spray Coatings: A Review *Journal of Thermal Spray Technology* **23(4)** 577–595
[9] Romero M, Liscano S, Palomo A & Padilla P 2014 Caracterización morfológica de polvos cerámicos para la fabricación de recubrimientos termorrociados *Iteckne* **11(1)** 93-98
[10] Vargas F 2010 *Development of thick ceramic layers with micrometric and nanometric structures by thermal spraying for tribological applications* (France: University of Limoges)
[11] Wang Y, Jiang S, Wang M, Wang S, Danny Xiao and Strutt P R 2000 Abrasive Wear Characteristics of Plasma Sprayed Nanostructured Alumina/Titania Coatings *Wear* **237(2)** 176-185
[12] Rosso M, Scrivani A, Ugues D, and Bertini S 2001 Corrosion Resistance and Properties of Pump Pistons Coated with Hard Materials *Int. J. Refract. Met. Hard Mater* **19(1)** 45-52

IOP Conf. Series: Journal of Physics: Conf. Series **935** (2017) 012022 doi:10.1088/1742-6596/935/1/012022

Projection parameters for zirconia-alumina-ceria coatings made by flame spraying from results of numerical simulation

L Rodríguez[1], M Ferrer[1], F Vargas[2] and G Peña[1]
[1] Universidad Francisco de Paula Santander, San José de Cúcuta, Colombia
[2] Universidad de Antioquia, Medellín, Colombia

E-mail: luisfranciscorw@ufps.edu.co

Abstract. A numerical simulation was performed with the software Jets et Poudres, the results let choose the parameters to deposit zirconia-alumina-ceria coatings of different composition on substrates of red clay, by thermal spraying with the oxyacetylene flame to obtain homogeneous coatings with good adhesion to the substrate. The effect of the projection distance (7, 10 and 12cm) between the substrate and the torch, the fusion percentage of particles and the K-Sommerfeld number was determined. This number is dimensionless and is affected by the projection distance and by the chemical composition of the particles. For a projection distance of 9cm, the fusion percentage of the particles varies between 83.8% and 100%, and the K-Sommerfeld number between 47.3 and 50 for the different compounds. This makes possible to obtain uniform coatings with good wettability, therefore, good adhesion to the substrate, while for the distance of 7cm the fusion percentage varies between 22% and 38%, due to the short time of the particles in the flame which causes low adhesion, when the projection distance is 12cm the particles do not have sufficient kinetic energy to reach the substrate and therefore the coating is not deposited.

1. Introduction

With the purpose to protect the surfaces of structures made of different materials alumina, zirconia and ceria coatings among others are used, which enhance their thermal, tribological and corrosive properties[1,2]. The zirconia ZrO_2 is used due to its stability to high temperatures and its low conductivity (0.8–1.5W/mK) [3]. At room temperature the zirconia is in its monoclinic phase, which does not allow it to be used in structural applications, that is why it is mixture with oxides from rare lands like the ceria which promote the retention in the cubic and tetragonal phases in room temperature. The aluminium oxide is used due to its low friction coefficient, that is why it is used in applications against wear and corrosion it also shows a low electrical and thermal resistance allowing it to be used in high temperatures [3].

One of the techniques used to obtain the coatings is by flame spraying (FS), which consists in melt a material (dust) and deposit it on a substrate. The oxyacetylene flame gets its energy of oxygen and acetylene combustion reaching temperatures close to 3100°C, which allows different of high-melting compounds to be used. The particles are adhered to the substrate by means of a mechanical anchor where the molten material in the form of droplets moistens the irregularities of the substrate and adheres while cooling simultaneously [4]. The morphology of the coatings obtained depends on the physical, chemical and morphological properties of the raw material and the projection parameters such as the type of neutral or oxidant flame, the distance between the torch and the substrate, the preheating temperature of the substrate and the flow of powders among others.

To predict the behaviour of the particles in the flame and his interaction with the substrate is rather difficult, that is why one resort is the numerical simulations that allow me to study the thermo kinetic behaviour of the particles in the flame. The software Jets et Poudres developed by the university of Limoges (France) allows to know the temperature, the melted percentage, and the dimensionless K-Sommerfeld number of the particles [5]. The K-Sommerfeld number was described by Mundo and Kappel, in the study of the impact of water drops and ethanol on cold surfaces, they found that the value of K is in the range from 3 to 57.7, where for minor values of 3 the particle bounces and does not stick fast to the substrate and if it is major of 57.7 the particle will splash on having struck with the substrate. This parameter is used in thermal projection since it allows to predict the behaviour of the particles when interacts with the substrate which determines the wettability and the homogeneity of the deposited splats, which is related to the anchorage of the coating [6].

The K-Sommerfeld value is given by the dimensionless numbers Weber (We) and Reynolds (Re), which relate the inertia, tension surface and viscosity of the molten particles and is determined for:

$$K = We^{1/2}Re^{1/4}\; K = [\rho v d/\mu]^{1/2}[\rho \mu^2 d/\sigma]^{1/4} \tag{1}$$

Where: ρ, v, d, μ and σ, are the density, the speed, the diameter, the viscosity and the superficial tension of the melted particle that strikes on the substrate[7].

2. Materials and methods

A numerical simulation is made with the software Jets et Poudres to determine the melted percentage and the K-Sommerfeld number of agglomerated particles of zirconia-alumina-ceria, after they have crossed a oxyacetylene flame, in order to know the conditions under which the melted or partially melted particles stick fast to the irregularities of the substrate and form a homogeneous coating, preventing the particles from bouncing or on the contrary they form splashes, aspects which prevent the proper ordering of splats, giving rise to porous coatings due to stacking defects and problems of adhesion or cohesion due to poor wettability.

The carried-out simulation of agglomerated particles of zirconia-alumina-ceria of 8, 10 and 12μm of diameter from different chemical composition, named as Composite 1 (C1), Composite 2 (C2) and Composite 3 (C3). A particle of powder of different chemical composition interacts with a oxidizer flame in the relation 1:3.2 and is projected to 7, 9 and 12cm, these distances are chosen according to [8,9], in order to evaluate how it affects the percentage in volume of the compounds in the k-Sommerfeld number and in the melted percentage of the particles [10]. The physical, chemical and thermal properties of the particles needed for the simulation are taken from the databases and calculated by the law of mixtures bearing in mind its composition in volume.

In order to validate the results of the simulation the coatings are done using commercial powders, one composite of zirconia-alumina ZrO_2, 36 % in weight Al_2O_3, from a commercial brand called Metaceram 25088 from Eutectic Castolin® and another of cerium oxide, which are placed in a ball mill to homogenize the size of the particles for his later agglomeration. The agglomeration is done in an agglomerating drum developed by the group GIPIMME of the university of Antioquia, like a binder a polyvinyl alcohol to 2% is used, which was selected in agreement to the results reported by [11], then they are left to dry during an hour in a stove to 70°C, finally they are sifted in a rotap, gathering the resulting particles in a 400 mesh and retaining them in a 500 mesh, the agglomerated powders are characterized morphologically by images SEM and his microstructure by refinement Rietvel DRX difractograms.

The coatings are done in the thermal projection chamber of the "Universidad de Antioquia", on substrate of red clay [12], in order to improve the structural properties of the tiles with the coating. The evaluation of the results is done by SEM, DRX and Pull Off.

3. Results and analysis

3.1. Composition of the raw material

In the Table 1 is named like C1, C2 and C3 to each one of composites of the powders obtained after the agglomeration, the concentration of every composite is given in percentage of volume, increasing the concentration of oxide of cerium. The results of diffraction of X-rays allows us to see three zones of amorphicity mainly between 10 and 20, 27-38 and 46-58 (2Θ) [13], as evidenced in Figure 1, for the named composite as C1. The refinement Rietveld allow find the percentage in weight of the present phases in each of the powders, as it appears in the Table 2, where from it is important to highlight that the zirconia is in its in monoclinic phase and the alumina in alpha or corundum.

Table 1. Composition of the particles for three powders of contribution.

Volume percent (%)				
Composite	Name	ZrO_2	Al_2O_3	CeO_2
Composite 1	C1	48.83	41.03	10.14
Composite 2	C2	46.85	39.37	13.78
Composite 3	C3	44.79	37.64	17.57

Table 2. Phases in the starting powders (%).

	Phase		
Compound	m $- Zr_4O_8$	α $- Al_2O_4$	c $- Ce_4O_8$
C1	59.2	26	14.8
C2	51.9	28.4	19.7
C3	47	24.4	28.6

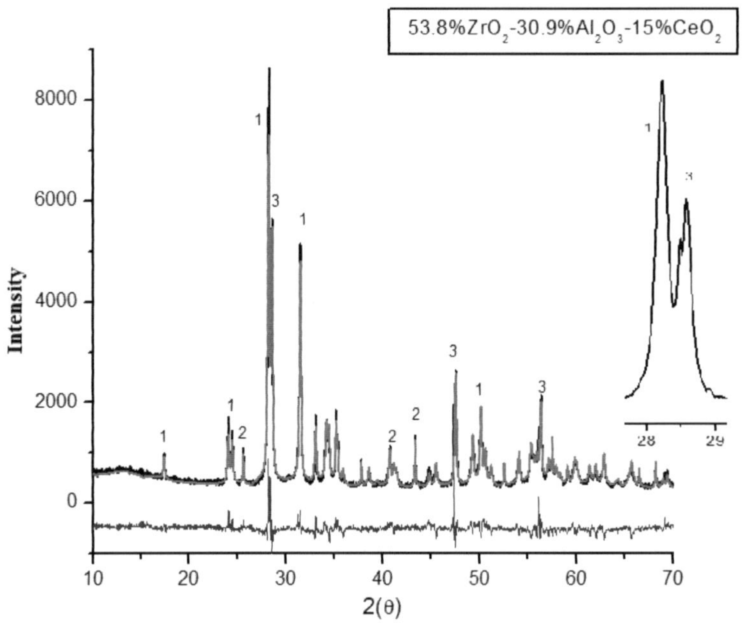

Figure 1. Phases presents in C1 $1. m - Zr_4O_8$ $2. \alpha - Al_2O_4$ $3. c - Ce_4O_8$.

3.2. Morphology of the powders

The SEM images (Figure 2), allow seeing that the powder particles are formed by agglomeration of smaller particles of nanometric order. The composite C2 shows a more uniform agglomeration where see oval particles with smaller added particles. In the composite C1 and C3 the agglomerated particles are smaller; nevertheless, they are formed also by nanometric particles.

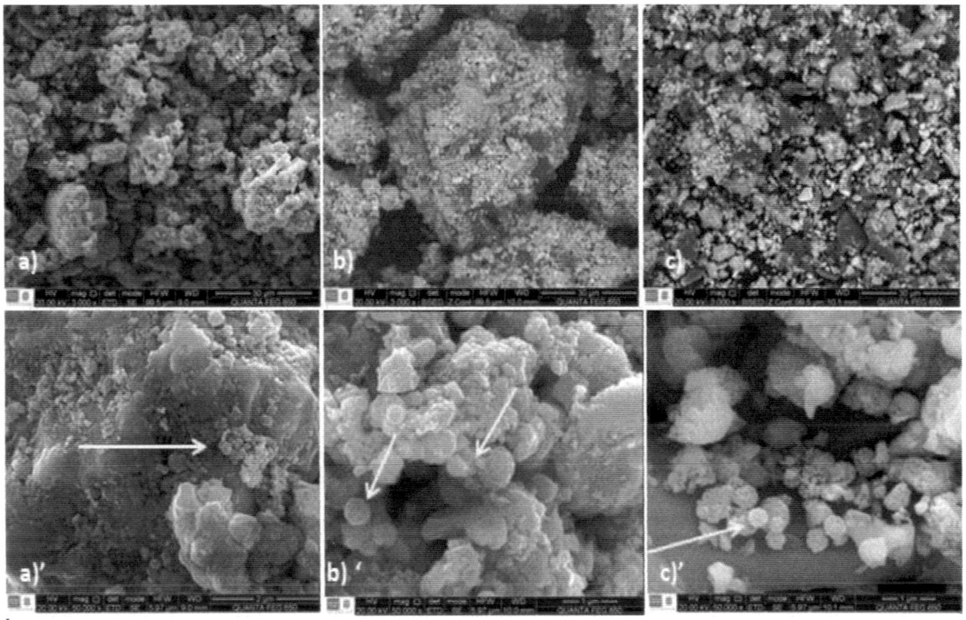

Figure 2. Morphology of the powders, where (a), (b) and (c) correspond to C1, C2 and C3 to a magnification of 3000X, (a'), (b') and (c') they are approximations to 50000X of them.

3.3. The simulation
The results of the simulation let us conclude that the parameter K-Sommerfeld depends on the projection distance and the size of the particles, as well as the melted percentage depends on the time of these in the warm zone of the flame. The parameter K-Sommerfeld, the melted percentage and the reach of the particles, for the different projection conditions are shown in the Table 3. For a distance of 9 cm conclude that the K - Sommerfeld vary between 43.7 and 50 and the melted percentage between 83.8 and 100%, whereas for a distance of 7cm the value of K overcomes the limit of 57.7 and finally for a distance of 12cm they do not have the sufficient energy to reach the substrate. In conclusion to obtain homogeneous coatings the particles must be projected to a distance of 9cm, with an approximate size of 8μm.

3.4. Projection parameters
According to the simulation results and the conclusion of some authors, the coatings are made with an oxidizer flame in a relation of 1:3.2 (acetylene: oxygen), to a distance of 9cm. The torch moves to a linear speed of 0.72cm/s, the sample holder is rotated at 116rpm. The flow of the powders was kept constant to 12.6g/min, three warm up and six projection passes are made.

3.5. Morphology and microstructure of the coatings
The coatings are named like R1, R2, and R3, obtained from the composite C1, C2 and C3 respectively. The superficial morphology of the coatings is see in SEM images. In the Figure 3(a), (b) and (c), there see the melted particles of a splat shaped like a disk, partially melted particles, splashes and cracks generated in the moment of solidification of the particles. In the images (a'), (b') and (c') the morphology of the coatings appears in a transverse cut, it can be seen that the particles in all the cases have good wettability with the substrate.

The coatings R1 and R3 have a uniform morphology with the presence of lamellas corresponding to fully melted particles, on the other hand in R2 where addition to the lamellas their unmelted particles due to the agglomerated particles of the composite C1 and C3 are smaller in comparison with C2, the results of the simulation show that the size particles influences the melting percentage.

Table 3. K- Sommerfeld, melting percentage, scope of studied particles.

Compound	Size Particle (μm)	Parameter	Projection distances		
			7cm	9cm	12cm
C1	8	K-Sommerfeld	80.78	47.9	41.35
		% Melted	27.1	100	100
		Scope (cm)	7	9	9.5
	10	K-Sommerfeld	151.1	100.7	76.88
		% Melted	10.7	59.2	100
		Scope (cm)	7	9	10.9
	12	K-Sommerfeld	220.8	167.3	126.5
		% Melted	1.4	26.4	100
		Scope (cm)	7	9	11.9
C2	8	K-Sommerfeld	89	50.0	44.6
		% Melted	15.8	100	100
		Scope (cm)	7	9	9.6
	10	K-Sommerfeld	84.6	104.9	79.9
		% Melted	20.3	44.5	100
		Scope (cm)	7	9	11
	12	K-Sommerfeld	227.18	173.6	110.2
		% Melted	1.1	19.3	91.2
		Scope (cm)	7	9	11.9
C3	8	K-Sommerfeld	90	47.3	44.6
		% Melted	15.8	83.8	100
		Scope (cm)	7	9	9.6
	10	K-Sommerfeld	162.2	110	76.79
		% Melted	6.1	34.5	100
		Scope (cm)	7	9	11.1
	12	K-Sommerfeld	234.45	180.1	115.7
		% Melted	0.8	15	70.8
		Scope (cm)	7	9	11.9

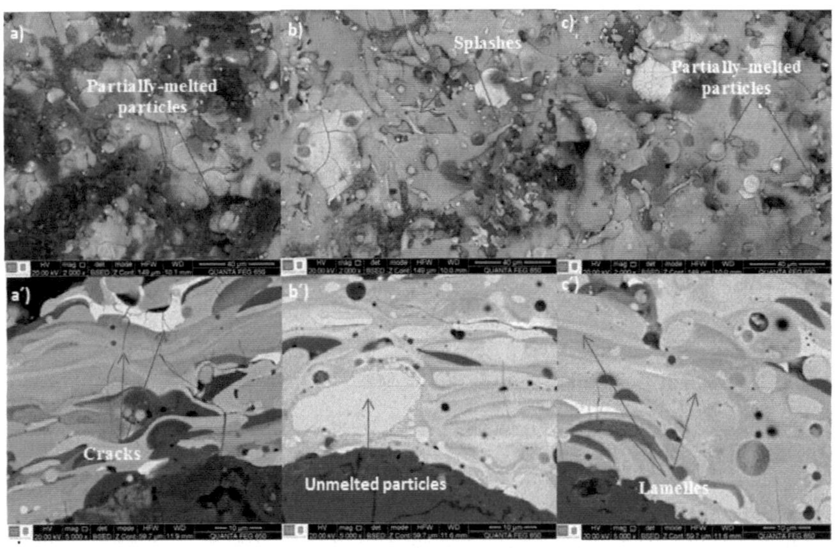

Figure 3. Morphology of the coatings. (a), (b) and (c) correspond to R1, R2 and R3 with a magnification of 2000X in the superficial part of the coatings. (a'), (b') and (c') it corresponds to a magnification of 5000X of the transverse surface of them.

The thickness of the coatings and his adhesion are recorded in the Table 4. The adhesion results show cohesive faults taking away part of substrate.

Table 4. Physical properties of the coatings.

Coating	Thickness (μm)	Adhesion Lb/in^2 (Psi)
R1	340 ± 27	1060 ± 183
R2	371 ± 87	1130 ± 60
R3	395 ± 66	1060 ± 162

4. Conclusions

The Software Jets et Poudres, allows a good approximation to obtain the coatings projection parameters of agglomerated powders of zirconia-alumina-ceria, for oxyacetylene flame. The conditions that predict the obtaining of homogeneous coatings and well adhered to the substrate correspond to a distance projection of 9cm, using an oxyacetylene flame oxidizer. The optimum K-Sommerfeld value for this type of coatings varies between 47.3 to 50 according to the results of the software and its validation.

Acknowledgments

GIPIMME-GIMACYR Group "Universidad de Antioquia".

References

[1] Fauchais P L, Heberlein JVR and Boulos M 2014 *Thermal spray fundamentals* (New York: Springer)
[2] Ferrer M, Moreno M, Vargas F, Peña G, Vera E and Pineda Y 2014 Depósito y caracterización de recubrimientos cerámicos de Al_2O_3 sobre acero 1020 por termorrociado *Rev. Col. Mat* **5** 317-322
[3] Muñoz J 2012 Una visión general de los cerámicos de circona – estructura, propiedades y aplicaciones *Rev. Colom. Mater.* **2** 1–18
[4] Pawlowski L 2008 *The science and engineering of thermal spray coatings, Second edition* (Great Britain: John Wiley & Sons, Ltd.) p 656
[5] Jets&Poudres 2009 *Software assistant for thermal spray coating* (France: Université de Limoges) (http://jets.poudres.free.fr/en/downloadupdates.html)
[6] Mundo C, Tropea C and Sommerfeld M 1997 Numerical and experimental investigation of spray characteristics in the vicinity of a rigid wall *Exp. Therm. Fluid Sci* **15(97)** 228–237
[7] Mundo C, Sommerfeld M and Tropea C 1995 Droplet-wall collisions experimental studies of the deformation and breakup process *Int. J. Multiph. Flow* **21**(2) 151–173
[8] Gonzalez A, Henao J, Diaz AF, López E and Vargas F 2013 Influencia de los parámetros de proyección térmica en la microestructura de los recubrimientos de circona-alúmina y circona-ceria usados como barreras térmicas *Rev. Latinoam. Metal. y Mater.* **33(2)** 272–81
[9] Cadavid E, Parra C, Soto D and Vargas F 2015 Estudio termo-físico de llamas oxiacétilenicas utilizadas en la proyección térmica *Memorias del VIII Congreso Internacional de Materiales (Paipa)* (Colombia: Universidad Pedagógica y Tecnológica de Colombia)
[10] Sommer F, Walcher H, Kern F, Maetzig M and Gadow R 2014 Influence of feedstock preparation on ceramic injection molding and microstructural features of zirconia toughened alumina *J. Eur. Ceram. Soc.* **34(3)** 745-751
[11] Arias J, Vargas F and Lopez E 2014 Aglomeración de partículas cerámicas nanométricas mediante peletización en disco para uso en proyección térmica por llama oxi-acetilenica *Rev. Col Mat.* **5** 290-295
[12] Araque M, Peña G and Vargas F 2015 Mechanical and tribological performance of coated ceramic tiles with alumina by thermal spraying process *TecnLógicas* **18(35)** 125–135
[13] Fauchais PL, Vardelle M, Vardelle A, Bianchi L 1995 Parameters controlling the generation and properties of plasma spray zirconia coatings *Plasma Chem Plasma Process* **16** 99–125

IOP Conf. Series: Journal of Physics: Conf. Series **935** (2017) 012023 doi:10.1088/1742-6596/935/1/012023

Optimization of the parameters for obtaining zirconia-alumina coatings, made by flame spraying from results of numerical simulation

M Ferrer[1], F Vargas[2] and G Peña[1]
[1] Universidad Francisco de Paula Santander, San José de Cúcuta, Colombia
[2] Universidad de Antioquia, Medellín, Colombia

E-mail: marthayasmidfp@ufps.edu.co

Abstract. The K-Sommerfeld values (K) and the melting percentage (% F) obtained by numerical simulation using the Jets et Poudres software were used to find the projection parameters of zirconia-alumina coatings by thermal spraying flame, in order to obtain coatings with good morphological and structural properties to be used as thermal insulation. The experimental results show the relationship between the Sommerfeld parameter and the porosity of the zirconia-alumina coatings. It is found that the lowest porosity is obtained when the K-Sommerfeld value is close to 45 with an oxidant flame, on the contrary, when superoxidant flames are used K values are close 52, which improve wear resistance.

1. Introduction

One of the most used materials to protect metal structures in aggressive temperature conditions and wear is zirconia (ZrO2) due to its low thermal conductivity $(0.8 - 1.5W/mK)$, however, the low thermal conductivity of zirconia limits to a great extent its tribological applications, because of it reaches a high temperature of contact with another material, which induces thermal stresses in the coating reducing its mechanical properties. For this reason zirconia is mixed with alumina, which has high hardness as well as high thermal conductivity making it excellent for applications where there is wear, in order to give better structural properties to the coating rare earth oxides such as yttria and ceria are added, they stabilize the zirconia in the tetragonal and cubic phases at room temperature [1], [2].

The most used method for the protection of structures is the application of a thermal spray coating, with this technique it is possible to make coatings and repair worn parts by applying a wide variety of materials on substrates of different composition, this is generally done by plasma [3], however, the flame spraying can be used to project high melting point materials, when the temperature and length of the primary zone are known by different $(C_2H_2{:}\,O_2)$ ratios (see Figure1) [4].

The software Jets et Poudres developed by the University of Limoges France, it allows to know all the thermal and kinetic history of the particle as it passes through the flame, by the values of temperature, melting percentage, and K-Sommerfeld number among others, data that facilitates the choice of parameters to obtain coatings [5-7].

K-Sommerfeld was described by Kappel and Mundo when they studied the impact of water droplets and ethanol on cold surfaces, the behaviour of a particle upon impact with a surface that may be the substrate or a previously deposited coating layer. Mundo et al, suggest that by number (K) it is possible to predict the interaction between the particle and the substrate, they define that if K is between 3 and 57.7 [8], the particle sticks on the substrate to form a disk-shaped splat and if K is less than 3, the particle will bounce after impact with the substrate, while if K exceeds 57.7 the particle will splashes after impact with the substrate, therefore, part of the material will not be deposited. The K-Sommerfeld is given by

$$K = We^{1/2}Re^{1/4} \quad (1)$$

Where We and Re are the dimensionless numbers of Weber and Reynolds, and which represent the relationship between the inertia of the molten particle with respect to its superficial tension and viscosity, expressed more specifically as:

$$K = [\rho v d/\mu]^{1/2}[\rho\mu^2 d/\sigma]^{1/4} \quad (2)$$

From where,ν, d, μ,y σ, are the density, velocity, diameter, viscosity and superficial tension of the drop (molten particle) that impacts the surface [8].

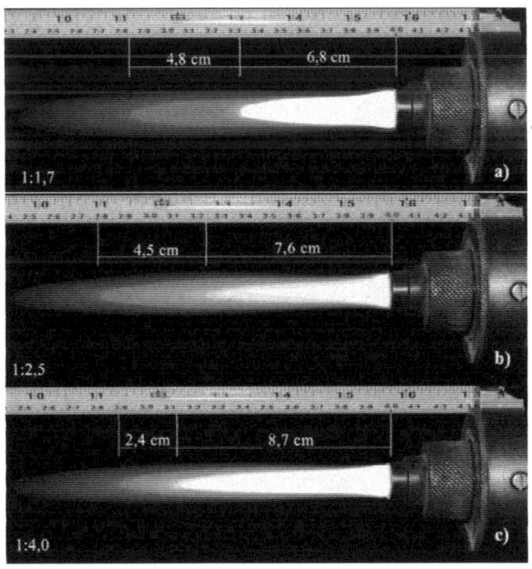

Figure 1. Length of the primary zone in a neutral oxyacetylene, oxidizing and super oxidant flame, respectively [4].

2. Materials and methods

Castolin-Eutectic commercial reference powders MetaCeram 25088™ are used to make of the coatings. A ceramic powder of ZrO_2-36 wt% Al_2O_3 is used to take the advantage of the good thermal properties of zirconia and the tribological properties of alumina. The chemical composition of the powders is obtained by X-ray fluorescence (XRF). The size distribution of the particles is determined by laser dispersion (Master Size 2000E); meanwhile, the porosity was determined by an optical microscopy images (OM) using Image-J software.

A numerical simulation was performed using the Jets et Poudres software [9], in order to find the parameters that allow to obtain homogeneous coatings with low porosity and good resistance to wear. A zirconia-alumina particle of less than 15μm to d_{10} of the distribution width of the powders used, which were made to interact with three flames, neutral, oxidant and superoxidant, for a projection distance of 8, 9 and 10cm. The thermal properties of the particles were obtained from the databases and taking into account the rule of the mixtures for compounds [10]. From the results of the simulation the parameter K-Sommerfeld and the melting percentage were given and they were used to choose the parameters to obtain the coatings.

The coatings are deposited using an Areste I chamber developed by the GIPIMME Group of the "Universidad de Antioquia (Colombia)", which is equipped with a Castolin-Eutectic Terodyn 2000 torch and a RAYTEK infrared pyrometer to measure the surface temperature of the substrate, as well as the electromechanical systems that control the speed of the torch and the sample holder.

3. Results and analysis

3.1. Characterization of raw material

According to the chemical analysis performed on the starting powder, it is established that it is composed mainly of 59.72wt% - ZrO_2, 35.7wt% - Al_2O_3 0.94wt% - HfO_2, 0.25wt% Y_2O_3 and others.

The particle size distribution obtained by dispersion laser for the zirconia-alumina particles between 15 and 95μm with an average size of 50μm, this shows a suitable average size to obtain ceramic coatings by flame spraying (FS) [11]. The analysis of the morphology of the powders is performed using SEM micrographs. In Figure 2, the powder morphology of zirconia (ZrO_2)-alumina (Al_2O_3) can be seen; it is made of spherical, micrometer, agglomerated and sintered particles, and also alumina nanoparticles (dark areas) and zirconia (light areas) can be seen [12], (see the cross section of a microsphere).

The analysis phase is performed by X-ray diffraction, (Figure 3) using the High Score Plus software, the presence of two phases is detected: the first with 63.5wt% of $m - Zr_4O_8$ and the second with 36.5wt% of $\alpha - Al_{12}O_{18}$.

Figure 2. Micrograph powders zirconia (ZrO_2)-alumina (Al_2O_3).

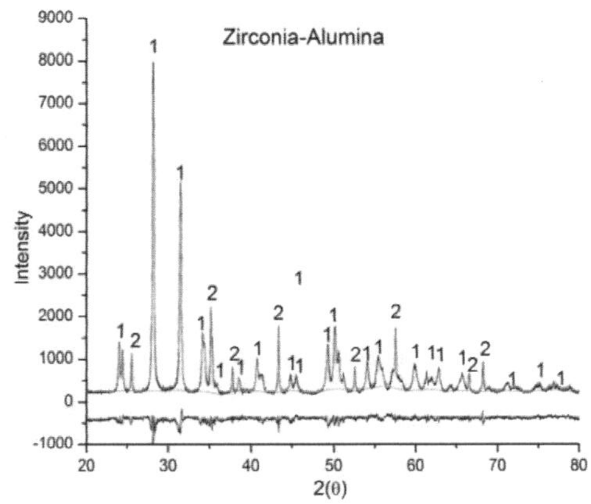

Figure 3. XRD spectrum of zirconia-alumina powders, used to make the coatings $1. m - Zr_4O_8$, $2. \alpha - Al_{12}O_{18}$.

3.2. Results of the simulation

The results of the simulation performed with the Jets et Poudres software for each of the conditions are shown in Table 1. As can be observed, for the neutral and oxidant flames at the projection distances of 8, 9 and 10cm and for the superoxidant flame at a distance of 10cm we obtain Sommerfeld values that are in the range between 3-57.7 condition in which the particle is deposited without bouncing, or forming splashes [5]. However, the melting percentage reached for the particles in the neutral flame is very low with respect to that reached for oxidant and superoxidant flame at a distance of 9 and 10cm. It is important to keep in mind that the particles are composed mainly of zirconia having a low coefficient of thermal conductivity of 0.8-1.5W/mK, making it difficult to diffuse the heat and its melting. In addition, in a neutral flame obtained from a ratio of oxygen and acetylene (C2H2: O2) equivalent to 1:1.9 by volume, a maximum temperature of 3113°C is reached, with a length of the primary reaction zone of about 6.8 cm, whereby the particles do not melt when the substrate is placed at 8, 9 or 10cm. On the other hand, to obtain an oxidizing flame an excess of oxygen is added in a ratio of 1:2.9 by reducing the flame temperature meanwhile having a longer primary reaction zone of about 7.6cm in length, then the particles spent a longer time in the hottest area of the flame and a short time in the secondary zone, reaching a higher melting percentage. In the case of a superoxidant flame where a greater amount of oxygen is added in the ratio 1:4.2, the flame temperature decreases and the length of the hot zone reaches a distance of about 8.7cm [4], thus, particles traveling 8cm in this flame do not reach a sufficient melting percentage to generate a homogeneous coating. According to this, it can be concluded that the condition in which the zirconia-alumina particles are best melted is when an oxidizing flame is used, where, the particles have a longer time in the hot zone, and this is why an increase in projection distance increases in the melting percentage of the particles. For the zirconia-alumina coatings it is evident that the K-Sommerfeld number is influenced by both the flame type and the projection distance.

Table 1. K- Sommerfeld number and melting percentage from the results of the simulation.

Flame	Parameters	Projection distance		
		8 cm	9 cm	10 cm
Neutral (1:1.9)	K-Sommerfeld	36.8	32.7	32.9
	Melting %	4.6	28	26
Oxidizing (1:2.9)	K-Sommerfeld	49	45	45
	Melting %	31	78.2	78.2
Super-oxidant (1:4.2)	K-Sommerfeld	75.2	63.3	52.1
	Melting %	37	100	83.4

3.3. Projection parameters
In order to create the coatings three flames were chosen, and three projection distances. The torch moves with a linear velocity of 0.59cm/s, the sample holder rotates at 116rpm, the flow of the powder remains constant at 9g/min. The coatings are called N8, N9 and N10, for those obtained with neutral flame at 8, 9 and 10cm in length, O8, O9 and O10 for those obtained with oxidizing flame and SO8, SO9 and SO10 to those obtained with superoxidant flame.

3.4. Validation
It is important to keep in mind that the materials given have wide distribution range in its size, which varies from about 15 to about 95μm, making it difficult to melt all the particles. Figures 4, 5 and 6 show SEM images of the surface of the coatings, where the obtained results in the simulation for both the K Sommerfeld and the melting percentage are show. The coatings N8, N9 and N10 (Figure 6) have a surface with non-melted particles and very few partially melted particles, which agrees with the melting percentage of the particles and the value of K-Somerfeld obtained in the simulation. The melted particles with an acceptable value within the range of Sommerfeld are the smallest that reach the melting temperature. The coatings O8, O9 and O10 (Figure 5) whose K-Sommerfeld is in the range of 45 to 49 and their melting percentage between 31 and 78.2% have a structure with well fused splats and some splashes due to excessive thermokinetic energy reached by the particles. The SO9 and SO10 coatings show a surface with melted particles and many splashes as a result of larger number of small particles achieved enough thermokinetic energy to melt and produce splashes at the time of impact. The SO8 coating, on the other hand, has large number of unmelted particles joined by smaller, well-fused particles. This is because the temperature and residence time of the particles in the flame are not enough for the larger ones to melt, these results are in agreement with those predicted in the simulation where the percentage of fusion is 37%.

The predominant phases are monoclinic zirconia and alpha alumina which are given from the starting powders, indicating that there were particles that did not reach the melting point, however, the cubic and tetragonal phases of zirconia are also there, which are retained due to the presence of yttrium and hafnium oxides. The alpha and beta phases of the alumina correspond to particles that melted and solidified (metatestable or transition eta phase). There are two areas corresponding to the amorphous part, the first in 22-37 (2θ) and the second in 52-65 (2θ). According to Fauchais *et. al.* [13], the amorphous phases are always presented and are formed by the high cooling rates up to $600 \times 10^6\ K/s$, in a conventional plasma spray process using micron sized particles (see Table 2).

Figure 4. The morphology of the surface view of the coatings obtained with neutral flame.

Figure 5. The morphology of the surface view of the coatings obtained with oxidizing flame.

Figure 6. The morphology of the surface view of the coatings obtained with superoxidizing flame.

Table 2. Phases in the zirconia (ZrO_2)-alumina (Al_2O_3) coatings.

Sample	Phases present in the coatings					
	$m-ZrO_2$	$t-ZrO_2$	$c-ZrO_2$	$\alpha-Al_2O_3$	$\gamma-Al_2O_3$	$\eta-Al_2O_3$
O8	51.7	6.4	21.9	15.4	3.2	1.4
O9	51.6	4.8	26.9	13.9	2.3	1.5
O10	49.0	15.6	9.6	15.2	9.8	1.7
SO8	55.3	14.7	4.8	16.8	2.3	6.1
SO9	48.7	9.2	23.7	11.3	6.2	0.8
SO10	47.1	7.0	27.2	6.1	12	0.7

The porosity percentage and the mechanical properties of hardness and wear resistance of the coatings were evaluated for samples O8, O9, O10, SO8, SO9 and SO10 which show a K-Sommerfeld value between 45 and 75.2. Table 3 shows the porosity of the coatings evaluated, where it can be observed that the samples O8, O9 and O10 corresponding to a K-Sommerfeld value between 45 and 49, they have a lower percentage of porosity. The SO8, SO9 and SO10 coatings with a K-Sommerfeld between 52.1 and 75.2 have a higher percentage of porosity due to the presence of splashes predicted by the value of Sommerfeld. With regard to microhardness it is important to keeping mind that this is affected by both the porosity and the presence of hard phases as the corundum [5,11], for example the coating SO8 is the most porous and is the one with the highest percentage of the phase $\alpha-Al_2O_3$, 16.8%. In the case of the SO10 coating, it shows the lowest porosity percentage, and has the lowest percentage of $\alpha - Al_2O_3$ of 6.1%, this makes the hard phase with porosity and the effective hardness similar in all the Cases studied.

Table 3. The mechanical properties of the coatings zirconia (ZrO_2)-alumina (Al_2O_3).

Sample	Porosity [%]	Microhardness Vickers [GPa]	Wear rate $\times 10^{-4}$ [mm^3/N.m]
O8	5.3 ± 0.9	7.8 ± 0.6	1.3 ± 0.07
O9	5.1 ± 0.9	8.4 ± 0.9	2.1 ± 0.20
O10	5.4 ± 1.2	8.6 ± 0.3	2.1 ± 0.46
SO8	9.1 ± 1.1	8.0 ± 0.4	1.2 ± 0.13
SO9	6.9 ± 1.1	8.3 ± 0.4	1.7 ± 0.31
SO10	5.8 ± 0.9	8.2 ± 0.6	1.3 ± 0.20

4. Conclusions

The simulation developed with the software Jets et Poudres allows to predict the thermokinetic behaviour of a particle of zirconia-alumina when interacting with an oxyacetylene flame, by means of the value K-Sommerfeld and the melting percentage. From these results it is possible to establish the most suitable projection parameters in order to obtain coatings with certain characteristics intended for specific applications, which makes the simulation a useful tool that optimizes the time, energy and materials needed to obtain homogeneous coatings with good mechanical properties.

Acknowledgments

A Sustainability Strategy 2014-2015 at the "Universidad de Antioquia" allowed GIPIMME group participation in the development of the coatings.

References

[1] Muñoz J 2012 Una visión general de los cerámicos de circona–estructura, propiedades y aplicaciones *Rev. Colomb. Mater.* **2** 1–18

[2] Garvie R C, Hannink RH and Pascoe R T 1975 Ceramic steel? *Nature* **258** 703–704

[3] Gonzalez A, Henao J, Diaz A F, López E, and Vargas F 2013 Influencia de los parámetros de proyección térmica en la microestructura de los recubrimientos de circona-alúmina y circona-ceria usados como barreras térmicas *Rev. Latinoam. Metal. y Mater.* **33** 272–281

[4] Cadavid E, Parra C, Soto D and Vargas F 2015 Estudio termo-físico de llamas oxiacétilenicas utilizadas en la proyección térmica *Memorias del VIII Congreso Internacional de Materiales (Paipa)* (Colombia: Universidad Pedagógica y Tecnológica de Colombia)

[5] Fauchais P L, Heberlein J V and Boulos M 2014 *Thermal spray fundamentals* (New York: Springer)

[6] Berndt E C 2001 Thermal Spray 2001 New surfaces for a new millennium *Int. Therm. Spray Conf.* **14**

[7] Mundo C, Tropea C and Sommerfeld M 1997 Numerical and experimental investigation of spray characteristics in the vicinity of a rigid wall *Experimental thermal and fluid science* **15(97)** 228–237

[8] Mundo C, Sommerfeld M and Tropea C 1995 Droplet-wall collisions: Experimental studies of the deformation and breakup process *Int. J. Multiph. Flow.* **21(2)** 151–173

[9] Jets&Poudres 2009 *Software assistant for thermal spray coating* (France: Université de Limoges) (http://jets.poudres.free.fr/en/downloadupdates.html)

[10] McWilliams A 2016 *High-performance ceramic coatings: Markets and technologies* (USA: BccResearch)

[11] Pawlowski L 2008 *The science and engineering of thermal spray coatings* second edition (Great Britain: John Wiley & Sons, Ltd.) p 656

[12] Fukumoto M, Tanaka Y, and Nishioka E 2004 Flattening Problem of Thermal Sprayed Particles *Mate. Sc. Forum* **449–452** 1309–1312

[13] Fauchais P L, Vardelle M, Vardelle A, Bianchi L and Léger A 1995 Parameters controlling the generation and properties of plasma spray zirconia coatings *Plasma Chem. and Plasma Process* **16** 99–125

IOP Conf. Series: Journal of Physics: Conf. Series **935** (2017) 012024 doi:10.1088/1742-6596/935/1/012024

Wear evaluation of flank in burins of high speed steel modified with titanium ions

J E Caballero[1] and E D V-Niño[2,3]
[1] Universidad Francisco de Paula Santander, Cúcuta, Colombia
[2] Universidad Industrial de Santander, Bucaramanga, Colombia
[3] Foundation of Researches in Science and Technology of Materials, Colombia

E-mail: deydannv@gmail.com

Abstract. This report shows the results obtained researching the flank wearing resistance performed by the high-speed steel (HSS) burins without any surface treatment (reference substrate) and others with surface treatment based on Titanium ions. The flank wearing was carried out by means of an industrial process by chip removal with repetitive tests of dry finished turning of AISI/SAE 1045 steel bars. The useful service life of the burins was evaluated according to ISO 3685:1993, and it was found that the burins treated with Titanium ions showed an increase in the flank wearing resistance with respect to the ones used as reference.

1. Introduction

The manufacturing industry generally uses one-point cutting tools and machine-tools in its production processes, which rapidly deteriorate under severe operating conditions, causing millions of losses that can account for up to 25% of production costs [1], so the application of coatings on the surface of cutting tools has had a constant development during the last decades, optimizing production without significantly increasing costs and especially improving surface properties such as wearing resistance, lower coefficient of friction, resistance to oxidation, among others [2].

According to the above, an alternative technique of surface modification of metallic materials capable of achieving unique physicochemical properties, which by traditional metallurgical methods cannot be obtained, is the three-dimensional ion implantation (3DII) of metal species [3-7].

In the present investigation, by means of a process of manufacture by chip removal of turning of dry finishing [8,9], according to ISO 3685: 1993 [10] is analysed the behaviour of the useful half-life in service of the burin of HSS steel not modified superficial (reference) with respect to superficially treated with Ti ions.

2. Experimental

The flank wearing tests was realized using Impact Tools on HSS burins of 3/8" square side for 4" long (see Table 1).

The surface of the steel burins was modified superficially with ions of Ti during 4 and 8 minutes by electric hybrid pulsed high voltage and electric arc to low pressures in the reactor JUPITER (Joint Universal Plasma and Ion Technologies Experimental Reactor) [11]. The burins before the surface treatment were subjected to a sputtering process (generated with a pulsed electric discharge of 5kV) with argon gas for 15 minutes.

Table 1. Total number of burins used in mechanizing testing.

Burin type	Number of burins
Of reference (untreated)	3
Modified with Ti for 4 minutes	3
Modified with Ti for 8 minutes	3

The flank wearing tests were carried out according to ISO Standard N° 3685:1993 [10] and the cutting parameters that define the criterion of the flank wearing of HSS Steel burins [12], are stated in the Table 2. The material selected to machining was bars of Carbon steel type AISI/SAE 1045 (CASH CK45) [8,9].

Table 2. Cutting parameters used in the machining repetitive tests.

Variable	Variable type	Measure
Cutting speed	Independent	20m/min on average
Advance	Independent	0.1mm/rev
Depth	Independent	1mm
Mechanized time	Independent	Intervals of 2 minutes up to determining that the wearing is lightly a minor to 0.300mm (criterion of uniform wearing according to ISO Standard 3685:1993)
Material of the tool	Independent	Burin of steel standard HSS not well-established and implanted with ions of Ti and N
Wearing of the tool	Dependent	Flank wearing measured in millimetres
Material to turning	Independent	Steel AISI/SAE 1045

Once the type of burin, the material and size of specimen to be machined, the surface treatment and the cutting parameters were selected, repetitive tests of orthogonal cutting in dry were made on lathe every 2 minutes using reference burins and every 8 minutes with metallic Titanium species modified burins.

3. Results and discussion

Figure 1 shows the burins before and after the process of superficial modification observing that those implanted with ions of Titanium got silvered violet coloration, due to the alloy formed by the metallic species into the atmosphere during the 3DII discharge process and the interaction the material components of the material of the substratum.

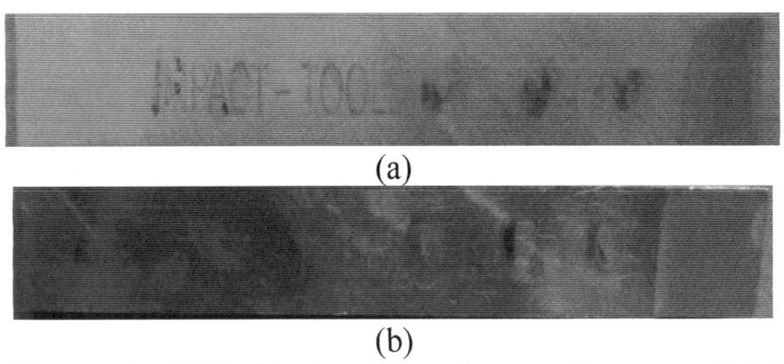

(a)

(b)

Figure 1. HSS Burin, (a) reference (no superficial modification and (b) modified with Titanium ions.

The results shows in the Table 3 and 4, of the flank wearing of burins not modified and modified superficially with Titanium ions during 4 minutes, were obtained making a maximum of seven (7) passes, whereas the results with Titanium ions during 8 minutes were achieved with a maximum of twelve (12) passes, with machining time each pass of four times bigger than the time established in the burins used as reference and equally to the established in the modified ones with Titanium during 4 minutes. The number of burin passes during the mechanized of the bar of AISI/SAE 1045 steel is determined by the instant in which the cutting tool came to the limit of the criterion VB_B = 0.300 mm that determines the end of the useful life in service according to the Standard ISO3685:1993 [8,10,12,14].

Table 3. Measures registered in the flank wearing test of reference burins (not modified superficially).

Time (min)	Wearing (mm) Burin 1	Burin 2	Burin 3	Average wearing (mm)
2	0.088	0.089	0.088	0.088
4	0.098	0.108	0.109	0.105
6	0.130	0.126	0.135	0.130
8	0.149	0.149	0.147	0.148
10	0.178	0.186	0.183	0.182
12	0.210	0.225	0.220	0.218
14	0.271	0.277	0.274	0.274

Table 4. Measures registered of the flank wearing tests of burins modified superficially with Titanium ions during 4 and 8 minutes.

Turning Time (min)	Wearing (mm) Burin Ti 4min			Average wearing (mm)	Wearing (mm) Burin Ti 8min			Average wearing (mm)
	Burin 1	Burin 2	Burin 3		Burin 1	Burin 2	Burin 3	
8	0.193	0.194	0.195	0.194	0.110	0.108	0.121	0.113
16	0.228	0.229	0.230	0.229	0.133	0.127	0.143	0.134
24	0.239	0.240	0.241	0.240	0.145	0.138	0.157	0.147
32	0.265	0.268	0.271	0.268	0.152	0.158	0.186	0.165
40	0.292	0.295	0.289	0.292	0.162	0.169	0.198	0.176
48	0.300	0.296	0.292	0.296	0.171	0.175	0.209	0.185
56	0.296	0.298	0.300	0.298	0.300	0.188	0.218	0.196
64		-		-	0.190	0.197	0.227	0.205
72		-		-	0.209	0.204	0.241	0.218
80		-		-	0.215	0.220	0.269	0.235
88		-		-	0.233	0.239	0.278	0.250
96		-		-	0.250	0.255	0.304	0.270

In the Figure 2 appears the measures of the flank wearing average depending on the time of mechanized, and is observed that the flank wearing average of the burins treated superficially with Titanium ions during 8 minutes show the best performance in relation to the flank wearing of the burins used as reference. The superficially modified ones with Ti ions during 4 minutes, had an increase of the useful life service of approximately 585.70% and 71.43%, respectively, so the flank wearing burins with Ti ions during 8 minutes achieved an approximate reduction of 28.47% according to the criterion of wear (VB_B = 0.300 mm) [8,10,12,13], defined by the number of passes in relation with the flank not superficially modified. The results show in the Figure 2 were obtained mechanizing a steel bar AISI/SAE 1045 of diameter 50.8mm×200mm, and making mediations with a Mitutoyo

Digital Caliper with an accuracy of 0.001mm, and they are the graphical representation of the Tables 3 and 4.

The flank wearing average of the burins used as reference showed an intensive growth as the time of turning increases. According to the total time of turning it was an increase of the useful life in service of 300% in the flank of the burins superficially modified with Titanium ions during 4 minutes with regard to the burins of reference. Nevertheless, according to the criterion of wearing about the number of passes, the flank wearing of the burins not modified registered an approximate reduction of 8.76% with regard to the flank of the burin modified with Ti ions 4 minutes, so the flank wearing with Ti ions 4 minutes was major. In spite of that the machined time established for a pass was four times more than the assigned to the burins not superficially treated.

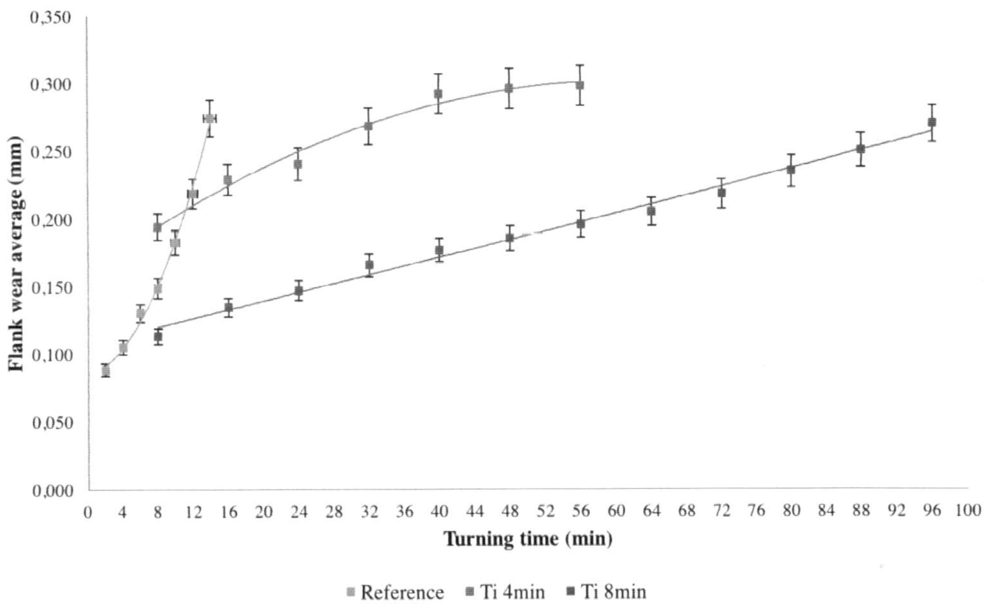

Figure 2. Flank wearing average with regard to turning time of reference and modified superficially with Titanium ions during 4 and 8 minutes burins.

The Figure 2 shows that the resistance of the flank wearing of the modified superficially burins with Titanium ions during 8 minutes improved significantly in comparison with the flank wearing burins with Titanium ions 4 minutes and the reference ones, which is why it becomes the most optimum and efficient surface treatment in the dry finishing turning process of AISI/SAE 1045 steel.

4. Conclusion

The repetitive tests on bars of Carbon steel type AISI/SAE 1045 getting in dry finished with machining on lathe, demonstrated that the flank wearing of high speed steel (HSS) burins superficially modified with Titanium ions during 8 minutes improved significantly the service useful life in comparison with the flanks of the burins treated with Titanium ions during 4 minutes and the ones used as reference (not modified), demonstrating that the covering with Titanium is a protection mechanism that might be implemented in the flank of the burins used in the manufacturing industry.

Acknowledgement

The authors express their gratitude to the Group of Investigation in "Física y Tecnología del Plasma y Corrosión (FITEK)" of the "Universidad Industrial de Santander", to the Laboratory of Machines and Tools of the "Universidad Francisco de Paula Santander" and to the Laboratory of Materials Resistance of the "Universidad EAFIT".

References

[1] Gómez Maryory Astrid, Gilberto Bejarano Gaitán, Osorio Jaime Alberto 2010 Mejoramiento de propiedades mecánicas y tribológicas en herramientas industriales mediante aplicación de recubrimientos multicapa de TiN/ZrN *Revista Facultad de Ingeniería* **54** 136-145

[2] Payan Harvey, Aperador William 2005 Estudio de la sinergia corrosión erosión de recubrimientos duros de TiN y CiN obtenidos sobre acero AISI 1045 *Sciencia Et Technica* **XIV(38)** 177-180

[3] Dougar V, Dulcé J, Tsygankov P A 2002 High voltage pulse discharge for ion treatment of metals *Rev. Sci. Instr.* **73(2)** 828

[4] V Niño Ely Dannier, Dulce Moreno Jaime, Garnica Hernán Alfonso, Dugar-zhabon Valeriy, Peña Rodríguez Gabriel y Tsygankov Petr 2012 Aplicación de descargas simultáneas de alto voltaje y arco eléctrico para el tratamiento superficial avanzado de metales *Revista Iteckne* **9(1)** 14-20

[5] Dugar-Zhabon Valeriy Dondokovich, Dulcé Moreno Hector Jaime, Garnica Villamizar Hernán Alfonso, Valbuena Niño Ely Dannier 2012 A new method for surface modifications of carbon steels and alloys *Materials Research* **15(6)** 969-973

[6] Dougar-Jabon V D, Dulce Moreno H J, Tsygankov P A 2003 Implantación iónica *Revista Respuestas* **8** 19-30

[7] Dulce H J 2015 Descargas eléctricas y sus aplicaciones *Eco. Mat.* **6(1)** 6-21

[8] Ely Dannier V-Niño, Jorge Enrique Caballero, Jesús Eduardo Contreras Naranjo 2016 Evaluación del desgaste de flanco en buriles de acero rápido implantado con iones de titanio y nitrógeno *Universidad, Ciencia y Tecnología* **20(80)** 106-113

[9] Hernández-González, W, Pérez-Rodríguez R, Zambrano-Robledo P, Guerrero-Mata M, Dumitrescu L 2011 Estudio del desgaste del flanco de carburos recubiertos y cermet durante el torneado de alta velocidad en seco del acero AISI 1045 *Revista de Metalurgia* **47(3)** 262-272

[10] Norma ISO 3685 1993 *Tool-life testing with single-point turning tools* (Switzerland: International Organization for Standardization)

[11] Dougar-Jabon, V D, Castro B J, Dulce Moreno H J, Tsygankov P A 1999 Device JÚPITER for ion implantation *Revista Colombiana de Física* **31(2)** 181-184

[12] Timings Roger 2005 Manual del ingeniero de taller (México: Limusa)

[13] Omveer Singh, Hitendra K Malik, Raj P Dahiya, P K Kulriy 2017 Tuning of mechanical and structural properties of 20 MC 5 steel using N ion implantation and subsequent annealing *Journal of Alloys and Compounds* **710** 253-259

[14] A Hartwig, M Decker, O Klein, H Karl 2015 Stoichiometric titanium dioxide ion implantation in AISI 304 stainless steel for corrosion protection (Germany: University of Augsburg)

IOP Conf. Series: Journal of Physics: Conf. Series **935** (2017) 012025 doi:10.1088/1742-6596/935/1/012025

Scanning electron microscopy of heat treated TiO_2 nanotubes arrays obtained by anodic oxidation

D I Naranjo[1], **S J García-Vergara**[1] **and S Blanco**[1]
[1] Universidad Industrial de Santander, Bucaramanga, Colombia

E-mail: siblanco@uis.edu.co

Abstract. Scanning electron microscopy was used to investigate the anatase-rutile transformation of self-organized TiO_2 nanotubes obtained on titanium foil by anodizing and subsequent heat treatment. The anodizing was carried out at 20V in an 1% v/v HF acid and ethylene glycol:water (50:50) electrolyte at room temperature. The anodized samples were initially pre-heat treated at 450°C for 4 hours to modify the amorphous structure of TiO_2 nanotubes into anatase structure. Then, the samples were heated between 600 to 800°C for different times, in order to promote the transformation to rutile structure. The formation of TiO_2 nanotubes is evident by SEM images. Notably, when the samples are treated at high temperature, the formation of rutile crystals starts to become evident at the nanotubes located on the originally grain boundaries of the titanium. Thus, the anatase - rutile transformation has a close relationship with the microstructure of the titanium, more exactly with grain boundaries.

1. Introduction

TiO_2 nanotubes are a material whose architecture offers great internal surface area and their ordered nature makes them an excellent medium for the transference of electrons between the different interfaces on which it is desired to work [1]. In addition, TiO_2 is an oxide with unique properties including good photocatalytic behaviour, ability to detect some substances in the medium, biocompatibility and could be used for photocurrent generation [1,2]. The synthesis of TiO_2 nanotubes can be performed by different routes such as electrochemical anodization, sol-gel synthesis, hydrothermal treatment and vapor deposition [3-6]. However, it is the electrochemical synthesis that has allowed better manipulation of the process parameters to improve the characteristics of the final product [1,8]. The application of TiO_2 nanotubes is closely related to their crystalline structure (anatase, rutile or bruquita) for this reason the understanding of the polymorphic transformations by post-anodized heat treatments has been explored widely [9-12]. The TiO_2 nanotubes obtained by anodizing are amorphous. The first transformation that takes place in the arrangement of nanotubes is from the amorphous structure to anatase. The anatase-rutile transformation takes place at temperatures between 600 to 900°C [13-15]. Regarding the transformation mechanism, Yu et. al. proposed a process of progressive transformation as a function of temperature, where rutile is initially formed at the metal-oxide interface at 600°C [13]. With the incrementation of the temperature to 700°C, the transformation occurs at the top of the nanotubes and finally the total transformation and collapse of the tubular structure occurs when the temperature rises to 900°C. On the other hand, Fang et al. suggested that the transformation starts at the metal-oxide interface at 600°C and progressively evolves from the base of the TiO_2 nanotubes to the upper part at 750°C [14]. Clearly, the phase transformation processes are affected by the temperature and duration of the heat treatment, since its parameters affect the thermodynamics and the kinetics of the transformation. The present work studies the effect of these parameters on the morphology of the TiO_2 nanotubes arrays formed by anodizing in a fluoride containing electrolyte.

2. Experimental

2.1. Sample preparation

TiO_2 nanotubes arrays were obtained by anodizing of commercially pure titanium foil with a surface area of 4cm^2. The titanium samples were polished with alumina paste to obtain a mirror-like surface finish. Subsequently, the specimens were washed with bi-distilled water and cleaned in ethanol using an ultrasonic bath for 15min. The anodizing was carried out at 20 V in an electrolyte comprising 1% v/v HF acid and ethylene glycol:water (50:50) for 90min at room temperature. The anodizing was performed in a cell with two (2) electrodes, where the titanium samples acting as the anode and a platinum wire as the cathode in a DC VOLT power supply GP5003.

2.2. Heat treatment

The anodized samples were initially pre-heat treated at 450°C for 4h in order to modify the amorphous anatase structure of TiO_2 nanotubes. The pre-heat treatment was done to ensure that subsequent heat treatments only promoted anatase - rutile transformation. Then, the anodized and pre-heat treated samples were heated to 600, 650, 700 and 800°C for 40, 60 120, 180, 240, 300, 360, 420 and 480min, in order to study the anatase-rutile transformation. The heat treatments were performed in a STableTemp Coler Parme furnace using an oxidizing atmosphere.

2.3. Characterization of the coatings

The morphology of TiO_2 nanotubes arrays obtained by anodizing and modified with subsequent heat treatments was characterized using a QUANTA 650 FEG-ESEM scanning electron microscope, operated at 25kV.

3. Results and discussion

3.1. SEM evaluation of morphology changes promoted by heat treatment

Figure 1 shows a homogeneous distribution of the TiO_2 nanotubes arrays formed by anodizing along the surface. The nanotubes have an average diameter of 100nm and a length between 300 and 400nm.

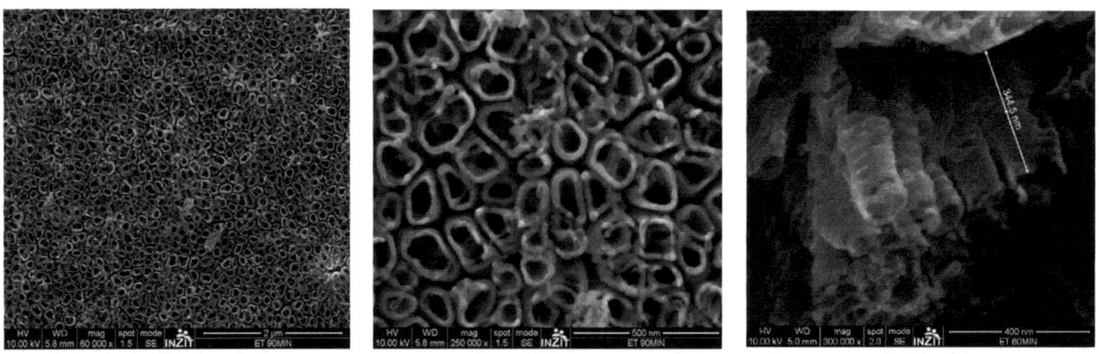

(a) (b) (c)

Figure 1. Scanning electron micrographs of titanium anodized in 1% v/v HF+(50/50% v/v) water/ethylene glycol electrolyte at 20V for 90min. (a)-(b) Surface and (c) Cross-section.

In our previous work [16] we confirm by SEM analysis that after heat treatment of the anodized samples at 450°C the morphology of the TiO_2 nanotubes arrays does not change compare with the

IOP Conf. Series: Journal of Physics: Conf. Series **935** (2017) 012025 doi:10.1088/1742-6596/935/1/012025

samples without heat treatment. Further, the heat treatment time does not affect the original morphology. At this temperature the amorphous TiO_2 generated by anodizing is transformed into anatase. When the calcination temperature increases to 800ºC, the nanotube arrays are completely destroyed and only dense rutile crystallites with size largest 100nm are observed [13]. The collapse of the tubular structure occurs because the anatase-rutile transformation is reconstructive, originating the rupture and reform of the bonds. The network parameters of the tetragonal crystalline network of the rutile are larger than the network parameters of the tetragonal crystalline network of the anatase [17].

(a) (b) (c)

Figure 2. Scanning electron micrographs of titanium anodized in 1% v/v HF+(50/50% v/v) water/ethylene glycol electrolyte at 20V for 90min, with or without heat treatment (HT) at 650ºC. (a) No HT, (b) HT for 60min and (c) HT for 480min.

The evolution of the anatase-rutile transformation was studied from the morphological changes of the TiO_2 nanotubes arrays after heat treatment at 650ºC. Figure 2(a) shows a homogeneous nanotube array before annealing. Figure 2(b) reveals a localized variation of the nanotube array in the shape of interconnected lines after 60min of heat treatment, which contrasts with a structure of collapsed nanotubes (Figure 2(c)), due to the reconstruction of its crystalline structure from anatase to rutile after of 480min of heat treatment. When comparing the pattern generated by the union of these interconnected lines, with the microstructure of the substrate, it is observed that these patterns are related with grain boundaries. This may suggest that the anatase-rutile transformation is favoured at the grain boundaries of the substrate.

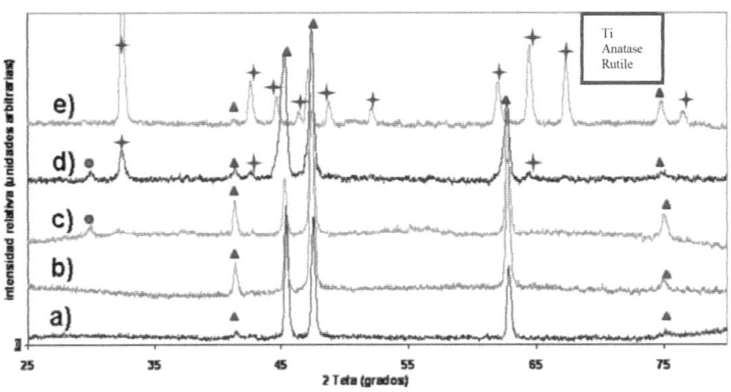

Figure 3. XRD patterns of titanium. Pure titanium without anodizing (a), titanium anodized in 1% v/v HF+(50/50% v/v) water/ethylene glycol electrolyte at 20V for 90min (b), titanium anodized as (b) and pre-heat treated at 450°C for 4 hours (c), titanium anodized as (b) and heat treated at 600°C for 4 hours and titanium anodized as (b) and heat treated at 700°C for 4 hours.

In order to support the above statement, regarding the appearance of rutile crystals in the TiO_2 nanotube array when subjected to different heat treatment times at 650°C, the X-ray diffractogram obtained for anodized and pre-heat treated samples at different conditions is presented in Figure 3. The

IOP Conf. Series: Journal of Physics: Conf. Series **935** (2017) 012025 doi:10.1088/1742-6596/935/1/012025

anatase - rutile transformation is confirmed above 450°C, as the signal corresponding to the anatase phase decreases as the signal of the rutile phase increases with increasing temperature.

3.2. Proposed evolution model

Based on the sequence of morphological changes that were evidenced at different times of heat treatment at 650°C on the TiO_2 nanotubes arrays obtained by anodizing, a mechanism for the anatase-rutile transformation is proposed from the analysis of the images obtained by SEM (Figure 4).

In the initial phase of heat treatment, a variation in morphology of the nanotubes array occurs preferentially in those located at the grain boundaries of the substrate (Figure 4(a)-(b)). This morphological variation is associated with the appearance of the first rutile crystals. As the heat treatment time increases, rutile crystals appear on the TiO_2 nanotubes located within the areas demarcated by the grain boundaries (Figure 4(c)-(d)), being clearly evident at 420min of treatment (Figure 4(e)-(f)). After 480min the anatase-rutile transformation reconstructs the morphology of the TiO_2 nanotubes arrays obtained by anodizing, shifting from amorphous TiO_2 nanotubes to rutile crystals [9,13,14].

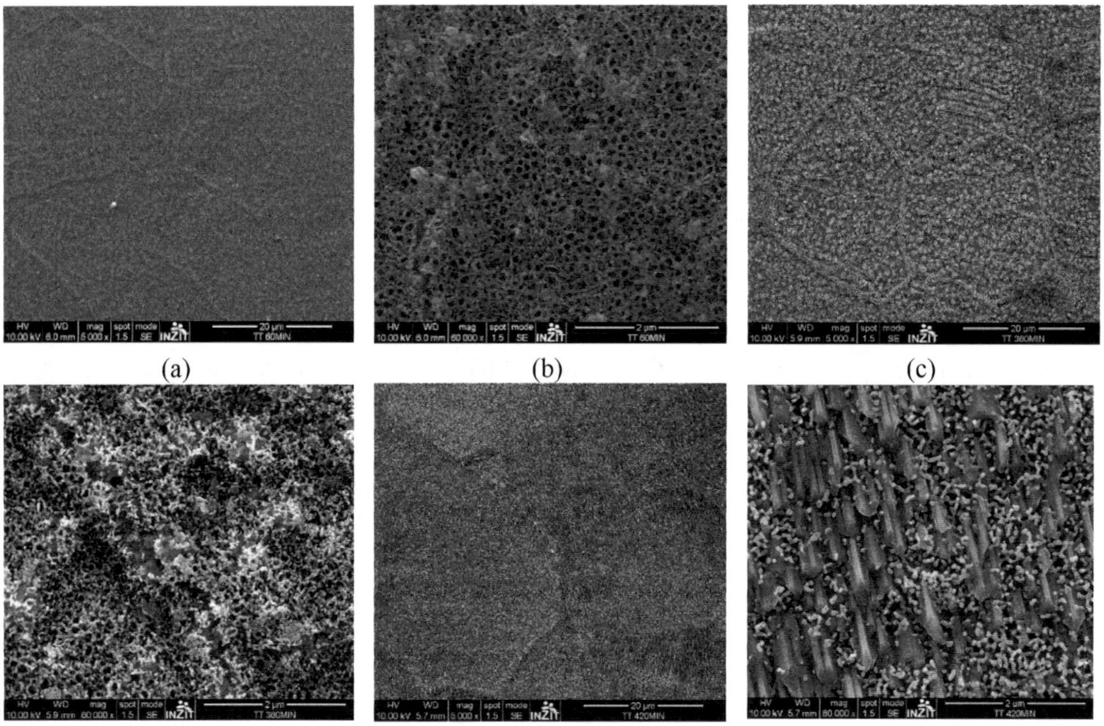

(d) (e) (f)

Figure 4. Scanning electron micrographs of titanium anodized in 1% v/v HF+(50/50% v/v) water/ethylene glycol electrolyte at 20V for 90min and then heat treated to 650°C for different times. (a)-(b) 60min, (c)-(d) 360min and (e)-(f) 420min.

Previous observations can be schematically represented in a 3-stage model presented in Figure 5.

a. Start of the anatase-rutile transformation on the TiO_2 nanotubes located at the grain boundaries of the substrate.
b. Progress of the anatase-rutile transformation on the TiO_2 nanotubes located inside the grains of the substrate.
c. Collapse of nanotubular structure and formation of rutile microcrystals.

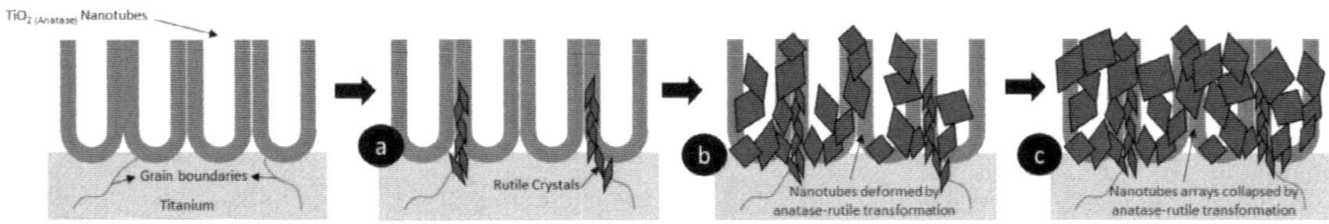

Figure 5. Squematic diagram of the proposed evolution model for anatase-rutile transformation in TiO_2 nanotubes at 650°C.

4. Conclusions

The thermal treatments of TiO_2 nanotubes arrays provided evidence that not only the crystal structure is modified with temperature, but also changes in the morphology of the anodic nanotubular layer. When the temperature does not exceed 450°C morphologically changes are not observed, but amorphous-anatase transformation is presented. At 650°C anatase and rutile phases are observed. This transformation is associated with a change in the network parameters, which generated degenerative morphological changes at the grain boundaries of the substrate. At higher temperatures, the TiO_2 undergoes a nearly complete transformation and the nanotubes are completely destroyed and well-defined rutile crystals are observed.

Acknowledgment

The authors are grateful for the financial support from Venezuelan: *Centro de Tecnología de Materiales, Fundación Instituto de Ingeniería,* and the contribution from researcher Vianney Carolina Quijada Marquez.

References

[1] Pang Y, Lim S, Ong H and Chong W 2014 *Applied Catalysis A: General* **481** 127-142
[2] Luo Q, Cai Q, Li X, and Chen X 2014 *J. Alloys and Compounds* **597** 101-109
[3] Fernandes J, Kohlrausch E, ShKhan S, Brito R, Machado G, Teixeira S, Dupont J and Leite Santos M J 2017 *Journal of Solid State Chemistry* **251** 217-223
[4] Brnardić I, Huskić M, Umek P and Holjevac Grgurić T 2013 *Ceramics International* **39(8)** 9459-9564
[5] Fan H, Zhang H, Luo X, Liao M, Zhu X, Ma J and Song Y 2017 *J. Power Sources* **357** 230-240
[6] Abdullah M and Kamarudin S 2017 *Renewable and Sustainable Energy Reviews* **76** 212-225
[7] Gong D, Grimes C, Varghese O, Hu W, Singh R, Chen Z and Dickey E C 2001 *J. Mater. Res.* **16** 3331-34
[8] Molchan T, Thompson G, Skeldon P, Zhong X and Molchan I 2015 *Electrochimica Acta* **176** 1233-1238
[9] Regonini D and Bowen C R, Jaroenworaluck A and Stevens R 2013 *Materials Science and Engineering R: Reports* **74(12)** 377-406
[10] Mo S and Ching W 1995 *Physical Review B* **51(19)** 13023
[11] Li J, Ishigaki T and Sun X 2007 *The Journal of Physical Chemistry C* **111(13)** 4969-4976
[12] Tang H, Prasad K, Sanjines R, Schmid P E and Levy F 1994 *Journal of applied physics* **75(4)** 2042-2047
[13] Yu J and Wang B 2010 *Applied Catalysis B: Enviromental* **94** 295-302
[14] Fang D, Luo Z, Huang K and Lagoudas D C 2011 *Appl. Surf. Sci.* **257** 6451-6461
[15] Ou H and Lo S 2007 *Separation and Purification Technology* **58** 179-191
[16] Suarez P, Blanco S, Camero S 2015 *Acta Microscopica* **24** 361-362
[17] Hanaor D and Sorrell C 2011 *Journal of Materials Science* **46** 855-874

IOP Conf. Series: Journal of Physics: Conf. Series **935** (2017) 012026 doi:10.1088/1742-6596/935/1/012026

Electric dipole moment of magnetoexciton in concentric quantum rings

L F García[1], I D Mikhailov[1] and S Yu Revinova[2]
[1] Universidad Industrial de Santander, Bucaramanga, Colombia
[2] Peoples' Friendship University of Russia, Moscow, Russia

E-mail: lgarciar@uis.edu.co

Abstract. We study properties of exciton in a weakly coupled concentric quantum rings, penetrated by an axially directed magnetic flux and subjected to an electric field in the ring's plane. To this end, we adopt a simple model of quasi-one-dimensional rotator, for which the wave functions and the corresponding energies we found by using the double Fourier series expansion method. Revealed multiple intersections of the energy levels provide conditions for abrupt changes of the radial and the angular quantum numbers, making possible the tunnelling of carriers between rings and allowing the formation of a permanent large dipole moment. We show that the electric and magnetic polarizability of concentric quantum rings with a trapped exciton are very sensible to external electric and magnetic fields.

1. Introduction

Recent progress in epitaxial techniques has resulted in the assembly of ring-like structures [1], which in the presence of the magnetic field applied along the ring axis exhibit the Aharonov-Bohm (AB) effect and other related quantum topological phenomena, such as the jumps of the magnetization, persistent currents, optical transparency, etc. [2-7]. Lately, there has been a great deal of attention to the theoretical study of double concentric quantum rings (DCQRs) with a single and few electrons [8-10], fabricated earlier [11]. In such structure, the external magnetic field, unlike the case a single ring, produces an entanglement of not only states with different magnetic momenta but also with different spatial distributions, providing a change of the dipole moment of trapped electron-hole pair. One can suppose additionally that an in plane, the external electric field could change the magnetic moment by means of increasing of their separation facilitating in such way the relative rotation. The analysis of oscillations of the electric dipole moment of the electron and the exciton, realized earlier, shows that the quantum ring, pierced by a magnetic flux and subjected to electric field in plane of the ring, allows a control of polarization properties associated to the terahertz radiation [12-13]. In this work, we attempt to extend the model, considered in Refs. [12-13], to study the similar effect in DCQR with a trapped electron-hole pair.

2. Theoretical model

We consider InAs/GaAs DCQR that consists of two very narrow circular rings of the same width w and of radii and heights respectively R_1, h_1 (inner) and R_2, h_2 (outer). In order to analyse properties of this structure associated to low lying energy levels we apply the valence bond theory in which the wave function is represented as a linear combination of weakly coupled orbitals with small overlap. These orbitals, represented in Figure 1 by means of four Lewis structures we mark by two indices $(i, j = 1,2)$, corresponding to a configuration, in which the electron is located within the ring of radius $R_e = R_i, (i = 1,2)$, and the hole within the ring of the radius $R_h = R_j, (j = 1,2)$.

In the effective mass approximation, the dimensionless Hamiltonian in cylindrical coordinates for the Lewis structure (i, j), in the presence of the magnetic field B applied along de symmetry axis and the electric field E directed over the plane, has in the adiabatic limit a following form (Figure 1):

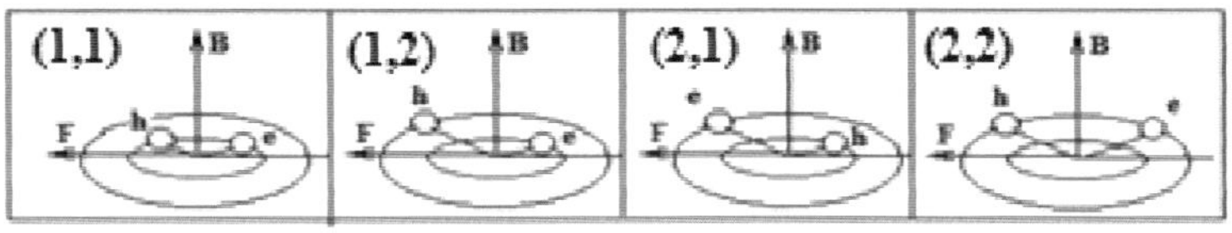

Figure 1. Valence bond structures for electron hole pair in DCQR.

$$H^{i,j} = H_{ei} + H_{hj} + U_{i,j};$$

$$H_{ki} = \frac{1}{\mu_k}\left[\frac{\pi^2}{h_i^2} - \frac{\delta^2}{R_i^2 \delta\varphi_k^2} + i\gamma\eta_k \frac{\delta}{\delta\varphi_k} + \frac{\gamma^2 R_i^2}{4}\right] + \eta_k \beta R_i \cos\varphi_k;\; k = e,h \quad (1)$$

$$U_{i,j} = -2/\sqrt{R_i^2 + R_j^2 - 2R_iR_j\cos\varphi_e - \varphi_h + w^2/12};\; i,j = 1,2;\; \eta_e = -1;\; \eta_k = +1$$

Here we use the effective Bohr radius $a_0^* = 4\pi\varepsilon_0\varepsilon\hbar^2/\mu e^2$, the effective Rydberg $R_y^* = e^2/8\pi\varepsilon_0\varepsilon a_0^*$, the parameters $\gamma = e\hbar B/2\mu^* R_y^*$ and $b = eFa_0^*/Ry^*$ as units of magnetic and electric fields strengths, being μ^* the reduced mass $\mu^* = m_e^* \cdot m_h^*/(m_e^* + m_h^*)$ and $\mu_e = m_e^*/\mu^*;\ \mu_h = m_h^*/\mu^*$. Unknown coefficients in double Fourier series of eigenfunctions $\Phi^{(i,j)}(\varphi_e,\varphi_h) = \sum_{m_1,m_2=-\infty}^{\infty} C_{m_1,m_2}^{(i,j)} e^{im_1\varphi_e} e^{im_2\varphi_h}$ of the Hamiltonian (1) satisfy the following secular equation:

$$\left[\frac{1}{\mu_e}\left(\frac{m_1}{R_i} - \frac{\gamma R_i}{2}\right)^2 + \frac{1}{\mu_h}\left(\frac{m_2}{R_j} - \frac{\gamma R_j}{2}\right)^2 + \frac{\pi^2}{\mu_e h_i^2} + \frac{\pi^2}{\mu_h h_j^2}\right] C_{m_1,m_2}^{(i,j)} - \frac{1}{2}\beta R_i\left(C_{m_1+1,m_2}^{(i,j)} + C_{m_1-1,m_2}^{(i,j)}\right) +$$

$$+\frac{1}{2}\beta R_j\left(C_{m_1,m_2+1}^{(i,j)} + C_{m_1,m_2-1}^{(i,j)}\right) + \sum_{\tilde{n},\tilde{m}=-\infty}^{\infty} V(m_1, m_1', m_2, m_2') C_{m_1',m_2'}^{(i,j)} = E^{(i,j)} \cdot C_{m_1,m_2}^{(i,j)}; \quad (2)$$

$$V(m_1, m_1', m_2, m_2') = -\frac{2\delta_{m_1-m_1', m_2'-m_2}}{\pi} \int_0^{\pi} \frac{\cos(m_1 - m_1')\vartheta \cdot d\vartheta}{\sqrt{R_i^2 + R_j^2 - 2R_iR_j\cos\vartheta}}$$

IOP Conf. Series: Journal of Physics: Conf. Series **935** (2017) 012026 doi:10.1088/1742-6596/935/1/012026

Eigenvalues $E_k^{(i,j)}$, $k=1,2,\ldots$ and eigenvectors $C_{nm}^{(i,j)}(k)$, $k=1,2,\ldots$, found by solving the equation (2) define energies and wave functions of the corresponding state of number k :

$$\Phi_k^{(i,j)}(\varphi_e,\varphi_h)=\sum_{m_1,m_2=-\infty}^{\infty} C_{m_1,m_2}^{(i,j)}(k)e^{im_1\varphi_e}e^{im_2\varphi_h} \tag{3}$$

3. Results

In order to analyse the effect of the electric and magnetic fields on the polarizability of the exciton in DCQRs, we have performed numerical calculations for low-lying energy levels as functions of the external magnetic field. For our calculations, we chose the geometrical parameters close to those of GaAs/Ga(Al)As DCQRSs fabricated earlier by using a droplet epitaxial technique and studied in Refs [14-16]. In the Figure 2 are shown results for DCQRs with fixed inner and outer rims radii $R_1=20nm$, $R_2=40nm$ and heights $h_1=6nm$, $h_2=7nm$, respectively.

Figure 2, we display lower energies as functions of the external magnetic field obtained by solving the eigenvalue problem (2) for three different values of the external electric fields applied along the symmetry axis.

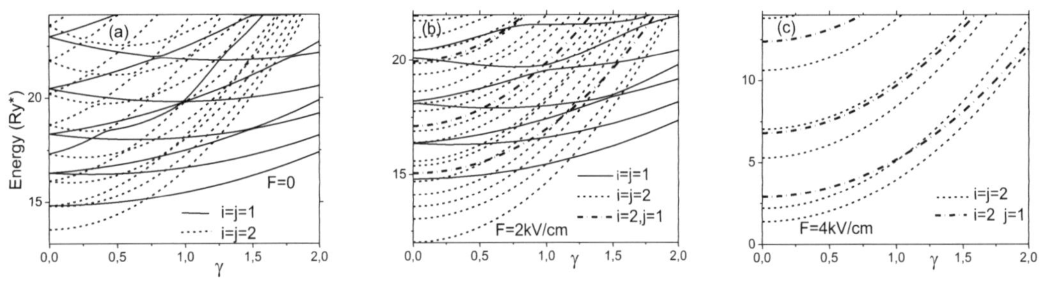

Figure 2. The energies of magnetoexciton in DCQR as functions of the external magnetic field under different external electric fields applied along the symmetry axis.

One can observe clearly a multiple intersections of energy levels of different Lewis structures for the zero electrics field case and as the electric field is relatively weak ($F=2kV/cm$). Almost all intersections corresponding to the points of the degeneration of the energy levels disappear as the electric field increasing becomes equal to 4kV/cm. To explain the comportment of the curves in Figures 2, one can to take into account three independent factors. First, the confinement energy inside the thinner inner ring is larger for both carriers, than one inside the outer ring and therefore ground state configuration in the absence of other two factors corresponds always to the Lewis structure (2, 2). Second, the external magnetic field provides, on the one hand the splitting the doubly degenerated levels, (paramagnetic effect), and, on the other hand, diamagnetic confinement, which pushes carriers toward the inner ring, increases their energy and restores periodically the degeneration of the energy levels. Finally, the electrostatic attraction between carriers forces to keep them close to each other, preferably within the same ring.

In Figures 2(a), (b) one can observe that in cases of zero- or weak electric fields there are threshold values of the magnetic field at which the ground state configuration is changed sharply from (2, 2) to (1, 1) due to dominant diamagnetic confinement. In addition, the major the external electric field, the larger is the critical value of the magnetic field, for which both carriers jump toward inner ring. It is because the electric field pushes them apart. The behaviour of the energies dependencies in Figure 4(c) for strong electric field is essentially different. In this case, the electric field breaks the axial symmetry of the structure, providing a strong localization of almost all lower states, and the magnetic field no longer able to induce circular electric current. This is the reason for the absence of the intersections between curves in Figure 4(c). On the other hand, the separation force, induced by a strong external electric field, becomes superior to the convergence force, generated by the electrostatic attraction and the magnetic confinement. Therefore, in Figure 4(c) there is no longer intersections between lower levels.

The change of the configuration at points of the intersection of energy levels are associated to an abrupt alteration of the electric and magnetic momenta and polarizabilities of the structure since these parameters are proportional to the first and the second derivatives of the corresponding energies with respect to the dimensionless external field γ . For example, the magnetic moment, $M_k^{(i,j)}$ and polarizability $\chi_k^{(i,j)}$ of the Lewis structure (i, j) at zero temperature for the state with number k one can find numerically by using the following relations

$$M_k^{(i,j)} = \partial E_k^{(i,j)}(\gamma)/\partial\gamma; \quad \chi_k^{(i,j)} - \partial^2 E_k^{(i,j)}(\gamma)/\partial^2\gamma \tag{4}$$

In Figure 3, we display the energies dependencies of the ground $(k=1)$ and two first excited $(k=2, 3)$ states on the magnetic field in the upper row and their first derivatives corresponding the magnetic moment according to the equation (4) in the lower row.

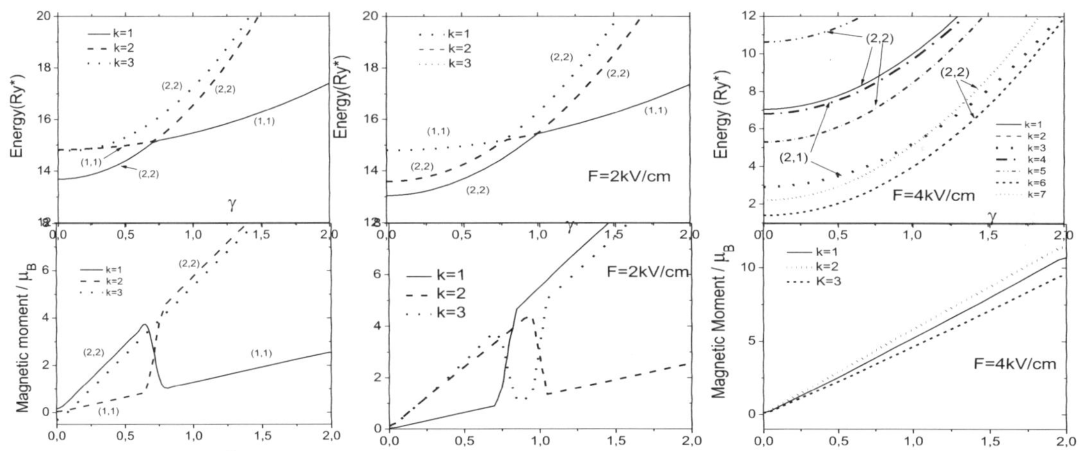

Figure 3. The energies (upper row) and magnetic momenta (lower row) of some magnetoexciton lower lying states in DCQR as functions of the external magnetic field under different external electric fields.

In general, the magnetic moment increases linearly with growth of the external field is linear typically for diamagnetic structures, which have no a proper magnetic momentum and have no Aharonov-Bohm oscillations. It indicates that the wave function for these lowest states are non-periodic, the angular carriers motion is restricted. Nevertheless, the linear dependency in the presence of a weak electric field is only sectional; at the points of the intersections, there are jumps of the magnetic moment related to the carriers' movement at radial direction between inner and outer rings. However, if the electric field is strong the intersections of energy levels and jumps of the magnetic moment disappear.

IOP Conf. Series: Journal of Physics: Conf. Series **935** (2017) 012026 doi:10.1088/1742-6596/935/1/012026

Similar analysis one can realize for the electric dipole moment and polarizability of the structure. The electric dipole moment $D_k^{(i,j)}$ and magnetic $\alpha_k^{(i,j)}$ for the Lewis structure (i, j) in the state with number k one can find by derivation the corresponding energies with respect the dimensionless electric field β :

$$D_k^{(i,j)} = \partial E_k^{(i,j)}(\beta)/\partial\beta; \quad \alpha_k^{(i,j)} = \partial^2 E(\beta)/\partial\beta^2 \tag{5}$$

In Figure 4(a), we present the energies dependencies of some lower energy levels on the electric field for structures (1, 1) with solid lines and for structures (2, 2) with dash lines. The energy dependencies for all states are strictly parabolic, which satisfy the relation $E_k = D_k F + \alpha_k F^2/2,\ k = 1,2,3\ldots$, being D_k and $\alpha_k F$ are the proper and induced by the external electric field dipole momenta of the structure, respectively. As the polarizabilities of different states coincide with curvatures of the energy dependencies, one can see in Figure 4(a) that polarizabilities of all states for the structure $(2,2)$ are superior to those for the structure $(1,1)$. The general formula for the polarizability of a quantum system [7] for analysed problem has a form:

$$\alpha_k = 2e^2 \sum_{k' \neq k} \left|\langle k|x|k'\rangle\right|^2 / (E_{k'} - E_k) \tag{6}$$

As in plane, matrix element is proportional to the ring radius, the polarizability of the Lewis structure in which the carriers are placed at the outer ring are larger. In Figure 4(b) we show curves of the electric dipole moment for the ground $(k=1)$ and two first excited states as functions of the electric field. The linear dependencies of curves for weak electric fields indicates that the structure in these states have no a proper dipole moment. Additionally, the jumps of the dipole momenta with an abrupt alteration of the slopes of the linear dependencies demonstrates an evidence of the structural transitions accompanied by a jump of the polarizability.

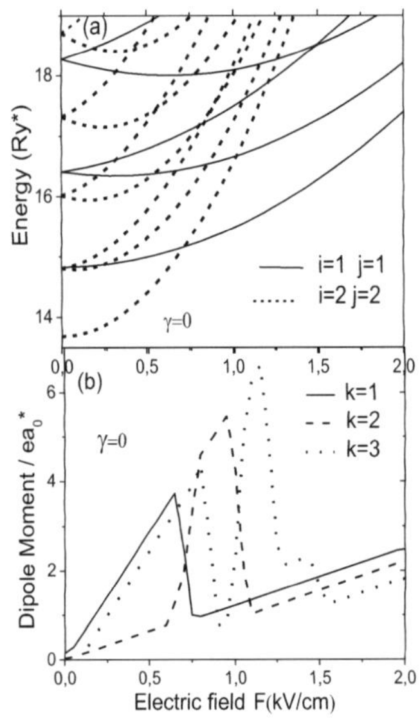

Figure 4. Energies and electric dipole momenta of exciton as functions of the external electric field

4. Conclusion

A simple model was proposed for theoretical analysis of possible transformations that undergoes electron-hole arrangement of the exciton trapped in concentric quantum rings in the presence axially directed magnetic and in transversally directed electric fields. The obtained results of the calculations, show the existence of abrupt changes in the curves of the dependencies of the electric and magnetic moment on external fields.

Acknowledgements

This work was financed by the «Universidad Industrial de Santander (UIS) through the «Vicerrectoría de Investigación y Extensión (VIE), DIEF de Ciencias (Cod. 5124)» and the «Patrimonio autónomo del Fondo Nacional de financiamiento para la ciencia, la tecnología y la innovación Francisco José De Caldas-Contrato RC– No. 275-2011» and the Cod. No. 1102-05-16923 subscribed with «Colciencias».

References

[1] A Lorke, R J Luyken, A O Govorov, J P Kotthaus, J M García, and P M Petroff 2000 *Phys. Rev. Lett.* **84** 2223
[2] S Olariu and I I Popescu 1985 *Rev. Mod. Phys.* **57** 339
[3] S Viefersa, P Koskinen, P Singh Deo, M Manninen 2004 *Physica* E **21** 1
[4] R A Römer and M E Raikh 2000 *Phys. Rev. B* **62** 7045
[5] F Ding, N Akopian, B Li, U Perinetti, A Govorov, F M Peeters, C C BofBufon, C Deneke, Y H. Chen, A Rastelli, O G Schmidt, V Zwiller 2010 *Phys. Rev. B* **82** 075309
[6] A V Chaplik 1995 *JETP Lett.* **62** 900
[7] A O Govorov, S E Ulloa, K Karrai and R J Warburton 2002 *Phys. Rev. B* **66** 081309(R)
[8] B Szafran and F M Peeters 2005 *Phys. Rev. B* **72** 155316
[9] Y Ch Guang, Y N Chen, D S Chuu 2007 *Solid State Communications* **143(11–12)** 515
[10] I Filikhin, S Matinyan, J Nimmo, and B Vlahovic 2011 *Physica E* **43** 1669
[11] L F García, S Yu Revinova, I D Mikhailov 2015 *Physica E* **71** 101
[12] T Mano, T Kuroda, S Sanguinetti, T Ochiai, T Tateno, J Kim, T Noda, M Kawabe, K Sakoda, G Kido, and N Koguchi 2005 *Nano Letters* **5(3)** 425-428
[13] M D Teodoro, V L Campo, Jr V Lopez-Richard, E Marega, Jr G E Marqués, Y Galvão Gobato, F Iikawa, M J S P Brasil, Z Y A Waar, V G Dorogan, Y I Mazur, M Benamara, and G J Salamo 2010 *Phys. Rev. Lett.* **104** 086401
[14] M Alexeev and M E Portoni 2012 *Phys. Rev. B* **85** 245419
[15] S Sanguinetti, T Mano, and T Kuroda 2014 *Physics of Quantum Rings* ed V. M. Fomin (Heidelberg: Springer) p 161
[16] T Kuroda, T Mano, T Ochiai, S Sanguinetti, K Sakoda, G Kido, and N Koguchi 2005 *Phys. Rev. B* **72** 205301

IOP Conf. Series: Journal of Physics: Conf. Series **935** (2017) 012027 doi:10.1088/1742-6596/935/1/012027

Effect of pH and chloroauric acid concentration on the geometry of gold nanoparticles obtained by photochemical synthesis

G R Conde Rodríguez[1], G H Gauthier[1], L O Ladeira[2], J A Sanabria Cala[1] and D Laverde Cataño[1]
[1] Universidad Industrial de Santander, Bucaramanga, Colombia
[2] Universidade Federal de Minas Gerais, Belo Horizonte, Brazil

E-mail: gerson.rafael.conde@gmail.com

Abstract. Due to their excellent surface properties, gold nanoparticles have been used in a wide range of applications from optics and catalysis to biology and cancer treatment by thermal therapy. Gold nanoparticles can absorb a large amount of radiation according to their geometry, such as nanospheres and nanorods. The importance of gold nanoparticles geometry is based on the electromagnetic spectrum wavelength where exists a greater absorption of radiation, which belongs to the visible region for nanospheres and ranges between visible and near infrared regions for nanorods, conferring greater biomedical applicability to the latter. When using photochemical synthesis method, which consists of reducing gold atoms to their metallic state with UV radiation, the geometry of gold nanoparticles depends on different variables such as: 1) pH, 2) concentration of chloroauric acid, 3) the surfactant, 4) concentration of silver nitrate, 5) temperature and 6) irradiation time. Therefore, in this study the geometry of the gold nanoparticles obtained by photochemical synthesis was determined as a function of solution pH and chloroauric acid concentration, using Spectrophotometry in the Ultraviolet Visible region (UV-vis) as characterization technique. From the analysis of the UV-vis spectra, it was determined that at an acidic pH the particles have two absorption bands corresponding to nanorods geometry, while at a basic pH only nanospheres are found and at a neutral pH the lower relative intensity of the second band indicates the simultaneous existence of the two geometries. The increase in the concentration of chloroauric acid produces a decrease in the amount of synthesized nanorods, seen as a decrease of the relative intensity of the second absorption band. Therefore, obtaining gold nanoparticles with nanorods geometry favours fields such as biomedicine, because they are capable of absorbing infrared radiation and can be used as photosensitive agents in localized thermal therapy against cancer.

1. Introduction

Nanoscale particles have been studied for years due to their physicochemical properties, which are different to those of their macroscopic equivalents [1]. This is due to the confinement of the electrons in the surface of a particle to a nanometric space, which leads to the appearance of the phenomenon known as Surface Plasmon Resonance (SPR) [2]. This phenomenon consists of the collective oscillation of the electrons at the nanoparticle surface, whose oscillation frequency is in resonance with frequencies corresponding to the electromagnetic spectrum [3].

Surface Plasmon Resonance depends mainly on the composition, size and shape of the nanoparticles, besides the medium in which they are dispersed [4]. Thus, gold is one of the most interesting metals for the synthesis of nanoparticles due to characteristics such as: 1) biostability, 2) biocompatibility, 3) absorption of large amounts of radiation at near-infrared frequencies due to the SPR and 4) the ability to form isotropic structures, such as nanospheres, or anisotropic structures such as nanorods [5-7].

Generally, the geometry of the gold particles corresponds to nanospheres because it is the most stable structure and relatively easy to synthesize. In these cases, the resonance phenomenon occurs at frequencies belonging to the visible region of the electromagnetic spectrum, which correspond to wavelengths around 520nm [8]. However, the wavelength at which the resonance occurs can range between the visible and the infrared regions for gold nanorods, making it possible to use this type of particles in a greater variety of applications ranging from optics and catalysis up to biology and cancer treatment by thermal therapy [9].

The geometry can be clearly identified from the absorption spectrum in the Visible Ultraviolet region (UV-vis), which is strongly associated with the SPR as shown in Figure 1 [10]. When the geometry of the nanoparticles is spherical, the spectrum contains a single high intensity band with maximum absorption in the range of 500 to 530nm wavelength. On the other hand, when the geometry corresponds to nanorods the spectrum contains two absorption bands: the first one with low intensity located in the wavelength range between 500 and 530nm, as it happens with the spheres, and the second one of high intensity localized between 600 and 1000nm, according to the length of nanorods [11].

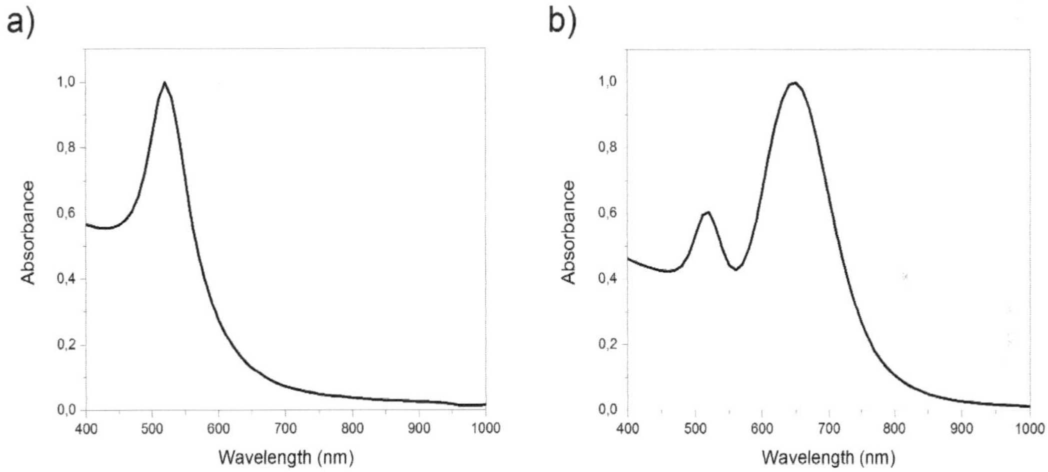

Figure 1. Absorption spectrum in the visible ultraviolet region of: (a) gold nanospheres, (b) gold nanorods.

One of the most recent methods used to obtain gold nanoparticles is photochemical synthesis, during which the gold atoms are reduced to their metallic state using ultraviolet (UV) radiation. One of the major advantages of this method is the time of synthesis, which oscillates between 30 and 90 minutes, increasing the possibilities of producing nanoparticles on a larger scale [12].

However, the geometry of the nanoparticles obtained by photochemical synthesis depends on different variables such as: 1) pH, 2) concentration of chloroauric acid, 3) surfactant, 4) concentration of silver nitrate, 5) temperature, 6) the irradiation time and 7) the presence of acetone and cyclohexane [12]. Therefore, in this study the geometry of the gold nanoparticles obtained by photochemical synthesis was determined as a function of solution pH and chloroauric acid concentration, using spectrophotometry in the Visible Ultraviolet (UV-vis) region as characterization technique.

IOP Conf. Series: Journal of Physics: Conf. Series **935** (2017) 012027 doi:10.1088/1742-6596/935/1/012027

2. Experimental

The reagents used during the photochemical synthesis are: 1) chloroauric acid 2) silver nitrate, 3) hexadecyltrimethylammonium bromide (CTAB), 4) ascorbic acid, 5) acetone, 6) cyclohexane, 7) hydrochloric acid and 8) ammonium hydroxide. All reagents are acquired with analytical grade, except for chloroauric acid which is produced in the laboratory of Nanomaterials–ICEx, UFMG. All solutions are prepared with milli-Q deionized water of 18.2MΩ resistivity.

The photochemical reactor consists of a quartz tube of 800cm long, 12mm inner diameter and 4mm thickness, and a UV lamp of 914mm long and 25mm diameter that generates light of 256nm wavelength and intensity of 15W/cm^2. All the laboratory instruments used during the synthesis are of glass, ceramic or plastic to avoid the reduction of the gold due to contact with metals. The glass materials were previously immersed in aqua regia for 2 hours, washed with distilled water and oven dried at 80°C for 2 hours to guarantee the quality and reproducibility of the results obtained [13].

The solution for photochemical synthesis is prepared from: 20mL of CTAB (120mM), 10mL of chloroauric acid solution with specific concentration, 2mL of silver nitrate (1.4mM), 650μL of acetone, 450μL of cyclohexane and 170μL of ascorbic acid (90mM). This solution is placed inside the quartz tube and irradiated with UV light for 1 hour. The suspension obtained is centrifuged 3 times at 10.000RPM for 15 minutes, collecting the precipitate concentrated in nanoparticles and dispersing in deionized water.

The pH of the irradiated solution takes values of 3, 5, 7 and 9, and is adjusted by adding hydrochloric acid or ammonium hydroxide right before being placed in the quartz tube. For its part, the solution of chloroauric acid takes concentration values 1.5, 3.0, 4.5, 6.0 and 7.5mM. The characterization of the gold nanoparticle suspensions is made from the absorption spectrum obtained by a Rayleigh VIS 723-G Spectrophotometer with scanning wavelengths between 330 and 1000nm, step of 1nm, in quartz cuvettes with optical path of 1 cm and using deionized water as the reference sample.

3. Results and discussion

The absorption spectra of the colloidal suspensions obtained by varying the pH during the photochemical synthesis are shown in Figure 2, which are normalized so the maximum absorbance value of the first band is equal to 1. It is possible to observe that the photochemical synthesis of gold nanoparticles is highly sensitive to the pH of the irradiated solution. Thus, when the pH is 3 or 5 it is possible to distinguish two absorption bands where the second one has a relatively higher intensity than the first one, indicating that the suspended particles correspond mostly to gold nanorods. On the other hand, when the pH is neutral the second band has a slightly higher intensity than the first one, which indicates the simultaneous presence of spherical and rods nanoparticles. Finally, when the pH is 9, only an absorption band is observed whose maximum is located at a wavelength around 515nm, indicating the exclusive presence of gold nanoparticles with spherical geometry.

The sensitivity of photochemical synthesis to pH is due to the rate at which ascorbic acid releases H^+ ions and donates the electrons that reduce the gold atoms to their metallic state. When the pH is basic the concentration of H^+ ions in solution is very low, causing fast dissociation of ascorbic acid and simultaneous reduction of a large amount of gold atoms, which leads to the formation of many nuclei. Because of this, the available concentration of reduced atoms decreases and the accelerated growth of the nuclei takes place to form nanospheres, avoiding the growth of anisotropic structures.

In the case of neutral pH, the rate at which ascorbic acid dissociates is lower, and so is the number of nuclei. This allows the available concentration of reduced atoms to be such that some of the nuclei have enough time to grow into anisotropic structures as nanorods, while the other nuclei produce spherical nanoparticles. On the other hand, when the pH is acid the concentration of H^+ ions in the solution is higher, causing a slow and progressive dissociation of ascorbic acid and giving time to most of nuclei to grow anisotropically, leading to a lot of gold particles with nanorods geometry.

IOP Conf. Series: Journal of Physics: Conf. Series **935** (2017) 012027 doi:10.1088/1742-6596/935/1/012027

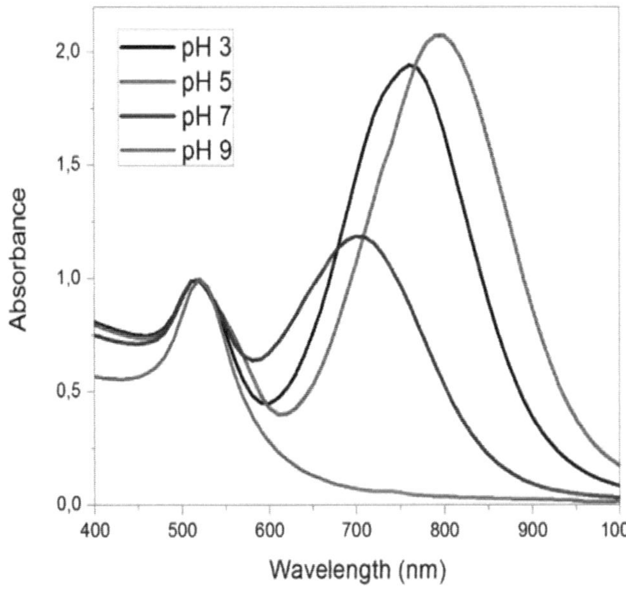

Figure 2. Absorption spectra of the suspensions obtained in the pH variation.

The absorption spectra of the colloidal suspensions obtained by varying the chloroauric acid concentration are shown in Figure 3, which are also normalized so the maximum absorbance value of the first band is equal to 1. It is possible to observe the decrease in the intensity of the second band of the absorption spectra as the concentration of chloroauric acid increases. For the concentrations of 1.5, 3.0 and 4.5mM the intensity of the second band is relatively high, which is associated with the presence of gold nanorods. However, the second band has a lower intensity for the concentrations of 6.0 and 7.5mM, related to the simultaneous presence of nanospheres and nanorods.

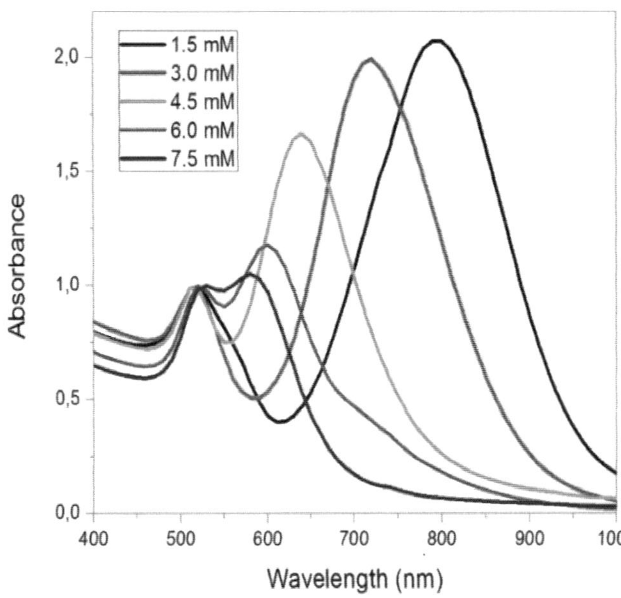

Figure 3. Absorption spectra of the suspensions obtained in the variation of chloroauric acid concentration.

One possible explanation for this behaviour is based on the way the gold nanoparticles are generated. Once the gold atoms are reduced to their metallic state, they are located inside micelles formed by the CTAB surfactant, which can behave differently according to the conditions of the solution. When the amount of CTAB is high, the excess of these molecules is added to the micelles, which lengthen and allow to generate gold nanorods. In contrast, when there are no excess molecules of CTAB, the micelles maintain their shape and this leads to nanospheres formation [14]. Thus, an increase in the concentration of chloroauric acid can cause lead to the absence of enough CTAB molecules to lengthen the micelles, which generates that only a small part of the formed particles are nanorods while the majority are nanospheres, which is evidenced from the absorption spectra.

4. Conclusions

The analysis of the absorption spectra in the Visible Ultraviolet region allowed to determine the great influence of the pH and chloroauric acid concentration on the geometry of the nanoparticles obtained by photochemical synthesis. The pH 3 and 5 promoted the majority formation of gold nanorods, neutral pH allowed the simultaneous formation of nanorods and nanospheres, and pH 9 produced the exclusive formation of gold nanospheres. This behaviour is due to the rate at which ascorbic acid dissociates in the presence of the H^+ ions of the solution. On the other hand, the increase in the concentration of chloroauric acid produces a decrease in the intensity of the second absorption band, associated to the lower presence of gold nanorods in relation to nanospheres, possibly due to the lower proportion of CTAB with respect to the golden ions. In this way, obtaining gold particles with nanorods geometry increases their capacity to be used in different applications such as biomedicine, since these absorb radiation in the infrared region of the electromagnetic spectrum and can be used as photosensitive agents in localized thermal therapy against cancer.

References

[1] Miranda O and Ahmadi T 2005 *The Journal of Physical Chemistry B* **109(33)** 15724–15734
[2] Huang X, Neretina S and El-Sayed M A 2009 *Advanced Materials* **21(48)** 4880–4910
[3] Kozek K, Kozek M, Wu W, Mishra S and Tracy, J 2013 *Chemistry of Materials* **25(22)** 4537–4544
[4] Hee Song J, Kim F, Kim D and Yang P 2005 *Chemistry A* **11(3)** 910–916
[5] Orendorff C and Murphy C 2006 *The Journal of Physical Chemistry B* **110(9)** 3990–3994
[6] Ni W, Kou X, Yang Z and Wang J 2008 *ACS Nano* **2(4)** 677–686
[7] Sau T and Murphy C 2004 *Journal of the American Chemical Society* **126(28)** 8648–8649
[8] Marin M, McGilvray K and Scaiano J 2008 *J. of the American Chemical Society* **130(49)** 16572–16584
[9] Vigderman L and Zubarev E 2013 *Chemistry of Materials* **25(8)** 1450–1457
[10] Hu M, Chen J, Li Z, Au L, Hartland G, Li X and Xia Y 2006 *Chem. Society Reviews* **35(11)** 1084–1094
[11] Haiss W, Thanh N, Aveyard J and Fernig D 2007 *Analytical Chemistry* **79(11)** 4215–4221
[12] Placido T, Comparelli R, Giannici F, Davide P, Capitani G, Striccoli M and Lucia M 2009 *Chemistry of Materials* **21(18)** 4192–4202
[13] Niidome T, Yamagata M, Okamoto Y, Akiyama Y, Takahashi H, Kawano T and Niidome Y 2006 *Journal of Controlled Release* **114(3)** 343–347
[14] Scarabelli L, Sánchez-Iglesias A, Pérez-Juste J and Liz-Marzán L 2015 *The Journal of Physical Chemistry Letters* **6(21)** 4270–4279

IOP Conf. Series: Journal of Physics: Conf. Series **935** (2017) 012028 doi:10.1088/1742-6596/935/1/012028

Detailed study of the magnetic behaviour at low scale in $La_{2/3}Sr_{1/3}MnO_3$

I C Arango[1], J E Ordoñez[1], C Dominguez[2], C Arango[1] and M E Gomez[1,3]
[1] Universidad del Valle, Cali, Colombia
[2] Université de Genève, Genève, Switzerland
[3] Centro de Excelencia en Nuevos Materiales, Cali, Colombia

E-mail: isabel.arango@correounivalle.edu.co

Abstract. The $La_{2/3}Sr_{1/3}MnO_3$ (LSMO) with Curie temperature above room temperature is the leading compound of the manganite perovskite family. Therefore, the physical properties are desirable for practical applications as magnetic sensors. However, when the dimensions are reduced the ferromagnetic properties of material are weakened. In this research, we have grown $La_{2/3}Sr_{1/3}MnO_3/SrTiO_3$ thin films by sputtering DC at high oxygen pressure at 830°C. X-Ray Diffraction (XRD) analysis reveals that only (0 0 2) LSMO peak are present, indicating a textured growth. The samples morphology was characterized by Atomic Force Microscopy (AFM). Additionally, LSMO microwires were patterned by UV lithography; the devices are a well-defined channel with current and voltage leads enabling four points resistance measurements. Resistivity versus temperature curves displays typical manganite behaviour with metal-insulator transition ~350K. We study the electric and magnetotransport properties in LSMO film and in wire channel and their dependence with size (width and length) for potential applications like magnetic sensors.

1. Introduction

The manganese perovskite has been studied extensively since it showed an intrinsic colossal magnetoresistance (CMR) near the Curie temperature T_C [1]. In the ferromagnetic state above to 300K, the $La_{2/3}Sr_{1/3}MnO_3$ (LSMO) behaves like a metal with electric properties; evince a ferromagnetic transition above room temperature, behaving as a good candidate to be used for spintronic devices [2,3]. At a certain temperature T_{MI} corresponding to metal–insulator transition they undergo simultaneously ferromagnetic (FM)–paramagnetic (PM) transition (Curie temperature, T_C) [1,4], even at temperature of 360–365K which is very closely to the highest Curie temperature of the bulk LSMO (T_C ~370K) [5]. Moreover, in a world following the ever demanding decrease of feature sizes for the fabrication of faster electronic circuits and devices, researches and development of tools allowing reproducible pattering at an ever decreasing scale gain increasing attention. Electronics manufacturers focus on reducing fabrication costs and addition of device functionalities [6]; the high-resolution the pattering technique of lithography is an excellent candidate [7,8]. In this work, we investigated the effect of volume minimization from thin film of LSMO to thin film microstructure on magnetotransport properties. By UV lithography, we patterned on LSMO films well defined channel (microwires) with current and voltage leads enabling four points resistance measurements. We evaluated the dependence of the resistivity with temperature, magnetization with the temperature and

IOP Conf. Series: Journal of Physics: Conf. Series **935** (2017) 012028 doi:10.1088/1742-6596/935/1/012028

isothermal hysteresis loops in LSMO films and microwires and study the dependence of magneto resistivity with microwires size, width and length.

2. Experimental details

We have grown LSMO on (001) $SrTiO_3$ (STO) substrate by sputtering technique at high oxygen atmosphere (O_2:99.999%). We deposited the LSMO film using DC sputtering at pressure of 2.7mBar with 830°C substrate temperature. The thin film was patterned by standard UV photolithography using a Microposit S1813 G2 positive resist. Pt contacts bonded by copper wire were used to connect the pads to the sample holder to determine the volume minimization effect. We structurally characterized LSMO thin films by X-Ray Diffraction (XRD), morphologically by Atomic Force Microscopy (AFM) and Scanning Transmission Microscopy (SEM) for determining the root mean square (RMS) roughness and the quality of microwires, respectively. It is studied the magnetic properties by magnetization measurements as function of temperature (10 to 400K) at zero field cooling (ZFC), and isothermal magnetization loops, M(H), for temperatures from 10K to 350K by means a Physical Property Measurement System (PPMS)® quantum design. We performed magnetotransport properties in microwires by resistivity as function of temperature measurements by the standard four-probe technique with current-in-plane geometry. For magnetoresistance measurements we applied magnetic field of 0.1, 1.0, 5.0, 10.0, and 30.0kOe in substrate plane along the microwire length direction.

3. Results and discussion

Figure 1(a) displays X-ray diffraction pattern in the 42 to 52 degrees range for LSMO(25.6nm)/STO film. θ-2θ scans indicates a (0 0 l) growth. No impurity or other additional phases was observed. The films present an out-of-plane lattice parameter c=(3.850±0.001) Å (substrate misfit 1.4%). Inset in Figure 1(a) shows the AFM micrograph for the same sample. We determined the RMS roughness of (1.9+0.1) nm, around 5 lattice parameters; the less roughness, the more enhanced the ohm contact efficiency due to increase of contact points. Figure 1(b) shows ZFC thermal demagnetization curves, from 10 to 400K, for the LSMO/STO film. To measure magnetization, we applied during heating a magnetic field of 1000Oe parallel to the plane of the film. The experimental data were fitted close to the transition temperature assuming a Gaussian distribution of domains, reported elsewhere [9] to determine the Curie temperature being 352K with and a transition width of 34K. Result indicate that samples exhibit magnetization at room temperature offering opportunities for the development of magnetic sensor and devices.

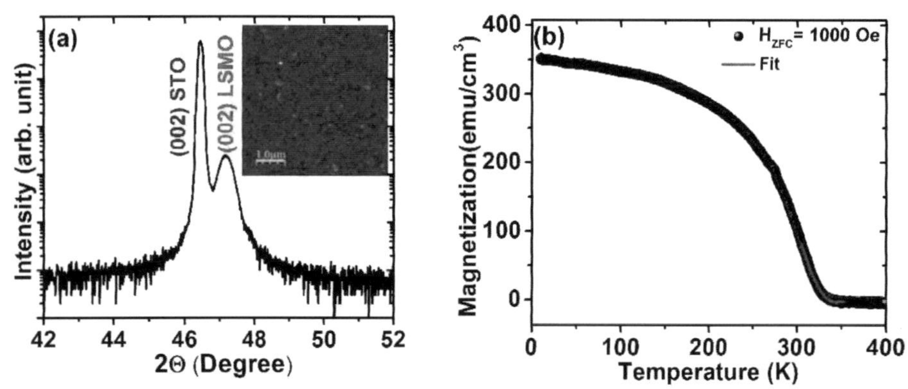

Figure 1. (a) X ray diffraction for a $La_{2/3}Sr_{1/3}MnO_3$ /$SrTiO_3$ thin film. Inset 5×5μm AFM image (b) Dependence of magnetization with the temperature for LSMO/STO.

The lithography process is based on etching the film with Ar^+ ions, removing the LSMO material in areas unprotected by the photoresist. We pattern a microwire connected to six electrodes as shown in Figure 2(a). The sample fabrication consisted of creating a well-defined channel with current and voltage leads enabling four-point resistance measurements. To avoid the influence of contact resistance, the device configurations have four-probe electronic transport measurements. Figure 2(a) shows that the current is applied along the wire (+I and -I), while the voltage is transverse (+V and -V). For the electrical contacts, 50nm of Pt was deposited. We have fabricated four microwires with different width (a)/length (l) ratio: W_1=10μm/300μm, W_2=10μm/150μm, W_3=25μm/300μm and W_4=25μm/150μm, Figure 2(b). The final microwires examined by SEM, shown in Figure 2(c), reveals that structures have 12.5μm and 25μm width, and 130μm and 300μm length.

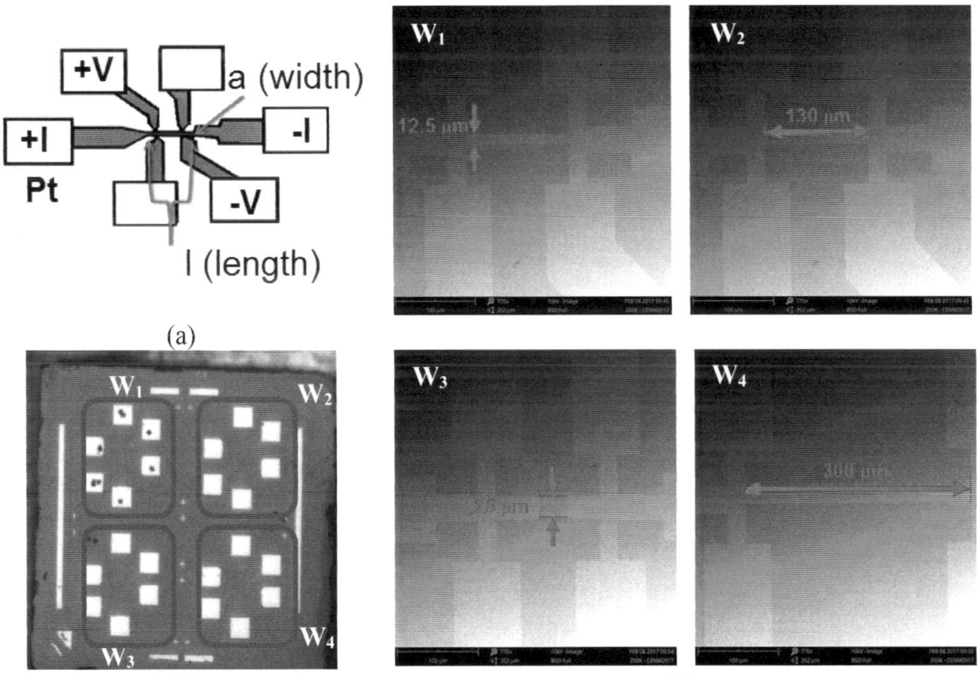

(b) (c)

Figure 2. (a) Sketch illustrating details of a microwire with width a, and length l. (b) SEM micrograph of a patterned film showing 4 microwires. (c) SEM images of the wires with the different width and lengths.

Figure 3(a) displays the resistivity ρ (H=0) as a function of temperature for a LSMO film (in Van der Pauw configuration), and for the microwires at different width/length (a/l) ratio: 0.033=10/300, 0.066=10/150, 0.083=25/300 and 0.166=25/150. All microwires exhibit M-I transition, where the peak shows the transition from the low-temperature metallic (ferromagnetic) to the high-temperature insulating (paramagnetic) phase; the maximum in curve ρ(T) corresponds to T_{MI}. It decreases with a/l ratio decrease, from T_{MI} ~365K for the thin film to ~344K for a/l=0.033 ratio microwire. Additionally, we observed an increase on resistivity when a/l ratio decreases. This behaviour has been associated to oxygen vacancies probably oxygen loss [9] during the ion milling etching process (Ar bombardment).

We have fitted the ρ(T) plots to the equation $\rho(T)=\rho_0+AT^{\alpha}$, where ρ_0 is the residual resistivity that we consider as a measure of the effective disorder, and AT^{α} a generic T-power law which can simulate different scattering processes [10]. At low temperature ($T<T_C$), the electron–electron and electron–magnon scattering and the polaronic effects are the key promoters of various concepts in electrical transport [11]. In previous investigations the deviation from the quadratic power law has been ascribed to a combination of terms due to different kinds of scattering [12,13]. Where α=2 is due to an electron-electron scattering, α=3 refer anomalous single magnons, two magnons α=4.5, spin-waves when α=3.5, acoustic phonons (α=4.5) [10,14]. Figure 3(c) shows the α exponent dependence with a/l ratios.

Our samples with high a/l ratio exhibit an electron-electron scattering, but when a/l ratio is decreasing, they exhibit an anomalous single magnons even a spin wave scattering.

Figure 4 plot the magnetoresistance as function of a/l ratio for applied field from H=100Oe to 30kOe. The magnetoresistance is defined as $MR(\%) = \rho_o - \rho_H/\rho_0 * 100$ [15]. It is calculated MR values at T_{MI}. It is observed that MR increases with the increase of the applied magnetic field:

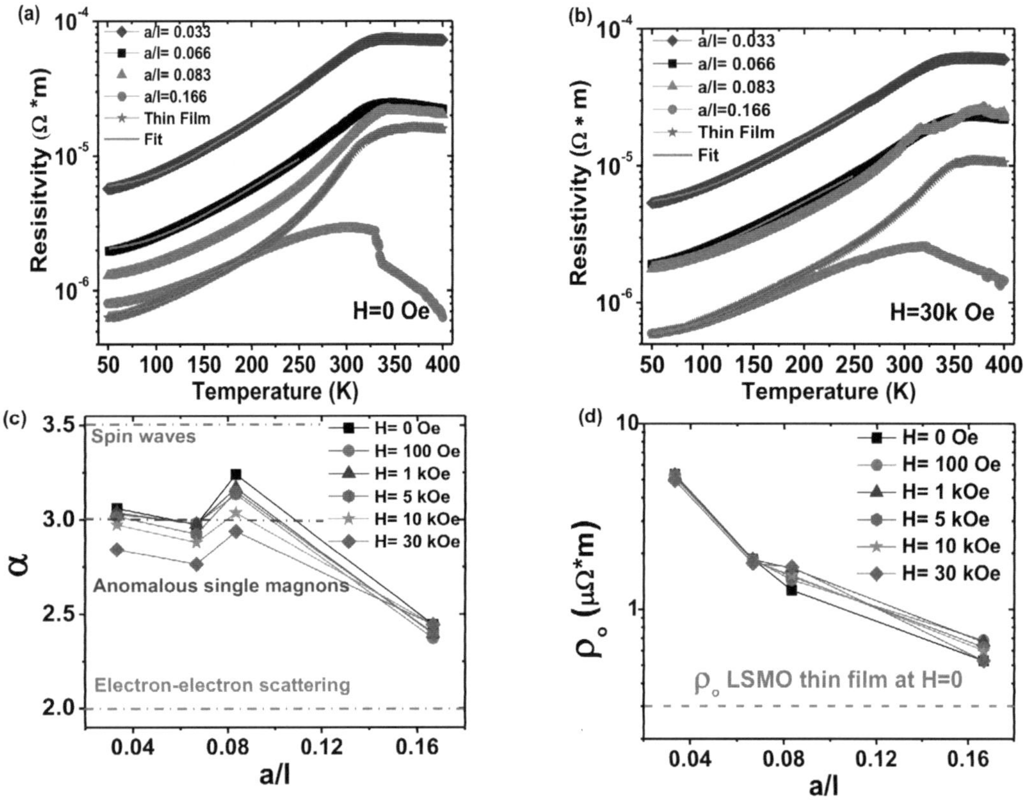

Figure 3. Temperature dependence of the resistivity at (a) H=0, (b) H=30kOe, for a LSMO film (green star) and for the wires with a/l ratio: 0.033 (blue diamond), 0.066(black square), 0.083 (grey triangle), 0.166 (dark yellow circle). (c) a/l-ratio vs of low-temperature exponent α, and (d) residual resistivity ρ_0, for applied magnetic fields of H=0Oe (black square), H=100Oe (red circular), H=1kOe (blue triangle), H=5kOe (dark cyan hexagon), H=10kOe (pink star) and H=30kOe (green diamond).

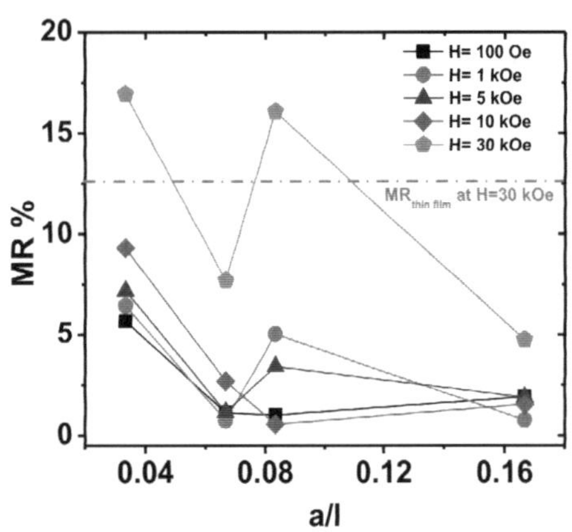

Figure 4. MR versus a/l ratio for different magnetic field applied: H= 100 Oe (black square), H= 1 kOe (red circle), H=5 kOe (blue triangle), H=10 kOe (green diamond), H=30 kOe (pink pentagon).

Compare MR for H= 100Oe and that for H= 30kOe. MR decreases in a non-monotonic way with the increase of a/l ratio, comparing, for example, the MR values for a/l=0.033 and 0.166. We obtain an 18%MR for sample under the 30kOe magnetic field and for the lower a/l ratio (0.033).

4. Conclusions

We studied magnetic and transport properties in 25 nm thick LSMO microwires in which we systematically varying the width (a)/length (l) ratio. We have found that the resistivity and the metal−insulator transition temperature strongly depend on a/l ratio of the wires and the applied magnetic field. Magnetoresistance measurements show that all samples satisfy the electron-electron scattering for higher a/l ratio, and an anomalous single magnons - spin waves for lower a/l ratio. Residual resistivity increases with the decrease of a/l ratio, indicating more density of defects in samples with lower a/l ratio, probably due oxygen loss during etching process. We show an enhancement of the magnetoresistance when decrease a/l ratio, and when increase the applied magnetic field, offering possible new opportunities for spintronic devices.

Acknowledgements

This work was sponsored by research projects No 110656933104, C.I. 7917 Colciencias-Univalle, and C.I. 7978 Univalle, “Centro de Excelencia en Nuevos Materiales (CENM)”. We thank University of Genève, Switzerland, for lithography process.

References

[1] Novickij V, Stankevic V, Grainys A, Novickij J, Tolvaisiene S and Stankevic T 2012 *Elektronika ir Elektrotechnika* **120(4)** 47-50
[2] Zurauskiené N, Balevicius S, Stankevic V, Kersulis S, Schneider M, Liebfried O, Plausinaitiene V and Abrutis A 2011 *IEEE Trans. Plasma Sci.* **39(1)** 411-416
[3] Sacanella J, Parisia F, Levya P and Ghivedler L 2004 *Physica B* **354(1)** 43-46
[4] Haghiri-Gosnet A M and Renard J P 2003 *J. Phys. D: Appl. Phys.* **36(8)** R127-R150
[5] Strbik V, Reiffers M, Dobrocka E, Soltys J, Spankova M and Chromik S 2014 *Applied Surface Science* **312** 212-215
[6] Bacells L, Calvo E and Fontcuberta J 2002 *J. Magn. Magn. Mat.* **242** 1166-1168
[7] Balla T, Spearing S and Monk A 2008 *J. Phys. D: Appl. Phys.* **41(17)** 174001
[8] Guo L. J 2007 *Advanced Materials* **19(4)** 495-513
[9] G. Campillo, M. Gómez, A. Hoffmann, R. Escudero y P. Prieto 2006 *J. Appl. Phys.* **99(08)** 08C106
[10] Nord M, Vullum P, Moreau M, Boscker J, Selbach S, Holmestad R and Tybell T 2015 *Appl. Phys. Lett.* **106** 041604
[11] Mercone S, Perroni C. A, Cataudella V, Adamo C, Angeloni M, Aruta C, De Filippis G, Miletto Z, Oropallo A, Perna P, Petrov A. Y, Scotti di Ucci U and Maritato L 2005 *Phys. Rev. B* **71** 064415
[12] Marín L, Morellón L, Algarabel P. A, Rodríguez L. A, Magen C, De Teresa J. M and Ibarra M. R 2014 *Nano Lettters* **14** 423-428
[13] Zurauskaite L, Balevicius S, Zurauskiene N, Kersulis S, Stankevic V, Simkevicius C, Novickij J and Tolvaisiene S 2012 *Lithuanian Journal of Physics* **52** 224-230
[14] Strbik V and Chomik S 2012 *J. Elect. Engin.* **63(4)** 270-272
[15] Ordoñez J.E, Gómez M. E and Lopera W 2016 *Rev. Mex. Fis.* **62** 543-548
[16] Zhang M, Li J, Peng Z, Li S, Zheng D, Jin A, Gu C, Li R and Liu C 2007 *Journal of Magnetism and Magnetic Materials* **316** L1-L4

IOP Conf. Series: Journal of Physics: Conf. Series **935** (2017) 012029 doi:10.1088/1742-6596/935/1/012029

Interfacial coupling in multiferroic $BiFeO_3$ and ferromagnetic $La_{2/3}Sr_{1/3}MnO_3$ thin films

C Dominguez[1,2], J E Ordoñez[1] and M E Gomez[1]
[1] Universidad del Valle, Cali, Colombia
[2] Université de Genève, Genève, Switzerland
[3] Centro de Excelencia en Nuevos Materiales, Cali, Colombia

E-mail: claribel.dominguez@correounivalle.edu.co

Abstract. Antiferromagnetic/Ferromagnetic coupling mechanics have been studying by growing successfully $BiFeO_3/La_{2/3}Sr_{1/3}MnO_3$ bilayers on $SrTiO_3$ single crystals by using rf and dc sputtering technique at pure oxygen pressures. We have investigated the magnetic behaviour of this samples, field cooling loops evidence interfacial coupling effect when antiferromagnetic ferroelectric $BiFeO_3$ is placed in contact with ferromagnetic $La_{2/3}Sr_{1/3}MnO_3$ indicate by the shift of the magnetization loop to negative values of the applied magnetic field. Our samples exhibited an exchange bias of 76Oe at 5K after field cooling the sample under 5000Oe. Temperature dependence of the exchange bias field showed exponential decay. The BFO/LSMO bilayer exhibits excellent ferroelectric behaviour (P_s=65μC/cm^2 at 4V and 100Hz). Coexistence of ferroelectric and ferromagnetic properties in the BFO/LSMO bilayer make it a promising candidate system for applications where the magnetoelectric behaviour is required.

1. Introduction

The increased demand for faster, smaller and non-volatile electronics is pushing the limits of current information systems. All these, point to the need to have ultralow power electronic for both information processing and storage systems [1]. One of the manifestations of the spintronics approach is to use a spin current to create a spin torque that can be transferred to a magnetic element (such as a domain wall) [2]; however, this requires the passage of significant amounts of current, which are prohibited in terms of energy consumption [3]. Thus, the challenge consists to explore the possibility to control the magnetic state of the ferromagnet by using electric fields instead of currents. A route to control the ferromagnetic state, for example, is the use of multiferroic materials in contact with ferromagnetic material. In this case, the antiferromagnetic order parameter in the multiferroic act as a medium that indirectly couples the ferromagnetic ordering of a ferromagnetic thin film and the ferroelectric ordering of the multiferroic. The coupling mechanism at the antiferromagnetic/ferromagnetic (AF/FM) interface is known as Exchange Bias (EB) effect [4]. Furthermore, significant efforts have been made to understand this mechanism in multiferroic/ferromagnetic heterostructures based on oxides with the purpose of extending the mechanism's operation at room temperature. The formation of novel magnetic states at the $BiFeO_3/La_{2/3}Ca_{1/3}MnO_3$ interface has been reported [5]; as well as multiferrocity at room temperature in $BaTiO_3$ ultrathin films with Fe or Co [6]. Chakhalian et al., has demonstrated an interface induced orbital ordering and FM driven by charge transfer in $YBa_2Cu_3O_7$ layer deposited onto $La_{2/3}Ca_{1/3}MnO_3$ [7]. An enhancement in magnetic moment and EB effect has been observed, when antiferromagnetic ferroelectric $BiFeO_3$ is placed in contact with ferromagnetic $La_{2/3}Sr_{1/3}MnO_3$ deposited on Si (1 0 0), associated with the formation of new FM phase in the $BiFeO_3$ sublattice brought on by magnetic reconstruction at the interface [8]. Although all these breakthrough investigations reveal interesting novel phenomena and the combination of ferromagnetism and ferroelectricity represent great potential for applications in novel devices, the number of materials for multifunctional applications with these properties at room temperature is quite limited. From this point of view, $BiFeO_3$ (BFO) is an excellent candidate as a room-temperature multiferroic material, known for its long-range ordering high Curie temperature (T_C ~1100K) and antiferromagnetic Néel temperature (T_N ~640K); this compound belongs to the perovskite family with lattice parameter of 3.96Å (in bulk) [9]. To study, the AF/FM coupling mechanics, we propose as strong FM candidate, half metallic perovskite manganite $La_{2/3}Sr_{1/3}MnO_3$ (LSMO), which has been widely used in spintronics with high spin polarization around 95% and possesses a T_C ~350K. The LSMO compound has a pseudocubic structure with a lattice constant of 3.87Å, similar as BFO, providing epitaxial growth of thin films. Here we investigate the coupling effect at the BFO/LSMO interface, by performing field cooling hysteresis loops, the interfacial interaction is observed when antiferromagnetic ferroelectric $BiFeO_3$ is placed in contact with ferromagnetic $La_{2/3}Sr_{1/3}MnO_3$. As consequence, a shift on the magnetization loop to negative values in the presence of magnetic field is observed. In turn, the ferroelectric hysteresis loop of the BFO/LSMO thin film shows that the LSMO buffer layer used as bottom electrode, greatly maintains the ferroelectric properties of the BFO thin film. These results will provide us some insight into the nature of the physical properties of the multiferroic/ferromagnetic heterostructures.

2. Experimental details

We have grown thin films based on AF-insulating BFO and ferromagnetic LSMO on (0 0 1) $SrTiO_3$ (STO) substrate. We used a multi-target dc-rf sputtering system, which permits in-situ growth of the BFO and LSMO materials. The LSMO layer was grown first via dc sputtering with oxygen-pressure at 3.0mbar, 850°C substrate temperature and power density of $2W/cm^2$. Then, the BFO layer was grown through rf sputtering at oxygen plasma pressure reduced at 1.0×10^{-1}mbar, and 600°C substrate temperature and rf power density of $3W/cm^2$; non-annealing process was required before cooling down to room temperature. In our studies, film thickness was chosen around 120±1nm and 10±1nm for the BFO and LSMO layers, respectively. Samples were structurally characterized via X-Ray Diffraction (XRD) in standard Bragg-Brentano configuration, using D'Xpert PRO diffractometer from PANalytical operating at 40kV and 80mA with Cu anode, λ=1.540Å and a hybrid monochromator to select the $K_{\alpha1}$ and $K_{\alpha2}$ radiation. Magnetic characterization used a DC magnetometer (Quantum Design PPMS VSM) performing thermal demagnetization curves in a range from 10 to 400K under zero field cooling (ZFC) (H=0Oe) and field cooling (FC) (H_{FC}=1000Oe); isothermal magnetization curves at ZFC and under field cooling FC (H_{FC}=5000Oe) for different temperatures (T=5 to 50K). In all samples, the magnetic field applied is parallel to the film plane. For the electrical measurements, circular Au electrodes ~15μm diameter were deposited on the BFO layer surface. We measured ferroelectric hysteresis loops at room temperature with the ferroelectric test system (Radiant Technologies) at 100Hz and until 4V.

3. Results and discussion

Figure 1 displays θ-2θ X-ray diffraction patterns for a LSMO film (a), BFO film (b) and (c) BFO/LSMO bilayer. All samples were grown on (0 0 1) STO substrate. The 2θ scans were performed between 20° and 80°. The (0 0 l) reflections of the STO substrate were identified in all patterns; Figure 1(a) identifies the Bragg peak for the LSMO layer; similarly, Figure 1(b) substrate peaks and (0 0 l) BFO reflections were found; whereas, Figure 1(c) shows the reflections for LSMO and BFO in the bilayer. There is no evidence of additional phases in the X-ray patterns, suggesting the epitaxial growth of the LSMO and BFO layers with (0 0 l) orientation perpendicular to the substrate surface. The LSMO peaks in the bilayer, maintain the position with respect to the single layer, while the BFO peaks are slightly displaced to high angle values, suggesting a decrease in the lattice parameter of the BFO material (Figure 1(c)). The respective lattice constant in bulk for the cubic STO, pseudocubic BFO, and cubic LSMO are 3.91Å, 3.96Å and 3.87Å, respectively. From the θ-2θ XRD data obtained around the (0 0 l) reflection, the out-of-plane lattice constant (*c*) was calculated for the BFO single layer 4.040±0.008Å and for the LSMO single layer 3.846±0.002Å. The lattice mismatch between LSMO and STO results in 1.5% in-plane tensile strain, causing a compression of the lattice constant in the out-of-plane direction for LSMO layers. On the other hand, the lattice mismatch between BFO and STO results in 5.3% in-plane compressive strain, causing an elongation of the lattice constant in the out-of-plane direction for BFO layers. However, the value of *c* for the BFO layer changes from 4.040±0.008Å to 3.951±0.008Å (2.2% reduction is obtained) when deposited onto the LSMO single layer. This reduction of *c* is accompanied by the expansion of the in-plane lattice constant, gives rise to large tensile strain on the LSMO, agreeing with other reports [10].

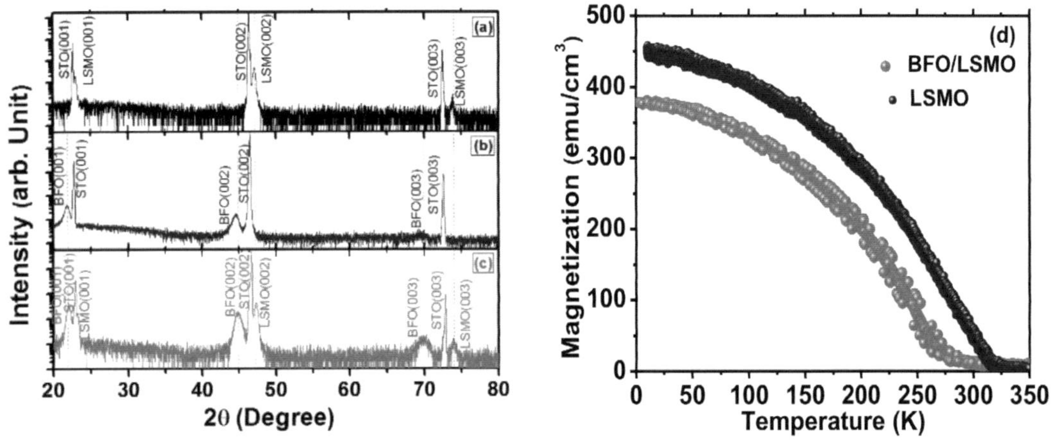

Figure 1. (a) XRD scans of the LSMO film (a), BFO film (b), and BFO/LSMO bilayer (c). All samples were grown on (0 0 1) oriented STO substrate. (d) FC magnetization versus temperature curves for the LSMO single layer (blue); and for the BFO/LSMO bilayer (red) with H_{FC}=1000Oe.

Figure 1(d) shows FC magnetization versus temperature curves of the LSMO single layer (blue) and BFO/LSMO bilayer (red). Data were collected during the warming cycle under the measuring field of 1000Oe after the sample was cooled under magnetic field of the same value (1000Oe). In all samples, the magnetic field was applied parallel to the film plane. The two curves exhibit the same typical behaviour of a FM manganite. Assuming a Gaussian distribution of domains, the FM-paramagnetic transition together with its width was calculated, as reported elsewhere on bilayers and single layers [5,11-13]. For the LSMO, we found a T_C ~340±1K, while for the bilayer it was 290±1K. The magnetization and T_C in the bilayer are consistently lower in comparison with the values obtained in the LSMO layer. In agreement with other reports, the magnetic properties strongly depend on parameters such as lattice strain, it has been shown that large values of tensile strain cause the T_C to decrease [14]. The smaller magnetization detected in the bilayer could be a result of the large tensile strain on the LSMO after the BFO is placed on top, through a Jahn-Teller effect [8,14]. You et al. observed a similar effect when LSMO was deposited on BFO, the decreasing of the T_C was explained in terms of a strain-induced layer formed in the BFO through a Jahn-Teller effect in response to the strain introduced by the top layer [15]. Additionally, remains in study if structural defects at the interface, oxygen vacancies, change in crystalline field of LSMO, spin canting at the interface, phase separation at the interface, strain-induced magnetic disorder in the film and other mechanisms can influence a decrease of the magnetization and T_C in these type of systems [16,17].

Magnetization as function of the magnetic field (M-H) have been collected for the BFO/LSMO bilayer under ZFC (H=0) and FC (H_{FC}=5000Oe) through different temperatures T=5, 10, 20, 40 and 50K (Figure 2(a)). The bilayers were warmed up to ~350K at the end of each return (the onset of the paramagnetic phase of LSMO), after cooling the sample under cycles of ±5000Oe at which the system was already saturated. The bilayer exhibits symmetric hysteresis under ZFC, inset Figure 2(a). However, we observed that the loop after FC exhibits a clear shift to negative values of the applied field, which shows strong temperature dependence. This exchange anisotropy (H_{ex}) created at the interface is known as exchange bias and is meanly attributed to AF/FM interfaces, in our case, when the AF-BFO layer is in contact with the FM-LSMO layer. Following the results reported in the literature, the observation of H_{ex} after the deposition of BFO on top of LSMO may be due to the strong coupling of the ferroelectric domains in BFO with the magnetic moment in LSMO, which occurs within the first few interface monolayers in the BFO sublattice [8]. To estimate the magnitude of the shift from these data, we defined H_{C1} as the location of the negative field and H_{C2} as the location of the positive field. The coercive field (H_C) is defined $H_C=-(H_{c1}-H_{c2})/2$ and the exchange bias (H_{ex}) is defined as the magnitude of the peak shift from zero as: $H_{ex}=(H_{c1}+H_{c2})/2$. The maximum value of H_{ex} was found at 5K (H_{ex}=76Oe), the clear evidence of interfacial effects and strong exchange coupling is manifest. We found strong temperature dependence on H_{ex} and H_C, both exhibit an exponential behaviour decreasing as function of temperature, Figure 2(b). Thus, the existence of frustration due to competing interactions is known to lead to an exponential decay of H_{ex} and H_C [18].

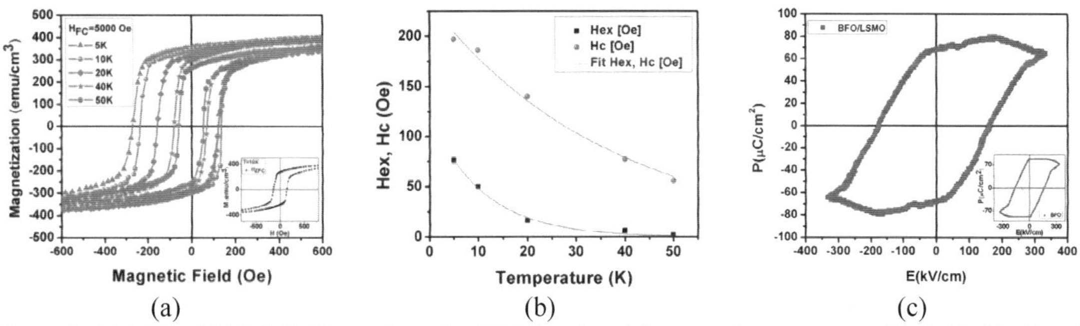

Figure 2. (a) M-H of BFO/LSMO sample under ZFC (inset) and for several temperatures (T=5, 10, 20, 40 and 50K) at H_{FC}=5000Oe. (b) Temperature dependence of H_C and H_{ex} obtained for the BFO/LSMO sample after cooled under 5000Oe from 400K. (c) P-E loops of the BFO/LSMO bilayer; inset shows the P-E loop for the BFO thin film deposited on Nb:STO substrate.

Hence, is reasonable to fit the observed H_{ex} and H_C with $H_{ex,C}=H_{ex,C}^0 \exp(-T/T_0)$, where H_{ex}^0 is the H_{ex} extrapolated value at T=0K, and T_0 is a parameter that decay exponentially, which multiplied by the Boltzmann constant can be associated to a characteristic energy of the exchange bias mechanism [18]. As the samples is cooled, H_{ex} decays (exponentially) showing a blocking temperature T_B ~50K. This is important because it sets an upper temperature limit where exchange bias can be observed and, thus, where the phenomenon can be operated. Our observations are consistent with the work done by S. Rao *et al.* [8], which reported strong temperature dependence of H_{ex}, obtaining a T_B ~53K. It has been reported by Schuller *et. al.* [4], and other authors [19-21], where the thicknesses of the AF layer play an important role in the blocking temperature of a system, that is, for thin films with thick AF layers T_B tends to be lower than the T_N. Considering that the anisotropy increases constantly below the Néel temperature, the variation of the AF layer thicknesses is a key to study the behaviour of the T_B in this type of systems.

Figure 2(c) plots the polarization-electric field (P–E) hysteresis loops for the Au/BFO/LSMO. The inset shows the P-E for the Au/BFO/Nb:STO capacitor structure at room temperature. A well-established hysteresis loop was observed for the BFO deposited directly onto the Nb:STO substrate, and for the BFO directly deposited on the LSMO layer. Polarization saturation values were found at ~73μC/cm^2 for the BFO and ~65μC/cm^2 for the BFO/LSMO bilayer, close to the values reported in the literature [22, 23]. It, therefore, shows that the LSMO buffer layer greatly maintains the ferroelectric properties of the BFO thin film, implying that can be used as bottom electrode to facilitate the domain reversal in BFO as have been reported previously by Ke *et. al.* [22]. Our studies are still at an early stage in the understanding of these heterostructures interfaces; therefore, the coexistence of ferroelectric and ferromagnetic orders occurring at the interface of the BFO/LSMO system, suggests that it may be a promising candidate system to pursuit the magnetoelectric effect.

4. Conclusion

In summary, BFO/LSMO was successfully grown on STO single crystals, deposited by rf and dc sputtering technique in oxygen atmosphere. Layers were grown textured on the STO substrate without presence of any measurable secondary phases. Our data evidence interface effects, in agreement with the JT effect, the slight relaxation of the BFO in the lattice plane after growing on the LSMO layer observed in the x-ray diffraction patterns, leads to a decrease of M_S and T_C, which are properties affected under larger tensile strain. Moreover, after field cooling the sample; the exchange bias when antiferromagnetic/ferroelectric BFO is placed in contact with ferromagnetic LSMO is detected. The exchange bias exhibits a stronger temperature dependence, an H_{EB} ~76Oe is noticed at 5K, which decays (exponentially) as function of temperature. Ferroelectric measurements showed that the LSMO buffer layer greatly maintains the ferroelectric properties of the BFO thin film, implying that the LSMO used as bottom electrode helps to facilitate the domain reversal in BFO.

Acknowledgements

This work was supported by "Colciencias" through research projects 1106-56-933104, "Colciencias-Universidad del Valle" C.I. 7917, "Universidad del Valle" C.I. 7978, CI 7999, and the "Centro de Excelencia en Nuevos Materiales (CENM)".

References

[1] Xu Q, Sheng Y, Xue X, Yuan X, Wen Z and Du J 2014 *Jpn. J. Appl. Phys.* **53** 08NM01
[2] Gross K, Westerholt K and Zabel H 2016 *New Journal of Physics* **18** 033007
[3] Ramesh R 2010 *Nat. Mater.* **9** 380-381
[4] Nogués J and Schuller I K 1999 *J. Magn. Magn. Mater.* **192** 203-232
[5] Dominguez C, Ordonez J E, Gómez M E and Lopera W 2014 *J. Phys. Conf Ser.* **480** 012029
[6] Valencia S, Crassous S, Bocher L, Garcia V, Moya X, Cherifi R O, Deranlot C, Bouzehouane K, Fusil S, Zobelli A, Gloter A, Mathur N D, A Gaupp, Abrudan R, Radu F, Barthélémy A and Bibes M 2011 *Nat. Mater.* **10** 753-758
[7] Chakhalian J, Freeland J W, Habermeier H -U, Cristiani G, Khaliullin G, Van Veenendaal M and Keimer B 2007 *Science* **318** 1114-1117
[8] Rao S S, Prater J T, Wu F, Shelton C T, Maria J P and Narayan 2013 *J Nano Letters* **13** 5814-5821
[9] Catalan G and Scott J F 2009 *Adv. Mater.* **21** 2463
[10] Wang L, Wang Z, Jin K J, Li J Q and Yang H X 2013 *Appl. Phys. Lett.* **102** 242902
[11] Campillo G, Berger A, Osorio J, Pearson J E, Bader S D, Baca E and Prieto P 2001 *J. Magn. Magn. Mater.* **237** 61-68
[12] Ordoñez J E, Gómez M E and Lopera W 2016 *Revista Mexicana de Física* **62** 543-548
[13] Ordoñez J E, Gómez M E, Lopera W and Prieto P 2015 *J. Phys.: Conf Ser.* **614** 012009
[14] Takamura Y, Chopdekar R V, Arenholz E, Suzuki Y 2008 *Appl. Phys. Lett* **92** 162504
[15] You L, Lu C, Yang P, Han G, Wu T, Luders U, Prellier W, Yao K, Chen L and Wang J 2010 *Adv. Mater.* **22** 4964
[16] Ordoñez J E, Gómez M E, Lopera W and Prieto P 2013 *IEEE Trans. Magn.* **49(8)** 4586-4589
[17] Ordoñez J E, Gómez M E, Lopera W and Prieto P 2014 *J. Phys.: Conf Ser.* **480** 012031
[18] Nalwa H 2001 *Handbook of Surfaces and Interfaces of Materials* (USA: Academic Press)
[19] Kiwi M 2001 *J. Magn. Magn. Mater.* **234** 584-595
[20] Berkowitz A and Takano K 1999 *J. Magn. Magn. Mater.* **200** 552-570
[21] Nogués J, Sort J, Langlais V, Skumryev V, Suriñach S, Muñoz J y Baró M 2005 *Phy. Reports* **422** 65-117
[22] Ke Q, Lu W, Huang X and Wang J 2012 *J. Electrochem. Soc.* **159** G11-G14
[23] Wang J, Neaton J B, Zheng H, Nagarajan V, Ogale S B, Liu B, Viehland D, Vaithyanathan Schlom D G, Waghmare U V, Spaldin N A, Rabe K M, Wuttig M and Ramesh R 2003 *Science* **299** 1719

IOP Conf. Series: Journal of Physics: Conf. Series **935** (2017) 012030 doi:10.1088/1742-6596/935/1/012030

In vitro electrochemical behaviour of Chitosan-PEG coatings obtained on Ti6Al4V by dip coating

A Sandoval Amador[1], S M Guerrero-Barajas[1], D R Hernandez-Salas[1], D K Sierra-Herrera[1], H A Estupiñan-Duran[2] and D Y Peña Ballesteros[1]
[1] Universidad Industrial de Santander, Bucaramanga, Colombia
[2] Universidad Nacional de Colombia, Medellín, Colombia

E-mail: anderson84f@outlook.com

Abstract. In the present study, coatings of Chitosan - Polyethylene Glycol was obtained by dip coating on Ti6Al4V from polymer mixtures (30-70, 50-50, 70-30). The aim of our study was analysed their ability to formation of apatite and their electrochemical behaviour in the presence of a simulated physiological fluid. The coating properties were evaluated by electrochemical impedance, atomic absorption, SEM and FTIR techniques. The results showed that surfaces coated with a 50-50 ratio of chitosan PEG stimulate the formation of calcium phosphates on the surface of the material. On the other hand, it was found that the number of layers generate a very significant effect on the resistance to the degradation of the alloy in the presence of SBF, generating an increase of up to 1 and 3 orders of magnitude in the impedance of the coating of 5 and 7 layers respectively compared to the Ti6Al4V alloy.

1. Introduction

In recent years, titanium and its alloys have developed into the key materials for orthopedic and dental implants surgery, owing to their excellent bulk of properties [1-3]. Nevertheless, these materials are associated with certain clinical challenges. For this reason, a wide number of works have recently been developed to improve the properties of the titanium and its alloys. Some of these studies have focused on improving the oseo-conductive capacity. While others studies have been focused on reducing the bacterial adhesion on the surface of the implant [4-7].

One of the ways to achieve this has been the modification of the layer of titanium oxide that is formed naturally on the surface of these materials. This modification has allowed improving the roughness of the surface thus favouring the cellular response and also the ability to integrate the material to the bone. [8-10]. Another alternative is the chemical treatment with solutions of NaOH or KOH, this type of modification has allowed to incorporate ions (OH^-) to the surfaces of titanium, which favour the nucleation of calcium phosphates, as is the case of hydroxyapatite [11]. Other authors have chosen to modify the composition of the titanium surface by the growth of calcium phosphates or other compounds that resemble the composition of the bone mineral matrix, in order to generate a surface whose composition favours osseointegration of the implant [12-14]. Likewise, there are works where the main materials of the superficial modification are the polymeric. These materials, when implanted and interact with the physiological environment, initiate a slow degradation process which allows the release of substances that are metabolized by the cells and thus allow to favour the conditions to improve cell adhesion and proliferation [15]. On the other hand, some polymeric materials manage to reduce the adhesion of bacteria, as is the case of chitosan and PEG [16,17].

IOP Conf. Series: Journal of Physics: Conf. Series **935** (2017) 012030 doi:10.1088/1742-6596/935/1/012030

The aim of this study was to evaluate the effect of the concentration and the number of layers of chitosan PEG coatings obtained by dip coating on Ti6Al4V in the capacity of apatite's formation and the electrochemical behaviour in simulated body fluid.

2. Experimental

Ti6Al4V cylindrical disc with 14mm diameter and 3 mm thick were ground with silicon carbide paper in successive grades from 120 to 1200 grit. Then were polished with a slurry of alumina (particle size 0.5μm) with cotton cloths. Subsequently the disc was cleaned by ultrasonically in ethanol for 15 minutes.

An alkaline treatment was performed in order to activate the surface, which was immersed in 10M NaOH solution for 24 hours at 60°C in a hydrothermal bath. The samples were washed in distilled water and dried at room temperature. In order to stabilize the solid formed, a heat treatment was carried out with a heating ramp of 25°C to 450°C in 43 minutes, maintaining this temperature for 1 hour.

The polymer solutions of Chitosan - polyethylene glycol was prepared from individual solutions of chitosan and PEG, dissolved in 0.5M acetic acid and water respectively. Subsequently they were mixed to form a single homogeneous solution. Three solutions were obtained with 30/70, 50/50 and 70/30 Chitosan-PEG ratios. The amounts of chitosan and PEG used for the preparation of the polymer ratios are shown in Table 1.

Table 1. Reagents to prepare different ratios of chitosan polyethylene glycol.

Chitosan (g)	Acetic acid (ml)	PEG (g)	Distilled water (ml)	Chitosan-PEG Ratio
0.09	9	0.21	21	30:70
0.15	15	0.15	15	50:50
0.21	21	0.09	9	70:30

The polymer solutions Chitosan-PEG were used to coat the discs of the alkaline treated Ti6Al4V alloy by immersion and extraction of the substrate using the Dip Coating technique. The immersion time of the samples in the polymer solution was 10 seconds, and then a drying was carried out in a preheated oven at 50°C for 30 minutes. The number of layers made on the specimens was 3, 5 and 7 layers.
The coated samples were immersed in polypropylene containers with 30ml of a Hank's solution (NaCl 8.0, KCl 0.4, $MgSO_47H_2O$ 0.2, $CaCl_2$ 0.185, Na_2HPO_4 0.046, KH_2PO_4 0.06, Glucose 1.0, $NaHCO_3$ 0.35g/L) for 180 hours to induce apatite formation in a hydrothermal bath at 37°C.

The coating morphology was analysed by scanning electron microscopy before and after than immersion on the Hanks solution. The functional groups present in the coating were analysed by Fourier transform infrared spectroscopy. The chemical reactivity of the coatings was evaluated by quantifying the amount of calcium adsorbed by the surfaces of Chitosan PEG by analysing the variation of the calcium concentration in the Hanks solution, using the atomic absorption technique with which it was possible determine the calcium present the Hanks solution after the immersion test. Finally, the electrochemical behaviour of the samples was evaluated simulating physiological conditions in hanks solution at 37°C. The tests were electrochemical impedance spectroscopy, for which a sweep of frequencies of 0.01 to 100 000Hz was applied applying $30mV_{rms}$. A potentiostat-galvanostat Gamry 600 was used for the realization of this tests, using a flat cell composed of three electrodes in which the Ti6Al4V discs were used as working electrode, graphite as counter electrode and Ag/AgCl 3M KCl as reference electrode. The exposure area of the sample was $0.63cm^2$. The samples were previously immersed in the solution for 5 minutes to allow stabilization of the system.

3. Results and discussion

Figure 1 shows the morphology of the coatings of chitosan polyethylene glycol. As can be seen the number of layers the coating becomes more homogeneous and decreases the number of cracks present on the surface. White spots observed as precipitates on surfaces are associated with chitosan particles that were not completely dissolved.

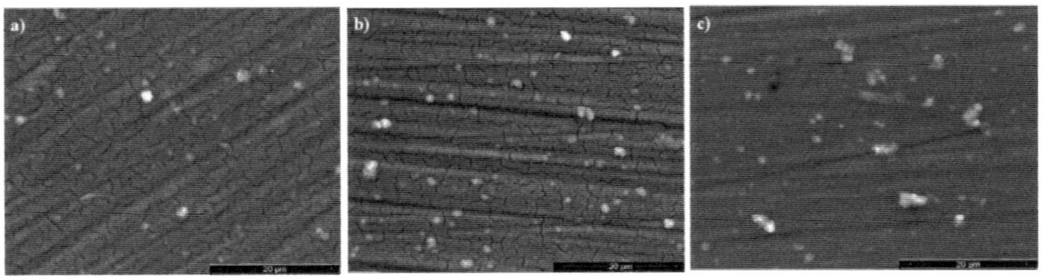

Figure 1. Morphology of surfaces Chitosan PEG. 5 coatings a) 30/70 b) 50/50 c) 70/30.

Figure 2 shows the analysis of the functional groups present on the developed chitosan polyethylene glycol surfaces. There is an asymmetric and flexion between 2860 and 3350cm^{-1} band, due to the hydroxyl and carboxyl groups present in chitosan rings, also NH deformation if primary amide associated with chitosan around 1617 and 1637cm^{-1}. The presence of these bonds favours primary amide between the calcium cations, the phosphates and the surface of the biomaterial, which could cause an increase in the nucleation and the precipitation of the calcium phosphates in the surface.

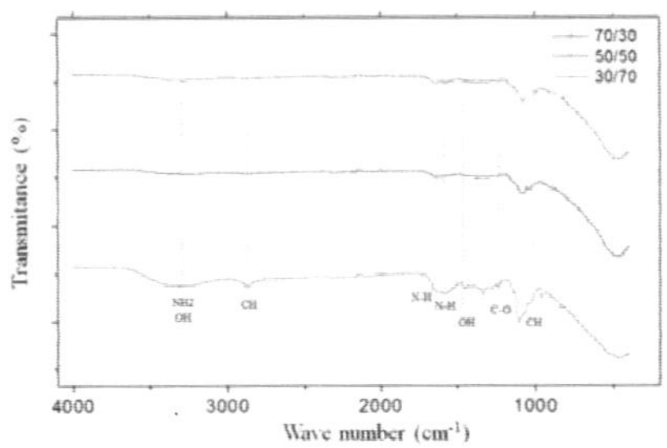

Figure 2. Infrared spectrum the ratio 30/70, 50/50, 70/30 Chitosan: PEG, 5 layers.

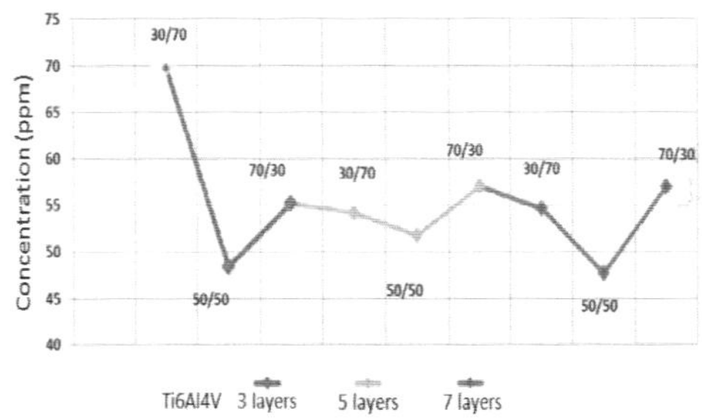

Figure 3. Concentration of calcium present in the fluid for different polymer ratios.

As can be seen in Figure 3, exist is a variation in the calcium concentration of aliquots with the Hank's solution, due to the phenomenon of adsorption on this metal ion by test samples, which reduces its concentration in each aliquot. It should be noted that the coatings that adsorbed the most calcium were 50/50 blend, this mean that it is the best combination of properties the polymer blend; By the chitosan having a porous matrix makes it more apt to interact with the fluid because it increases its surface area, favouring the nucleation and growth of phosphates [18]. In turn the PEG is a water-soluble polymer, which readily allows the activation of oxygen species which could provide an electron-bonding network to adsorb Ca^{2+} by electrostatic interaction to form apatite nucleation sites [19].

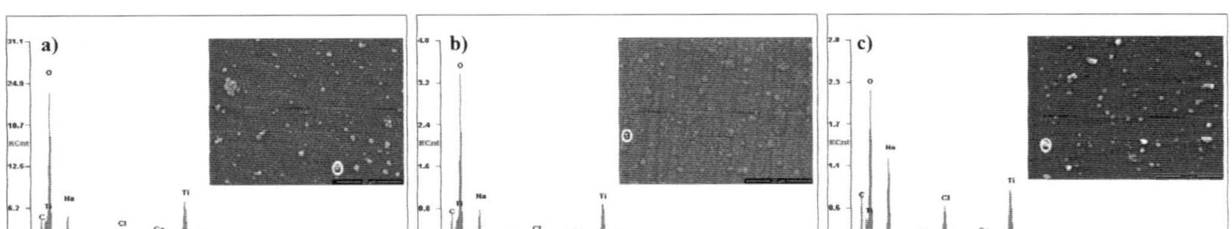

Figure 4. Compositional analysis of chitosan polyethylene glycol coatings. (a) 30/70, (b) 50/50, and (c) 70/30.

The composition and morphology of the chitosan polyethylene glycol coatings are shown in Figure 4. It can be seen that after 7 days of immersion in Hanks the coatings show precipitates whose composition indicates a possible formation of calcium phosphates on the surface. In addition, it can be observed that other elements appear in the composition which are associated with the precipitation of other salts that are present in the Hanks solution. It is noteworthy that the coatings with the 50/50 and 70/30 ratios are those with the highest number of calcium phosphate precipitates, because chitosan by its cationic nature allows it to interact with negative ions as well as phosphate groups (PO43-) present in Hank's solution.

In Figure 5(a), at high frequencies (10^4-10^5Hz), the solution resistance presents a pseudocapacitive behaviour corresponding. In very low frequencies (0.01-1Hz) the 50/50 and 70/30 ratios have the highest electrochemical impedance associated with a highly resistive behaviour, this behaviour corresponds to the interface coating-material base Ti6Al4V. For the 30/70 ratio a very stable behaviour is observed in low and medium frequencies, which indicates the presence of a passive, compact and very stable layer [20].

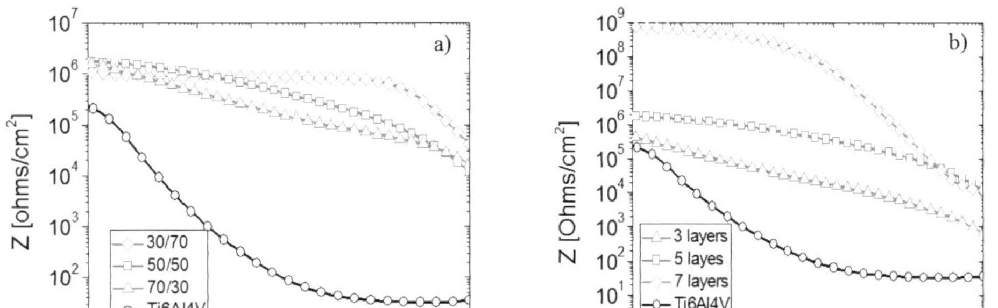

Figure 5. Bode diagram. (a) different Chitosan / PEG ratios, (b) different number layers.

In Figure 5(a) behaviour solution is observed at high frequencies (10^4-10^5Hz), also a resistive behaviour for 3 layers and capacitive one for 5 and 7 layers. This behaviour happens in the interface coating-physiological solution where the resistivity for 3 layers refers to a compact layer, and the capacitive effect for 5 and 7 layers, it is due to a greater interaction of the polymer layer with Hank's solution due to a larger area of exposure by the porous chitosan structure for a 50/50 ratio. In addition, it is observed that at low frequencies (0.01-1Hz) it is mainly observed the behaviour of the base material with the coating where the resistance of coating increases as the number of layers increases, where the highest is for 7 layers and the lowest for 3 layers.

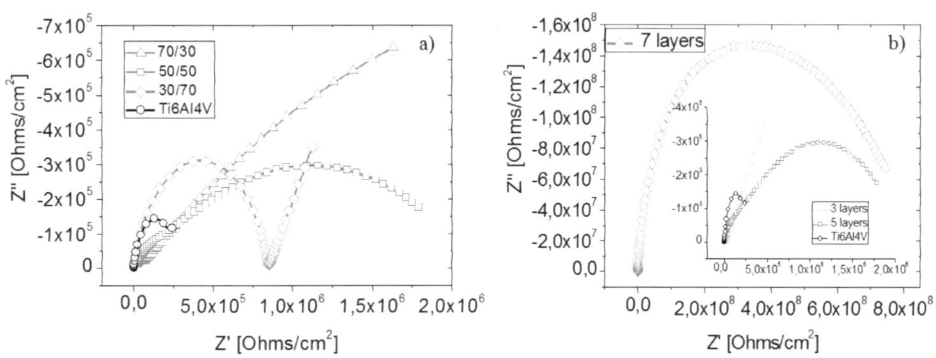

Figure 6. Nyquist diagram (a) different Chitosan/PEG ratios, (b) different number layers.

Figure 5 shows the bode plots for the chitosan-PEG coatings. It can be observed that in the high frequency range (10^4-10^5Hz) the resistance to the solution presents different values depending on the type of coating being evaluated. This change in resistance behaviour is associated with the solubility of PEG in the electrolyte. On the other hand, in the low frequency range (10^{-2}-10^0Hz) it is observed that the variation in the different PEG chitosan ratios can generate an increase of up to one order of magnitude in the impedance of the Ti6Al4V alloy (Figure 5 (a)), while increasing the number of layers of the coating, the impedance increases up to 4 orders of magnitude (Figure 5(b)).

As can be seen in Figure 6(a), the diameters of the semicircles in the Nyquits diagram increase with increasing concentration of chitosan, this behaviour can be associated with the low solubility of chitosan. The effect of increasing the number of coating layers on the impedance is shown in Figure 6(b), where we can see that the 7-layer coating has a larger dome diameter, which is associated with a high resistance to degradation. In addition, the increase in the height of the domes indicates that the coating of chitosan polyethylene glycol tends to be highly capacitive, which could favour the processes of electrical interaction between the surface and the anionic and cationic species of the Hanks solution.

4. Conclusions

The coatings of PEG chitosan showed morphology in function of the number of layers, where the highest homogeneity was achieved after depositing 7 layers on the surface of the Ti6AL4V alloy. Under the conditions of study, it was achieved after depositing 7 layers surface of the Ti5AlV alloy. Under the conditions of study, it was possible to determine that the coatings with the 50/50 ratio favour the formation of the adsorption o calcium phosphates on the surface, besides these two coatings presented highest resistance to the degradation in the presence of the Hanks solution.

Acknowledgment

The authors wish to thank Jaimes Rueda & Cia Ltda, Universidad Industrial de Santander, Vicerrectoria de Investigación y Extension and Grupo de Investigaciones en Corrosión for funding this Project under internal code 1888.

References

[1] Cai K, Hu Y, Jandt K D and Wang Y 2008 *J. Mater. Sci: Mater. Med.* **19** 499
[2] Fang K, Saong W, Wang L, Wei H, Ren S, Xu X and Song Y, 2014 *Int. J. Nanomedicine* **9** 4649
[3] Sandoval Amador A, Miranda Vesga L J, Pérez Martínez J S, Peña Ballesteros D Y and García Vergara S J 2016 *Revista Materia* **21(3)** 606
[4] Sandoval Amador A, Montañez Supelano N D, Vera Arias A M, Escobar Rivero P and Peña Ballesteros D Y 2017 *J. Phys.: Conf. Ser.***786** 012010
[5] Chua P H, Neoh K G, Kang E T and Wang W 2008 *Biomaterials* **29** 1412
[6] Buxadera-Palomero J, Calco C, Torrent Camarero S, Gil F J, Mas Moruno C, Canl C and Rodriguez D 2017 *Colloids surf B Biointerfaces* **152** 367
[7] Kumeria T, Mon H, Aw M S, Gulati K, Santos A, Griesser H J and Losic D 2015 *Colloids surf B Biointerfaces* **130** 255
[8] Lim S S, Chai C Y and Loh H S 2017 *Mater. Sci. Eng. B.* **76** 144
[9] Beltrán Partida E, Valdez Salas B, Curiel Álvarez M, Castillo Uribe S, Escamilla A and Nedev N 2017 *Mater. Sci. Eng. B.* **76** 59
[10] Mansoorianfar M, Tavoosi M, Mozafarinia, Ghasemi A and Doostmohammadi A 2017 *Surf. Coat. Technol.* **321** 409
[11] Yu P, Ning C, Tan G, Zhang Y, Liao J, Sun J, Peng Q, Zhong M, Yu Z and Ni G 2013 *Surf. Coat. Technol.* **235** 267
[12] Carrado A, Perrin-Schmitt F, Le Q V, Giraudel M, Fischer C, Koening G, Jacomine L, Behr L, Chalom A, Fiette L, Morlet A and Pourroy G 2017 *Dent. Mater.* **33** 321
[13] Liu S, Li H, Su Y, Guo Q and Zhang L 2017 *Mater. Sci. Eng. B* **70** 805
[14] Abd El-Aziz A M, Ranian M, E B, Nahla A T, Azza E M and Sherif H K 2017 *Mater. Sci. Eng. B* **76** 1188
[15] Liu X, Chu P L and Ding C 2004 *Mat. Sci. Eng. R* **47** 49
[16] Gebhardt F, Seuss S, Turhan M C, Hornberger H, Virtanen S and Boccaccini A R 2012 *Mater. Lett.* **66** 302
[17] Hüsnügül Yilmaz A, and Erdal C 2017 *Prog. Org. Coat.* **102** 194
[18] Pallab D, Santanu D and Jyotirmoy C 2012 *Carbohyd. Polym.* **87** 1354
[19] Jegatheeswaran S and Sundrarajam 2015 *Mater. Sci. Eng. C* **51** 174
[20] Mina Escobar A, Aperador Chaparro W, Caicedo Angulo j C, Tirado Mejia L S and Aguilar Castro Y 2013 *Rev. Fac. Ing.* **22** 61

IOP Conf. Series: Journal of Physics: Conf. Series **935** (2017) 012031 doi:10.1088/1742-6596/935/1/012031

Formation of Ca/P ceramic coatings by Plasma Electrolytic Oxidation (PEO) on Ti6Al4V ELI alloy

Y Rodriguez-Jaimes[1], **D I Naranjo**[1], **S Blanco**[1] **and S J García-Vergara**[1]
[1] Universidad Industrial de Santander, Bucaramanga, Colombia

E-mail: sagarver@uis.edu.co

Abstract. The formation of PEO ceramic coatings on Ti6Al4V ELI alloy was investigated using a phosphate/calcium containing electrolyte at 300 and 400V at 310K for different times. The Plasma Electrolytic Oxidation (PEO) coated specimens were then heat treated at 873 and 1073K for 2 hours. Scanning electron microscopy, Energy Dispersive X-Ray Spectroscopy (EDS) and X-ray diffraction analysis were used to study the composition and the morphology of the ceramic coatings. The corrosion behaviour of the coatings was studied by Electrochemical Impedance Spectroscopy (EIS) in Simulated Body Fluid (SBF). The PEO-treated specimens primarily revealed a porous structure with thickness between 4 and 12μm, according to the voltage and process time used. The coatings are mainly composed of hydroxyapatite; however, as the voltage and anodizing time increase, the Ca/P ratio decreases. Generally, the corrosion resistance of the alloy was improved by the PEO-treated coatings, although the specimens treated at 1073K showed the presence of cracks that reduced the protective effect of the coatings.

1. Introduction

Titanium and its alloys are widely used as medical, orthopedic and dental implants due to their high biocompatibility, high strength to weight ratio, high load capacity, chemical stability, low toxicity and excellent corrosion resistance [1]. However, titanium alloys release small amounts of metal ions once they are exposed to corporal fluids, causing cytotoxic, genotoxic, carcinogenic and mutagenic effects [2,3]. Several methods have been used to modify the surface condition of the alloy, in order to produce a stable barrier layer that reduces metal ion release, while increasing alloy bioactivity [4,5]. The PEO is a method for surface finishing of important light alloy metals. It results in a multifunctional ceramic coating, which is relatively thick and contains containing amorphous and crystalline phases with enhanced corrosion, wear, dielectric and thermal barrier properties [6-8]. One of the major advantages of the PEO process is the possibility of incorporating Ca and P ions into the titanium alloy surface by controlling process parameters such as composition of the electrolyte, applied voltage, current density and treatment times [9-11]. The high temperatures in the micro discharge channels during the PEO process allow for the combinations of Ca and P ions to form a hydroxyapatite phase in the coatings [12]. The presence of hydroxyapatite enhances the bioactivity of titanium alloys [13]. This study explores the influence of voltage and treatment time on the formation of Ca and P rich PEO coating on a Ti6Al4V alloy, in order to improve the corrosion resistance of the alloy in a SBF solution.

2. Experimental

2.1. Materials and surface treatment

Samples of Ti6Al4V-ELI alloy of 13.7 mm diameter and 1.55 mm thick were grounded by 80, 120, 180, 220, 320, 400 and 600 grid SiC paper and cleaned using ultrasonic agitation in acetone, followed by rinsing in deionized water and drying in warm air. The PEO treatment was carried out using a voltage controlled DC power supply (Kepco BHK 500-0.4 MG) in an electrolyte comprising 0.06M calcium glycerophosphate (Ca-GP) and 0.3M calcium acetate monohydrate (CA). Ti6Al4V alloy samples connected to the positive output of the power supply served as a working electrode, whereas a stainless-steel sheet was used as a counter electrode. The cell was immersed in a stainless-steel vessel with water to maintain the cell temperature at 309K. The process was carried out at 300 and 400 V for 180 and 420s. The treated samples were washed with deionized water and dried with cold air. Finally, they were subjected to a crystallization heat treatment at 873 and 1073K for 2h in a Carbolite CWF 1200 furnace.

2.2. Coating characterization

Surface morphologies of the coatings were investigated using a Quanta 650 FEG Scanning Electron Microscope (SEM) equipped with microwave Energy Dispersive X-Ray Spectroscopy (EDS). The phase structure of the PEO coatings was detected by X- Ray Diffraction (XRD) using a Bruker D8 Discovery equipment. Coating thickness was evaluated using an eddy current gauge.

2.3. Electrochemical experiments

Electrochemical Impedance Spectroscopy (EIS) and Linear Polarization Resistance (LPR) measurements were carried out at room temperature in an aqueous SBF solution described by Kokubo [14]. A standard three-electrode cell arrangement was used, in which the coated samples served as the working electrode, a platinum wire acted as the counter electrode, and an Ag/AgCl electrode was employed as reference. Electrochemical measurements were performed using a GAMRY 650 potentiostat/galvanostat, with a 23mV amplitude perturbation and a frequency scan from 10^4 to 0.01Hz.

3. Results and discussion

3.1. Coating morphology and composition

The coating surfaces revealed micropores of different sizes and shapes. The porous structure strongly depends on the voltage and time involved in the PEO process (Figure 1) [9]. It can be clearly seen that at low voltage, fine micropores with sizes less than 1 μm in diameter are randomly distributed on the surface (Figure 1 (a) and (b)). On the other hand, micropores with diameters larger than 1 μm are observed on the samples treated at high voltage (Figure 1 (c) and (d)). The number of micropores decreases as the time and applied voltage increase. However, the diameter of the micropores is bigger for higher voltages. On the other hand, cracks of dielectric rupture were observed due to the rupture of the titanium oxide layer formed at 400V (Figure 1 (c) and (d)). The cracking is caused by the high energy values released during the PEO process, which could have increased the ionic transport towards the metal/coating interface, and therefore, could have also increased the conductivity of the coating [9]. The Ca/P ratio was obtained by EDS analysis for each PEO condition before heat treatment. It was observed that by increasing the voltage and the treatment time the Ca/P ratio decreases. The coatings are constituted of non-stoichiometric hydroxyapatite (HAp), because the Ca/P ratio is less than 1.67. However, Ca/P ratio values can be associated with other calcium phosphates. For example, in the coatings obtained at 420s and 300V, the Ca/P ratio is close to dicalcium phosphate dehydrate ($CaHPO_4.2H_2O$) or brushite (Ca/P: 1.0), while for the coatings obtained at 180 s and 300V the Ca/P ratio is close to tricalcium phosphate ($Ca_3(PO_4)_2$, Ca/P: 1.5) [15]. The previous indicated that crystals of brushite and tricalcium phosphate could be formed during PEO treatment [16].

The effect of heat treatment temperature on the morphology and Ca/P ratio of the coating was studied using the PEO treated samples at 300V for 180s. Figure 2 shows the surface morphology of PEO coatings heat treated for 2h at 873 and 1073K. At 873K, the surface morphology showed no appreciable changes (Figure 2 (b)) with respect to the untreated sample (Fig 2 (a)). On the other hand, the sample treated at 1073K shows a rough surface (Figure 2 (c), (d)), probably due to the crystallization of amorphous species, allotropic transformations or formation of crystalline species [17]. Rutile crystals are formed above 873K [18] and tricalcium phosphate is formed on the surfaces of the PEO coatings after heat treatment at 993 K [16].

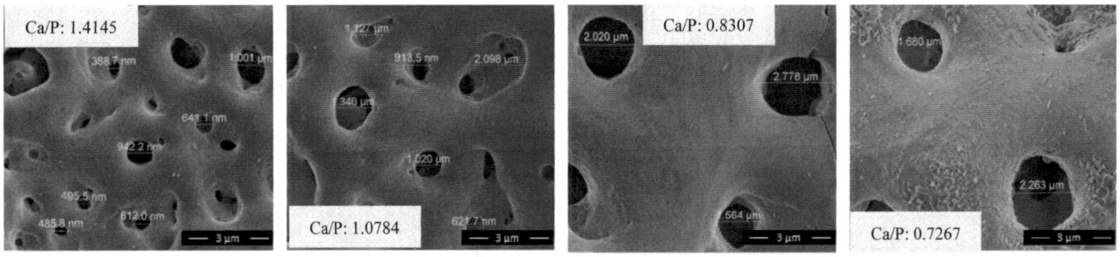

(a) (b) (c) (d)

Figure 1. Scanning electron micrographs of PEO coatings formed on Ti6Al4V ELI alloy in 0.06M Ca-GP+0.3M CA solution. (a) 180s, 300V, (b) 420s, 300V, (c) 180s, 400V and (d) 420s, 400V.

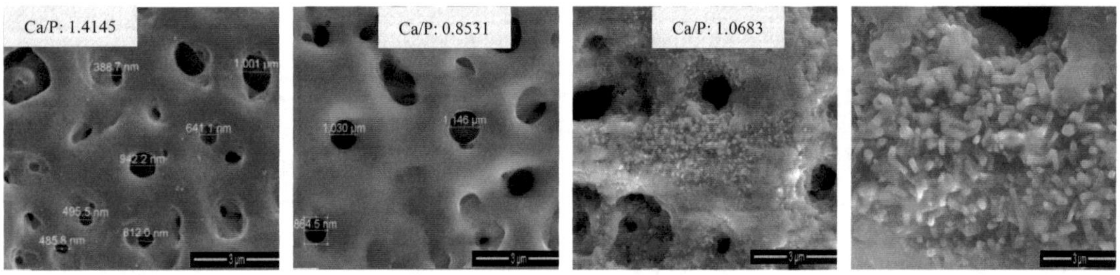

(a) (b) (c) (d)

Figure 2. Scanning electron micrographs of PEO coatings formed on Ti6Al4V ELI alloy in 0.06M Ca-GP+0.3M CA solution for 180s at 300V, before and after heat treatment (HT). (a) No HT, (b) HT to 873K, (c) HT to 1073K and (d) detail of condition (c).

XRD diffraction analysis indicates that the coatings are mainly composed of titanium dioxide (TiO_2), as anatase or rutile (Figure 3). The anatase phase decreases with the increase of heat treatment temperature due to its polymorphic transformation into rutile. Importantly, the presence of hydroxyapatite in the coating is confirmed. Hydroxyapatite enhanced the bioactivity properties of the coatings. Also, other phases are evident from the XRD patterns. Peaks corresponding to β-tricalcium phosphate or Whitlockite (β-$Ca_3(PO_4)_2$) are observed in the samples heat treated at 1073K. Whitlockite is one of the crystalline calcium phosphates and has a Ca/P ratio equal to 1.5 [15]. Brushite ($CaHPO_4 \cdot 2H_2O$) peaks are observed for the sample heat treated at 1073K in good agreement with the SEM and EDS results. The heat treatments improve the crystallinity of TiO_2 and lead to the crystallization of amorphous calcium phosphates.

3.2. Coating thickness

Table 1 presents the average thickness measurement of the PEO coatings. The coatings formed at 400V show higher values (>10μm) than the ones formed at 300V (<5μm). At higher voltage, the energy released in the process is great enough to break the barrier layer initially formed and continue the oxidation process [18]. The potential difference applied between the cathode and the anode exceeds the dielectric limit of the coating more easily, the rupture of the formed layer occurs for longer time resulting in a thicker coating. However, the thickness of the coatings does not increase significantly as the treatment time increases.

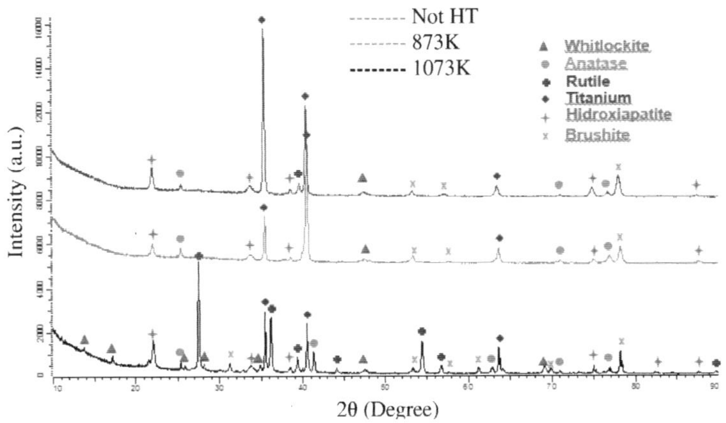

Figure 3. XRD patterns for Ti6Al4V ELI alloy following PEO treatment in 0.06M Ca-GP+0.3M CA solution at 300V for 180s, before and after heat treatment (HT).

Table 1. Average thickness of the PEO coatings formed on Ti6Al4V ELI alloy in 0.06M Ca-GP+0.3M CA solution.

PEO condition	180s–300V	420s–300V	180s–400V	420s–400V
Thickness (μm)	4.75	5.00	10.63	12.13

3.3. Electrochemical behaviour of PEO coatings in SBF

The polarization resistance values of the PEO coatings are presented in Table 2. The coatings obtained at 300V for 180s and heat treated at 1073K showed the highest polarization resistance values. The heat treatment appears to have a significant effect on the polarization resistance. There is an increase of two orders of magnitude relative to the sample without heat treatment. The PEO coatings obtained showed a capacitive behaviour at high frequencies, except for the condition obtained at 180s, 300V and heat treated at 1073K, which presented a resistive behaviour. The Nyquist spectra (Figure 4) shows that the samples without heat treatment and those heat treated at 1073K, except for the 180s at 300V condition, exhibited charge transfer control (represented by a dome) and diffusion control (represented by a straight line). On the other hand, diffusion control predominated in all systems heat treated at 873K.

Table 2. Polarization resistance values of Ti6Al4V ELI alloy following PEO treatment in 0.06M Ca-GP+0.3M CA solution.

PEO condition	Not HT	HT 873 K	HT 1073 K
(a) 180s–300V	1.5E+05	8.3E+04	6.7E+07
(b) 420s–30 V	2.6E+05	8.3E+05	2.2E+05
(c) 180s–400V	2.5E+05	4.4E+07	3.6E+04
(d) 420s–400V	2.3E+05	6.0E+08	1.2E+05

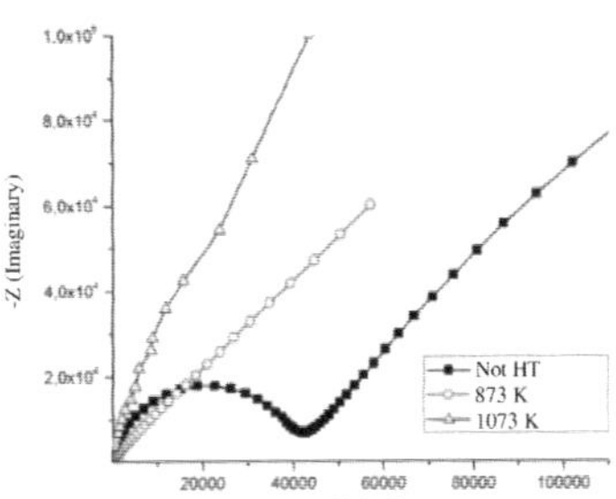

(a)

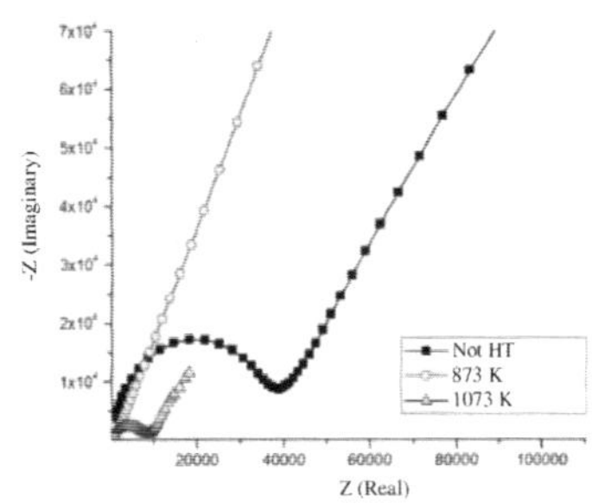

(b)

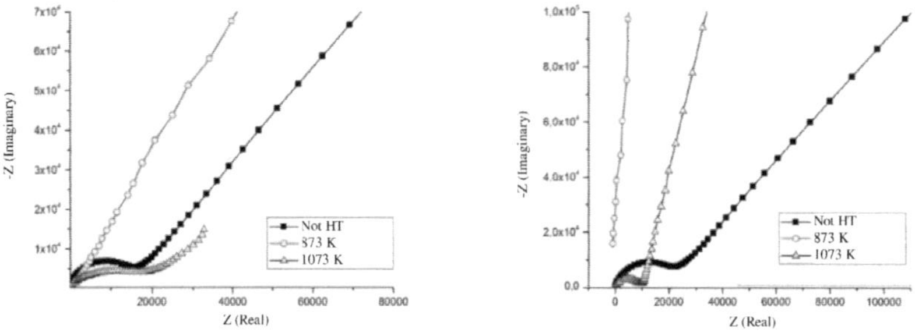

(c) (d)

Figure 4. Nyquist diagrams of Ti6Al4V ELI alloy following PEO treatment in 0.06M Ca-GP+0.3M CA solution, before and after heat treatment. (a) 180s, 300V, (b) 420s, 300V, (c) 180s, 400V and (d) 420s, 400V.

4. Conclusions

PEO coatings rich in Ca and P were obtained on a Ti6Al4V alloy using a calcium glycerophosphate (Ca-GP) and calcium acetate monohydrate (CA) electrolyte at 300 and 400V. The Ca/P ratio remains greater than one, indicating the formation of stoichiometric or non-stoichiometric hydroxyapatite. The coatings are porous, continuous and thick. The most suitable condition to study bioactivity of the alloy is PEO treatment at 300V for 180s, which also showed high corrosion resistance in a SBF solution. Heat treatment of the coatings results in the formation of calcium phosphates such as HAp, Brushite and Whitlockite.

Acknowledgment

The authors are pleased to acknowledge the financial assistance provided by Universidad Industrial de Santander and Quirúrgicos Especializados S.A.

References

[1] Durdu S, Usta M and Berkem A S 2016 *Surf. Coat. Technol.* **301** 85-93
[2] Finet B, Weber G and Cloots R 2000 *Mater. Lett.* **43** 159-165
[3] Matykina E, Arrabal R, Mingo B, Mohedano M, Pardo A and Merino MC 2016 *Surf. Coat. Technol.* **307** 1255-1264
[4] Liu X, Chu P K, and Ding C 2004 *Mater. Sci. Eng. R. Rep.* **47** 49-121
[5] Puelo A, Nanci A 1999 *Biomaterials* **20** 2311–2321
[6] Du K, Guo X, Wang O and Tian Y 2013 *Mater. Lett.* **91** 45-49
[7] Matykina E, Arrabal R, Pardo A, Mohedano M, Mingo B, Rodriguez I, Gonzales J 2014 *Mater. Lett.* **127** 13-16
[8] Quintero D, Galvis O, Calderón J A, Gómez M A, Castaño J G, Echeverría F and Habazaki H 2015 *Surf. Coat. Technol.* **283** 210-222
[9] Wang YM, Guo J W, Zhuang J P, Jing Y B, Shao Z K, Jin MS, Zhang J, Wei DQ, and Zhou Y 2014 *Appl. Surf. Sci.* **299** 58-65
[10] Nan K H, Wu T, Chen J H, Jiang S, Huang Y and Pei G X 2009 *Materials Sci. Enginner. C* **29** 1554-1558
[11] Song W H, un Y K, Han Y and Hong S H 2004 *Biomaterials* **25** 3341-3349
[12] Durdu S and Ustan M 2014 *Ceramics International* **40** 3627-3635
[13] Hanawa T 2010 *Japanese Dental Science Review* **46** 93-101
[14] Kokubo T and Takadam H 2006 *Biomaterials* **27** 2907-2915
[15] Sadat-Shojaj M, Khorasani MT, Dinpanah-Khoshdargi E and Jamshidi A 2013 *Acta Biomaterialia* **9** 7591-7621
[16] Wei D, Zhou Y, Jia D and Wang Y 2007 *Surf. Coat. Tech.* **201** 8723-8729
[17] Li H, Khor K A and Cheang P 2002 *Biomaterials* **23** 2105-2112
[18] Wang Y, Lei T, Jiang B and Guo L 2004 *Appl.Surf. Sci.* **233** 258-266

The effect of pulsed current electrodeposition parameters of calcium phosphates coating on Ti6Al4V ELI

D K Sierra-Herrera[1], A Sandoval-Amador[1], N D Montañez-Supelano[1] and D Y Peña-Ballesteros[1]
[1] Universidad Industrial de Santander, Bucaramanga, Colombia

e-mail: diana.sierra4@correo.uis.edu.co

Abstract. Pulse current electrodeposition is a technique of special interest, due to the advantages it has, like easy operation, high control in the amount, homogeneity and purity of the deposited material, and low cost. This work studies the influence of the pulsed electrodeposition parameters variation on the characteristics of calcium phosphates coatings, including the composition, crystallinity and morphology. The influence of the current density and pulse on and off time on the physicochemical properties of the obtained coatings were evaluated. The coatings were electrodeposited on Ti_6Al_4V using $Ca(NO_3)_2 \cdot H_2O$ and $NH_4H_2PO_4$ with a Ca/P molar ratio of 1.67. The coatings were characterized by X-ray diffraction (XRD), Fourier transform infrared spectroscopy (FTIR) and Scanning electron microscopy (SEM). The analysis of DRX confirmed the formation of HAP. The results revealed that the variation of the current density modified the morphology of the coating. Also, the amount of material deposited increases as the off-time pulse increases, allowing the diffusion of the ions in the solution towards the working electrode.

1. Introduction

Titanium and its alloys are materials widely used in orthopedic implants due to its excellent corrosion resistance and biocompatibility, however, these materials show low chemical bonding with osseous structures [1]. Bioactive materials, such as calcium phosphates, can improve the osteoconductive and osseointegration of orthopedic prosthesis, and have good adaptation with the human body, high crystallinity and biocompatibility [2]. The calcium phosphates have been widely used as coating on bio-inert materials such as the Ti_6Al_4V alloy. Since this one is present in the mineral bone phase, it encourages the regeneration of the surrounding osseous tissue in the implant.

In order to obtain calcium phosphates coatings on metallic surfaces, studies have been carried out applying techniques such as Electrophoretic deposition [3], Sol-Gel [4], plasma spraying [5,6], pulsed laser deposition [7], and electrochemical deposition [8-10]. Several authors have a great interest in electrochemical deposition due to the advantages of this technique, since it is easy to apply on pieces with irregular shapes and the pieces can be coated at low temperature and low cost of operation. The amount of material deposited in the coatings can been modified by controlling the process variables such as time, temperature, current density and number of pulses applied. In the present text we studied the influence of current density and pulse on/off ratio on the morphology and structure of coatings on Ti_6Al_4V obtained by means of pulsed current electrodeposition.

IOP Conf. Series: Journal of Physics: Conf. Series **935** (2017) 012032 doi:10.1088/1742-6596/935/1/012032

2. Materials and methods

Ti_6Al_4V cylindrical discs with 14 mm diameter and 3 mm thick were used as the substrate for the pulsed electrodeposition. The specimens were polished with sandpaper with different grades from 400 to 1200 grit, after polishing, the substrate was pre-etched in an acidic mixture (HF:HNO3:H2O=1:3:10) for 30s, washed with distilled water and cleaned by sonication in ethanol for 15 minutes.

The electrolyte for deposition was prepared by mixing 0.042mol/L of $Ca(NO_3)_2 \cdot H_2O$ and 0.025mol/L of $NH_4H_2PO_4$ with a Ca/P molar ratio of 1.67. The pH of the electrolyte was adjusted to 4.5 using NH_4OH and the temperature was maintained at 65°C, the magnetic stirring was of 450rpm to keep the uniform concentration of the electrolyte. The electrodeposition was performed in a three-electrode cell using a Gamry reference 600 potentiostat; Ti_6Al_4V was used as a working electrode, a platinum mesh as a counter (anode) electrode and the reference electrode was a standard silver-sliver chloride (Ag/AgCl in saturated KCl). The deposition process was carried out in a galvanostatic mode and the deposition time was 30min. The current density values were i_{on}=10mAcm^{-2}, i_{off}=0mAcm^{-2} and i_{on}=20mAcm^{-2}, i_{off}=0 mAcm^{-2}, the on time (t_{on}) pulsed was 1s and the off time (t_{off}) pulsed were varied as 4s and 8s. After deposition, the specimens were removed from the electrolyte solution, washed with distilled water and dried at 70°C for 5h and annealed at 550°C for 2h.

The samples of pulsed electrodeposition of calcium phosphates on Ti_6Al_4V with 1s constant pulse on time will refer: P1—10mA.cm^{-2}, 4s pulse off time, P2—10 mA.cm^{-2}, 8s pulse off time, P3—20mA.cm^{-2}, 4s pulse off time and P4—20mA.cm^{-2}, 8s pulse off time.

The surface characterization of the coatings was carried out by (Quanta FEG-650 field emission scanning electron microscope (FESEM)). It was used to observe the morphology of the calcium phosphates coating deposited by electrochemical deposition method, and the Bruker D8 advance X-ray diffractometer (XDR) with Cu Kα radiation generated at 40kV and 40mA was used for studying the crystallinity and the present phases on the coating. The chemical composition of the coating was analyzed for Fourier transform infrared spectroscopy (FT-IR) using Nicolet iS50 Spectrometer with ATR technique. The spectrum range was from 1600cm^{-1} to 400cm^{-1} with a number of 32 scans at the 4cm^{-1} resolution.

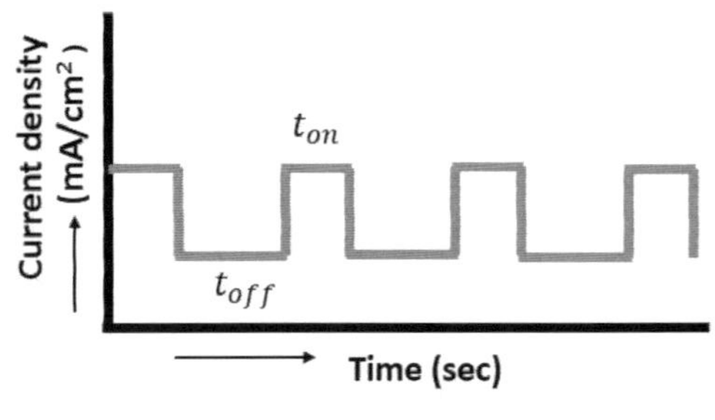

Figure 1. Pulsed current scheme.

3. Results

Figure 2 shows SEM images of the calcium phosphates coating on Ti_6Al_4V by pulsed electrodeposition for 30min at constant pulse on time of 1s, varying the current density and off time pulse. The surface morphology obtained from the coating changes a little with increasing current density, mostly needle-like and plate-like structures were formed. This structure difference could be attributed to a small amount of Brushite ($CaHPO_4 \cdot H_2O$) and calcium pyrophosphate ($Ca_2P_2O_7$) contained in the as-deposited coating, A needle-like structure may be beneficial for bone growth, as reported Seyedraoufi *et al* and Liou *et al* [11,12].

The needle-like coating deposited crystal sizes were uniform. Nevertheless, there is a decrease in the size of the crystals with the increasing current density Figure 2(a1, b1 and c1). In the same way, it is observed that when the pulse off time is at 8s, the crystal sizes are bigger than the registers for off time pulse at 4s as show Figure 2(b1, c1 and d1). This can be related to a change in the transport mechanisms that could affect the nucleation and growth mechanisms of the deposited crystals [13]. On the other hand, Figure 2(d1). Shows a coating with more porosity, the reason for this may be that, when current density increases at 20mAcm^{-2}, the cathodic hydrogen evolution reaction increases too, and the escaping H_2 gas leads to a porous structure in the layer. Cracking is not observed on the coating obtained by pulsed electrodeposition.

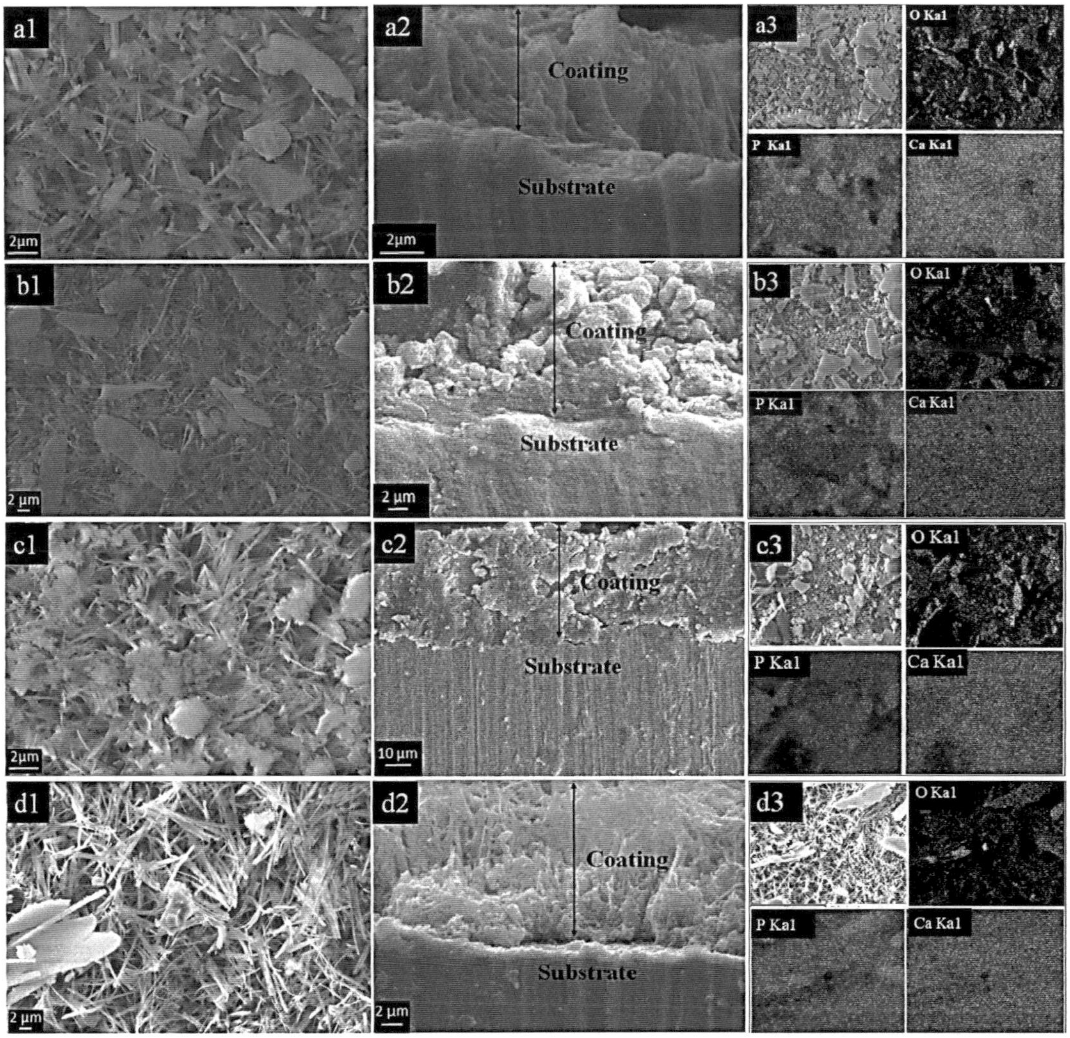

Figure 2. SEM morphologies and cross-sectional image and EDS elemental mappings of coating on Ti6Al4V by pulsed electrodeposition with 1 s constant pulse on time and (a) P1 (b) P2 (c) P3 (d) P4.

The cross sections of the coating along with substrates material have been presented in Figure 2(a2, b2, c2 and d2) in order to assess the coating thickness at various current densities and off time pulsed. The thickness of the calcium phosphates coating on Ti_6Al_4V at different conditions ranged from 4 to 31μm. The thickness measurement for coated sample at pulsed time of 4s with low current density 10mAcm^{-2} resulted in 4.75μm (Figure 2(a2)). When the current density increased to 20mAcm^{-2}, the thickness was 31.39μm (Figure 2(c2)), in the same way, the sample at pulsed time of 8s with 10mAcm^{-2} presented a thickness of 13.465μm (Figure 2(b2)) and for 20mAcm^{-2} it was 14.83μm (Figure 2(d2)). Hence, the coating obtained by the pulsed electrodeposition method with increases in the current density produces a thicker coating than the one that can be obtained from a low current density and it's further supported by the results from Chakraborty *et al.* and Gopi, *et al.* [14,15].

The mapping was carried out to evaluate the distribution of calcium and phosphorus on Ti_6Al_4V. EDS elemental mappings of the coating are shown in Figure 2 and it can be inferred that the Ca, P and O are uniformly distributed on the substrate.

The XRD patterns of the layer deposited on the Ti_6Al_4V substrates after annealing obtained at constant 1s pulse on are presented in Figure 3(a-d). The results indicate that Calcium phosphates crystals were successfully fabricated by pulsed electrodeposition. The patterns indicated that during the electrodeposition process a mixture in the coating whose contents were hydroxyapatite, brushite and calcium pyrophosphate was obtained. The diffraction peaks are observed at 2θ values for HA at 25.8°, 32.2° and 40.5°, for calcium pyrophosphate at 26.7°, 38.5, 47.05° and brushite 29.2° and 35.4°. Similar results were observed in other research by Gopi *et al* and Sameri *et al.* [15,16].

Figure 4(a-d) represents the FTIR spectra of the calcium phosphates coating. The bands characteristics for HA, $CaHPO_4{\cdot}H_2O$ and $Ca_2P_2O_7$ have been reported in the literature [17–19]. The bands characteristics for PO_4^{3-} appear at 497-557, 1032-1188cm^{-1} and for P_2O_7 appear at 719 and 923cm^{-1} respectively. The trace at 497cm^{-1} is attributed to the band PO_4^{3-} bending O—P—O (υ_2), the bending mode for the phosphate group O—P—O (υ_4) is reflected at 533, 557cm^{-1} and stretching mode P—O (υ_3) of the PO_4^{3-} 1032, 1075, 1135, 1168 and 1188cm^{-1}. On the other hand, the bands assigned to pyrophosphate groups P_2O_7 are attributed to the conversion of $CaHPO_4$ into $Ca_2P_2O_7$ during annealing, as reported by Raynaud [17]. In fact, the results of FTIR spectra corroborate the results of XRD regarding the obtainment of calcium phosphates coating by the pulsed electrodeposition.

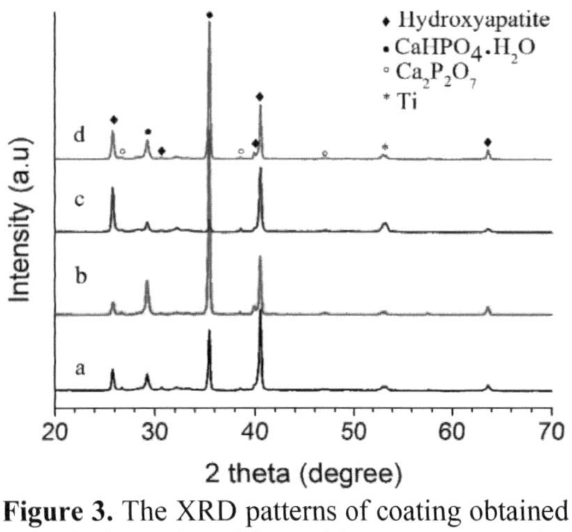

Figure 3. The XRD patterns of coating obtained pulsed electrodeposition with 1s constant pulse on time and (a) P1 (b) P2 (c) P3 (d) P4.

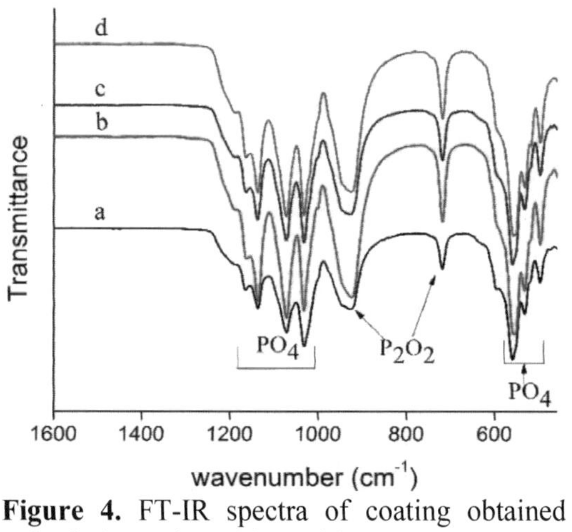

Figure 4. FT-IR spectra of coating obtained pulsed electrodeposition with 1s constant pulse on time and (a) P1 (b) P2 (c) P3 (d) P4.

4. Conclusion

In this study, calcium phosphate coatings were produced by pulsed electrodeposition at different current densities and pulse off time. Calcium phosphates coating with mostly needle-like partially and plate-like structures morphology were formed by the pulse electrodeposition method. The patterns of XDR and FTIR show that the as-deposited coating contained calcium phosphates such as HA, $CaHPO_4{\cdot}H_2O$ and $Ca_2P_2O_7$. Results showed that pulse electrodeposition parameters have an effect on the morphology and crystal size of the coatings. The SEM results revealed that the size of the crystals varies with the increase of the current density. In the same way, the coating presents a higher porosity when it is electrodeposited with a current density of 20mAcm^{-2}. The increment of pulse off time effectively influenced the coating. The thickness of the deposited coating increases as the off-time pulse increases; the range of the thickness was about 4 to 31μm. Furthermore, it is believed that the off time pulsed in the electrodeposition affects the nucleation and growth mechanisms of the deposited crystals maintaining more favorable conditions for deposition.

Acknowledgment

The authors want to thank Jaimes Rueda & Cia Ltda, Universidad Industrial de Santander, Vicerrectoria de Investigación y Extension and Grupo de Investigaciones en Corrosión for funding this Project under internal code 1888.

References

[1] Geetha M, Singh A K, Asokamani R and Gogia A K T 2009 *Prog. Mater. Sci.* **54** 397
[2] Hench L L 2005 *J. Am. Ceram. Soc.* **81** 1705
[3] Drevet R, Jaber N B, Fauré J, Tara A, Larbi A B C and Benhayoune H 2016 *Surf. Coat. Technol.* **301** 94
[4] Anjaneyulu U and Vijayalakshmi U 2017 *Mater. Lett.* **189** 118
[5] Gu Y W, Khor K A and Cheang P 2003 *Biomaterials* **24** 1603
[6] Shamray V F, Sirotinkin V P, Smirnov I V, Kalita V I, Fedotov A Y, Barinov S M and Komlev V S 2017 *Ceram. Int.* **43** 9105
[7] Dinda G P, Shin J and Mazumder J 2009 *Acta Biomaterialia* **5** 1821
[8] Blackwood D J and Seah K H W 2009 *Mater. Sci. Eng.* **29** 1233
[9] He D H, Wang P, Liu P, Liu X K, Ma F C, and Zhao J. 2016 *Surface Coatings Technology* **301** 6
[10] Benea L, Danaila E and Ponthiaux P 2015 *Corros. Sci.* **91** 262
[11] Seyedraoufi Z S, Mirdamadi S and Rastegari S 2013 *Int. J. Eng.* **27** 939
[12] Liou S C, Chen S Y and Liu D M 2003 *Biomaterials* **24** 3981
[13] Seyedraoufi Z S and Mirdamadi S 2014 *Mater. Chem. Phys.* **148** 519
[14] Chakraborty R, Sengupta S, Saha P, Das K and Das S 2016 *Mater. Sci. Eng.* **69** 875
[15] Gopi D, Indira J and Kavitha L 2012 *Surface Coatings Technology* **206** 2859
[16] Saremi M and Golshan B M 2006 *Iran. J. Mater. Sci. Eng.* **3** 1
[17] Raynaud S, Champion E, Bernache-Assollant D and Thomas P 2002 *Biomaterials* **23** 1065
[18] Xu J, Butler I S and Gilson D F R 1999 *Spectrochim. Acta - Part A Mol. Biomol. Spectrosc.* **55** 2801
[19] Yu Y D, Zhu Y J, Qi C and Wu J 2017 *Ceram. Int.* **43** 6511

IOP Conf. Series: Journal of Physics: Conf. Series **935** (2017) 012033 doi:10.1088/1742-6596/935/1/012033

Corrosion of Ti6Al4V ELI surfaces texturized by CO_2 laser in SBF

A Sandoval-Amador[1], J J Daza Rodriguez[1], N D Montañez Supelano[1], J A Sanabria Cala[1] and D Y Peña-Ballesteros[1]
[1] Universidad Industrial de Santander, Bucarmanaga, Colombia

E-mail: anderson84f@outlook.com

Abstract. The purpose of our study was to evaluate the corrosion behavior of Ti_6Al_4V ELI surfaces textured with linear geometry using a CO_2 laser. The linear geometry of the textures was obtained using a laserpro X380RX device with a 40W power supply. The characterization of the textured surface was performed by scanning electron microscope and X-ray dispersive energy spectroscopy. The effect of the separation distance of the linear textures on the corrosion resistance of the Ti_6Al_4V ELI alloy was evaluated by resistance to linear polarization, electrochemical impedance spectroscopy and potentiodynamic polarization curves in SBF. The results showed that the separation distance of the linear textures does not generate a significant effect on the corrosion resistance of the material. However, all the textured surfaces presented less resistance to corrosion than the unmodified material. This decrease in the corrosion resistance is attributed to the inclusion of elements such as molybdenum, silicon and phosphorus that are present in the lines due to the implementation of the cermark coating prior to the laser textured process.

1. Introduction

The Ti_6Al_4V ELI alloy is a material widely used as a biomedical implant due to its high strength weight ratio, good corrosion resistance, and relatively low Young modulus [1]. However, this alloy has a bioinert performance, which can be evidenced in a low cellular adhesion due to titanium oxide pore formation that in the long term could trigger a serious problem in the implant [1,2]. For this reason, several studies developed in the last decades intend to transform the bio-inert nature of this material into a biological active surface [3-9]. Laser texturing is an inexpensive method, easy to implement that allows the modification of the surface of metal materials generating active sites that favor cellular adhesion as has been widely reported in literature [10-14]. These studies reveal that laser texturing is a technique with high potential that may be implemented in the biomedical industry. However, few studies have focused on the effect that this technique generates on the corrosion resistance of titanium and its alloys [2]. In the present study, laser surface texturing of Ti_6Al_4V ELI alloy has been undertaken with linear geometry using a CO_2 laser. The effect of the separation distance between lines textured on corrosion resistance has been evaluated in SBF.

2. Methodology

Ti_6Al_4V ELI discs of 14mm in diameter and 3mm thick were used. Samples were polished with silicon carbide paper (180 to 1200 grit) and ultrasonically cleaned in ethanol during 15 minutes. The laser texturing process was performed using a Laserpro X380RX which is equipped with a CO_2 laser and operated with 40W. In Figure 1 the designs that were generated over alloy surfaces can be observed. The line width was approximately 100μm in these designs.

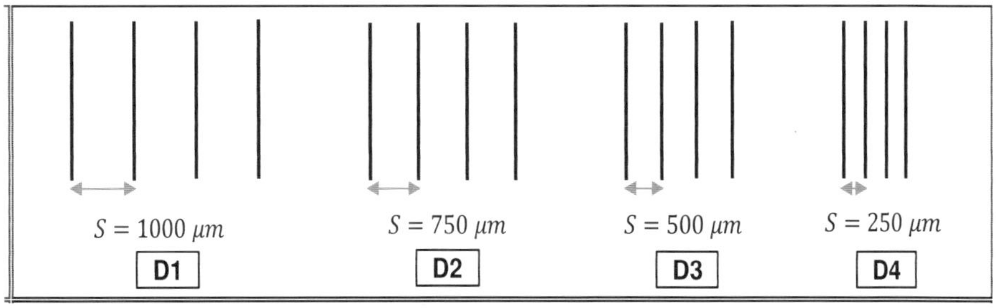

Figure 1. Layout of textured designs over the Ti_6Al_4V surface.

After the laser texturing, the surfaces were observed using Scanning Electron Microscope (SEM) and a punctual elementary composition analysis was performed over the textures generated by EDX analysis. A Gamry 600 Potentiostat-Galvanostat and a three-electrode flat cell were used in order to evaluate corrosion resistance of the textured surfaces. For this purpose, samples of Ti_6Al_4V with an exposure area of $0.5cm^2$ as working electrode and Ag/AgCl 3M KCl as reference electrode were used together with the high purity graphite bar as counter electrode. Kokubo SBF [15] was used as electrolyte. All tests were performed at 37°C. Furthermore, open circuit potential of the system was determined for one hour and based on these results, linear polarization resistance tests were carried out applying potential sweep of +/-30mV around the OCP. The electrochemical impedance spectroscopy tests were performed applying 10mVAC, running a frequency sweep between 0.01Hz and 100000Hz. Finally, a potential sweep from -500mV to 4000mV at a scanning speed of 1mV/s was used for potentiodynamic polarization tests.

3. Results

Figure 2 shows the scanning electron micrographs of the Ti_6Al_4V textured surfaces with linear geometry. Speckled material in the shape of droplets and amorphous microstructures can be seen in clear areas that were ejected due to foundry produced by the laser texturing process [16]. Additionally, molten material can be seen in dark areas caused by thermal gradient produced by the laser texturing that includes evaporation and sublimation processes of surface materials [17].

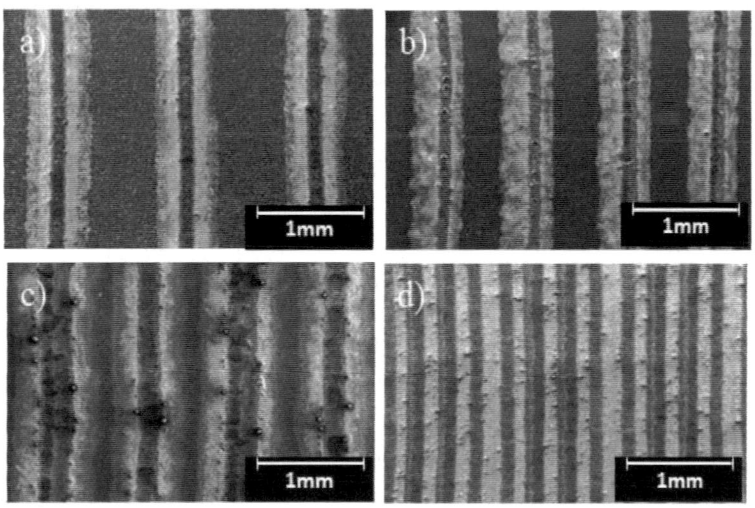

Figure 2. Scanning electron micrographs of the textured Ti_6Al_4V (a) D1, (b) D2, (c) D3, (d) D4.

The point composition analysis of Figure 3 shows a change in the surface composition. The presence of oxygen in the textured area indicates that the process of laser texturing under environmental conditions produces oxidation in the Ti_6Al_4V surface [17]. On the other hand, it is important to highlight presence of high amount of molybdenum in this area as well as small amounts of silicon, phosphorus and carbon. The incorporation of these new elements is attributed to Cermark coating implementation during the surface texturizing process [11].

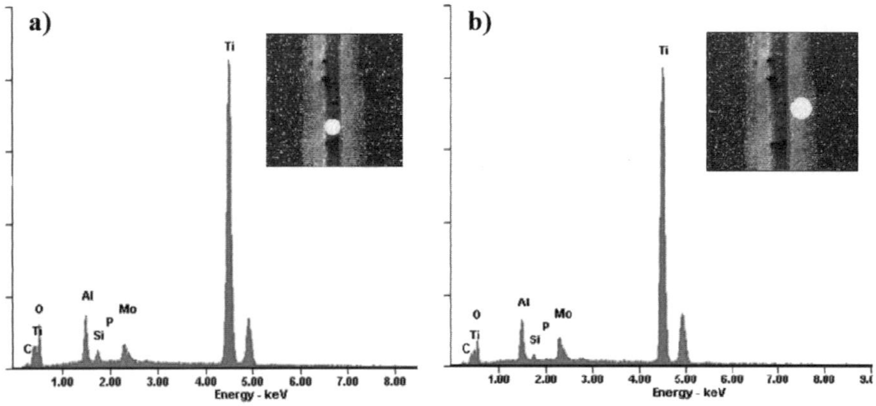

Figure 3. EDS point analysis of a line texturized by laser (a) inner part and (b) outer part.

The Nyquist diagram (Figure 4) shows the electrochemical performance of the Ti_6Al_4V texturized surfaces. One can see that the unmodified Ti_6Al_4V sample shows the highest impedance, which is attributed to the larger dome diameter that forms in this diagram. This behavior is linked to the formation of a homogeneous and stable titanium oxide layer. On the other hand, surfaces with laser texturing show smaller dome diameters, which could possibly be attributed to the chemical reactivity in SBF of the elements that compose the Cermark coating. However, it is worth mentioning that the D1 sample exhibits a performance that is similar to the unmodified material between medium and high frequencies. This may be due to the fact that this design has a lower number of lines on the surface that generate less interaction sites for the elements of Cermark with the electrolyte.

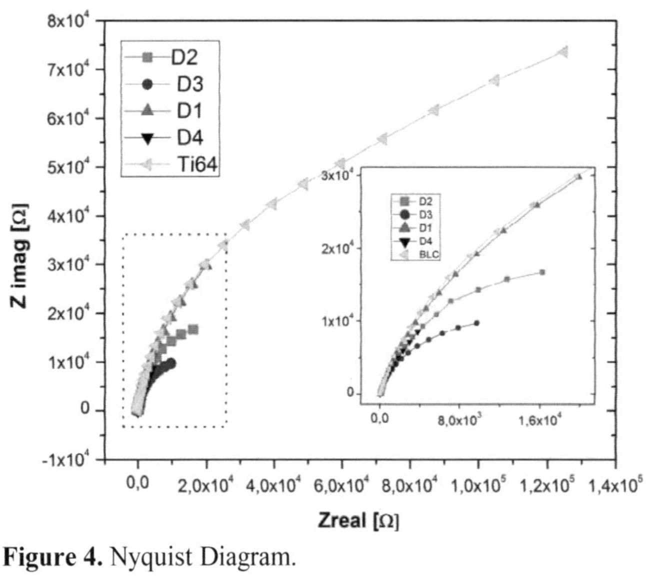

Figure 4. Nyquist Diagram.

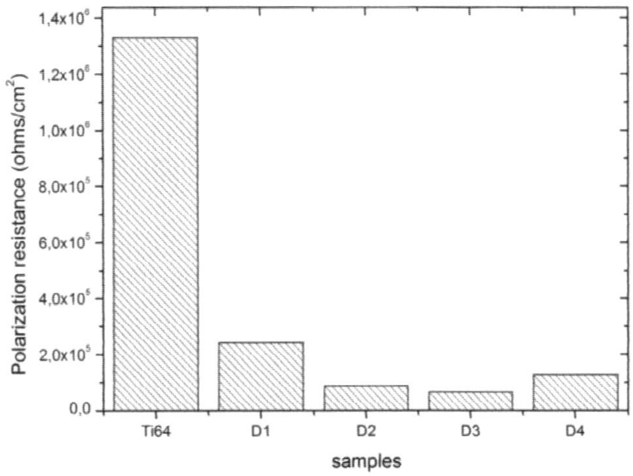

Figure 5. Polarization resistance of texturized Ti6Al4V surfaces.

Linear polarization resistance data (Figure 5) reveals that textured samples show lower polarization resistance values with respect to the unmodified Ti_6Al_4V surface. This performance is consistent with the results in the Nyquist diagram. On the other hand, Figure 6 shows potentiodynamic curves obtained for laser-textured samples which confirm that unmodified and D1 samples display the highest corrosion resistance which can be associated to the low currents of corrosion and passivation current reported in these curves. Additionally, it can be noted that in each of the curves the passive zone starts around 0.5V and ends around 2V. This situation is due to the formation of a passive layer of titanium oxide. The area from 2V to 3V is an area where oxygen detachment occurs associated with the regeneration of the passive layer over the alloy surface.

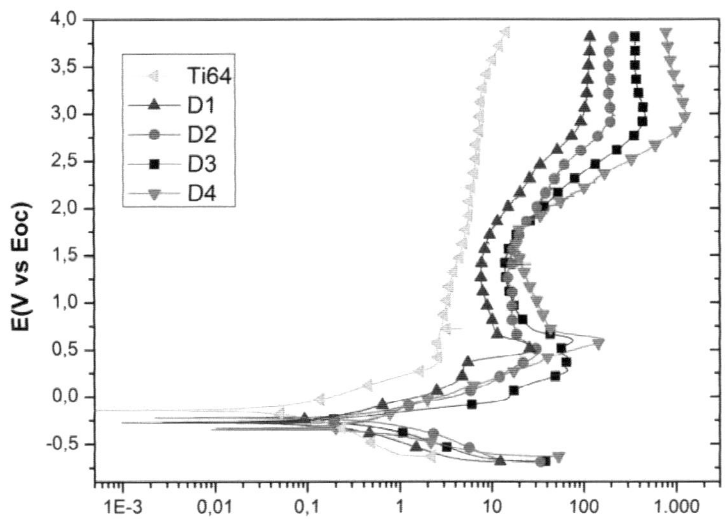

Figure 6. Potentiodynamic polarization curves of Ti6Al4V textured surfaces.

Corrosion rates showed in Figure 7 were calculated using the ASTM G-102 international standard [18]. The results determined that the unmodified Ti_6Al_4V and D1 sample produced the best performance against corrosion, while other study samples reported higher corrosion rates than those of D1samples. Resistance to damage of the unmodified sample can be attributed to the formation of a homogeneous titanium oxide layer which can protect the material when it is in contact with the SBF. On the other hand, D1 sample performance is associated with the lower number of lines generated by laser texturing which generated less Cermark coating interaction sites with the SBF.

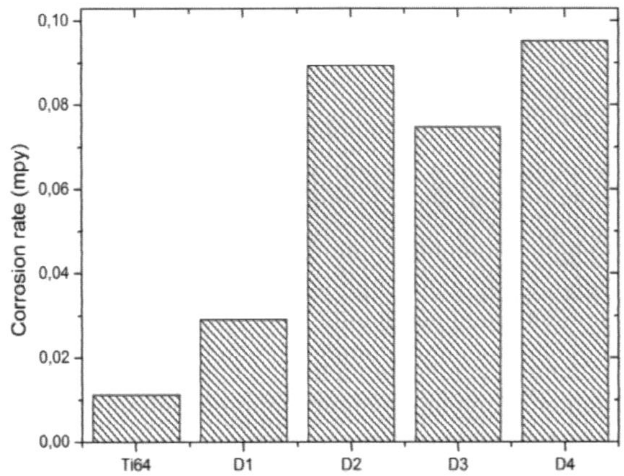

Figure 7. Corrosion rate of texturized Ti6Al4V surfaces.

Results showed that laser-textured line separation distance on the Ti_6Al_4V ELI alloy generates morphology and chemical composition changes in surface of the alloy, which is reflected in the decrease of the corrosion resistance of the material. However, this change is mainly attributed to use of Cermark coating in the texturing process. Therefore, under the study condition used, Cermark coating is not recommended for textures line of a 100μm width.

4. Conclusions

Laser engraving geometry defined by parallel slots leads to morphology changes that affect the Ti_6Al_4V alloy composition, structure and electrochemical performance. SEM analysis revealed presence of splashed material on the slot topography and the EDS chemical microanalysis reflected that this re-deposited material includes new compounds that are induced during the laser texturing process. Ti_6Al_4V alloy electrochemical evaluation using RPL, EIS techniques and potentiodynamic polarization curves showed that textured samples have lower resistance to corrosion with respect to the unmodified material, due to new compounds included by laser and Cermark coating. When these compounds are immersed in SBF, it produces a higher chemical reactivity, causing an ion release, which decreases the corrosion resistance of the alloy.

Acknowledgment

The authors wish to thank Jaimes Rueda & Cia Ltda, Universidad Industrial de Santander, Vicerrectoria de Investigación y Extension and Grupo de Investigaciones en Corrosión for funding this Project under internal code 1888.

References

[1] Pfleging W, Kumari R, Besser H, Scharnweber T and Majumdar J D 2015 *Appl. Surf. Sci.* **355** 104
[2] Kumari R, Scharnweber T, Pfleging W, Besser H, and Majumdar J D 2015 *Appl. Surf. Sci.* **357** 750
[3] Barrere F, Mahmood T A, de Groot K, and van Blitterswijk C A 2008 *Mater. Sci. Eng. R.* **59** 38
[4] Bacakova L, Filova E, Parizek M, Ruml T and Svorcik V 2011 *Biotechnol. Adv.* **29** 739

IOP Conf. Series: Journal of Physics: Conf. Series **935** (2017) 012033 doi:10.1088/1742-6596/935/1/012033

[5] Bis was A, Bayer I S, Biris A S, Wang T, Dervishi E and Faupel F 2012 *Adv. Colloid. Interf. Sci.* **170** 2
[6] Estupiñan Duran H A, Peña Ballesteros D Y and Vasquez Quintero C 2016 *Dyna* **79** 99
[7] Montañez Supelano N D, Peña Ballesteros D Y and Estupiñan Duran H A 2016 *Ingeniare* **24** 215
[8] Reyes Jaimes L J, González Romero H A, Sandoval Amador A, Peña Ballesteros D Y and Estupiñan Duran H A 2015 *Revista Ciencia en Desarrollo* **6(1)** 33
[9] Sandoval Amador A, Miranda Vesga L J, Pérez Martínez J S, Peña Ballesteros D Y and García Vergara S J 2016 *Revista Materia* **21(3)** 606
[10] Sandoval Amador A, Montañez Supelano N D, Vera Arias A M, Escobar Rivero P and Peña Ballesteros D Y 2017 *J. Phys.: Conf. Ser.* **786** 012010
[11] Sandoval Amador A, Carreño García H, Escobar Rivero P and Peña Ballesteros D Y 2016 *J. Phys.: Conf. Ser.* **687** 012012
[12] Györgyey A, Ungvári K, Kecskeméti G, Kopniczky J, Hopp B, Oszko A, Pelsöczi I, Rakonczay Z, Nagy K and Turzo K 2013 *Mater. Sci. Eng. C* **33** 4251
[13] Lawrence J, Hao L and Chew H R 2006 *Surf. Coat. Technol.* **200** 5581
[14] Chen J, Ulerich J P, Abelev E, Fasasi A, Arnold C B and Soboyejo W O 2009 *Mater. Sci. Eng. C* **29** 1442
[15] Kokubo T and Takadama H 2006 *Biomaterials* **27** 2907
[16] Fiorucci M P, Lopez A J and Ramil A 2014 *Int. J. Adv. Manuf. Technol.* **75** 515
[17] Fiorucci M P, Lopez A J and Ramil A 2015 *J. physc.: Conf. Ser.* **605** 012022
[18] Annual Book of ASTM Standards 2015 *ASTM G102-89* (USA: ASTM International)

IOP Conf. Series: Journal of Physics: Conf. Series **935** (2017) 012034 doi:10.1088/1742-6596/935/1/012034

HOS cell adhesion on Ti6Al4V ELI texturized by CO_2 laser

A Sandoval-Amador[1], Y M Bayona–Alvarez[1], H Carreño Garcia[1], P Escobar-Rivero[1] and D Y Peña-Ballesteros[1]
[1] Universidad Industrial de Santander, Bucaramanga, Colombia

E-mail: anderson84f@outlook.com

Abstract. In this work, the response of HOS cells on Ti_6Al_4V ELI textured surfaces by a CO_2 laser was evaluated. The test surfaces were; smooth Ti_6Al_4V, used as the control, and four textured surfaces with linear geometry. These four surfaces had different separation distances between textured lines, D1 (1000 microns), D2 (750 microns), D3 (500 microns) and D4 (250 microns). Toxicity of textured surfaces was assessed by MTT and the cellular adhesion test was performed using HOS ATCC CRL 1543 line cells. This test was done after 5 days of culture in a RPMI 1640 medium supplemented with 10% fetal bovine serum and 1% antibiotics. The results showed that the linear textures present 23% toxicity after 30 days of incubation, nevertheless, the adhesion tests results are inconclusive in such conditions and therefore the effect of the line separation on the cell adhesion cannot be determined.

1. Introduction

Due to their excellent properties, titanium and its alloys have been established as the most widely used materials for manufacturing medical devices. This type of material has a clearly defined market in bone fixation and replacement applications. Due to the bioinert behaviour of the titanium surface, the need to implement different surface modification techniques have been generated in order to improve the osseointegration process of this material [1-4]. Thermal and chemical oxidation, cathodic electrodeposition, plasma electrolytic oxidation and laser modification are some of the most widely used surface modification methods.

Recently some authors have suggested that the osseointegration process of the implant material can be favoured by increasing surface roughness or texture [5-9]. An easy-to-apply method for the surface texturizing of these materials is laser texturizing, since it creates a great variety of surface patterns accurately, quickly and with excellent repeatability [5,10]. Therefore, the purpose this study is to analyse the effect of laser texturizing distance separation patterns using a CO_2 laser in the adhesion of HOS line osteoblasts on Ti_6Al_4V ELI surfaces.

2. Methodology

Ti_6Al_4V discs with 14mm diameter and 3mm thickness were used. The disks were chipped with silicon carbide paper (1200 grit) and cleaned ultrasonically with ethanol for 10min. Afterwards, disks were modified by laser texturing using a laserpro X380-RX equipped with a CO_2 laser. The patterns generated on the discs of the titanium alloy had 100 microns wide lines, with 4 different distances between lines (1000μm, 750μm, 500μm, and 250μm). The power used for the laser texturing was 40W. After the texture process, the samples were viewed in an Olympus BX41 optical microscope, to observe the morphology of the surface, in addition to the surface reconstruction in 3D with the aid of the Zen lite program.

The biologic response of the specimens was determined by the quantitative adhesion of human osteosarcoma cells (HOS, ATCC CRL-1543), following the methodology previously described in [9,10]. The HOS cells were kept in RPMI supplemented with 10% fetal bovine serum and 1% penicillin / streptomycin in an atmosphere of 5% CO2, 95% humidity and a temperature of 37°C. Cells were sub-cultured once they had a 90% cell confluency (3 days culture) by trypsin-EDTA peeling. Textured surfaces D1, D2, D3 and D4, previously sterilized in autoclave at 121°C, for 30 minutes at 15psi, were placed inside 24 well plates and 1.5×104cells/mL were grown on their surface. After 5 days of incubation the adhered cells were detached with trypsin-EDTA and centrifuged at 1500rpm for 5 minutes. The quantification was carried out re-suspending the pellet with yellow eosin and counting the viable cells by optic microscopy with the Neubauer chamber. Surfaces without texturizing Ti_6Al_4V were used as control.

3. Results

As can be seen in the micrographs of Figure 1, notice that the textured lines are homogeneous along the surface of the Ti_6Al_4V alloy. The clear areas present in the micrograph are the areas of the alloy that were not exposed to the texturing process; there are some smooth lines that are product of the surface preparation process. The dark areas are due to the laser-textured process. This process induces a black coloration of the material due to the implementation of the Cermark coating, which must be applied to the surface following current standard protocols to generate textures with the laserpro X380-RX equipment. Additionally, as the separation distance between lines decreases, an increase in the apparent width of the lines can be seen because the Cermark coating generates black marks outside the lines generated.

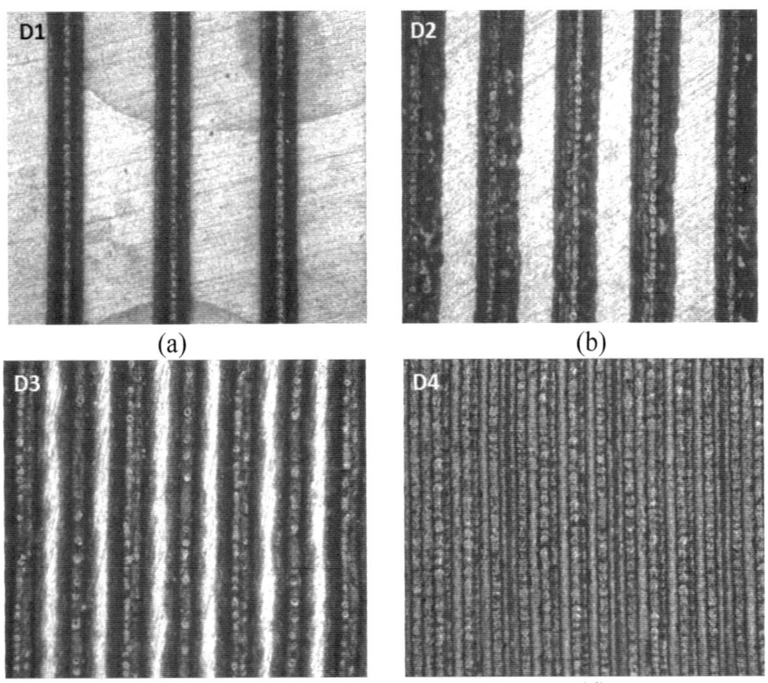

(a) (b)

(c) (d)

Figure 1. Optic micrograph of the Ti6Al4V texturized surfaces (a) 1000µm, (b) 750µm, (c) 500µm and (d) 250µm at 5X.

With the implementation of Zen lite software, 3D reconstruction of the textured surfaces was carried out. As can be seen in Figure 2 the textured process generates an increase in surface roughness, which could favour the adhesion of the bone cells.

IOP Conf. Series: Journal of Physics: Conf. Series **935** (2017) 012034 doi:10.1088/1742-6596/935/1/012034

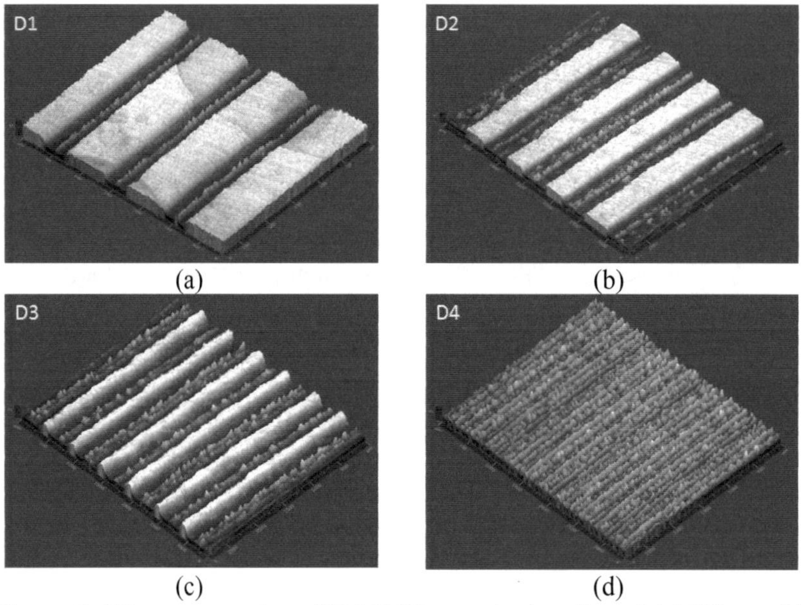

Figure 2. 3D reconstruction of Ti6Al4V texturized surfaces (a) 1000μm, (b) 750μm, (c) 500μm and (d) 250μm.

The toxicity tests show that line textured surfaces generate a toxicity of 23%. This result is attributed to the fact that the texturing process is causing the inclusion of new elements to the surface as has been previously reported in [10]. This material incorporated into the surface of the Ti_6Al_4V alloy can affect the toxicity of the alloy. On the other hand, when performing the cell adhesion test it was found that the textured titanium alloy surfaces have high affinity with HOS cells adherence. This increase in adhesion could be explained by an increase in roughness due to the texture of the material. In most cases it was possible to quantify the number of adhered cells by this method. However, surface D3 showed the best adhesion result with a value of 6×103 cells. Figure 3 shows individual results of the cells adhered in each of the replicates of the surfaces of the modified material. On the other hand, Figure 4 shows the percentage of viable cells when exposed to the vital yellow dye eosin. The results of each of the samples are shown independently.

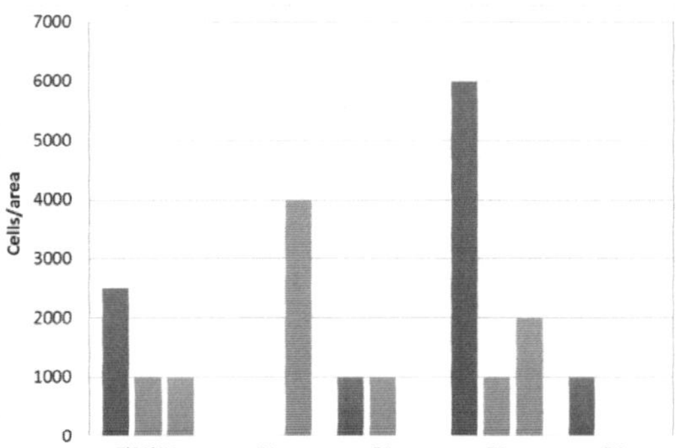

Figure 3. Cell adhesion on Ti6Al4V texturized surfaces, D1=1000μm, D2=750μm, D3=500μm and D4=250μm.

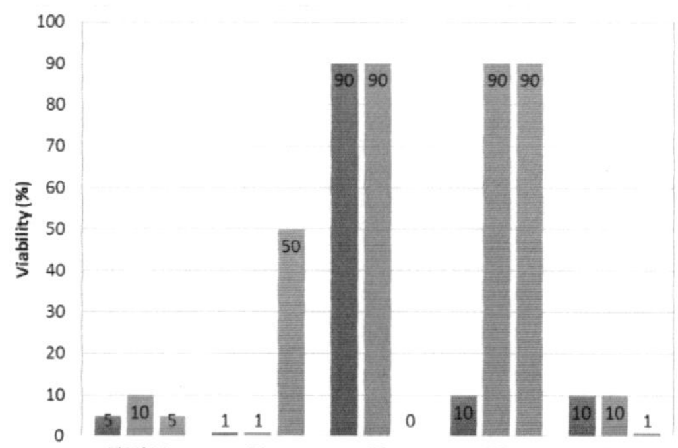

Figure 4. Cell viability on Ti6Al4V texturized surfaces, D1=1000μm, D2=750μm, D3=500μm and D4=250μm.

The results showed that the textured Ti_6Al_4V alloy samples have good affinity with HOS cells. Besides, due to the high toxicity percentage found by the MTT assay and the low viability reported in the adhesion tests, it is not possible to conclude the effect the separation distance of the textured lines by laser on the titanium alloy. When comparing the results of this study with projects previously developed by Sandoval *et al.* [9-10] where circular morphology textures were used, it is possible to conclude that due to the size of the lines and to the inclusion of a large number of elements resulting from the Cermark surface coating an unfavourable response occurs by the lined textured surfaces to the cell adhesion process.

4. Conclusions

Patterns obtained by laser texturizing using a CO_2 laser do not produce a cytotoxic response, nevertheless, the results of the adhesion tests are inconclusive and do not allow the effect of the line separation on the cell adhesion to be determined. Therefore, it is recommended to continue the study of the texture geometry effect on the cellular response.

Acknowledgment

The authors wish to thank Jaimes Rueda & Cia Ltda, Universidad Industrial de Santander, Vicerrectoria de Investigación y Extension and Grupo de Investigaciones en Corrosión for funding this Project under internal code 1888.

References

[1] Pfleging W, Kumari R, Besser H, Scharnweber T and Majumdar J D 2015 *Appl. Surf. Sci.* **355** 104
[2] Hu T, Hu L and Ding Q 2012 *Surf. Coat. Technol.* **206** 5060
[3] Qin Y, Xiong D and Li J 2015 *Surf. Coat. Technol.* **269** 266
[4] Fasasi A Y, Mwenifumbo S, Rahbar N, Chen J, Beye A C, Arnold C B and Soboyejo W O 2009 *Mater. Sci. Eng. C* **29** 5
[5] Kumari R, Scharnweber T, Pfleging W, Besser H, and Majumdar J D 2015 *Appl. Surf. Sci.* **357** 750
[6] Chen J, Ulerich J P, Abelev E, Fasasi A, Arnold C B and Soboyejo W O 2009 *Mater. Sci. Eng. C* **29** 1442
[7] Györgyey A, Ungvári K, Kecskeméti G, Kopniczky J, Hopp B, Oszko A, Pelsöczi I, Rakonczay Z, Nagy K and Turzo K 2013 *Mater. Sci. Eng. C* **33** 4251
[8] Liang C, Wang H, Yang J, Cai Y, Hu X, Yang Y, Li B, Li H, Li H, Li C and Yang X 2013 *Appl. Mater. Interfaces* **51** 8179
[9] Sandoval Amador A, Montañez Supelano N D, Vera Arias A M, Escobar Rivero P and Peña Ballesteros D Y 2017 *J. Phys.: Conf. Ser.***786** 012010
[10] Sandoval Amador A, Carreño García H, Escobar Rivero P and Peña Ballesteros D Y 2016 *J. Phys.: Conf. Ser.* **687** 012012

IOP Conf. Series: Journal of Physics: Conf. Series **935** (2017) 012036 doi:10.1088/1742-6596/935/1/012036

Mechanical properties of ceramic structures based on Triply Periodic Minimal Surface (TPMS) processed by 3D printing

S Restrepo[1], S Ocampo[1], J A Ramírez[1], C Paucar[1] and C García[1]

[1] Universidad Nacional de Colombia, Medellín, Colombia

E-mail: cpgarcia@unal.edu.co

Abstract. Repairing tissues and organs has been the main goal of surgical procedures. Since the 1990s, the main goal of tissue engineering has been reparation, using porous scaffolds that serve as a three-dimensional template for the initial fixation of cells and subsequent tissue formation both in vitro and in vivo. A scaffold must have specific characteristics of porosity, interconnectivity, surface area, pore volume, surface tortuosity, permeability and mechanical properties, which makes its design, manufacturing and characterization a complex process. Inspired by nature, triply periodic minimal surfaces (TPMS) have emerged as an alternative for the manufacture of porous pieces with design requirements, such as scaffolds for tissue repair. In the present work, we used the technique of 3D printing to obtain ceramic structures with Gyroid, Schwarz Primitive and Schwarz Diamond Surfaces shapes, three TPMS that fulfil the geometric requirements of a bone tissue scaffold. The main objective of this work is to compare the mechanical properties of ceramic pieces of three different forms of TPMS printed in 3D using a commercial ceramic paste. In this way it will be possible to clarify which is the TPMS with appropriate characteristics to construct scaffolds of ceramic materials for bone repair. A dependence of the mechanical properties with the geometry was found being the Primitive Surface which shows the highest mechanical properties.

1. Introduction

The concept of additive manufacturing has expanded the possibilities of manufacturing highly complex parts since its introduction in the 80's by Charles Hull [1, 2]. Different additive manufacturing techniques have emerged over this time, including stereolithography (SLA), selective laser sintering (SLS) and inkjet printing [3, 4]. However, it has been found that parts manufactured from extrusion-based processes have better mechanical properties and higher density in their microstructure compared to those made from processes based on powder deposition [5].

Due to the emergence of additive manufacturing techniques and the relative ease they provide to construct highly complex parts, their use has been explored in tissue engineering applications such as the design and fabrication of functional scaffolds for repair, regeneration or reconstruction of tissues or organs [6]. The fundamental objective in the design and construction of a scaffold for tissue engineering is to be able to mimic the extracellular matrix (ECM) of the tissue that will be replaced or modified [7]. A mechanically stable environment must be provided to house the necessary cells, growth factors and other biological factors in a porous structure that facilitates the migration, adhesion, proliferation and vascularization of the growing tissue [8]. In particular, in bone regeneration, the scaffolds must have specific architectural characteristics in order to support the biological and mechanical requirements for the application. The structure of the scaffold plays a fundamental role, since it must have a distribution of interconnected pores and high porosity in order to allow the cellular penetration, vascular growth and a good diffusion of nutrients [9]. It is well known that the porosity percentage of the trabecular bone is in the range of 50% to 90% while the compact bone presents a percentage of porosity less than 10% [10]. Regarding the mechanical properties of spongy bone tissue, it has been found that its maximum compressive strength is between 2 and 12MPa while its Young's modulus is between 0.02 and 0.5GPa [11].

Modelling of structures for scaffolds represents a challenge because it is necessary to obtain geometries with interconnected pores in a controlled manner. The triply periodic minimal surfaces (TPMS) are interesting geometries in the design of scaffolds since the use of 3D printing became popular [12]. The TPMS have been studied by differential geometry for more than 250 years. One of the pioneers in the study of these surfaces was Lagrange, who postulated that there should be a surface of minimal area having a closed curve as a boundary [13]. The solids that are obtained when thickness has been given to these surfaces constitute a unit cell which has cubic symmetry. The periodic repetition in space of these unit cells form a porous structure whose pores are found interconnected. This is ideal for the design of scaffolds [10].

In the present work the mechanical behavior of three different TPMS processed by 3D printing were compared using a conventional commercial ceramic paste in order to determine which of these geometries is most suitable for the manufacture of scaffolds for bone tissue.

2. Experimental

2.1. Design of triply periodic minimal surfaces

The Triply Periodic Minimal Surfaces (TPMS) selected for comparing their mechanical properties were: Gyroid, Schwarz Primitive Surface and Schwarz Dyamond Surface. This TPMS were selected according to their outstanding strength to compression and toughness [14]. These surfaces have interconnected pore circuits. TPMS can be approximated by means of an implicit surface represented by trigonometric combinations of sinusoidal functions [13]. To model these equations, the software MathMod® was used. The unit cell of each TPMS was obtained, which was replicated in three dimensions, 5 times in each XYZ coordinate axis so that the resulting surfaces consisted of 125 unit cells total. From the modelled surfaces, Blender® CAD software was used to impart thickness to these. The solids were designed with a wall thickness of 2.4mm and overall dimensions of $50 \times 50 \times 50mm^3$. Table 1 summarizes the equations, surfaces, and modelled solids.

2.2. Printing of the triply periodic minimal surfaces

From the generated models, Slicing Software Cura 3D Printing software version 15.04.2 was used to configure the printing parameters and obtain the G-Code necessary to control the movements of the printer. A layer height of 0.45mm, a density of 100% were determined and the print speed was set between 15 and $25mm/s$ to ensure good print quality.

For printing, a Wasp printer model Delta 20×40 was used. A conventional commercial ceramic paste was used for the printing of the chosen minimal surfaces. In order to facilitate the printing process the paste was adjusted with an approximate humidity of 23% to improve its plasticity and viscosity. Since the parts undergo a shrinkage due to drying and heat treatment, a compensation of 8% is made in the dimensions of the models.

The paste was positioned on the feed cylinder of the printer and extruded with a pressure of 101PSI (Approximately, 7Bars). The paste was deposited, following the established design, by means of an extruder with a nozzle of 1.2mm diameter. After print, the pieces were dried at

IOP Conf. Series: Journal of Physics: Conf. Series **935** (2017) 012036 doi:10.1088/1742-6596/935/1/012036

100°C in a Binder stove until the pieces reached constant weight. Subsequently a heat treatment was carried out at 1050°C for two hours in a furnace Sentro Tech Corp SA 1700X.

Table 1. Equations, surfaces and models of TPMS-based solids designed from trigonometric approximations.

Schwarz Primitive	Gyroid	Schwarz Diamond
$cos(x) + cos(y) + cos(z) = 0$	$cos(x)sin(y) + cos(y)sin(z)$ $+cos(z)sin(x) = 0$	$sin(x)sin(y)sin(z)$ $+sin(x)cos(y)cos(z)$ $+cos(x)sin(y)cos(z)$ $+cos(x)cos(y)sin(z) = 0$

	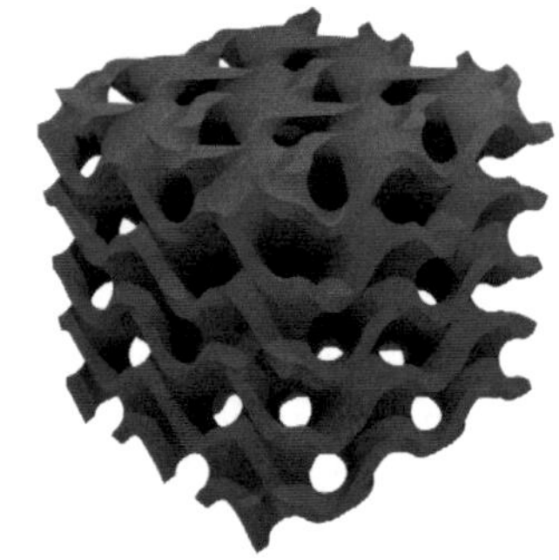	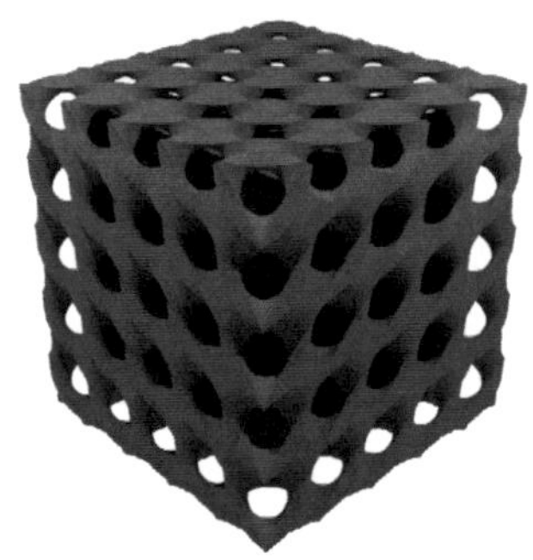

2.3. Characterization of triply periodic minimal surfaces

Using the software CAD Rhinoceros 5.0, the percentage of porosity of the pieces of the different geometries was determined making the relationship between the obtained pieces and a body of cubic shape completely dense of the same dimensions. For crystalline phase identification, X-Ray Diffraction with a double circle multipurpose Xpert-Pro PANanalytical Radiation Cu-Ka (λ=0.15406nm) diffractometer in a range of diffraction of 0-80° (2θ) was used. The printed parts were measured superficially with a Mitutoyo digital caliper with a resolution of 0.01mm and their faces were rectified by means of 400 grid sandpaper to ensure flat-parallel faces and dimensions.

2.4. Mechanical characterization of triply periodic minimal surfaces

The sintered bodies were subjected to the maximum compressive strength test using a Shimadzu universal machine AG-X model. Five tests were performed for each type of geometry at a constant speed of $0.5mm/min$ with a load cell of 100kN.

IOP Conf. Series: Journal of Physics: Conf. Series **935** (2017) 012036 doi:10.1088/1742-6596/935/1/012036

3. Results

Figure 1 shows the photographs of the pieces obtained. There are differences in the orientation of the main channels generated by the pores of each of the TPMS. In the Schwarz Primitive the main channels are interlaced perpendicularly and oriented in direction (0 0 1). In the Gyroid, the main channels are oriented in the direction of the diagonals of the cube (1 1 1). In the Diamond the main channels are oriented in direction (1 1 0).

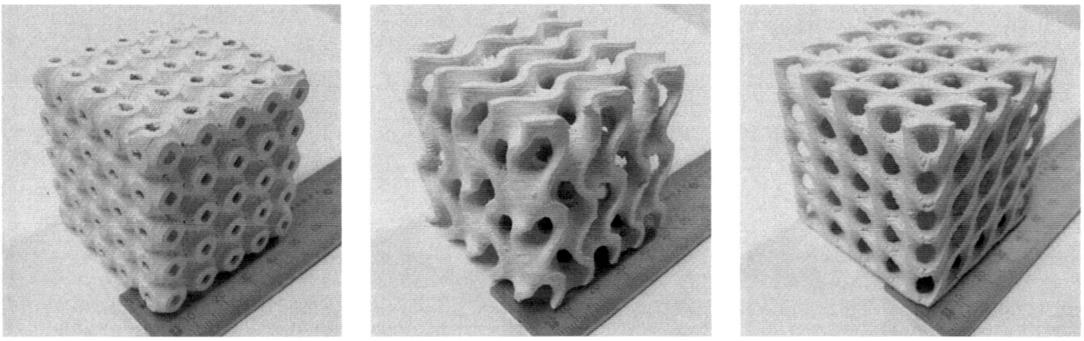

(*a*) (*b*) (*c*)

Figure 1. TPMS obtained by 3D printing. (*a*) Schwarz Primitive. (*b*) Gyroid. (*c*) Schwarz Diamond.

Figure 2 shows the X-Ray diffractograms of the ceramic paste before and after thermal treatment at 1050°C for 120 minutes.

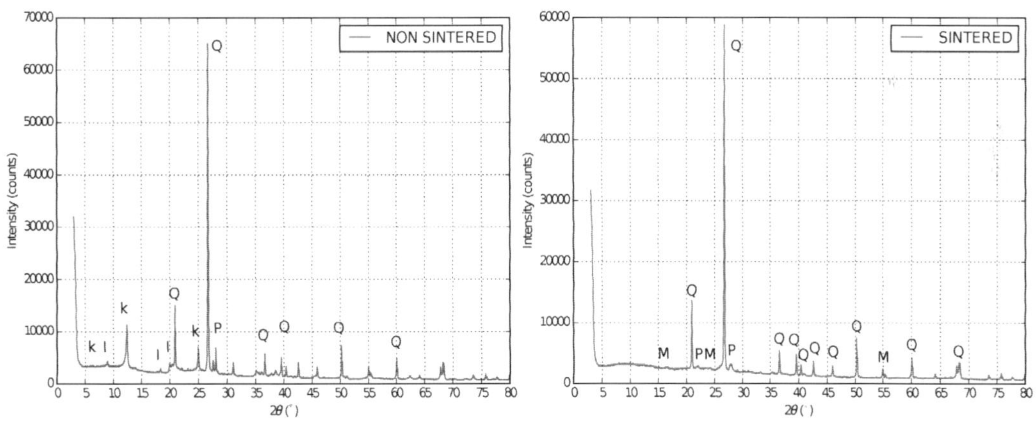

(*a*) (*b*)

Figure 2. X-Ray diffractograms of the ceramic paste. (*a*) Before thermal treatment. (*b*) After thermal treatment.

In the paste without thermal treatment, it is possible to recognize illite (I) (2θ=9°, 18° and 20°), kaolinite (K) (2θ=6.3°, 12.5° and 25°) in addition to feldspar (P) (2θ=27.5°) and quartz (Q) (2θ=20.8°, 26.6°, 36.5°, 39°, 50°, and 60°) as main crystalline phases. After calcination the intensity of reflections at 2θ=9° and 2θ=12,5° (illite and kaolinite) decreased as a result of the formation of partially amorphous metakaolinite (broad band between 2θ=5° and 2θ=15°) and mullite (2θ=16.5°, 26.6°, 36.5°, 39°, 50°, and 60°). After firing, signals of illite, kaolinite and

IOP Conf. Series: Journal of Physics: Conf. Series **935** (2017) 012036 doi:10.1088/1742-6596/935/1/012036

feldspar decreased and finally disappeared leaving quartz, feldspar and mullite (M) as the main crystalline components.

Figure 3 shows the results of the maximum compressive strength of the tested pieces. Table 2 summarizes the data of the calculated porosity of the pieces, the measured porosity, the maximum compressive strength and the Young's modulus.

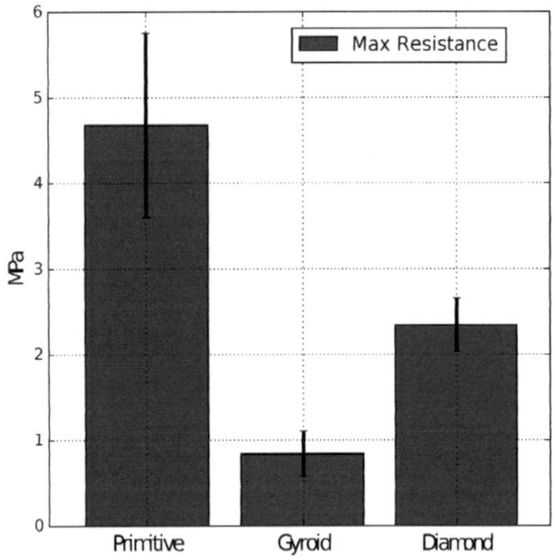

Figure 3. Maximum compressive strength (MPa).

Table 2. TPMS's Young modulus.

TPMS	Porosity (%)	Young modulus (MPa)
Schwarz Primitive	$72.29441 \pm 1 \times 10^{-5}$	335.0
Gyroid	$81.73618 \pm 1 \times 10^{-5}$	51.7
Schwarz Diamond	$78.52382 \pm 1 \times 10^{-5}$	238.3

The maximum compressive strength of the trabecular bone is in the range of 2 to 12MPa [11]. Of the TPMS tested, Schwarz Primitive and Schwarz Diamond have maximum compression strengths, compatible with those of the bone, being Schwarz Primitive the one with the greatest mechanical resistance (4.7MPa). A direct relationship between the porosity and the mechanical behavior of the parts is evidenced. As the pieces are more porous, the compressive strength and Young's modulus are smaller. This may be associated with the orientation of the main channels defined by the porosity. In the Schwarz Primitive the direction in which the load is applied in the tests of resistance to compression is parallel to the columns of support of the geometry, whereas in Gyroid and Schwarz Diamond the orientation of the columns that support the load show inclination with respect to the direction of the applied load generating a torque in the walls of the structure. Sintered Ceramic materials containing glass [15] have a good mechanical behavior in the case of compression, however when strain appear as in the case of Gyroid and Schwarz Diamond, the mechanical behavior of the pieces may be affected.

4. Conclusions

It was possible to manufacture ceramic pieces based on TPMS by means of 3D printing. All geometries studied show porosities within the range of those of the trabecular bone. A direct relationship between the percentage of porosity and the mechanical behavior of the studied geometries is evidenced. At lower porosity, greater compressive strength and Young's modulus. The Schwarz Primitive geometry presented the best mechanical behavior associated with the orientation of the columns that support the load.

IOP Conf. Series: Journal of Physics: Conf. Series **935** (2017) 012036 doi:10.1088/1742-6596/935/1/012036

References

[1] Gross B C, Erkal J L, Lockwood S Y, Chen C and Spence D M 2014 Evaluation of 3D printing and its potential impact on biotechnology and the chemical sciences *Anal. Chem.* **86** 3240–3253

[2] Bamford R 2013 3D printing in clay building objects in coiled layers *Ceramics Monthly* **61(2)** 60

[3] Wilkinson S and Cope N 2015 *3D Printing and sustainable product development green information technology* ed Mohammad Dastbaz *et. a.l* (Boston: Morgan Kaufmann) chapter 10 pp 161-183

[4] Noguera R, Ljeune M and Chartier T 2005 3D fine scale ceramic components formed by ink-jet prototyping process *J. Eur. Ceram. Soc.* **25(12)** 2055-2059

[5] Thomas A, Kolan K C, Leu M C and Hilmas G E 2017 Freeform extrusion fabrication of titanium fiber reinforced 13–93 bioactive glass scaffolds *J. Mech. Behav. Biomed. Mater.* **69** 153–162

[6] Afshar M, Anaraki A P, Montazerian H and Kadkhodapour J 2016 Additive manufacturing and mechanical characterization of graded porosity scaffolds designed based on triply periodic minimal surface architectures *J. Mech. Behav. Biomed. Mater.* **62** 481–94

[7] Berman B 2012 3-D printing: The new industrial revolution *Bussiness Horizons* **55(2)** 155

[8] Migliaresi C and Motta A 2014 *Scaffolds for tissue engineering: Biological design, materials, and fabrication* (Singapore: Pan Stanford Publishing Pte. Ltda.)

[9] Roseti L, Parisi V, Petretta M, Cavallo C, Desando G, Bartolotti I and Grigolo B 2017 Scaffolds for bone tissue engineering: State of the art and new perspectives *Materials Science and Engineering: C* **78** 1246-1262

[10] Fantini M, Curto M and De Crescenzio F 2017 TPMS for interactive modelling of trabecular scaffolds for bone tissue engineering *Advances on Mechanics, Design Engineering and Manufacturin* (USA: Springer International Publishing) pp 425–35

[11] Velasco M A, Narváez-tovar C A and Garzón-alvarado D A 2015 Design, materials, and mechanobiology of biodegradable scaffolds for bone tissue engineering *BioMed Research International* **2015(729076)** 1-21

[12] Abueidda D W, Bakir M, Abu Al-Rub R K, Bergström J S, Sobh N A and Jasiuk I 2017 Mechanical properties of 3D printed polymeric cellular materials with triply periodic minimal surface architectures *Mater. Des.* **122** 255–267

[13] Sánchez E M 2015 *Superficies minimales. Historia, desarrollo y aplicaciones a otras ciencias* (España: Universidad Complutense de Madrid)

[14] Montazerian H, Davoodi E, Asadi-Eydivand M, Kadkhodapour J and Solati-Hashjin M 2017 Porous scaffold internal architecture design based on minimal surfaces: A compromise between permeability and elastic properties *Mater. Des.* **126** 98–114

[15] Zhang Z F, Eckert J and Schultz L 2003 Difference in compressive and tensile fracture mechanisms of Zr59Cu20Al10Ni8Ti3 bulk metallic glass *Acta Mater.* **51** 1167–79

IOP Conf. Series: Journal of Physics: Conf. Series **935** (2017) 012037 doi:10.1088/1742-6596/935/1/012037

A novel biphasic calcium phosphate derived from fish otoliths

N D Montañez-Supelano[1], A Sandoval-Amador[1], H A Estupiñan-Durán[2] and D Y Peña-Ballesteros[1]
[1] Universidad Industrial de Santander, Bucaramanga, Colombia
[2] Universidad Nacional de Colombia, Medellín, Colombia

E-mail: ing.nerly.montanez@hotmail.com

Abstract. Calcium phosphates are bioceramics that have been widely used as bone substitutes because they encourage the formation of bone on their surface and can improve the healing of the bone. Hydroxyapatite HA (calcium/phosphorus ratio of 1.67) and tricalcium phosphate TCP (calcium/phosphorus ratio of 1.50) are the most common calcium phosphates. Natural materials have begun to be tested to make HA or TCP such as shells of cardiidae (family of mollusks) and eggshells. The calcium phosphate obtained has a high ability to precipitate apatite. In this work, the mixed phase ceramic of beta-Tri-calcium phosphate / hydroxyapatite (β-TCP/HA) was synthesized by aqueous precipitation from fish otoliths, which are monomineralic species composed of aragonite. Otoliths of the specie Plagioscion squamosissimus, commonly called the river croaker, were used. Techniques such as DRX, Raman spectroscopy and SEM-EDS were used to characterize the raw material and the obtained material. X-ray diffraction analysis revealed the presence of two crystalline phases of calcium phosphates with 86.2% crystallinity. SEM micrographs showed agglomeration of particles with porous structure and submicron particle sizes.

1. Introduction

Calcium phosphates (CaPs) that were used as coatings or bulk implants improved bonding between bone and implant. The most known CaPs are Beta–Tri–calcium phosphate ($Ca_3(PO_4)_2$, β-TCP) and calcium hydroxyapatite ($Ca_{10}(PO_4)_6(OH)_2$, HA), which have been used as corrective therapies for bone defects and diseases [1-3]. HA has excellent biocompatibility and its components are similar to those of natural bone and teeth. However, it has been found that dense hydroxyapatite hinders bone growth due to its low biodegradability, so in these cases β-TCP works better, the latter has been proposed for orthopedic applications because its bioresorption rate is higher compared to pure hydroxyapatite. A mixture of HA and β-TCP called "biphasic calcium phosphate (BCP)" was first introduced by Moore and Anuta in 1985. The BCP consists of a mixture of CaPs phases with similar Ca/P molar radius [4,5].

However, there are controversies in the appropriate amount of each of the present phases and the physico-chemical properties in the BCP. The higher relative biological solubility of tricalcium phosphate in biological medium and the partial dissolution of HA, increases the local concentration of phosphate (HPO_4^{2-}, PO_4^{3-}) and calcium (Ca^{2+}) ions, followed by a precipitation of calcium carbonate phosphate, these precipitates induce the growth of bone to the implant by attraction of osteoblasts, followed by osteoclasts that dissolve the carbonaceous precipitates in native bone in the defects related to low resorption of HA. A high concentration of TCP results in the rapid formation of bone at the expense of long-term osseointegration. In addition, a high amount of HA can occlude the porosity of the metal surface so that adequate bone apposition is not achieved. Resorption kinetics governs the fixation of the implant and therefore the formation of bone [1,5,6].

Calcium phosphates are synthesized by chemical techniques such as aqueous precipitation, sol-gel, hydrothermal method and thermal deposition, however raw material costs are high and yield is small. Other naturally occurring materials have begun to be tested to make HA or β-TCP such as shells of cardiidae (family of mollusks) and egg shells. The HA obtained has a high ability to precipitate apatite [4].

Natural materials not widely studied in this type of applications are the otoliths. Otoliths are a material found in the inner ear cavity of fish, located in the saccule, utricle and canals, they have a few grams or milligrams in weight and its morphology depends on the specie. An X-ray diffraction analysis shows that otoliths are monomineralic species composed of aragonite. The otolith grows from a central core of organic material called primordium. The organic matrix provides a support that favors the nucleation of aragonite crystals which grow on its c axis, so apparently the aragonite is accompanied by proteins called otoline [7-9]. In a biocompatibility assay in mice, polyethylene tubes containing otoliths were implanted subcutaneously into the back of Wistar mice. The results showed that after 30 days the tissue surrounding the implants was composed of fibroblasts and new corrugated collagen fibers [10]. In this research fish otoliths of the Sciaenidae family of the Plagioscion squamosissimus specie, commonly called Curvinata, river croaker or stone head were used. From otoliths a BCP was synthesized with physicochemical characteristics suitable to be used in the application of orthopedic implants.

2. Experimental

To synthesize the biphasic calcium phosphate, the otoliths were washed with ethanol and deionized water by sonication for 1 hour for each solution and could dry in a horizontal laminar flow biosecurity chamber for 30 minutes. The natural material was calcined at 1000 Celsius degrees for 2 hours at a heating rate of 2°C/min to produce calcium oxide (CaO). Since CaO is highly hygroscopic, calcium hydroxide ($Ca(OH)_2$) was formed according to equation (1).

A solution of phosphoric acid 0.6M (H_3PO_4 85%, Merck) is mixed with a solution of calcium hydroxide 1M to pH 4.5, at a rate of 5ml/min with continuous stirring at 90°C for one hour and another hour more without temperature. The mixture is left without stirring for 48 hours at room temperature. The precipitated calcium phosphates are washed with deionized water and dried for 24 hours at 105°C. The reaction carried out is shown in equation (2), after that, the calcium phosphates were calcined at 1050°C for 3 hours.

$$\mathrm{CaCO_3} \xrightarrow[1000°C]{} \mathrm{CaO} \xrightarrow[H_2O]{} \mathrm{Ca(OH)_2} \tag{1}$$

$$10Ca(OH)_2 + 6H_3PO_4 \xrightarrow[1050°C]{} Ca_{10}(PO_4)_6(OH)_2 + 18H_2O \tag{2}$$

The phase formation in the ceramic was carried out by X-ray diffraction (XRD, Bruker D8 advance, Germany) in 2θ=20-70°. The qualitative analysis of the phases present in the calcium phosphates was performed by comparing the profile observed with the diffraction profiles reported in the PDF-2 database of the International Center for Diffraction Data ICDD (PDF-2 010-80 -3956 for HA, and for β-TCP 010-70-2065). The quantitative analysis of the phases was performed by refinement by the Rietveld method of the observed profile, adding a known internal standard quantity (Aluminum oxide, -100 mesh, 99%, Corundum, α-phase. Aldrich No. 23.474-5) corresponding to 20.28%.

The morphology of the otolith and the sample obtained was analyzed by scanning electron microscopy (SEM Evo Ma 10 Carl Zeiss Detector X-act oxford instruments). To observe the otolith sample, it was covered with gold (equipment Quorum Q1SORES) in the vacuum for 100 seconds, producing a layer of thickness of approximately 6nm. The functional groups of CaPs were verified by Raman spectroscopy (Labram HR Evolution HORIBA Scientific) with a 532nm laser.

3. Results and discussion

The X-ray diffraction pattern of otolith powder and Raman spectrum of fish otolith are shown in Figure 1. It is evident that among the calcium carbonate polymorphs, the aragonite peaks (PDF number 000-75-2230 (ICDD)) coincide with the diffraction pattern shown in Figure 1 and 2. According to Pattanaik, the growth of the calcic matrix is mediated by kinetic factors such as the degree of super-saturation of calcium carbonate, which favored the formation of aragonite [11]. Besides that, others environmental conditions like temperature, pollution and biological activity affect the formation of the biomineral structure [12].

The quantitative analysis of the XRD resulted in 82.5% aragonite and 17.5% amorphous. This biomineral structure cannot be formed in a biological environment and is accompanied by acidic glycoproteins, part of the percentage of amorphous is due to the content of acidic groups in the proteins, which are involved in biomineralization processes resulting in a higher capacity of attract Ca^{2+} ions, Additionally, this structure can construct water-insoluble gel structures in the biomineral matrices. Because the proteins are not easily separable, the identification is very complicated [9,13,14].

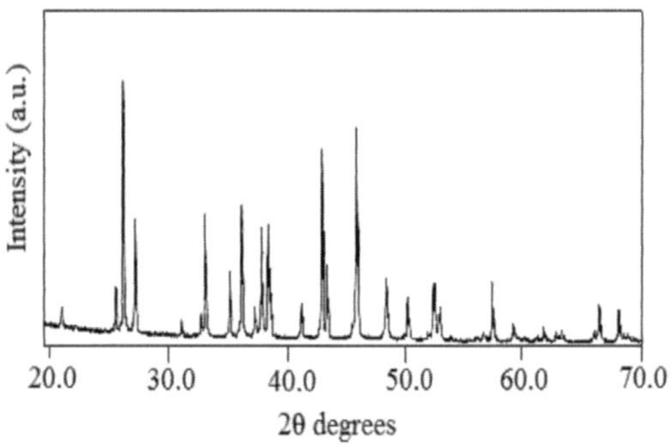

Figure 1. XRD pattern of fish otolith.

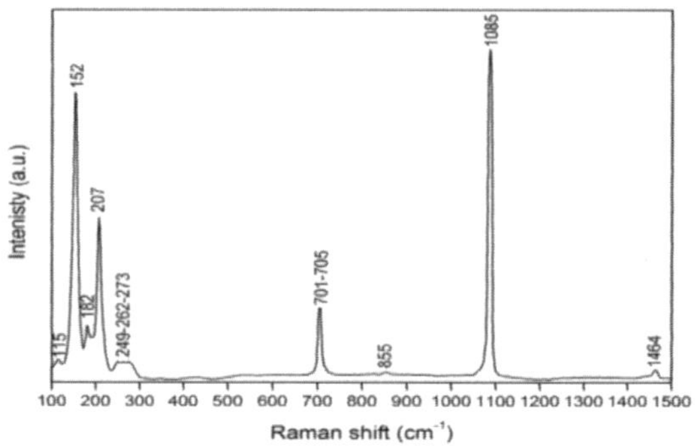

Figure 2. Raman spectrum of fish otolith.

Figure 2 shows the Raman spectrum of fish otolith. The crystallization of calcium carbonate in its aragonite phase is shown marked by the principal peaks 152, 207, 701-705 y 1085cm^{-1}. The bands between 115 and 273cm^{-1} were assigned to the vibrational modes of the cell, two bands of 701 and 705cm^{-1} were attributed to CO_3 in plane bending, 855cm^{-1} to δ(CCH) ring and C-C stretching, 1085cm^{-1} to symmetrical stretching of C-O and the band 1464cm^{-1} to CO_3 asymmetrical stretching. Both the 855 and the 1464cm^{-1} bands are also attributed to the vibrational modes of organic fractions in the otolith as proline, glutamic acid and serine for the first case [15].

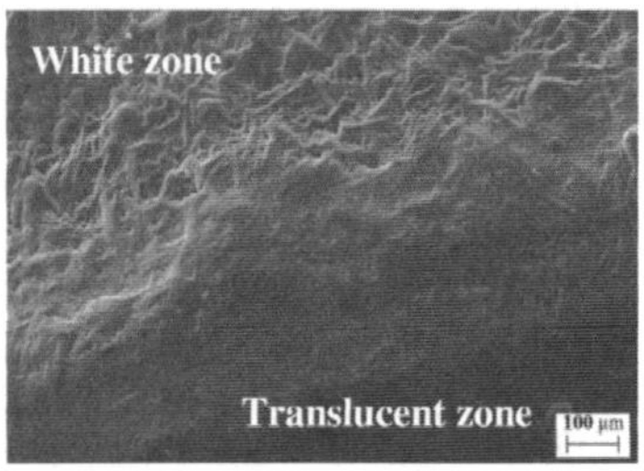

Figure 3. SEM micrograph of fish otolith.

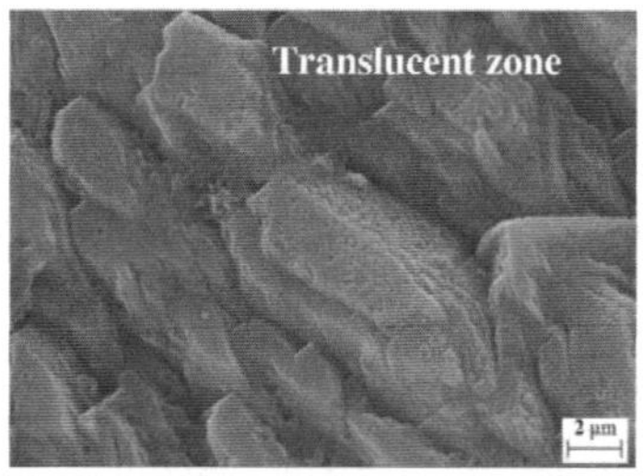

Figure 4. SEM micrograph of translucent zone.

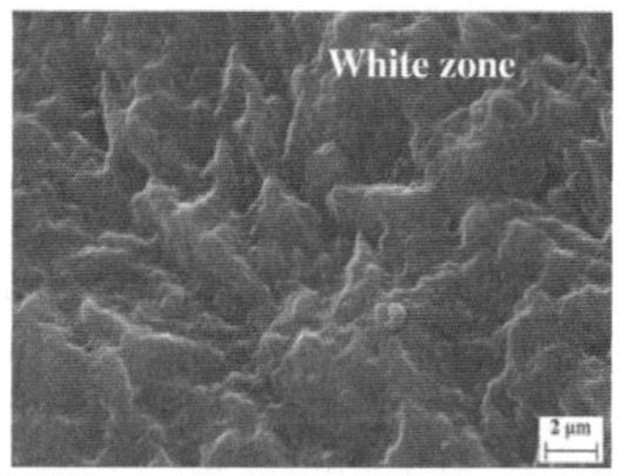

Figure 5. SEM micrograph of white zone.

SEM micrograph of fish otolith shows a different morphology on its outer surface. Two areas are clearly distinguished, which we will call "white zone" and "translucent zone" (Figure 3). The translucent area shows an aggregate structure in the form of flakes formed from small crystals of aragonite (Figure 4). These crystals are similar in appearance to otoliths of the Hoplostethus Atlanticus fish, which were described as aragonite [16]. The white zone has aggregates in the same direction as those described above, however, the morphology is not the same, and the aragonite crystals are not so evident (Figure 5). Despite this, there was no significant difference in the percentage of calcium in the EDS analysis of the two zones, these percentages oscillated between 18 and 22% (atomic %) of calcium.

The X-ray diffraction patterns of BCP powder synthetized from otoliths of River Croaker are shown in Figure 6. The presence of two crystalline phases of hydroxyapatite and tricalcium phosphate in proportions 18.3% and 67.9% respectively is evident. A total of 13.8% of amorphous was found. Figure 6 shows the main characteristic peaks of HA at 2θ diffraction angles relevant to each peak: 25.9°, 31.7°, 32.2°, 32.8°, 34.1°, 46.9° and 49.4°. For β-TCP, the main peaks are; 25.9°, 26.5°, 29.6°, 31.0°, 32.8° and 34.3°, corresponding to the results shown by Ebrahimi [5].

The Raman spectrum of BCP powder synthetized from otoliths is shown in Figure 7. According to the literature, two peaks at 948 and 970cm^{-1} were assigned to the symmetric vibration P–O stretching mode ($\nu_1 PO_4^{3-}$) of β-TCP. Between 500 and 650cm^{-1}, the peaks 548, 611 and 630cm^{-1} were assigned to the asymmetric vibration bending mode ($\nu_4 PO_4^{3-}$) of β-TCP and the peak 595cm^{-1} were assigned to the asymmetric vibration P–O of HA. Typical HA vibrations ($\nu_2 PO_4^{3-}$) bending mode (O–P–O) at about 430-450cm^{-1} were presented, as well as, asymmetric stretching ($\nu_3 PO_4^{3-}$) mode at 1015-1089cm^{-1}. The vibrations ($\nu_2 PO_4^{3-}$) bending mode to 406cm^{-1} was identified for β-TCP [17-19]. The peak 276cm^{-1} could not be identified.

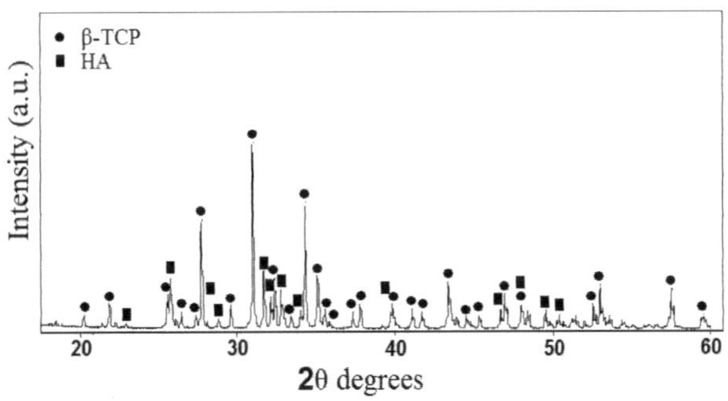

Figure 6. (a) XRD pattern of BCP derived from fish otolith (■: HA peaks, ●: β-TCP peaks).

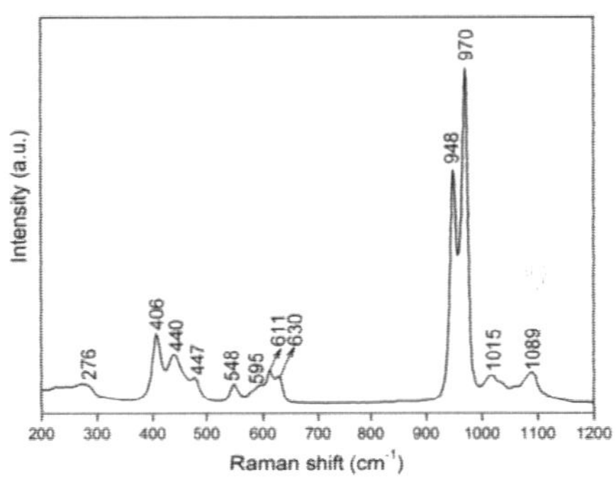

Figure 7. Raman spectrum of BCP.

SEM micrograph and chemical mapping for oxygen, phosphorus and calcium are shown in Figure 8 and Figure 9. The BCP particles were larger spherical, in porous structure, with a particle size up to 2μm. This morphology is similar to that reported by Gao *et al* [20] for 100% β-TCP with a particle size of 1μm. The percentage of HA and amorphous causes the particle size to increase in the BCP. The distribution of oxygen, phosphorus and calcium elements is homogeneous throughout the sample and no precipitates are found (Figure 9).

Bioceramics with a high dissolution rate encourage important pH changes that can result in inflammatory processes [21], while bioceramics with very slow dissolution rates can hamper bone growth. Therefore, manufacturing and evaluating BCP with different proportions of HA and β-TCP is of scientific interest, since the results regarding kinetics of degradation can be varied. According to the results found the obtained proportion of HA and TCP can provide a surface suitable for cell proliferation, especially if they come from natural materials.

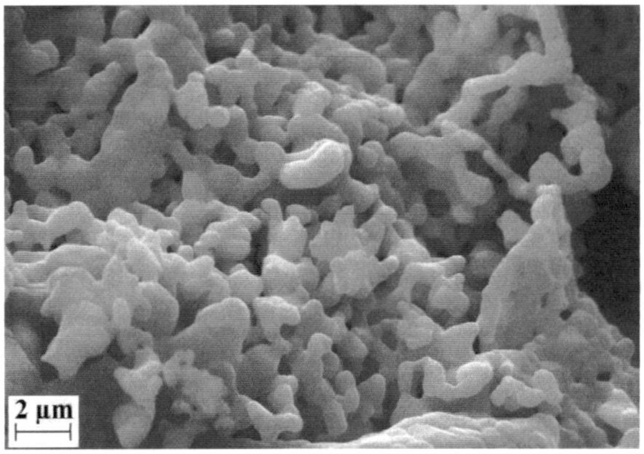

Figure 8. SEM micrograph of BCP

Figure 9. Chemical mapping corresponding to micrograph, oxygen, phosphorus and calcium.

4. Conclusions

Otoliths of the Plagioscion squamosissimus specie are monomineral species from the aragonite type accompanied by organic fractions. The XRD and Raman analysis of BCP synthesized from fish otoliths confirmed the presence of two crystalline phases HA and β-TCP and the different modes of vibration of the PO_4^{3-} ion. On the other hand, the morphology shows that there was more crystallization of the TCP phase with respect to the HA phase and the percentage of amorphous that modifies the particle size. The material synthesized 18% HA/67.9% β-TCP is a novel BCP derived from a natural material, with physicochemical characteristics similar to synthetic BCP, since they are materials that can encourage both bone replacement and bone regeneration.

Acknowledgment

The authors would like to thank Jaimes Rueda & Cia Ltda, Universidad Industrial de Santander, Vicerrectoría de Investigación y Extension and Grupo de Investigaciones en Corrosión for funding this project under internal code 1888.

References

[1] McEntire B, Bal B, Rahaman M, Chevalier J and Pezzotti G 2015 *J. Eur. Ceram. Soc.* **35** 4327
[2] Ergun C, Evis Z, Webster T and Sahin F 2011 *Ceram. Int.* **37** 971
[3] Kumta P, Sfeir C, Lee D, Olton D and Choi D 2005 *Acta Biomater.* **1** 65
[4] Lertcumfu N, Jaita P, Manothan Sm Jarupoon P, Eitssayeam S, Pengpat K and Rujijanagul G 2016 *Ceram. Int.* **42** 10638
[5] Ebrahimi M, Botelho M and Dorozhkin S 2017 *Mater. Sci. Eng. C.* **71** 1293
[6] Ebrahimi M and Botelho M 2017 *Data in Brief.* **10** 93
[7] Degens E, Deuser W and Hardrich R 1969 *Mar. Biol.* **2** 105
[8] Mulcahy S, Killingley J, Phleger C and Berger W 1979 *Oceanologica Acta* **2** 423
[9] Wojtas M, Holubowicz R, Poznar M, Maciejewska M, Ozyhar A and Dobryszycki P 2015 *Biochemistry* **54** 6525
[10] Santos T, Rodrigues C, Calado G, Cleveilton J, Rodrigues S, Filho X and Cavalcanti R 2013 *Interfaces Científicas - Saúde a Ambiente* **1** 71
[11] Pattanaik S 2005 *Nucl. Instrum. Methods Phys. Res., Sect. B* **229** 367
[12] Yan L, Shengrong L, Fengquin D, Wenjie L, Na Z and Xin C 2012 *Afr. J. Biotechnol.* **11** 6847
[13] Tohse H, Saruwatari K, Kogure T, Nagasawa H and Tagaki Y 2009 *Cryst. Growth Des.* **9** 4897
[14] Wojtas M, Wolcyrz M, Ozyhar A and Dobryszycki P 2011 *Cryst. Growth Des.* **12** 158
[15] Jolivet A, Bardeau J, Fablet R, Paulet Y and Pontual H 2013 *Can. J. Fish. Aquat. Sci.* **70** 711
[16] Gauldie R 1993 *J. Morphol.* **218** 1
[17] Kim D, Chun H, Lee J and Yoon S 2014 *Ceram. Int.* **40** 5145
[18] Khan A, Awais M, Khan A S, Tabassum S, Chaudhry A and Rehman I 2013 *Appl. Spectrosc. Rev.* **48** 329

IOP Conf. Series: Journal of Physics: Conf. Series **935** (2017) 012037 doi:10.1088/1742-6596/935/1/012037

[19] Sidane D, Rammal H, Beljebbar A, Gangloff S, Chicot D, Velard F, Khireddine H, Montagne A and Kerdjougj H 2017 *Mater. Sci. Eng. C* **72** 650
[20] Gao C, Yang B, Hu H, Liu J, Shuai C and Peng S 2013 *Mater. Sci. Eng, C* **33** 3802
[21] Garrido C, Lobo S, Turíbio F and LeGeros R 2011 *Int. J. Biomater.* **2011** 1

IOP Conf. Series: Journal of Physics: Conf. Series **935** (2017) 012038 doi:10.1088/1742-6596/935/1/012038

Study of a magnetorheological fluid submitted to a uniform magnetic field

H A Fonseca[1], **E Gonzalez**[1] **and J Restrepo**[2]

[1] Universidad Pedagógica y Tecnológica de Colombia, Tunja, Colombia
[2] Universidad de Antioquia, Medellín, Colombia

E-mail: hugoalexander.fonseca@uptc.edu.co

Abstract. In this work, the rheological and hyperfine properties of a magnetorheological fluid (MRF) under the action of a uniform external magnetic field are analysed. Powders of native mineral magnetite of micrometric particle size, after a pulverization process, form the solute of these fluids. The sizes of these samples are selected by sieving in order to obtain sizes of around 20μm and 45μm. The powders are characterized by means of Mössbauer spectroscopy to analyse their stoichiometry giving rise to a non-stoichiometric magnetite $Fe_{2.96}O_4$ in addition to a hematite component. Result of viscosity and shear stress in the low-speed regime were analysed using the Hershel Buckley method. In particular, the case of surface tension it decreases with the application of a uniform magnetic flux density, which is understood in terms of a phase separation due to the formation of mesoscopic structures, thus decreasing the cohesion force and increasing the adhesion force.

1. Introduction

Magnetic fluids or magnetic colloids are two-phase systems formed by a liquid (carrier) and a solid phase (ferromagnetic or ferrimagnetic powder), which act as dynamic fluids and they can be controlled by an external magnetic flux density. Magnetic fluids can be classified in ferrofluids (size nm), magnetorheological (MRF) fluids (μm size) and composite magnetic fluids. The solid or dust particles are coated by a tensoactive surfactant adhered to the liquid-solid interface to mediate the interaction between these phases by avoiding agglomeration by electrostatic interaction. On the other hand, thermal agitation holds particles in suspension by steric repulsion. Magnetic fluids form a new class of magnetic materials whose novelty lies in the way as they behave changing their properties under the influence of a magnetic fields without losing the character of fluid, changing rheological properties such as local density, viscosity, stress and surface tension.

Magnetic fluids are materials of paramount importance due to their potential for technological applications. They are used in impact (suspensions) or vibration (seismic) suspension systems [1], in optics for lens polishing or for improving lens capability [2], in microelectromechanical (MEMs) technology [3] and for mineral cleaning [4] among others.

In particular, magnetoviscosity and shear stress properties of MRF as well as the formation of mesoscopic structures under an external applied magnetic field is still a subject of interest and debate upon which we focus our attention in this work.

2. Experimental

Native magnetite was grinded up in a vibrating cylinder mill at 140rpm after which two set samples of sizes around 20μm labelled as T5 and 45μm labelled as T2 respectively, were obtained by sieving.

These values were then confirmed by applying statistical methods to micrographs taken by optical microscopy [5].

The magnetite powders were then covered with oleic acid and dissolved in SAE oil in percentages of 20%-30%-50% and 15%-22.5%-62.5% respectively. Mixtures were then mechanical stirred in a thermal bath at 56°C during periods of 0.5, 1, 1.5 and 2 hours.

Sample were characterized by means of room temperature Fe^{57} Mössbauer spectroscopy for stoichiometric purposes by taking into account spectral areas of the tetrahedral (A) and octahedral (B) crystallographic sites of magnetite. Surface tension was characterized by measuring the surface breaking stress made by an aluminium ring and viscosity was measured in a BrokField 115 viscometer in parallel plate mode. The measurements in the viscometer were established under two different modes. First, by fixing the value of the magnetic flux density and varying the speed of the spindle, and second by fixing the speed of spindle and varying the magnetic flux density.

3. Results and discussion

After fitting the room temperature Mössbauer spectrum of the native sample, two components, one ascribed to hematite (11%) and the other one to magnetite (89%), were evidenced as can be observed in Figure 1. As concerns to the magnetite component, the corresponding $Fe^{2.5+}$ y Fe^{3+} spectral areas allow an estimation of the degree of stoichiometry via the oxidation parameter (x) in the framework of a model of vacancies distributed in either tetrahedral or octahedral sites.

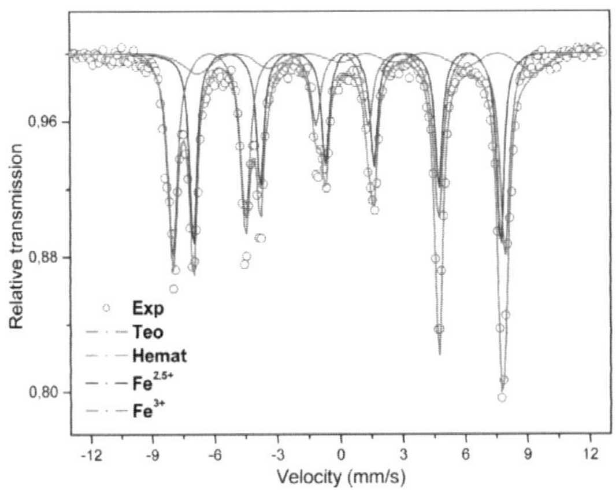

$\Gamma_1(Fe^{2.5+})$	$\Gamma_2(Fe^{3+})$	R	x	$Fe_{3-x}O_4$
0.44	0.56	0.79	0.12	$Fe_{2,88}O_4$

Figure 1. Room temperature Fe^{57} Mössbauer spectrum with its respective fit. The corresponding line widths (□) of tetrahedral and octahedral sites, the ratio of the spectral areas (R) and oxidation (x) parameters are included.

To do that, we employ the following equation [6], from which the oxidation parameter x was extracted and where the ratio $R = \mathrm{A(Fe_B^{2.5+})/A(Fe_A^{3+})}$, obtained from the fitting process, was used,

$$R(\phi_B, x) = \frac{2-6x}{1.06(1-(1-\phi_B)x)+(6-\phi_B)x} \quad (1)$$

In equation (1), ϕ_B represents the fractional concentration of vacancies in B sites, which ranges between 0 and 1. However, as it has been already shown [6], the oxidation parameter x is practically insensitive to ϕ_B and differences are only noticed within a 0.8% of precision. In our case, the best estimate for x is 0.12 giving rise to a non-stoichiometric magnetite with formula $Fe_{2.88}O_4$. The effects of the hematite on the native sample is also reflected in the magnetic properties, more concretely, coercive forces and saturation values are given by: Hc_{T2}=132.75Oe, Hc_{T5}=144.5Oe, Ms_{T2}=81.59emu/g, Ms_{T5}=77.53emu/g [5].

IOP Conf. Series: Journal of Physics: Conf. Series **935** (2017) 012038 doi:10.1088/1742-6596/935/1/012038

To analyse the rheological properties, it must be stressed that the viscosity of the MRF is established according to the conditions of laminar flow and its value must be in principle independent of the method used to measure it. However, the rheological information implies using a methodology that depends in practice upon some variables such as the container and spindle. The corresponding results are shown in Figure 2 were a sweep of low shear speeds by instrumentals limitations was considered.

The behaviour of the viscosity is ruled by the magnetic properties of the hydrodynamic volume of the magnetite, showing a special rearrangement or rotation of its magnetic moments to become aligned according to the magnetic flux density consistent with the Hershel-Buckley model as endorsed by the linear fitting shown in Figure 2. In this way, more resistance to the laminar movement is expected.

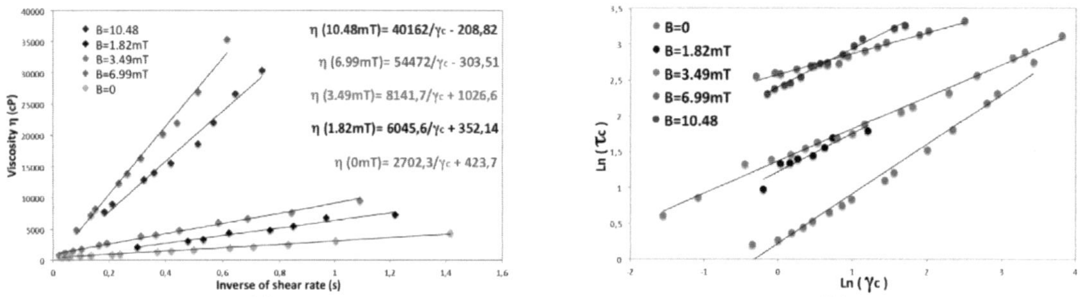

(a) (b)

Figure 2. Behaviour of viscosity (a) and shear stress (b) for different values of the magnetic flux density and shear rate for percentages 20% of magnetite with size T2.

From Figure 2, η is the viscosity, γ_c shear rate, τ_c shear stress the slopes represent the effort required to move the MRF at each speed and the cut-off point indicates the limit of Newtonian behaviour. As is observed, slopes increase with the magnetic flux density until a critical value of the field of B=6.99mT, from which phase the separation occurs by gravitational and magnetic effects due to the formation of mesoscopic structures, so the information recorded does not belong to a uniform MRF. The shear stress curves are given in the Figure 3, which are in agreement with those obtained by J. M. Linke and S. Odenbach [7].

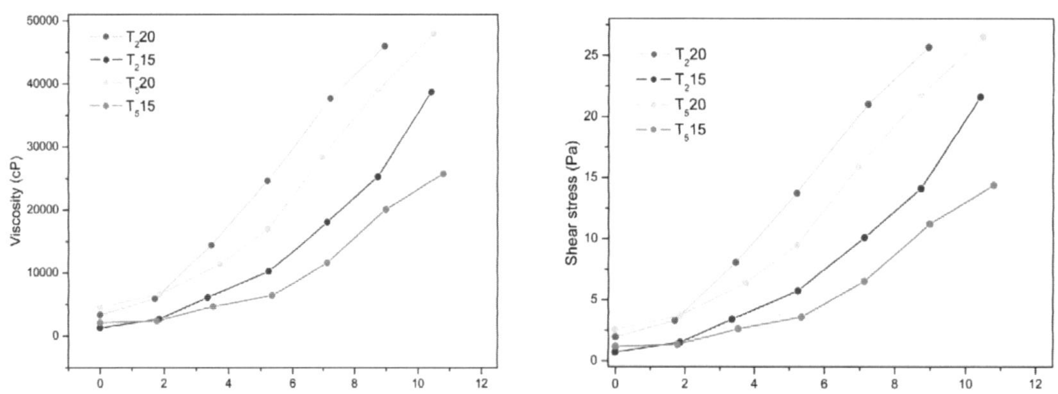

(a) (b)

Figure 3. Dependence of viscosity (a) and shear stress (b) with the magnetic flux density at a constant shear rate and for two different percentages of magnetite (15% and 20%).

On the other hand, by varying the magnetic flux density and fixing the speed of the spindle at 2rpm we obtain the results shown in Figure 3. As can be observed, magnetoviscosity is evidenced through

IOP Conf. Series: Journal of Physics: Conf. Series **935** (2017) 012038 doi:10.1088/1742-6596/935/1/012038

the remarkable variations of viscosity as the magnetic flux density increases, which is attributed to the formation mesoscopic structures. Our results are consistent with those found by E. V. Korobko, *et. al.* [8] for shear stress and R. P. Pant, *et. al.* [9] for viscosity.

The increase of viscosity and shear stress with magnetic flux density can be explained as due to the long-range magnetic dipole interaction between agglomerates causing a spatial rearrangement of solute in liquid carrier. Moreover, the magnetic dipole moment μ of the agglomerate is proportional to the hydrodynamic volume given by $\mu = V\chi B/\mu_0$, where V is the hydrodynamic volume, χ the magnetic susceptibility and μ_0 magnetic permeability [9,10].

Rotation and rearrangement of the magnetic dipole moments are ruled by the surface functionality provided by the electric character of the oleic acid layer around the aggregate; which can in turn gives rise to some preferred orientations of the magnetic moments and changes in surface tension T_σ.

The effects of surface tension allow us to consider whether the MRF can be used in cleaning native minerals that may be with unwanted material on the surface of others like as coal [4]. The results of the surface tension measurements are shown in Figure 4.

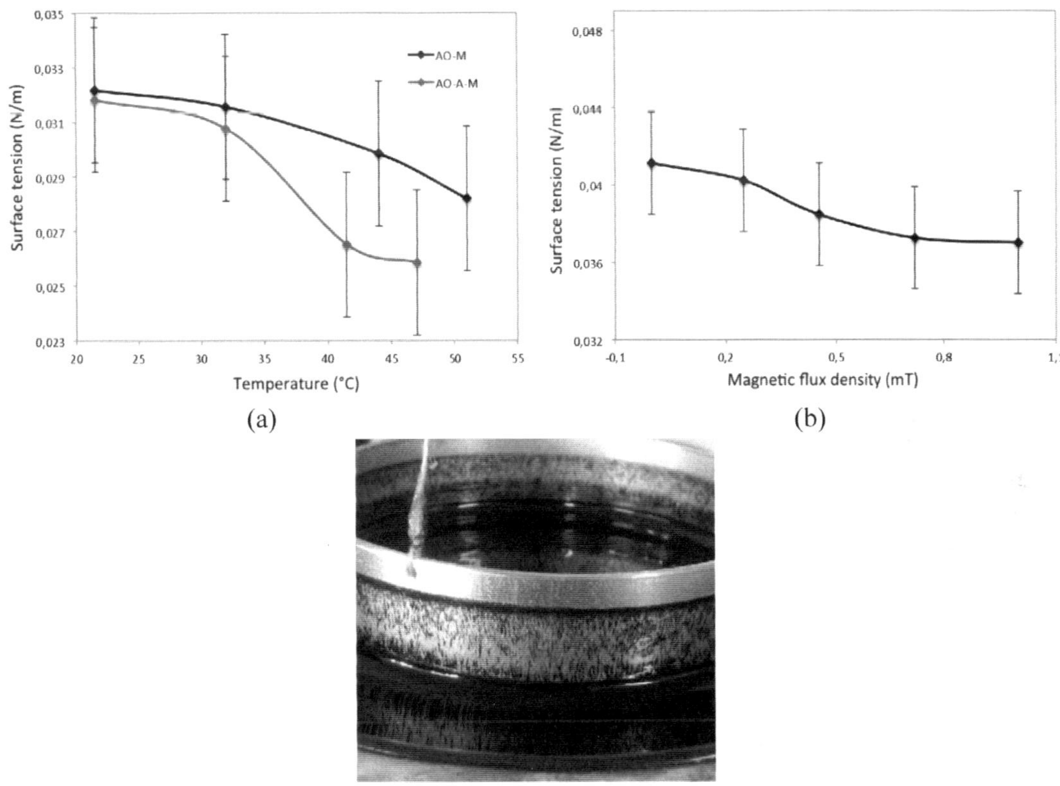

Figure 4. Behavior of surface tension as function of temperature for the liquid residue of oleic acid (OA) – magnetite (M) and oleic acid (OA) – oil (O) – magnetite (M) (a), and dependence with the magnetic flux density (b). At the bottom the formation of mesoscopic structures due to the field is visualized for percentage 20% of magnetite with size T5 (c).

Figure 4 shows also the formation of mesoscopic columnar structures and of fibrous-type for which a magnetostatic interaction between them, as a consequence of the radial asymmetry and the diameters involved, is expected [10]. Additionally, magnetite powders are not physically bound to each other due steric repulsion as a consequence of the oleic acid molecules, which are coordinated and attached of the Fe atoms of the bi-dentate surface formed by covalent oxygen bonds. Besides, the oxidation

IOP Conf. Series: Journal of Physics: Conf. Series **935** (2017) 012038 doi:10.1088/1742-6596/935/1/012038

parameter favours bonding by increasing adhesion to the molecule. For concentrations of oleic acid higher than the approximate equivalent of two monolayers there is an appreciable steric repulsion as it has been reported elsewhere [10-12].

4. Conclusions

Results of the rheological properties of the MRF show characteristics that are strongly dependent on the concentration and size of magnetite, temperature and the magnetic flux density. This MRF was characterized from its nature of micrometric sized aggregates and non-uniform geometry, which exhibit a pseudoplastic behaviour consistent with the Herschey Bulkley model.

The study of the MRF, gives information of our understanding about how this fluid reacts under certain circumstances or how to control some reaction, as it would be the case of damping systems by effects of instability by an applied field; or in transport systems where magneto-viscosity play an importance role.

It can be established also that the effective surface of the magnetic powders increases with size reduction allowing a better colloidal stability by lowering the gravitational energy $E_g = \Delta\rho V_{T_5} g L$ as compared to the thermal one $E_T = kT$ [10].

Finally, in MRF, its knowledge about the rheological behaviour under an applied magnetic flux density is currently incomplete and models of constitutive equations have not been provided still in general form [13].

Acknowledgements

Support provided by the CODI-UdeA project 2016-10085 and the exclusive dedication UdeA program to one of the authors (J. R) is acknowledged.

References

[1] F Donato et al. 2007 Estudio experimental de la viscosidad de un fluido magneto - reológico a base de magnetita mineral *Rev. Mex. Fís. E* **53(1)** 3140

[2] Shin-Tson Wu, et al. 2011 Adaptive mechanical-wetting lens actuated by ferrofluids *Opt. Commun.* **284** 2118

[3] R Pérez-Castillejos 2000 The use of ferrofluids in micromechanics *Sensor Actuat* **84** 176

[4] M. Pilar Triviño, et al. 2010 Implementación de fluidos magnetorreológicos para beneficio de minerales *Prospect* **8(1)** 77

[5] Fonseca H A, Gonzalez E, Restrepo J, Parra C A and Ortiz C 2016 Magnetic effect in viscosity of magnetorheological fluids *Journal of Physics: Conference Series* **687(1)** 012102

[6] J D Betancur, et al. 2003 Thermally driven and ball-milled hematite to magnetite transformation *Hyperfine Interactions* **148/149** 163

[7] J M Linke and S Odenbach 2015 Anisotropy of the magneto viscous effect in a ferro- fluid with weakly interacting magnetite nanoparticles *J. Phys.: Condens. Matter.* **27(176001)** 7

[8] E V Korobko, Z A Novikova and M A Zhurauski 2015 Magnetorheological fluids with two-component dispersed phase *J. Phy.: Conf. S.* **602** 012033

[9] R P Pant, et al. 2013 The size induced effect on rheological properties of Co-ferrite based ferrofluid *J. Non-Cryst. Sol.* **361** 3842

[10] R E Rosensweig 1997 Ferrohydrodynamics (New York: Dover publications Inc.)

[11] X Batlle, et al. 2007 Surfactant effects in magnetite nanoparticles of controlled size *J. Magn. Magn. Mater.* **316** e756

[12] M T López, et al. 2005 Stability go magnetizable colloidal suspensions by addition of oleic acid and silica nanoparticles *Colloids and Surfaces A: Physicochem. Eng. Aspects* **264(1-3)** 75

[13] R Y Hong, et al. 2007 Rheological properties of water-based Fe_3O_4 ferrofluids *Chem. Eng. Sci.* **62** 5912

IOP Conf. Series: Journal of Physics: Conf. Series **935** (2017) 012039 doi:10.1088/1742-6596/935/1/012039

Electrical and structural behaviour of the perovskite $LaCr_{0.4}Co_{0.4}Fe_{0.2}O_3$

W Rativa-Parada[1], J A Gómez-Cuaspud[1], E Vera-López[1] and J B Carda-Castelló[2]

[1] Universidad Pedagógica y Tecnológica de Colombia, Tunja, Colombia

[2] Universitat Jaume I, Castellón de la Plana, España

E-mail: wilson.rativa@uptc.edu.co

Abstract: The electrical and structural properties of the $LaCr_{0.4}Co_{0.4}Fe_{0.2}O_3$ perovskite are investigated. The oxide is synthetized by polymerization-combustion method, using citric acid as a chelating agent and low calcination temperature. The X-ray diffraction, Raman spectroscopy and transmission electron microscopy analysis show conformation of a pure phase with rhombohedral *(R-3c)* structure and confirmed high structural crystallinity facilitated by synthesis method. The characterization by means of impedance spectroscopy is performed at room temperature. It is observed that the oxides behave as materials of the semiconductor type and that the conductivity increase in accordance to a thermal excitation phenomenon.

1. Introduction

The perovskite $LaCrO_3$ had been extensively studied as electrode material in solid oxide fuel cells because it has shown good stability in both oxidizing and reducing atmospheres. It has been found that this oxide presents an orthorhombic lattice at room temperature, rhombohedral lattice at 260°C, and cubic lattice at 1600°C, which gives it diverse properties [1]. Furthermore, it has been observed that modified $LaCrO_3$ materials show a high resistivity towards carbon deposition and an excellent electrochemical activity to hydrocarbon reforming reactions, which can be significantly enhanced by introduction of transition cations in the structure as Fe and Co. Several methods have been employed as solid state reaction [2], hydrothermal [3], sol-gel [4] and co-precipitation [5] to try controlling Cr volatilization, which is one of the main drawbacks of these oxides, but the polymerization-combustion has shown the best results for their control, due to the use of citric acid in the formation of metal-citrate compounds and the low calcination temperatures [6-8]. In this paper the perovskite $LaCr_{0.4}Co_{0.4}Fe_{0.2}O_3$ is synthesized by the polymerization-combustion method and it is study with the aim of evaluating doping effect on the electrical and structural properties.

2. Experimental

The starting materials were $La(NO_3)_3.6H_2O$, $Fe(NO_3)_3.9H_2O$, $Cr(NO_3)_3.9H_2O$ and $Co(NO_3)_3.9H_2O$ (99.99%) (1.00M). Citric acid (2.00M) was used as chelating agent in a 0.5:1 molar ratio with respect to total metal cation content. The dissolution was drying to 90°C by 120min with continuous stirring and reflux. It was continued with heating at 350°C for 2 hours until the consolidation of a gel, which was calcinated for 12 hours at 850°C. The structural characterization was carried out in a Panalytical X´pert PRO-MPD diffractometer from 10 to 80° (2Θ) using Cu radiation and Bragg-Brentano configuration. The transmission electron microscopy observations were performed with a JEOL 2100 microscope employing an acceleration voltage of 200kV and a LaB_6 thermionic gun. An HR-UV 800 infinity microprobe (Jobin-Yvon) spectrometer with a laser of 785nm was employed to collect the Raman shift. The electrical characterization was performed in an Agilent 4294A spectrometer using pellets of solids compacted and coated with platinum on the surface.

3. Results and discussion

Figure 1 reveals the formation of a rhombohedral *R-3c (167)* single phase perovskite with cell parameters a=5.4619Å, c=13.5284Å, volume of cell=349.511Å^3, unlike what happens with the un-doped lanthanum chromite that tends to be orthorhombic [1]. The Fe and Co doping generated a reduction on the BO_6 octahedron and smaller B-O bond distances than $LaCrO_3$, for which a rhombohedral crystalline structure was formed, according to previous reports in where the doping at the site B of the lattice generated a lower distortion of the BO_6 octahedron [9]. No additional peaks were found, suggesting that all metal elements were integrated within the perovskite structure. The crystalline size was calculated with Debye-Scherrer equation [10], using a constant of 0.89 and the strongest peak (1 2 1). The estimated crystallite size was 31nm.

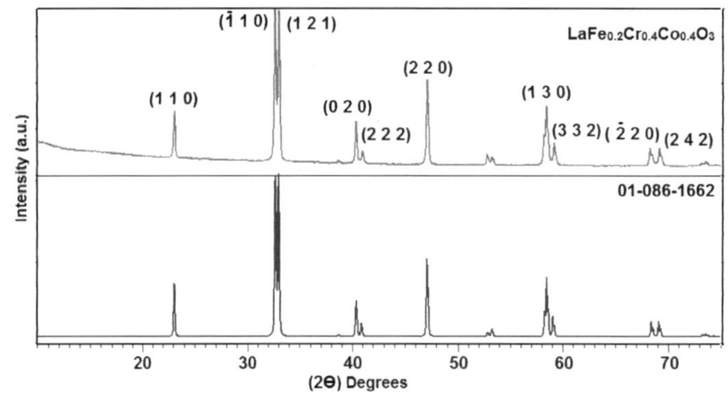

Figure 1. XRD pattern of $LaCr_{0.4}Co_{0.4}Fe_{0.2}O_3$.

TEM images reveal the crystalline that the materials reached. Figure 2(a) shows an interplanar spacing of 3.15Å corresponding to the (1 2 1) plane, ratifying the results obtained by means of X-ray diffraction about the nanometric size of the sample. Figure 2(b) confirms high crystallinity obtained by polymerization-combustion method without the formation of planar defects, moreover ratifying the existence of a rhombohedral crystalline structure.

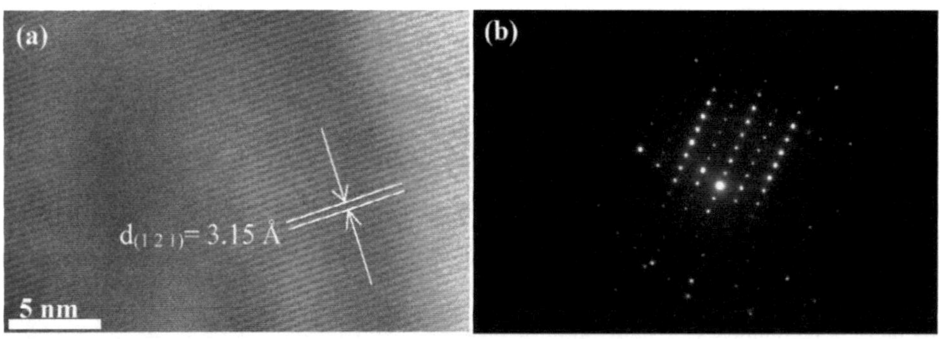

Figure 2. TEM images of $LaCr_{0.4}Co_{0.4}Fe_{0.2}O_3$.

Figure 3 shows the existence of signals at 553 and 647cm^{-1}, which are associated to B_{1g} stretch modes of a rhombohedral structure [8]. The Co and Fe doping generated a displacement of the peaks of the $LaCrO_3$ original perovskite toward a lower Raman shift and doing the peaks located below

300cm^{-1} more diffuse [11], since in other reports $LaCrO_3$ have been seen that the same vibrational modes appear in 590 and 719cm^{-1} [12].

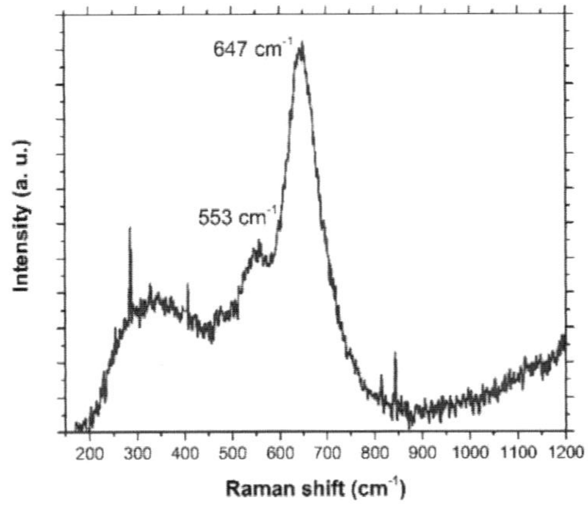

Figure 3. Raman shift of $LaCr_{0.4}Co_{0.4}Fe_{0.2}O_3$.

The electrical properties were measured as a function of the temperature. Figure 4(a) exhibits the impedance spectra in where an only semi-arc is observed without the appearance of a charge transfer resistance, where resistivity decreases with increasing of the temperature, besides at high temperatures an ohmic resistance is observed, indicating that the oxide has a semiconductor type behaviour [13]. Figure 4(b) confirms that the perovskite is formed by resistance inside (bulk) and at the grain boundary, since the peaks appear at both low and high frequencies [8].

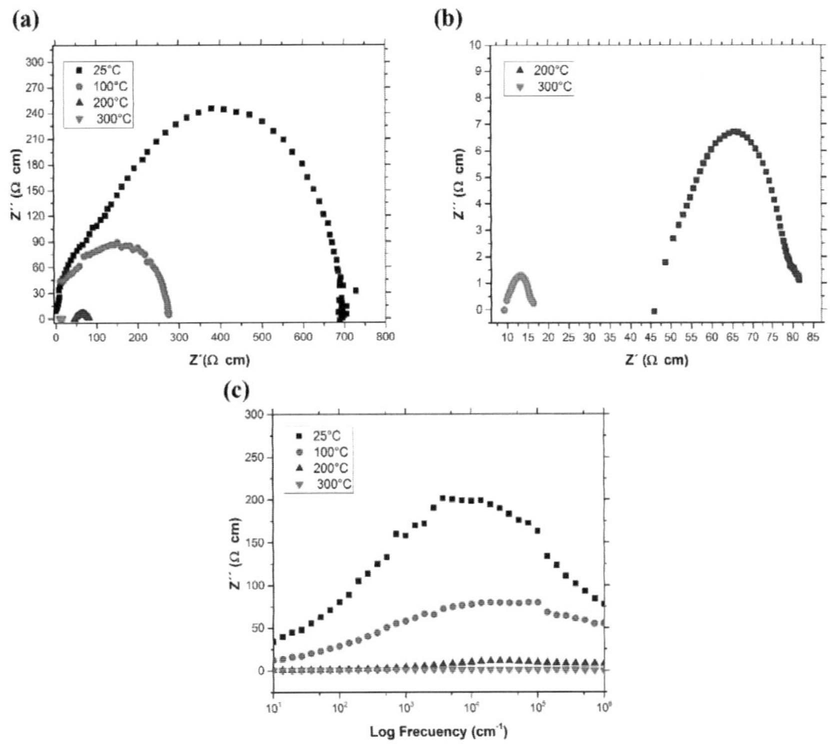

Figure 4. (a) (b) Nyquist and (c) Imaginary part of impedance (Z´´) vs Frequency logarithm plots of $LaCr_{0.4}Co_{0.4}Fe_{0.2}O_3$.

Figure 5 shows that conductivity increase with an increment of the temperature, which indicate that oxide present a thermal excitation phenomenon that enhances the mobility of the charge carries between B-O-B bonds. This behaviour allows linear slope of the Arrhenius plot to be the activation energy value. The value of the activation energy was 1.05 eV. Both the linear tendency and low value of the slope of the Arrhenius plot indicate that the conductivity mechanisms are mainly caused by electronic species, according to the Nyquist plots [14,15].

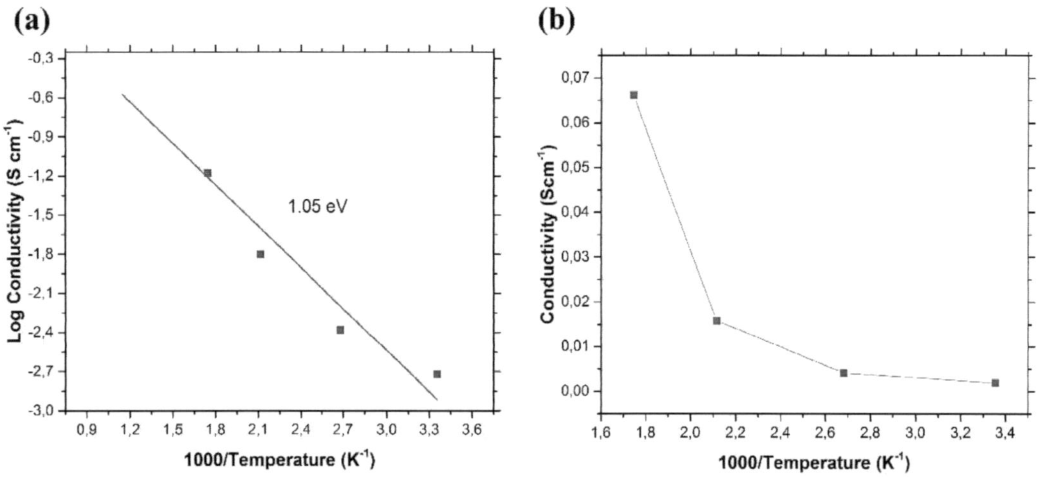

Figure 5. (a) Conductivity and (b) Arrhenius plots of $LaCr_{0.4}Co_{0.4}Fe_{0.2}O_3$.

4. Conclusions

The perovskite $LaCr_{0.4}Co_{0.4}Fe_{0.2}O_3$ synthesized through polymerization-combustion method reached outstanding structural and electrical properties to be applied as electrode in solid oxide fuel cells. The oxide showed a rhombohedral *(R-3c)* single phase with the deformation of the BO_6 octahedron caused by incorporation of Co and Fe, which also induced a displacement of the Raman peaks toward lower Raman shift with respect to un-doped $LaCrO_3$. The material presented a semiconductor type electrical behaviour with a thermal excitation phenomenon and an activation energy of 1.05eV.

References

[1] Gupta S, Mahapatra M K, Prabhakar Singh P 2013 Phase transformation, thermal expansion and electrical conductivity of lanthanum chromite *Materials Research Bulletin* **48** 3262–3267

[2] He S, Chen H, Li R, Ge L, Guo L 2014 Effect of $Ce_{0.8}Sm_{0.2}O_{1.9}$ interlayer on the electrochemical performance of $La_{0.75}Sr_{0.25}Cr_{0.5}Mn_{0.5}O_3$ de $Ce_{0.8}Sm_{0.2}O_{1.9}$ composite anodes for intermediate-temperature solid oxide fuel cells *Journal of Power Sources* **253** 187-192

[3] Wang S, Huang K, Zheng B, Zhang J, Feng S 2013 Mild hydrothermal synthesis and physical property of perovskite Sr doped $LaCrO_3$ *Materials Letters* **101** 86–89

[4] Ratha M K, Lee K T 2015 Investigation of aliovalent transition metal doped $La_{0.7}Ca_{0.3}Cr_{0.8}X_{0.2}O_{3-\delta}$ (X=Ti, Mn, Fe, Co, and Ni) as electrode materials for symmetric solid oxide fuel cells *Ceramics International* **41** 10878–10890

[5] Choi B H, Park S A, Park B K, Chun H H, Kim Y T 2013 Controlled synthesis of $La_{1-x}Sr_xCrO_3$ nanoparticles by hydrothermal method with nonionic surfactant and their ORR activity in alkaline medium *Materials Research Bulletin* **48** 3651–3656

[6] Lee B I, Ravindra, Gupta R K, Whang C M 2008 Effects of solvent and chelating agent on synthesis of solid oxide fuel cell perovskite, $La_{0.8}Sr_{0.2}CrO_{3-\delta}$ *Materials Research Bulletin* **43** 207–221

[7] Kolisetty A, Fu Z, Koc R 2017 Development of $La(CrCoFeNi)O_3$ system perovskites as interconnect and cathode materials for solid oxide fuel cells *Ceramics International* **43** 647–7652

IOP Conf. Series: Journal of Physics: Conf. Series **935** (2017) 012039 doi:10.1088/1742-6596/935/1/012039

[8] Nithya V D, Immanuel R J, Senthilkumar S T, Sanjeeviraja C, Perelshtein I, Zitoun D, Selvan R K 2012 Studies on the structural, electrical and magnetic properties of $LaCrO_3$, $LaCr_{0.5}Cu_{0.5}O_3$ and $LaCr_{0.5}Fe_{0.5}O_3$ by sol–gel method *Materials Research Bulletin* **47(8)** 1861–1868

[9] Silva Jr. R S, Barrozo P, Moreno N O, Aguiar J A 2016 Structural and magnetic properties of $LaCrO_3$ half-doped with Al *Ceramics International* **42** 14499–14504

[10] Hammouda S B, Zhao F, Safaei Z, Babu I, Ramasamy D L, Sillanpää M 2017 Reactivity of novel Ceria–Perovskite composites CeO_2-$LaMO_3$ (MCu,Fe) in the catalytic wet peroxidative oxidation of the new emergent pollutant 'Bisphenol F': Characterization, kinetic and mechanism studies *Applied Catalysis B: Environmental* **218** 119–136

[11] Selvadurai A P B, Pazhanivelu V, Jagadeeshwaran C, Murugaraj R, Muthuselvam I P, Chou F C 2015 Influence of Cr substitution on structural, magnetic and electrical conductivity spectra of $LaFeO_3$ *Journal of Alloys and Compounds* **646** 924-931

[12] Tompsett G A, Sammes N M 2004 Characterisation of the SOFC material, $LaCrO_3$, using vibrational spectroscopy *Journal of Power Sources* **130** 1–7

[13] Venkataramana K, Madhuri C, Reddy Y S, Bhikshamaiah G, Reddy C V 2017 Structural, electrical and thermal expansion studies of tri-doped ceria electrolyte materials for IT-SOFCs *Journal of Alloys and Compounds* **719** 97-107

[14] Benali A, Bejar M, Dhahri E, Graça M F P, Costa L C 2015 Electrical conductivity and ac dielectric properties of $La_{0.8}Ca_{0.2-x}Pb_xFeO_3$ (x=0.05, 0.10 and 0.15) perovskite compounds *Journal of Alloys and Compounds* **653** 506-512

[15] Omri A, Bejar M, Dhahri E, Es-Souni M, Valente M A, Graça M P F, Costa L C 2012 Electrical conductivity and dielectric analysis of $La_{0.75}(Ca,Sr)_{0.25}Mn_{0.85}Ga_{0.15}O_3$ perovskite compound *Journal of Alloys and Compounds* **536** 173–178

IOP Conf. Series: Journal of Physics: Conf. Series **935** (2017) 012040 doi:10.1088/1742-6596/935/1/012040

Characterization of fly ash, slag and glass hull for the obtaining of vitreous materials

D M Ayala Valderrama[1] and J A Gómez Cuaspud[1]
[1] Universidad Pedagógica y Tecnológica de Colombia, Tunja, Colombia

E-mail: diana.ayala01@uptc.edu.co

Abstract. This article presents the structural and thermal characterization of fly ash, the waste from blast furnace slag and the glass hull, generated as common residues in industry, which cannot be recycled easily or destroyed in a simple and fast way. In the particular case of fly ash, at present are being used as a lightweight aggregate in the production of cement, concrete and additive in the production of glass and glass ceramics. As far as the slag and hull, are being used as additives for the asphalt and concretes, however its use still is restricted, reason why its use in alternative ways are necessary. Initially the chemical composition of residues was established, determining that the fly ashes contains SiO_2, Al_2O_3 and Fe_2O_3 oxides; 90% of the total composition, was confirmed by X-ray diffraction analysis. As minor constituents, small percentages of Mg, P, S, K, Na and Ti were found. For the slag case, the phases of Fe_3O_4, $Ca_3Mg(SiO_2)_4$ and $Ca(MgAl)(Si,Al)_2O_6$ were identified, observing the presence of amorphous phase higher than 94% of the total phase of the system. Meanwhile, the glass hull sample showed a higher percentage of 95% amorphicity, mainly identifying a weak signal associated with silicon oxide SiO_2. The thermal analyses of the samples, exhibit a decrease in mass for samples between 25-1000°C was observed, which can be attributed to different physical-chemical events that occur in the materials. The heat flow for each sample is related with the removal of the water retained by the physisorption processes around 92-110°C in all cases. With this previous characterization of the precursors, a sample was composed using 70% fly ash, 10% slag and 20% of glass hull was composed and treated at 1200°C/1.5 hours, obtaining a dense black glassy material for potential applications in field of the glass ceramics.

1. Introduction

The fly ash is a raw waste material mainly generated by the coal combustion process for energy production [1]. This material has received special attention in recent years, due to the increased research related to the study of its physicochemical properties, focused on the use as an aggregate in the manufacture of concrete and reinforced cements, due to its crystalline mineral compounds, carbon particles and glass aluminosilicates, each with levels of reactivity, stability and specific composition that allow to achieve diverse properties [2,3]. This material can provide a strong support for the obtaining of applications in construction field as low-cost adsorbents for the removal of organic compounds, gas treatment, light aggregates, mine fillings, road sub-base, zeolite synthesis, solidification of waste and the manufacture of geopolymers among others [3-5], demonstrating that the chemical composition of these materials is satisfactorily robust to be used in design of value-added products [6]. The main components of a fly ash are the SiO_2, Al_2O_3 and Fe_2O_3 oxides (with values around 70%, classified as class F ashes), the presence of different concentrations of MgO, CaO, Na_2O and K_2O as minor components could be an important factor in reducing costs in the glassmaking process due to the possibility of reducing evolving temperatures in the synthesis of glass materials. In addition, this effect of reducing energy costs can be enhanced by the addition of secondary components related to slag and glass hull, which allow the production of a wide variety of vitreous materials for thermal induction systems, and decorative materials between others according to the literature [6-9]. In this sense, the manufacture of vitreous derivatives related to the recycling of industrial waste presents a series of benefits such as the reduction of the electric consumption, the improvement of the quality of the material, the reduction of the emissions of greenhouse gases and the cost of Transport of raw materials [10], Reason justifying the development of the current work in the development of an experimental methodology for the production of vitreous material with the use of fly ash, slag and glass hulls from the region of Boyacá, Colombia, Materials that have been characterized to ensure the homogeneity of the process and the possibility of obtaining value-added materials [11,12].

2. Methodology

The design of the synthesis process was initiated with the physical processing and standardization of the raw materials of fly ash, slag and glass hull by an initial stage of milling using a standard 200 US mesh. The morphology and surface characteristics of materials were evaluated by scanning electron microscopy (SEM-EDX) on a Q600 SDT V20.9 Build 20 equipment. For the analysis, the samples were coated with platinum for a better resolution images in a sputtering QUORUM Q150R ES instrument. The elemental composition of the raw materials was analysed by X-ray fluorescence (XRF), using a PANalitical MiniPal 2 spectrometer operated at 20KeV. The structural analysis was performed in an X-ray diffraction equipment, PANanalytical X-Pert PRO 2.2, using the Cu Kα radiation at (1.54Å), with steps 0.020° (2θ) in a continuous mode from 10-90°. Based on the characterization, a percentage composition of the starting materials 70% ash, 10% slag and 20% glass hull were employed to evaluate the characteristics of the vitreous material in accordance with previous works. The results of thermal stability and transition temperature (Tg) were validated by thermal analysis (TGA-DSC), which were performed on a SDT Q600 20, DSC-TGA Standard instrument, using a heating rate of 10°Cmin^{-1} under argon flow conditions (100 mLmin^{-1}), from room temperature to 1100°C. The results were analysed by the enthalpy values associated to the main thermodynamic transitions shown by the obtained vitreous material.

3. Results discussion

Initially, the scanning electron microscopy (SEM) images, show the presence of particles of all possible morphologies, related to the basic conformations of these materials as shown in Figure 1.

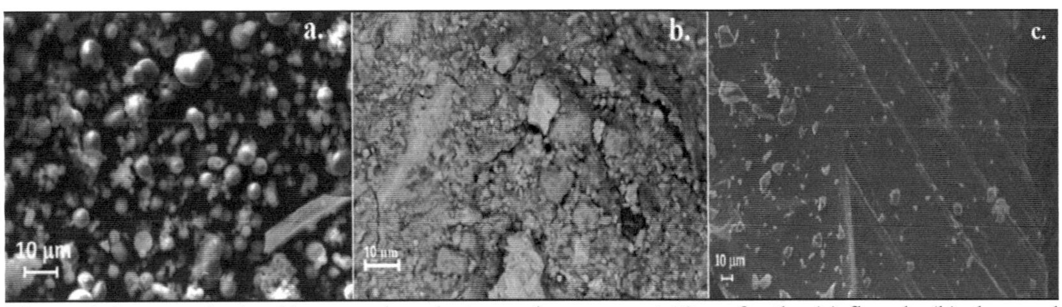

Figure 1. Images obtained by scanning electron microscopy at 10µm for the (a) fly ash, (b) slag, and (c) Glass hull samples.

The micrographs reveal the existence of heterogeneous particles separated from each other, on which it was possible to perform the morphological identification of a total of 10000 particles, which were selected in the automatic configuration of the scanning electron microscope, satisfying the statistical criteria in the analysed samples. The particle size evaluation in all cases using the Image J software, permitted to establish a variable distribution as indicated in the histograms of Figure 2. In the case of the fly ash sample, the results showed a homogeneous granulometric distribution around 100-150μm, whose composition is mainly associated with the presence of aluminosilicates derived from coal combustion processes, which confers to this material a low density and a high mechanical resistance as has been established by Żyrkowski *et. al.* [12]. In the case of the slag sample, the results showed the presence of particles behaving as agglomerates with a size of about 125-150μm, indicating condensation of droplets by vapours' and collection of liquid or powder in the process of the blast furnace of the steel industry [13]. On the other hand, the glass hull sample shows the presence of embedded glass silica, which demonstrates its low tendency to be incorporated into the glass material with high variability in particle sizes between 70-225μm showing a high heterogeneity without a tendency defined, taking into account that the source is not unique, making this type of distribution in size and composition very variable. According to the work of García-Fraile *et. al.*, Alabama. *et. al.*, and Shaheen. *et. al.* [4,14], the characteristics of the raw materials can be exploited in favour of a better incorporation of the individual components for the manufacture of glass products, allowing the improvement of the synthesis process to consolidate an optimal physical interaction with low porosity and excellent levels of compaction. Accordingly, the work by Matsunaga *et. al.* [15], indicate that the quality of glass materials obtained from raw materials having a particle size distribution very close to about 150±50μm, allow the generation of a large number of applications for structural purposes and of sieving for the adsorption of gases, treatment of organic compounds and of mercury among others [3,16,17], confirming the potential of these raw materials for the development of value-added vitreous materials with excellent results.

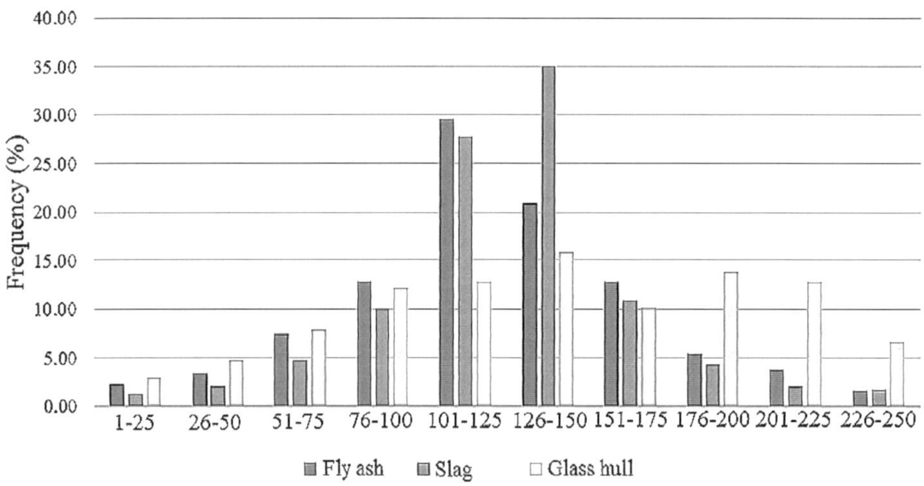

Figure 2. Percentage distribution histograms particle size for fly ash, slag and glass hull samples.

The basic composition results measured by X-ray dispersive spectroscopy (EDX), show that in the case of fly ash, the elements of highest contribution by weight are Si and Al represented in about 85% of the total composition, while the other 15% corresponds to minor elements of K, Fe and Ca oxides as shown in Figure 3. For the slag sample, the results indicate a very varied composition related with transition elements, however the greatest contributions are related with the presence of the SiO_2, CaO and Al_2O_3 oxides, which conform at least the 85% of the total weight, the remaining percentage is discriminated in transition elements, relevant to the process of steelwork, this result indicates a relevant pozzolanic reactivity that allows it to be considered as an effective additive in the production of cement mixtures in accordance with Wang *et. al.* [17].

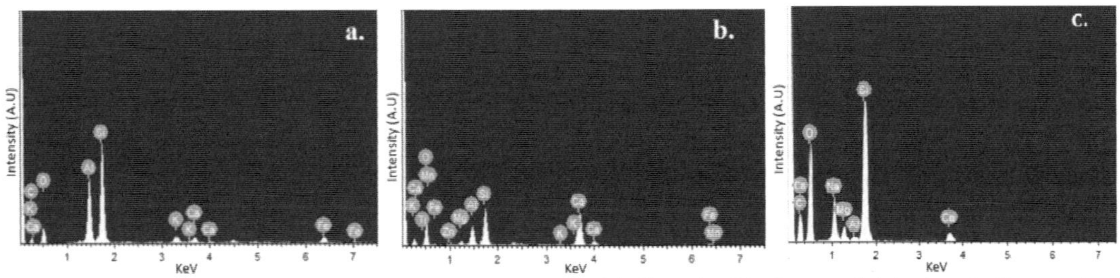

Figure 3. X-ray dispersive spectroscopy patterns for (a) fly ash, (b) slag, and (c) glass hull samples.

The glass hull sample, exhibit a marked presence of calcium and aluminium oxides with minor transition elements related to the typical dyes of this raw material as indicated in Table 1. Although the high content of silica in two of three precursors of raw materials (fly ash and glass hull), represent a high viscosity in the final vitreous material, this represent the possibility of obtention of low-porosity materials with a high bulk density in accordance with Żyrkowski *et. al.* [12], although other oxide precursor components may generate a decrease in viscosity, this can be compensated by the effect of alumina (Al_2O_3), which reduce the concentration of oxygen in the final material, enabling the devitrification processes and increasing the chemical stability to sulfides attack and the mechanical strength of vitreous materials [3,6,8,11,13,16,19].

Table 1. Percentage values of elemental composition of precursor materials associated with fly ash, slag and glass hull.

					Fly ash						
Element	O	Al	Si	K	Ca	Fe	Mg	Ti	Mn	Zn	Na
Weight (%)	29.39	18.25	37.3	2.88	3.12	9.06	-	-	-	-	-
					Slag						
Element	O	Al	Si	K	Ca	Fe	Mg	Ti	Mn	Zn	Na
Weight (%)	9.06	6.55	13.38	0.19	19.33	1.65	1.81	0.22	0.91	1.44	-
					Glass hull						
Element	O	Al	Si	K	Ca	Fe	Mg	Ti	Mn	Zn	Na
Weight (%)	45.87	0.46	36.18	-	6.32	-	2.39	-	-	-	8.78

The results of the structural analysis performed by X-ray diffraction in the precursor materials are shown in Figure 4, where the characteristic signals associated to the main identified crystalline phases are identified, which were analysed and processed by the X¨Pert High Score software in the ICDD databases.

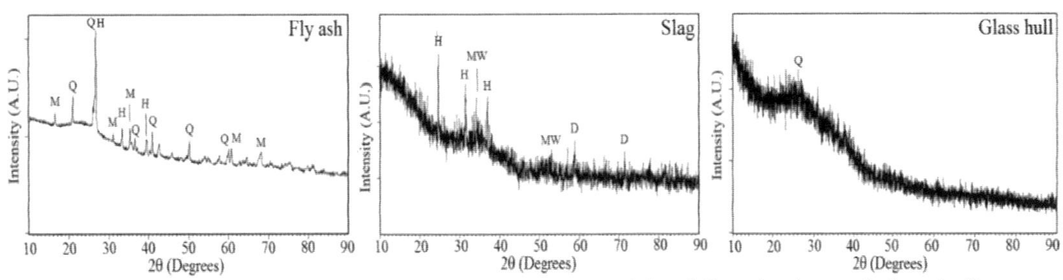

Figure 4. X-ray diffraction patterns of the three raw materials of fly ash, slag and glass hull.

In the case of fly ash, the characteristic signals associated with silicon oxide as quartz SiO_2 (Q), mullite $Al_{4+2x}Si_{2-2x}O_{10-x}$ (M) and hematite Fe_3O_4 (H), which are common materials to this type of samples have been identified extensively in the literature [15,19]. Detailed analyses of the slag sample, identify the presence of a hematite Fe_3O_4 phase, like the final stage of oxidation processes of iron in the steel industry. Additionally, were identified the crystalline phases related to the merwinite $Ca_3Mg(SiO_2)_4$ (MW) and diopside $Ca(Mg,Al)(Si,Al)_2O_6$ (D) phases, being clear that the presence of amorphous material exceeds the 94% of the total phase, indicating a strong vitrification process in the sample. These features, allow to establish the prevailing presence of MgO_6 and SiO_4 compositions, which favours the obtaining of vitreous material with application in the adsorption of gaseous pollutants NOx, SOx, organic compounds, mercury in air, dyes and organic compounds in wastewater, in accordance with the development of multifunctional vitreous materials [3,16,20,21]. Meanwhile, the glass hull sample showed a high percentage amorphicity around 95%, in which was possible to identify a weak signal associated with silicon oxide SiO_2 as shown in Figure 4.

The thermal analysis of the raw materials shown in Figure 5, indicate a decrease in mass of samples between 25-1200°C. This mass loss can be attributed to different physicochemical events that arise in these materials as have been well-known by Pavas and Molina *et. al.* [20]. Initially, the heat flow for each sample present several endothermic and exothermic signals related with the removal of water retained by the physical adsorption process around 92-110°C in all cases.

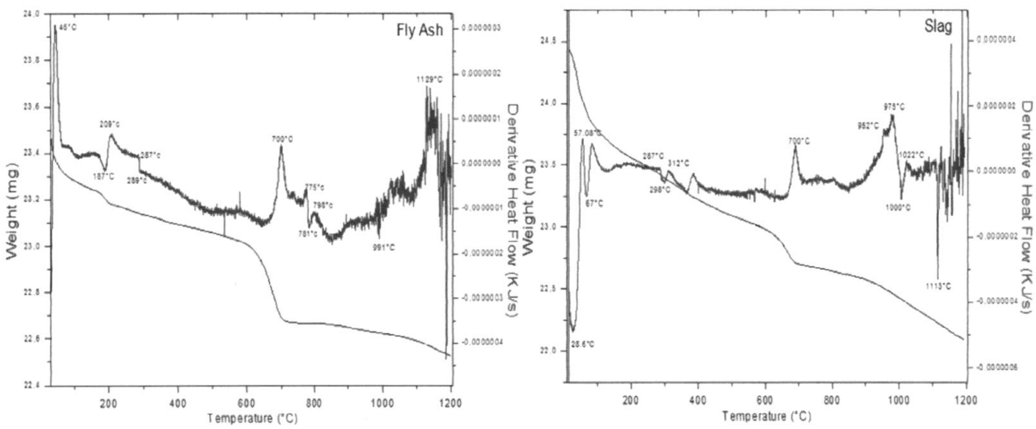

Figure 5. Thermal analysis between 25 and 1200°C for raw materials of (a) Fly ash, and (b) Slag.

The region between 120-200°C, can be attributed to elimination of organic matter mainly represented by oils and other substances derived from the physical manage of raw materials. The zone between 200-600°C, represent the elimination of carbonates and oxo-carbonates species, which are common contaminants of Na, K, Ca, and Mg oxides that represent at least a 70% of weight in all samples [22]. Finally, the zone around 600-1200°C represent different stages of melting and volatilization of minor species of low melting point oxides, characterized by strong exothermic signals and a weight loss around 65% in all cases. The exothermic signal observed at 700°C can be attributed to the oxidation of iron and magnesium compounds present in the three materials [23]. These materials present several exothermic signals that allow us to see the nucleation and crystallization points when subjected to thermal treatments, and which could contribute to the production of vitreous materials.

With the results of the physicochemical characterization of the raw materials, a composition of 70% of fly ash, 10% of slag and 20% of glass hull could be established, with the objective of forming a glass material for potential applications and obtaining of a value added product. To this end, the mixture was subjected to a thermal process according to the thermogravimetric analysis performed in the sample (Figure 6(a)) at 1200°C for 90 minutes using a heating chute of $2°Cmin^{-1}$, with a slow cooling step, obtaining A black and dense vitreous material according to the process developed by Savante *et. al.*, Alabama and Oziel Méndez [23,24]. As indicated in Figure 6(a) the mixture prior to heat treatment presented endothermic and exothermic signals between 1000°C and 1100°C, which indicate the temperature of nucleation and crystallization of the mixture. The corresponding structural characterization by X-ray diffraction shown in Figure 6(b), confirm the obtaining of an amorphous material of at least (95.3%) in phase content, without evidence of surface crystallization in accordance with the obtaining of a vitreous material [23-26].

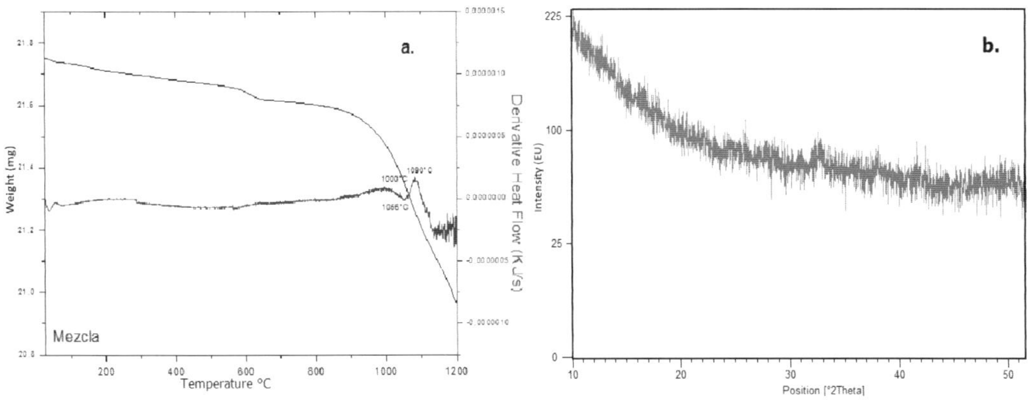

Figure 6. TGA-DSC prior to thermal treatment and curves X-ray diffraction pattern of vitreous material obtained at 1200°C for 90 min.

This shows that coal fly ash, blast furnace slag and glass hull waste have a wide range of industrial and technological applications.

4. Conclusions

The three characterized raw materials of fly ash, slag and glass hull, shown from point of view of surface, composition and structural conformation an optimal characteristic for use as components in design of vitreous materials for potential applications. The XRD results of mixture in percentage (70% fly ash, 10% slag and 20% glass hull), confirm the obtaining of an amorphous material without presence of the main diffraction signals associated with the raw materials confirming the evolution of a chemical reaction between main components in accordance with the objective of current work. The thermal analysis of vitreous sample, confirm the presence of an exothermic signal around 1200°C related with a vitreous transition in accordance with the obtaining of a material for potential applications in field of vitreous add-value products. With current results, is clear that is necessary to complement present work with the aim to reveal the effectiveness of proposed methodology to obtaining of new and advanced vitreous materials for potential applications.

Acknowledgements

The authors thank the Government of Boyacá and Colciencias for the financial support, according to call 733 of 2016.

References

[1] Valadares Folguera M, Novaes de Oliveira P and Estevam Alarcon O 2005 *American Ceramic Society Bulletin* 9201–9205
[2] Hyup Lee, Ho Sung Ha, Chang Hoon Lee, Yong Bok Lee, Pil Joo Kim 2006 *Bioresource Technol.* 97 1490–1497
[3] Ahmaruzzaman M 2010 *Prog. Energy Combust. Sci.* **36** 327–363
[4] Shaheen S M, Hooda P S and Tsadilas C D 2014 *Journal of Environmental Management* **145** 249–267
[5] Hanif A, Lu Z and Li Z 2017 *Constr. Build. Mater.* **144** 373–384
[6] Ribeiro M J, Labrincha J A and Ferreira J 2003 *Researchgate* 1-19
[7] Ossa M. and Jorquera H 1984 *Mater. Construcción* **34** 3–17
[8] Zhu m, Ji R, Li Z, Wang H, Liu L and Zhang Z 2016 *Constr. Build. Mater.* **112** 398–405

[9] Bakharev T 2005 *Cem. Concr. Res.* **35** 1224–1232
[10] Andreola F, Barbieri L, Lancellotti I, Leonelli C and Manfredini T 2016 *Ceram. Int.* 42 13333–13338
[11] Hongyu Liu, Hongxia Lu, Deliang Chen, Hailong Wang, Hongliang Xu, and Rui Zhang 2009 *Ceram. Int.* **35** 3181–3184
[12] Zyrkowski M, Costa Neto R, Santos L, and Witkowski K 2016 *Fuel* **174** 49–53
[13] Guézennec A G, Huber J C, Patisson F, Sessiecq P, Birat J P and Ablitzer D 2005 *Pow. Technol.* **157** 2–11
[14] García-fraile D J 2015 *Caracterización de cenizas de centrales térmicas del norte de españa para su aplicación en usos alternativos* (España: Universidad de Oviedo) pp 1-87
[15] Matsunaga T, Kim J K, Hardcastle S and Rohatgi P K 2002 *Mater. Sci. Eng. A* **325** 333–343
[16] Ferreira C, Ribeiro A and Ottosen L 2003 *J. Hazard. Mater* **96** 201–216
[17] Wang S, Zhang C and Chen J 2014 *J. Mater. Sci. Technol.* **30** 1208–1212
[18] Ji R, Zhang Z, Yan C, Zhu M and Li Z 2016 *Constr. Build. Mater.* **114** 888–895
[19] Liu H, Wei G, Liang Y and Dong F 2011 *J. Cent. South Univ. Technol.* **18** 1945–1952
[20] Pavas E G and Molina C S 1998 *Revista Colombiana de química* **27(1)** 89–97
[21] Melero F 2000 *Tecnología y gestión de la innovación* 194,195,196.
[22] Mangialardi T 2001 *J. Hazard. Mater.* **87** 225–239
[23] Oziel Méndez Guerrero D, Alicia Vázquez Méndez B and Álvarez Méndez A 2001 *Bol. la Soc. Esp. Ceram. y Vidr.* **50** 143–150
[24] Savante B 1995 *Bol. Soc. Esp. Cerám. Vidr.* **34** 119–134
[25] Ercenk E, Sen U, Bayrak G and Yilmaz S 2014 *Acta Phys. Pol. A* **125** 626–628
[26] Barbieri L, Lancellotti I, Manfredini T, Rincon J and Romero M 1999 *Fuel* **78** 271–276

IOP Conf. Series: Journal of Physics: Conf. Series **935** (2017) 012041 doi:10.1088/1742-6596/935/1/012041

Characterization of films formed by the aluminizing of T91 steel

J A Sanabria Cala[1], G R Conde Rodríguez[1], D Y Peña Ballesteros[1], D Laverde Cataño[1] and L S Quintero Rangel[2]
[1] Universidad Industrial de Santander, Bucaramanga, Colombia
[2] Multiservicios de Ingeniería 1A S.A., Barrancabermeja, Colombia

E-mail: sanabriacalaj@gmail.com

Abstract. The aluminizing of a T91 martensitic ferritic steel was carried out by a novel modification to the traditional technique of packed cementation, with the objective of producing a diffusion coating of aluminum in a shorter time and operating cost, from a technique that allows the reuse of powder packaging and which the coating of metal parts with complex shapes can be secured. As an aluminum source, commercial foil is used to wrap the piece to be coated, while the powder packaging contains aluminum oxide Al_2O_3 and an activating salt, ammonium chloride NH_4Cl. During the deposition process of the coating, the NH_4Cl is decomposed by reacting with foil, and thus, aluminum halides can be transferred to the metallic substrate, which deposit aluminum on the T91 steel surface while Al_2O_3 can be recycled for subsequent processes. The results of the diffractograms and micrographs indicated the strong influence of temperature, exposure time and ammonium chloride concentration in the formation and growth evolution of a stable coating of iron-aluminum and iron-aluminum-nickel on the T91 steel surface, which was effectively deposited at a temperature of 700°C and an exposure period of 9 hours. The coating formed on the T91 steel surface could play a protective role towards the material by acting as a physical barrier between the alloy and other corrosive species in high temperature operated systems.

1. Introduction

The aluminizing of steels is widely used in industry because it can give low chromium alloys a protection against corrosion at high temperatures. One of the most used methods of aluminizing is the hot immersion of the parts to be covered in an aluminum bath. During this process, the piece is annealed in a pure hydrogen atmosphere at a temperature between 675 and 785°C. Subsequently, the piece is heated in a furnace with a non-oxidizing atmosphere to an approximated temperature of the aluminizing bath, with the objective of obtaining an aluminum coating with improved wettability and minimum uncoated points [1]. Finally, the hot piece is transferred to the aluminum and silicon bath, where the coating is obtained. Thus, is possible to realize the aluminizing of steel sheets that are commercialized by industries ranging from electrical appliances to the automotive industry, which manufacture multiple parts for high temperature applications such as ovens, mufflers, exhaust pipes, etc. The parts made from these aluminized sheets have zones that are sensitive to corrosion at high temperatures in the seams or adhesives of the sheets, as well as uncoated points, generating future damage to the material.

Another technique used for the aluminizing of steels that are subjected to high temperature corrosion conditions is the packed cementation, which is used in components for turbines, power generation plants, chemical applications, aerospace and petrochemical industries [2-3]. According to the process proposed

IOP Conf. Series: Journal of Physics: Conf. Series **935** (2017) 012041 doi:10.1088/1742-6596/935/1/012041

by Bianco et al. [3-5], the process of aluminizing by packed cementation can be carried out at 900ºC using sodium chloride and a powder alloy with 42% Al-58% Fe by weight, immersed in an argon atmosphere and a reforming gas (5% H_2+Ar). During heating in the furnace, the system is continuously rotated to eliminate the chemical depletion zone of the alloy powder within the package, wrapping the substrate with a ceramic sheet to produce a particle-free surface, and thereby, introducing a diffusion barrier for the transport of the gaseous species. This method has been used to obtain coatings by diffusion of aluminum, after 6 hours of reaction, of approximately 100 microns of thickness.

Other researchers proposed the use of CVD-FBR technique (Chemical vapor deposition by fluidized bed reactor) for aluminizing martensitic ferritic steels using 99.5% pure aluminum powder. The mixture was fluidized with analytical argon as an inert gas, hydrogen chloride with purity of 99.99% as the reactive gas, and analytical type gaseous hydrogen as reducing gas. With this process, Fe_2Al_5 coatings of 4μm thickness could be obtained at 550ºC and 2 hours of material exposure time [5-7].

In the present study, a new aluminizing process was standardized, in which, commercial foil is used as an aluminum source and it wraps the piece to be covered, providing a concentrated source of aluminum throughout the process for growth and formation of the coating. This is in the closest area to the substrate surface, making the aluminum deposition process more efficient. The aluminizing by applying packed cementation in T91 steel, using commercial foil as a source, contributes in the industry because more efficient and economic processes are standardized, which are developed at low temperatures and short operating times in comparison with other techniques that are used nowadays.

2. Experimental

The coupons for the gravimetric tests are machined in rectangular form with dimensions of: 2×1×0.5cm from a martensitic ferritic steel (Fe-9Cr-1Mo), which are characterized based on ASTM E-415 standard [8] using an optical emission spectrometer. The surface of the coupons is prepared metallographically by using silicon carbide sandpaper, treatment initiated with sandpaper No. 120 up to No. 2400, based on ASTM G31-72 standard [9]. Prior to the tests, the coupons were cleaned with ethanol and then with acetone in an ultrasonic bath.

For all the tests, a horizontal tubular type batch furnace is used. Powder pack mixtures are prepared with a composition of 94% Al_2O_3 and 6% NH_4Cl by weight, which are carried to a ball mill for 24 hours to obtain a homogenization of the mixture. The gravimetric coupons are wrapped in commercial foil pre-treated by oxidation at 500°C, and then soaked in the powder package into alumina crucibles. The crucibles are introduced into the horizontal tubular furnace and exposed to a continuous flow of high purity analytical nitrogen. This procedure is carried out at temperatures of 500, 600 and 700°C, and operating times of 3, 5 and 9 hours.

3. Results and discussion

The oxidation pre-treatment at 500°C of the commercial foil is carried out in order to promote the growth of an aluminum oxide thin layer on its surface, and thereby, avoid that the aluminum of the foil melts on the metallic substrate at temperatures above the aluminum melting temperature, equivalent to 660°C, which would produce an irregular and non-homogeneous surface. The result of pre-treatment is presented in Figure 1 where it is possible to observe, in addition to the large pores, the formation of a fine porosity along the foil entire surface. This commercial foil pre-treatment process leads to the formation of a number of fine pores, which increases with the oxidation time and provides the diffusion paths for the volatile species generated by the breakdown of the salt, which subsequently react with the aluminum inside the foil to produce the volatile aluminum halides that effectively deposit the aluminum for forming the coating. Therefore, the smooth surface of the coated coupons shows that the coating was formed by gaseous species and not by aluminum which melted and deposited during the coating process, since on this last case, metallic deposits would be observed completely adhered to the surface of the steel substrate as reported in other cases where aluminum powder was used [3].

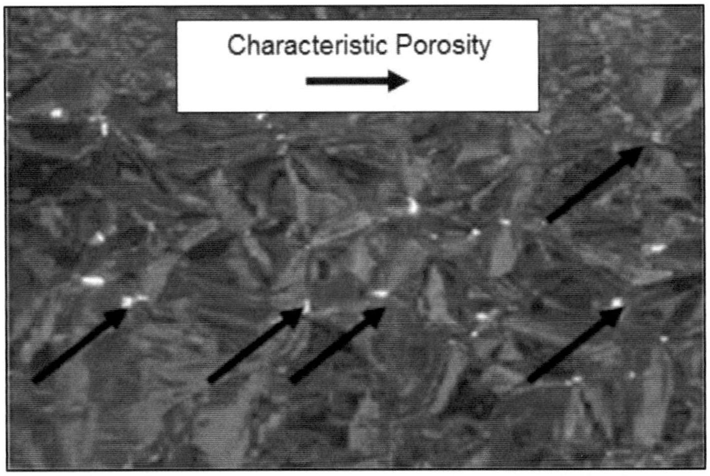

Figure 1. Characteristic porosity of commercial foil after pre-treatment.

The thickness of the coatings obtained by the aluminizing of the martensitic ferritic T91 steel coupons exposed to system temperature conditions of 500, 600 and 700°C, times of 3, 5 and 9 hours, and pressure of 5psi in N_2 atmosphere are detailed in Figure 2. It is observed that at longer exposure times, the thickness of the coating formed by the diffusion of aluminum on the T91 steel surface has an increasing behavior, where the process is carried out with an optimum concentration of 6% by weight of NH_4Cl in the system. On the contrary, it was not possible to aluminize T91 steel by this technique at a temperature of 500°C, regardless of the operating time.

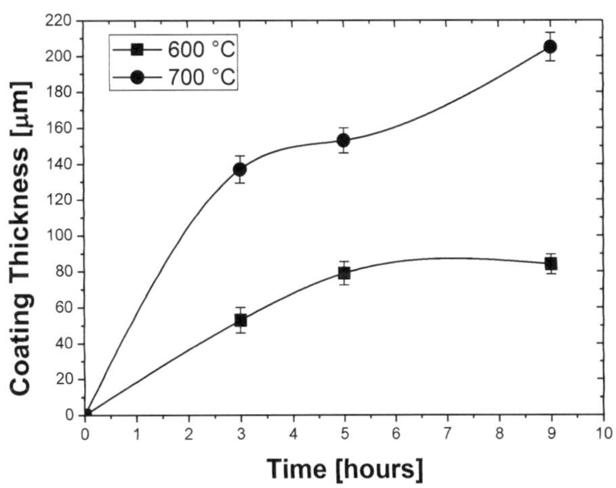

Figure 2. Variation of the coating thickness at different operating conditions.

In the micrographs of Figure 3, the thickness of the coatings obtained on the T91 steel surface is evident at system temperatures of 600 and 700°C, and exposure times of 3, 5 and 9 hours. A smooth light gray surface is detailed in the micrographs in all the coatings formed under the above conditions. Moreover, it was evident that the obtained coatings are continuous, of homogeneous thickness, thermodynamically stable, with pore-free morphology and without evidence of unstuck parts, obtaining an increase in the coating thickness formed on the T91 steel surface in a range from 53 to 205μm.

IOP Conf. Series: Journal of Physics: Conf. Series **935** (2017) 012041 doi:10.1088/1742-6596/935/1/012041

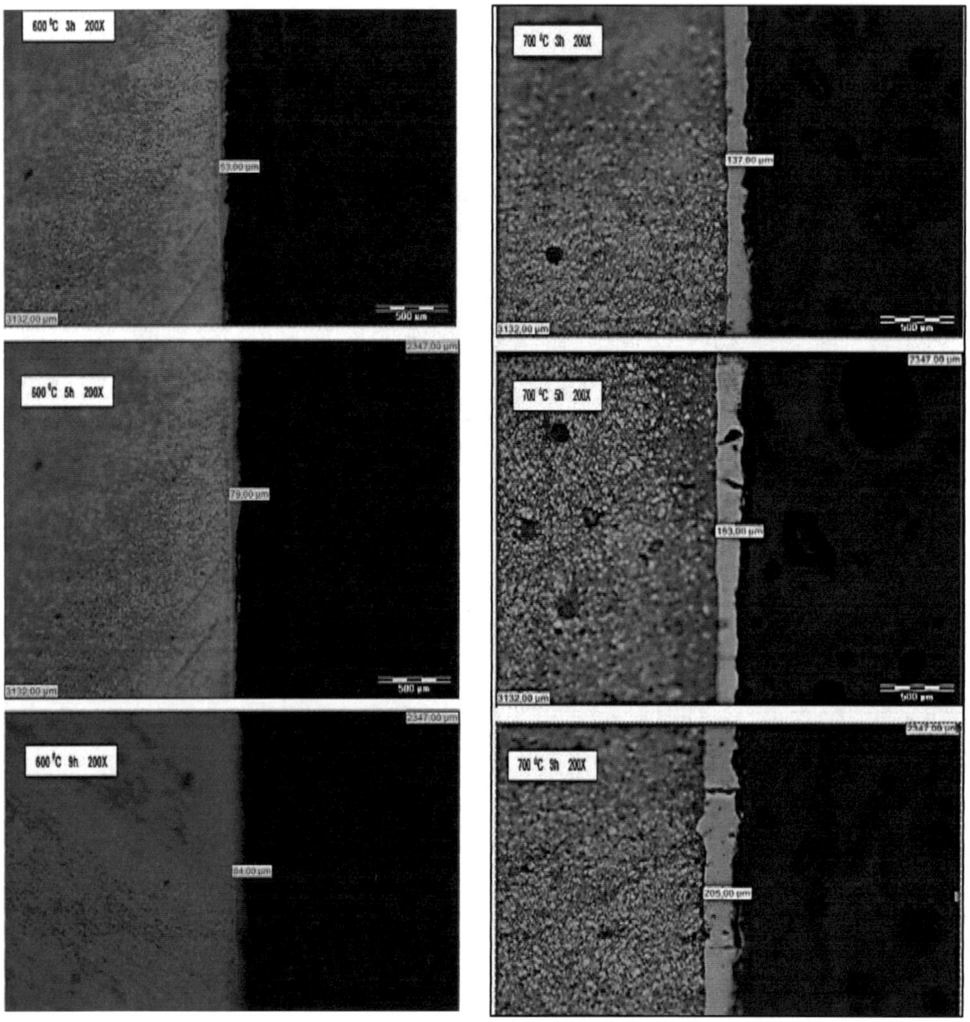

Figure 3. Micrographs of the coating formed on the T91 steel at different operating conditions.

X-ray diffraction was performed on the exposed coupon at 700°C for 9 hours, which shows that the crystalline phases present with greater intensity correspond to compounds of FeAl and $Fe_{0.99}Al_{0.99}Ni_{0.02}$ on the aluminized T91 steel surface, which are listed in Table 1. Likewise, in the micrograph of Figure 4 is observed that the aluminum coating on the T91 steel surface consists of two layers, which are identified from the diffractograms as Fe-Al (layer 1) and Fe-Al-Ni (layer 2).

Table 1. Phases found in the coating formed on the steel surface exposed to 700°C, 9 hours and NH_4Cl 6%w/w.

Crystalline phases	Name
Fe_3O_4	Magnetite
Fe_2O_3	Hematite
FeCr	Iron-Chrome
FeAl	Iron-Aluminum
$Fe_{0,99}Al_{0,99}Ni_{0,02}$	Iron-Aluminum-Nickel

IOP Conf. Series: Journal of Physics: Conf. Series **935** (2017) 012041 doi:10.1088/1742-6596/935/1/012041

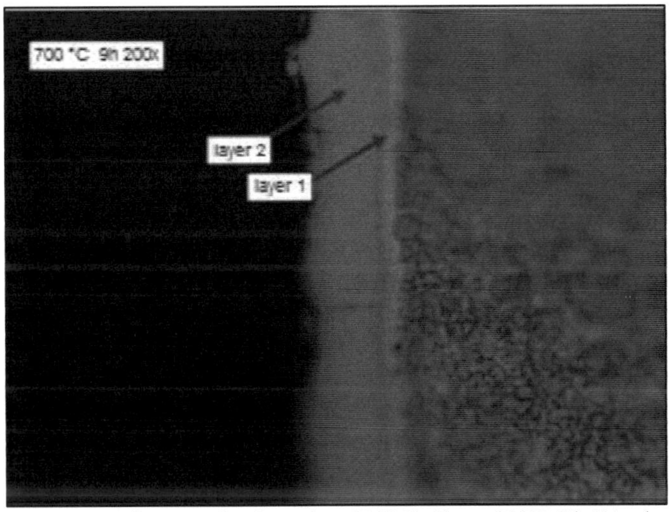

Figure 4. Micrograph of steel at 700°C, 9h and NH_4Cl 6%w/w.

4. Conclusions

It is possible to generate a coating on a martensitic ferritic steel by diffusion aluminizing from commercial foil at temperatures of 600 and 700°C, and exposure times from 3 to 9 hours. The oxidation pre-treatment applied to the commercial foil is enough to prevent the aluminum from melting on the steel surface during the coating process. The technique can be applied at temperatures above the melting temperature of aluminum, which allows to obtain coatings in shorter times compared to the traditional ones. By the application of the X-Ray Diffraction technique, the main compounds present in the coating were identified as FeAl and $Fe_{0,99}Al_{0,99}Ni_{0,02}$. The use of commercial foil instead of aluminum powder as a source makes the T91 steel aluminizing process efficient and cost-effective because it considerably reduces the costs of raw material, in addition to reducing energy consumption by performing the process in a shorter time than traditionally required to achieve coating of more than 100 microns in thickness.

References

[1] Bayer G 1995 *Materials Performance* **2** 34-38
[2] Bates B, Wang Y and Zhang Y 2009 *Surface and Coatings Technology* **204** 766-70
[3] Bates B, Zhang Y, Dryepondt S and Pint B 2014 *Surface and Coatings Technology* **240** 32-9
[4] Xing X, Han Z, Wang H and Lu P 2015 *Journal of Rare Earths* **33** 1122-8
[5] Zhang H and Jianfeng S 2015 *Rare Metal Materials and Engineering* **44** 2628-32
[6] Bianco R 1993 *Journal of the Electrochemical Society* **4** 1191-203
[7] Bianco R 1993 *Journal of the Electrochemical Society* **4** 1181-90
[8] Annual Book of ASTM Standards 2014 *ASTM E415-14* vol 03.05 (USA: ASTM International)
[9] Annual Book of ASTM Standards 2011 *ASTM G1-03* vol 03.02 (USA: ASTM International)
[10] Kim S 1998 *Materials Science and Technology* **14** 822-5
[11] Wang Q, Zhang Y and Wilson D 2010 *Surface and Coatings Technology* **204** 2737-44
[12] Grüters J and Galetz M 2016 *Intermetallics* **60** 19-27

IOP Conf. Series: Journal of Physics: Conf. Series **935** (2017) 012042 doi:10.1088/1742-6596/935/1/012042

Determination of the coefficient of dynamic friction between coatings of alumina and metallic materials

A Santos[1], E Córdoba[1], Z Ramírez[2], C Sierra[1] and Y Ortega[1]
[1] Universidad Pontificia Bolivariana, Floridablanca, Colombia
[2] Unidades Tecnológicas de Santander, Bucaramanga, Colombia

E-mail: alfonso.santos@upb.edu.co

Abstract: This project aims to determine the coefficient of dynamic friction between micrometric size coatings of alumina and metallic materials (Steel and aluminium); the methodology used to achieve the proposed objective consisted of 4 phases, in the first one was developed a procedure that allowed, from a Pin on Disk machine built based on the specifications given by the ASTM G99-05 standard (Standard test method for wear tests with a Pin on Disk machine), to determine the coefficient of dynamic friction between two materials in contact; subsequently the methodology was verified through tests between steel-steel and steel-aluminium, due to these values are widely reported in the literature; as a third step, deposits of alumina particles of micrometric size were made on a steel substrate through thermal spraying by flame; finally, the tests were carried out between pins of steel of aluminium and alumina coating to determine the coefficients of dynamic friction between these two surfaces. The results of the project allowed to verify that the developed methodology is valid to obtain coefficients of dynamic friction between surfaces in contact since the percentages of error were of 3.5% and 2.1% for steel-steel and aluminium-steel, respectively; additionally, it was found that the coefficient of friction between steel-alumina coatings is 0.36 and aluminium-alumina coating is 0.25.

1. Introduction

Friction is defined as the resistance to relative displacement between two surfaces that are in contact and depends on the characteristics of the surfaces that are interacting with each other and this phenomenon is quantified through the friction force which prevents or delays the displacement of one body with respect to another or on the surface with which it is in contact. Friction exists even when there is no relative movement between the two bodies and this is known as static friction; on the contrary if it is present the relative movement, it is designated by the name of dynamic friction.

The friction force has a magnitude equal to the product between the coefficient of friction and the resulting normal force due to the proper weight of the elements and the external forces acting on the system. The coefficient of friction is a dimensionless parameter and is characteristic of each pair of materials that are in contact and it also depends on many other factors such as temperature, roughness, relative speed between surfaces, etc. [1,2].

In mechanical systems where the elements interact with each other through sliding contacts, friction plays a very important role [3,4]; The determination of the friction force is generally modelled by Coulomb´s La, established experimentally in 1781, and defines that the frictional force that can exist in the bodies in contact is directly proportional to the value of the normal force of contact between them; the constant of proportionality is the coefficient of friction. Many experimental studies determine the coefficient of dynamic friction through the consideration of only a translation movement between the surfaces in contact [5-7], tests that allowed to estimate the normal and frictional forces present between materials; the dependence of the friction force of the displacement and velocity; and the modelling of the friction interface under conditions of nonlinear dynamic analysis. Other researchers have used rotational systems to experimentally determine the coefficient of friction, among which stand out the pin on disk system [8,9], circular spin plates [10] and roller systems and radially loaded discs [11]. It is also used the measurement of the angular displacement and the torque used in the test for the determination of the variables involved in the friction phenomenon, for different applications [12,13].

The pin on disk equipment are designed to perform adhesive wear tests, their specifications and the procedure for the execution of the tests are listed in the ASTM G99-05 (Standard test method for wear tests with a Pin on Disk machine) which is a standard method for obtaining valid data. For wear testing through this equipment are required two specimens, the first is a pin which is positioned perpendicular to the other specimen which is usually a circular disk. The machine causes a relative displacement between these specimens which result in the formation of a wear path on the disk; the pin-shaped specimen is pressed onto the disk with a specific load and that specimen is fastened to a positioning device with counterweights. Wear reports are made in terms of loss of volume.

Despite the simplicity of the measurement and calculation, there are practical challenges to accurately quantify these basic tribological parameters. Friction is an atypical process of non-equilibrium and slippage often leads to wear which is highly stochastic. The values of the coefficient of friction and wear rates reported in the literature typically show a wide variation even for nominally identical tests; the origin for these variations is often unknown [14-16].

This paper reports the results obtained of the dynamic coefficient of friction for alumina coatings deposited through thermal spray by flame process from particles of micrometric size and two different metals (steel and aluminium) obtained from tests in a pin on disk equipment.

2. Methodology

2.1. Development of a procedure that determines the coefficient of dynamic friction

The pin-on-disk equipment are used to quantify the loss of material originated during wear tests; however, the determination of the coefficient of friction must be performed indirectly through the measurement of some parameters during the test to find the friction force originated and from this value and based on Coulomb´s law determine the searched parameter.

2.2. Validation of the procedure for the determination of the coefficient of dynamic friction

The verification of the procedure developed for the calculation of the coefficient of dynamic friction was carried out through the realization of tests for two different configurations (steel-steel and steel-aluminium), since they are data quite reported in the literature. Tests were performed with an applied load of 10N, the time of each test was 5 minutes and the data were taken at constant time intervals of 10 seconds. A total of 10 tests were done for each pair of materials tested.

2.3. Deposit and characterization of alumina coatings

The alumina coatings were made through the thermal spraying by flame process, using an Eutalloy 85 BX gun, which is adapted to a conventional oxyacetylene combustion equipment. The distance between the torch and the base material was approximately 30cm, with an inclination to the horizontal of 60°. The oxygen pressure used was 25psi and that of the acetylene was 5psi.

2.4. Realization of tests on alumina-coated specimens

The coated specimens were tested in the pin-on-disc equipment using steel and alumina pins to determine the coefficient of dynamic friction between the coating and these two metals. The procedure and test parameters used for these tests was the same as that followed for the validation of the methodology.

3. Obtained results

3.1. Procedure for the determination of the coefficient of dynamic friction

The procedure developed is based on the determination of the electrical power consumed in the test (equation (1)), which is calculated as the subtraction between the value obtained during the test and the power consumed under vacuum conditions (without specimen). For the determination of the electrical power, the current (I) and the voltage (V) being supplied to the motor through the network were measured.

$$P_e = \left(\sqrt{3} * I * V * \cos\emptyset\right)_{test} - \left(\sqrt{3} * I * V * \cos\emptyset\right)_{vacuum} \quad (1)$$

The cosine of the angle ϕ, is the power factor of the motor used to carry out the tests.

The mechanical power is obtained by multiplying the electric power consumed by the motor efficiency (η) and this magnitude is the product of the torque (T) used in the test by the angular velocity that the specimen must have (ω), 300rpm defined by the standard. From the above, it is defined that the torque consumed in the test can be obtained from equation 2.

$$T = \frac{P_e * \eta}{\omega} \quad (2)$$

The multiplication of the friction force (F_f) by the distance of the pin to the centre of rotation (r) of the specimen determines the torque that is applied for the test (T); in addition, Coulomb`s law states that there is a proportionality between the normal (N) and the friction force by the coefficient of friction (μ) present between the two materials. From the above it is deduced that through equation 3 the coefficient of dynamic friction can be quantified between the material of the coating and the one of manufacture of the pin.

$$\mu_K = \frac{T}{N*r} = \frac{P_e*\eta}{N*r*\omega} = \frac{\left[\left(\sqrt{3}*I*V*\cos\emptyset\right)_{test} - \left(\sqrt{3}*I*V*\cos\emptyset\right)_{vacuum}\right]*\eta}{N*r*\omega} \quad (3)$$

The value of the normal is obtained from the static equilibrium conditions of the system where is installed the pin, which is shown on Figure 1.

Figure 1. Free body diagram of the system where the pin is installed.

By means of a summation of moments at the point where R_y is found, there is obtained the equation 4 that allows to calculate the normal that appears in the specimen in a disc shape.

$$N = 16.01 + W_{pin} \; [N] \tag{4}$$

Replacing the value of normal in equation 3 gives the equation 5 which allows to calculate the coefficient of dynamic friction between two materials.

$$\mu_K = \frac{\left[(\sqrt{3}*I*V*\cos\emptyset)_{test} - (\sqrt{3}*I*V*\cos\emptyset)_{vacuum}\right]*\eta}{(16.01+W_{pin})*r*\omega} \tag{5}$$

3.2. Validation of the procedure for the determination of the coefficient of dynamic friction

Using the equation 5, the coefficient of dynamic friction between steel - steel and steel - aluminium was determined; the requested values from the equation are defined by the characteristics of the motor arranged in the equipment (cos ϕ=0.65 y η=0.602), the test conditions defined by the standard (r=0.02m y ω=10πrad/s) and the measurements of current, voltage and weight of the pin. The results obtained from the tests are listed on Figure 2.

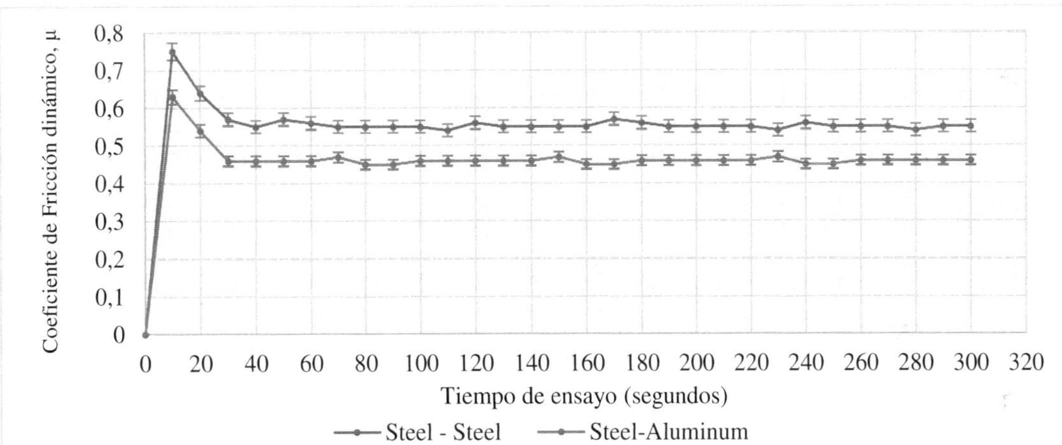

Figure 2. Coefficient of dynamic friction between steel-steel and steel-aluminium obtained through the developed methodology.

The obtained coefficient of dynamic friction between steel-steel was 0.553, while for steel-aluminium was 0.459; which represents a 3.6% and a 2.2% of error respectively regarding the values reported on the literature. Additionally, the classical behaviour of this variable is observed when a static coefficient of friction greater than the dynamic coefficient is obtained from the tests.

3.3. Realization of test on alumina-coated specimens

The results of the coefficient of dynamic friction obtained from the tests performed on alumina-coated specimens from particles of micrometric size and deposited through a thermal spray by flame process and using steel and aluminium pins are listed in Figure 3. The coefficients of dynamic friction between Steel-alumina coating and aluminium-alumina coating was 0.35 and 0.25 respectively.

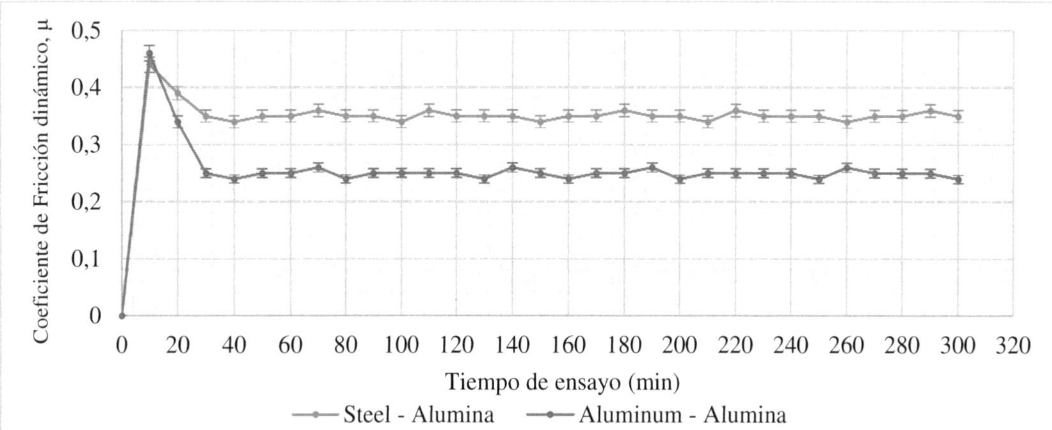

Figure 3. Coefficient of dynamic friction between steel-alumina and aluminium-alumina.

4. Conclusions

The developed methodology allows to determine the coefficients of dynamic friction between two materials using a pin on disk wear equipment

The percentages of error in the magnitudes of the coefficients of dynamic friction between steel-steel and Steel-aluminium was 3.6 % and 2.2% respectively.

The coefficients of dynamic friction between Steel-alumina coating and aluminium-alumina coating was 0.35 and 0.25 respectively.

References

[1] Maegawa S, Itoigawa F and Nakamura T 2015 Effect of normal load on friction coefficient for sliding contact between rough rubber surface and rigid smooth plane *Tribology International* **92(1)** 335-343

[2] Popov V L 2010 *Contact mechanics and friction: physical principals and applications* (Heidelberg: Springer)

[3] Dweib A H and D´Souza A F 1990 Self-excited vibrations induced by dry friction, part 1: Experimental study *J. Sound Vib.* **137(2)** 163-175

[4] Kang J and Krousgrill C M 2010 The onset of friction – induced vibration and spragging *J. Sound Vib.* **329(17)** 3537-3549

[5] Hoskins E R, Jaeger J C and Rosengren K J 1968 A medium-scale direct friction experiment *Int. J. Rock Mech. Min. Sci. & Geomech. Abstr.* **5(2)** 143-152

[6] Worden K, Wong C X, Parlitz U, Hornstein A, Engster D, Tjahjowidodo T, Al-Bender F, Rizos D D and Fassois S D 2007 Identification of pre-sliding and sliding friction dynamics: Grey box and black-box models *Mech. Syst. Signal Process.* **21(1)** 514-534

[7] Schwingshackl C W, Petrov E P and Ewins D J 2012 Measured and estimated friction interface parameters in a nonlinear dynamic analysis *Mech. Syst. Signal Process.* **28(1)** 574-584

[8] Lim S C, Ashby M F and Brunton J H 1989 The effects of sliding conditions on the dry friction of metals *Acta Metall.* **37(3)** 767-772

[9] Wakuda M, Yamauchi Y, Kanzaki S and Yasuda Y 2003 Effect of surface texturing on friction reduction between ceramic and steel materials under lubricated sliding contact *Wear* **254(3-4)** 356-363

[10] Dai H-L, Dai T and Yan X 2015 Thermoelastic analysis for rotating circular HSLA steel plates with variable thickness *Appl. Math. Comput.* **268** 1095-1109

[11] Shon S, Kahraman A, LaBerge K, Dykas D and Stringer D 2012 Influence of surface roughness on traction and scuffing performance of lubricated contacts for aerospace and automotive gearing *Proceedings of the ASME/STLR International Joint Tribology Conference* (USA: ASME) IJTC2012-61212 pp 87-89

[12] Kang Y S 1997 Friction identification in a sight stabilization system at low velocities *Mech. Syst. Signal Process.* **11(3)** 491-505

[13] Povey T and Paniagua G 2012 Method to improve precision of rotating inertial and friction measurements in turbomachinery applications *Mech. Syst. Signal Process.* **30** 323-329

[14] Novak R and Polcar T 2014 Tribological analysis of thin films by pin on disk: Evaluation of friction and wear measurement uncertainty *Tribol. Int.* **74** 154-163

[15] Velkavrh I, Lüchinger M, Kern K, Klien S, Ausserer F, Voyer J, Diem A, Schreiner M and Tillmann W 2017 Using a standard pin on disk tribometer to analyses friction in a metal forming process *Tribol. Int.* **114** 418-428

[16] Torre Pérez A, García-Atance Fatjó G, Hadfield M and Austen S 2011 A model of friction for a pin on disk configuration with imposed pin rotation *Mech. Mach. Theory* **46** 1755-1772

IOP Conf. Series: Journal of Physics: Conf. Series **935** (2017) 012043 doi:10.1088/1742-6596/935/1/012043

Analysis of the stormwater drainage of the historic walls of Cartagena de Indias between the bastions of San Lucas, Santa Catalina and Santa Clara

A Utria[1], M Saba[1] and E Quiñones-Bolaños[1]
[1] Universidad de Cartagena, Cartagena, Colombia

E-mail: manuelsaba1987@gmail.com

Abstract. The city walls of Cartagena between the Baluarte of San Lucas and Santa Clara reflects the consequences of poor stormwater drainage. In the present investigation the current drainage state of this area have been evaluated, assessing its sizing and conditions to be then modelled through SWMM 5 software for return periods of 5, 10, 15 and 20 years respectively. Finally, it has been determined that the physical deterioration of the drainage elements justifies the malfunction of the whole system, therefore cleaning and periodic maintenance of the drainage elements is strongly suggested.

1. Introduction

Batteries, fortifications, defensive walls, forts and fortified ports are common in the list of world heritage of America, and the fortifications of Latin America and the Caribbean are not stranger to it. At the present time, all these scenarios are cultural expressions of the same continent, therefore protect and preserve theme it is of the extreme importance [1]. Some fortifications have a high degree of authenticity in terms of location and configuration, shapes and designs, materials and constituents, however, they are generally subjected to severe tropical climates (including tropical storms and hurricanes), which may threaten the authenticity of the property, as in the case of the fortifications of Old Havana [2]. Among the most significant monuments there are undoubtedly the walls of Cartagena [3-6], World Heritage Sites according to UNESCO since 1984, whose purpose was to protect the City from the wartime continuous attacks [7], which are nowadays seriously deteriorated according to research done in the matter [8].

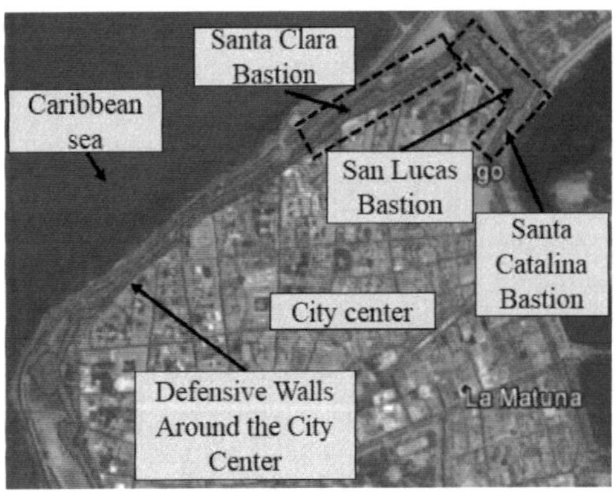

Figure 1. Area and bastions of interest.

Water is one of the factors that have a greater incidence in the process of deterioration of this type of fortifications, since its passage attrition causes erosion, furthermore the infiltration of water in the pores and micropores of the stones, reacts with the stone materials of the wall generating as end products substances vulnerable to be dragged by wind or overflow, in addition to vegetation and plants that in the growth process of its roots expand the joints between the elements of the wall. Being the walls extensive in its totality, the sector of the bastion of San Lucas, Santa Catalina and Santa Clara have been chosen as part of the study (Figure 1). A characterization of the pluvial drains of the mentioned sector was made, and a hydraulic simulation and a description of the problematic is presented [9].

2. Methodology

2.1 Field work
In the field work a study of the interested area was made, and the typologies of drainage were identified on which the rainwater is evacuated in the structure. In addition, dimensions, afferent areas and conditions of each drainage were defined.

2.2 Calculation of drainages hydraulic capacity
After obtaining all the field data, the hydraulic modelling has been done, using the Uniform Flow Method to calculate the flow rate of each drainage element, Horton's Method [10] for infiltration control and the Rational Method for calculating the runoff rate have been chosen; Manning's coefficient of permeability and coefficient values were obtained from the Technical Regulation of Drinking Water and Basic Sanitation of Colombia (RAS 2000) [11], according to the characteristics of each drainage element.

Precipitation data were obtained from the Rafael Núñez station of the city of Cartagena defined in the Master Plan of Pluvial Drainage [12]. Using the software SWMM 5, the different objects were created in the plane of the study area, like drainage sub-basins of the system and the afferent area of each drainage element, and the different properties of the same were fixed. Then initial and final points of each conduit were added, called nodes, and later the conduits were joined with the nodes.

Subsequently, through the rain icon, the time series of data, the value of precipitation in millimetres and the time intervals of these precipitations were defined (Figure 2, model of the channel 1).

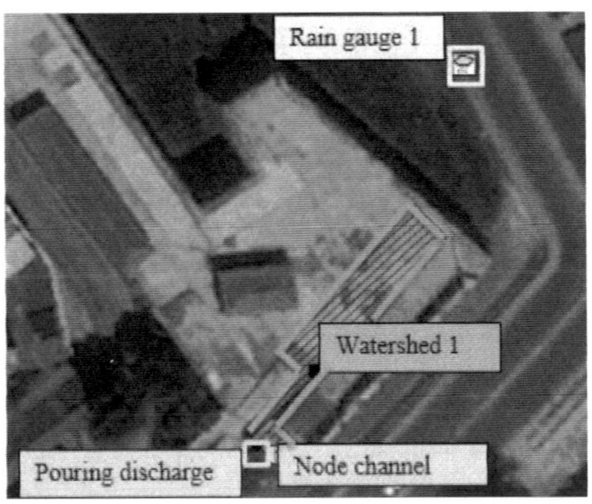

Figure 2. Channel 1 modelling.

With all the aforementioned data, the flow rate that hydraulically works in each drainage element was calculated. At the hydrological level, for each drainage element was evaluated the flow rate of runoff that must support for the return periods of 5, 10, 15 and 20 years.

3. Discussions and results
For the initial diagnosis of the pluvial drainages of the study area, the elements were classified by typology (Figure 3).

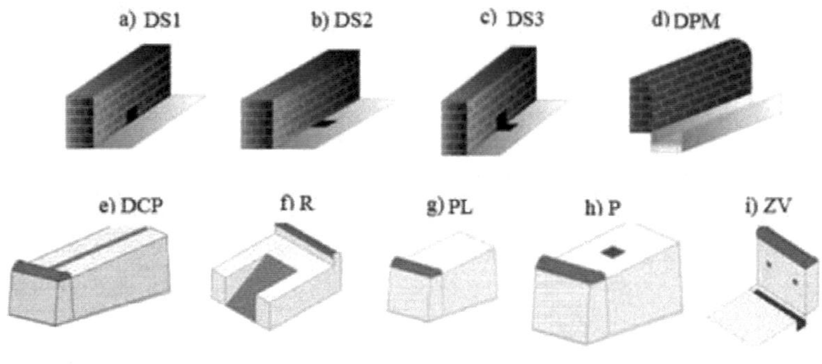

Figure 3. Drainage elements, adapted from [8].

Each of them is described below:

- DS1 – Drainage of Parapet: channels located at the top of the wall, to evacuate rainwater falling on the embankment, its opening is in the parapet;
- DS2 - Platform Drainage: channels located at the top of the wall, to evacuate rainwater falling on the embankment, its opening is on the platform;
- DS3 - Combined Drainage: channels located at the top of the wall, to evacuate rainwater falling on the embankment, they have an opening in the platform and in the parapet;
- DPM - Wall Foot Drainage: rectangular channels located at the bottom of the wall, more precisely at the foot of the wall, they collect rainwater draining from their afferent areas (green areas, road, etc.) and of the upper drains (DS1, DS2, DS3);
- DCP - Drainage Platform Channel: channels of rectangular form located in the embankment; (DS1, DS2, DS3), or to other continuous DCPs. In order to achieve this, it is necessary to conduct the water from its concentration points to another type of drainage (DS1, DS2, DS3);
- R - Ramp: inclined structures that start in the upper embankment of the walls and end in the ground floor of the same. They allow a part of the rainwater that falls on the embankment to drain into the drainage of the Historic Centre of the city;
- PL - Embankment: the upper platforms or platforms of each of the stretches of the walls. It could be said that it is the element with which the rainy waters make their first contact. The rain waters that precipitate on it, are transported by means of slopes towards the DS1, DS2, DS3, DCP or Ramps;
- P - Rainwater Well or Reservoir: rectangular holes located on the platform of the wall panels, it is believed that these holes had the function of storing rainwater in the past;
- ZV-Green Zones: These are areas with vegetation, located in front of the cliffs or contraescarpas of the walls. The rainwater that precipitates them is evacuated towards the MPD.

Once the drains of the structure were numbered and catalogued, traditional distances, slopes and areas were measured for each individual drainage. In general, the rainwater drainage of the walled cordon between the bastions of San Lucas, Santa Catalina and Santa Clara of the city of Cartagena is constituted by a total of eighty-six (86 channels), most of all DS1, DS2 and DS3, of which are normally operating 16% (14 channels), partially operating 28% (24 channels) and non-functional 56% (48 channels).

Particularly the ramps of the sector of the San Lucas bastion are physically deteriorated, with many cracks and presence of vegetation, the reason for this condition could be the lack of periodic maintenance of these structures, nevertheless its slope allows to evacuate raining waters coming from a part of the embankment of that section. As well the ramps of the sector of the Santa Catalina Basin and Santa Clara

IOP Conf. Series: Journal of Physics: Conf. Series **935** (2017) 012043 doi:10.1088/1742-6596/935/1/012043

have good conditions and their slopes allow to efficiently evacuate the water evacuating them to the drainage of the Historic Centre. The flow data is digitized in Table 1.

Table 1. Hydraulic calculations of the most representative channels.

Dimension	DS1 n°1	DS1 n°12	DS2 n°31	DS3 n°61
Canal elevation (m)	0.190	0.210	0.100	0.175
Canal base (m)	0.120	0.090	0.120	0.170
Canal grade	0.020	0.020	0.021	0.495
Manning Coefficient	0.016	0.016	0.016	0.016
Caudal (l s^{-1})	25.72	18.56	11.88	39.04

Runoff flow results according to the return period (Table 2):

Table 2. Runoff flow according to the return period of the most representative channels.

Return period (years)	Flow DS1 n°1 (l s^{-1})	Flow DS1 n°12 (l s^{-1})	Flow DS2 n°31 (l s^{-1})	Flow DS3 n°61 (l s^{-1})
5	6.432	3.195	0.990	0.683
10	7.287	3.617	1.122	0.774
15	7.838	3.890	1.207	0.833
20	8.255	4.097	1.271	0.877

On the other hand, the hydraulic flows obtained in each drainage element do not exceed the total capacity of these elements, however the presence of external factors such as solid residues, vegetation and/or sediments that interrupt the route of rainwater precipitating in the sector studied, before reaching their final disposal point, preventing the normal operation of most drainage elements. For the green areas, have been found that permeability is very low and penetration of water is irrelevant. Consequently, according to the results obtained by the SWIMM 5 for the different return periods of 5, 10, 15 and 20 years, almost none of the channels are flooded, with the exception of two of them.

According to the height of the water sheet inside the critical elements, 87% of these do not exceed the height of the channel, which represents the maximum height of the water level. In terms of flow velocities, have been found that the vast majority of the modelled channels exceeded the parameter of the minimum speed established by the Basic Water and Sanitation Regulation RAS, mentioned above, whose value is 0.75m s^{-1}.

4. Conclusions

In the analysis of all the drainage elements modelled, none of the channels show flooding for the return periods of 5, 10, 15 and 20 years, with the exception of two of them because their cross-sectional dimensions are relatively smaller compared to the other drainage elements. According to the height of the water sheet inside the shaped elements, only 13% of these exceeds the height of the channel, which means that those elements work under pressure and increase the cross-sectional area would be necessary.

The main causes of malfunction found were: lack of maintenance and internal collapses due to degradation of the same wall material, presence of vegetation and solid residues, accumulated sediments, corrosion, inverted slopes and specifically in the wall foot drains (DPM) were observed to have obstructed outlets, consequently the water remains stagnant causing foul odors and proliferation of vectors and microorganisms.

A campaign of revision and cleaning as well as enlargement of the undersized channels is suggested, in order to allow the structure to evacuate the water properly, safeguarding its lifetime and its historicity.

Acknowledgment

This project has been funded with support of the European Commission. This publication reflects the view only of the author, and the Commission cannot be held responsible for any use which may be made of the information contained therein - Elarch Program (project reference number: 552129-em-1-2014-1-it-era mundus-ema21).

References

[1] Camara Muñoz A 2005 *Los ingenieros militares de la monarquia hispana de los siglos XVII y XVIII* Section 1-45 (Madrid: Fernando Villaverde Ediciones)

[2] Castañeda A, Corvo F and Rivero C 2006 Influencia de la agresividad atmosférica en el deterioro de las estructuras en ciudad de la habana *Proc. IX Congreso Iberoamericano de Metalurgia y Materiales* (Cuba: IBEROMET)

[3] Dorta E M 1951 *Cartagena de Indias, Puerto y Plaza Fuerte* (Colombia: Fondo Cultural Cafetero)

[4] Segovia R 1998 *The Fortifications of Cartagena de Indias: Strategy and History* (Cartagena de Indias: El Ancora Editores)

[5] Zapatero J M 1979 *History of the Fortifications of Cartagena de Indias* (Colombia: Ediciones Cultura Hispánica, del Centro Ibero-Americano de Coop. y Dirección Gen. de Relac. Cultur. del Minist. de Asuntos Ext)

[6] Cabellos Barreiro E 1991 *Cartagena de Indias, Magica agropolis de America* (Madrid: Ministerio De Fomento)

[7] UNESCO 1984 *World Heritage Committee Eighth Ordinary Session, Convention concerning the protection of the world cultural and natural heritage* (Argentina: UNESCO)

[8] Fondo financiero de proyectos de desarrollo 2009 *Estudio y diseños tecnicos para la restauracion del lienzo de muralla entre el momumento de la india Catalina y India Catalina y el museo de la Marina en Cartagena de Indias* (Cartagena de Indias: Fonade)

[9] Herrera A D 2009 *Factores ambientales y climáticos en la restauración de los lienzos de muralla en la ciudad de Cartagena de Indias* (Cartagena de Indias: Fonade)

[10] Marbello Pérez R 2005 *Flujo Uniforme Y Determinación De La Rugosidad En Canales* (Medellín: Universidad Nacional de Colombia) pp 159-190

[11] RAS 2000 *Reglamento técnico del sector de agua potable y saneamiento basico* Minist. Desarro. Económico, Dir. Agua Potable y Saneam. Básico p 45

[12] Alcaldia de Cartagena 2008 Plan Maestro de Drenajes Pluviales (Cartagena de Indias: Alcaldia de Cartagena)

Formation and decomposition of siderite for CO_2 treatment

E Y Mora[1,2], A Sarmiento[1], E Vera[1], V Drozd[2], A Durigyn[2] and S Saxena[2]
[1] Universidad Pedagógica y Tecnológica de Colombia, Tunja, Colombia
[2] Florida International University, Miami, USA

E-mail: eduin.mora@uptc.edu.co

Abstract. In this research work, we studied the conditions for formation and decomposition of siderite $FeCO_3$ from hematite Fe_2O_3 along with carbon dioxide CO_2 at suitable thermodynamic conditions. As reductant agents were used mixtures of two elements, metallic iron and graphite. Best levels of carbonation were found in mixtures with bigger amounts of metallic iron. It was demonstrated that CO_2 capture capacity by hematite depends of temperature, CO_2 pressure, and reaction time. Temperatures between 100 and 150°C, pressures between 10 and 30bar and reaction times between 1 and 4h were adjusted for analyse the carbonation behaviour; siderite formation was improved by increases of these three variables. There was no carbonation without water in the mixtures, due to kinetic limitations. CO_2 capture capacity was calculated from Rietveld refinement results. Using vacuum system and Dielectric Barrier Discharge, DBD plasma, the siderite was decomposed at 300°C, and 320°C respectively. Techniques as X-ray diffraction, and surface area analysis were employed to study the material.

1. Introduction

Carbon dioxide CO_2 is a pollutant gas and the main cause of global warming with contribution close to 65%, reaching over 30.000 million metric tons in 2010, which has allowed generated high levels of CO_2 atmospheric concentration, around 400ppm in 2015 [1,2]. The CO_2 emissions have been increased in the last decades due to mainly generation energy, the manufacture of products at industrial level, fuel combustion and rainforest burning. Some agreements between countries have been subscript and although have helped to develop technologies and strategies to reduce the global warming, however these are still away to be a real solution [3,4]. Among others possible solutions are exchange from coal to renewable energy sources, improvements in energy efficiency in the combustion of fossil fuel, and replacement of raw material for waste materials rich in CaO. Amine in chemical absorption, is the mature technology mainly used in industrial processes, which has the disadvantage of high spend of energy for regeneration; physical adsorption can be implemented using different substances as zeolites or activated carbon but depends strongly of material features as pore size which increases the cost; membranes, cryogenics and micro algal are new promisors options, but are still in laboratory scale [5-7].

A very interesting technology to capture CO_2 that has gained interest in researches is use metal oxides that along with CO_2 at determined conditions of pressure and temperature can generate carbonates by mean of exothermic reaction. After that, those carbonates are heated in order to liberate the CO_2 and hence, metal oxides are regenerated inside an endothermic reaction [4,8,9]. Iron oxides are a good candidate to capture CO_2 in steel or non-steel industries; iron oxides or iron ore become

IOP Conf. Series: Journal of Physics: Conf. Series **935** (2017) 012044 doi:10.1088/1742-6596/935/1/012044

iron carbonate and can be regenerated at temperatures around 400°C [10] in order to be used in new cycle carbonation- calcination until active sites of material disappear.

Here, it is presented a study about some conditions of pressure, temperature and reaction time for carbonation of mixtures from hematite, iron and graphite; there were studied these materials because hematite is the iron oxide most frequently found inside iron ore, besides iron and graphite are materials available and employed in industrial processes as steel industry.

2. Methods and experimentation

The carbonation reaction was carried out using a closed cylindrical vessel of length 1.25 inches and internal diameter of 0.35 inches; inside this vessel were added 0.25g of hematite (99.945% metal basis) mixed with metallic iron (Good Fellow, 99% purity, <60mm) and graphite powder (Alfa Aesar, crystalline - 325 mesh, 99% in different molar radios. Approximately 0.15ml of water was added to the vessel wall to improve the kinetics of carbonation reaction. High purity CO_2 gas (99.99% purity, Airgas) was introduced into the system at desired pressures. Before the experiments, CO_2 gas was flushed three times to ensure an inert atmosphere inside the cylindrical vessel. After the desired reaction time for each experiment, the system was cooled by environment. The CO_2 sorption capacity was calculated for all experiments, using the results generated in Rietveld refinement from XRD patterns. The calcination temperature of $FeCO_3$ and the generated products were analysed by mean of two methods, heating the sample in vacuum environment and plasma generated in a dielectric barrier discharge configuration, using Argon atmosphere; to get vacuum was used a pump Franklin Electric MOD 1101006418, 0.5HP, 1725rpm, and to get plasma was used a power supply operating at 12Kv and 18KHz.

The product characterization was performed using X-ray powder diffraction method. Bruker GADDS/D8 is equipped with Apex Smart CCD Detector and direct-drive rotating anode. The MacSci rotating anode (Molybdenum) operates with a 50kV generator and 20mA current. X-ray beam size can vary from 50 to 300μm. The usual collection time is 1200s. An isothermal gas adsorption was employed to quantify internal surface areas of the reactant powder particles; this information is needed to get the N_2 adsorption–desorption isotherm. Micromeritics Tristar II 3020 (surface area and porosimetry analyser instrument) was used with N_2 as the adsorptive gas at 77K (liquid nitrogen bath). The samples were first degassed under 200°C using a N_2 gas flow for 2h to remove the moisture and other adsorbed gases before analysis.

3. Results

3.1. Iron oxide carbonation

Siderite (JCPDS #00-083–1764) was identified as product of carbonation reaction at different conditions. Figure 1 shows the XRD pattern as result of the carbonation reaction using hematite, iron and graphite in equimolar proportion at 30bar CO_2 pressure, 150°C and 4 hours of time reaction; here, experiments performed without water did not present carbonation, due to kinetics limitations. Based on [11] water on the surface of hematite facilitates the reaction with CO_2, which results to the formation of CO_3^{2-} and H^+ ions. Free Fe^{+2} ions can further react with CO_3^{2-} to form $FeCO_3$.

Additionally, keeping constant the molar amount of hematite, there were used different molar radios between metallic iron and graphite in order to identify the best conditions for CO_2 capture. Table 1 presents the values for CO_2 capture capacity of the different mixtures at 100°C, 30bar and 2h.

According the values of CO_2 capture capacity, better results are accomplished using bigger amounts of metallic iron; this is due to graphite has lower activation energy. The best CO_2 capture capacity of the mixture is obtained using only metallic iron, reaching 6.3636mmolCO_2/gsorbent, which is translated to 45.74% of conversion.

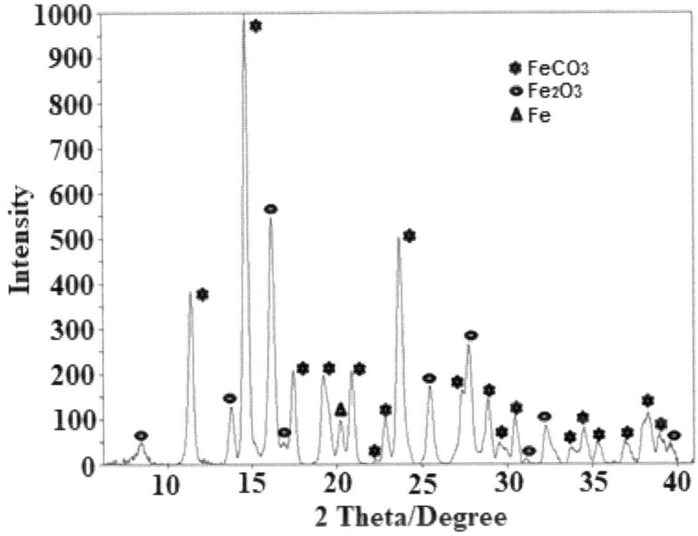

Figure 1. XRD pattern in siderite formation at 30bar CO_2 pressure, 150°C and 4 hours using hematite, iron and graphite in equimolar proportion.

Table 1. CO_2 capture capacity of several mixtures of hematite, iron and graphite at 30bar CO_2 pressure, 100°C and 2 hours.

Molar Radio (Fe/C)	CO_2 capture capacity (mmolCO2/gsorbent)
1:0	6.3636
0.75:0.25	5.8752
0.5:0.5	5.1834
0.25:0.75	2.9087
0:1	0.0987

3.2. Dependence of temperature, pressure, reaction time

It has been demonstrated that factors as temperature, pressure and time reaction influence strongly the carbonation of metallic oxides [8,12,13]. Table 2 shows several values for CO_2 capture capacity at different conditions, keeping equimolar radios between hematite, iron and graphite.

Table 2. CO_2 capture capacity at different combined effects of pressure, temperature and reaction time.

Pressure (bar)	Temperature (°C)	Reaction time (h)	CO_2 capture capacity (mmolCO_2/gsorbent)
10	100	2	2.0978
20	100	2	3.4759
20	150	1	3.8902
20	150	4	4.9882
30	100	4	6.1751
30	150	4	6.3347

It can be evidenced from Table 2 that with the increases of pressure, keeping constant the temperature and reaction time, CO_2 capture capacity by mixture increases; hence, with the enough CO_2 pressure, molecules can react with the solid mixture at suitable conditions to form $FeCO_3$. Longer reaction times increase the CO_2 capture regardless of conditions of pressure and temperature. Increase the temperature allows CO_2 molecules diffuse easier into the pores of hematite, iron and graphite and these CO_2 molecules are trapped in order to react with the mixture to form $FeCO_3$.

3.3 Thermal regeneration and recyclability

The decomposition temperature and generated products were studied from the sample processed in carbonation, it means at 30bar of CO_2 pressure, 150°C and 4h. As explained above, two methods were used to calcination, vacuum and plasma. The XRD patterns obtained from the decomposed samples are shown in Figure 2 and Figure 3, for vacuum and plasma respectively. Siderite was completely decomposed at 300°C in vacuum and at 320°C in plasma generated from 12Kv, 18KHz and electrical power delivered of 28w.

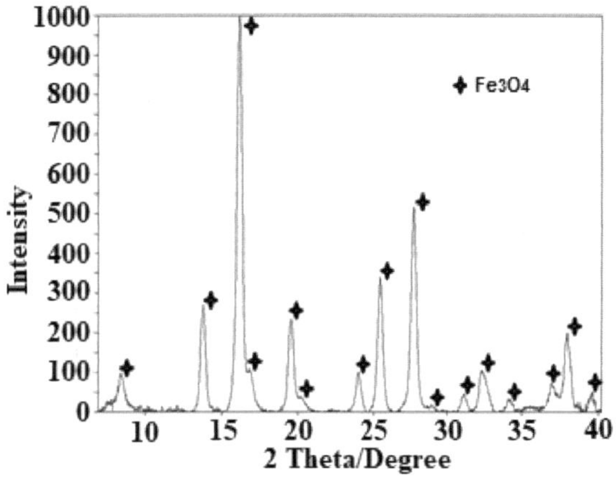

Figure 2. XRD pattern of calcinated samples from vacuum.

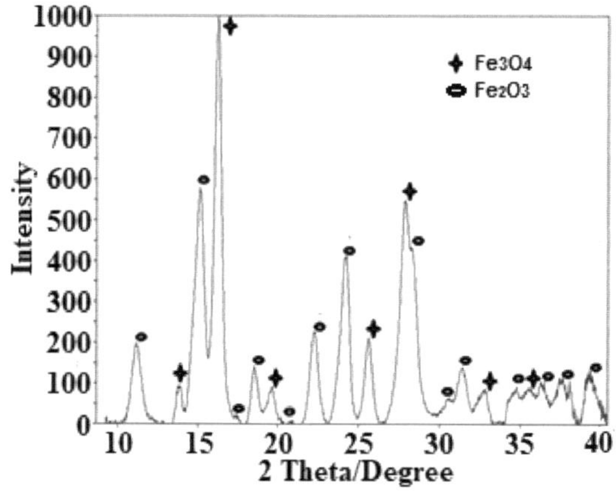

Figure 3. XRD pattern of calcinated samples from plasma.

According both XRD patterns, the calcinated products found are magnetite (JCPDS #00-089-2355), and hematite (JCPDS #00-033-0664). As can be inferred magnetite and hematite can be used along with carbon dioxide in a new cycle carbonation-calcination. Figure 4 depicts the N_2 adsorption–desorption isotherm curves for material calcinated in vacuum.

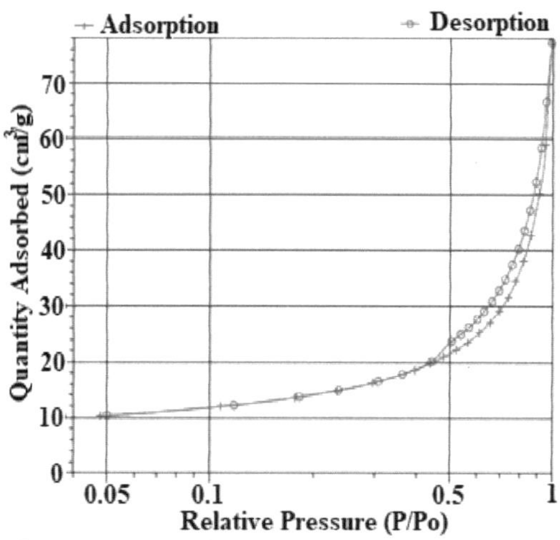

Figure 4. N_2 adsorption-desorption isotherm curve for sample calcinated in vacuum.

IOP Conf. Series: Journal of Physics: Conf. Series **935** (2017) 012044 doi:10.1088/1742-6596/935/1/012044

This shape corresponds to type IV isotherms according to IUPAC classification. Mesoporous materials employed to capture gas, follow this typical isotherm curve; these kinds of patterns present hysteresis indicating the presence of slit-like pores.

4. Conclusions

Capture of carbon dioxide by hematite, iron and graphite was studied as an alternative to avoid emissions of this pollutant gas to atmosphere. Carbonation at different molar radios of metallic iron and graphite was analysed. Best levels of CO_2 capture capacity were found in mixtures where iron amount was higher. There was no carbonation in experiments without water, because water acts as an efficient catalyst, overcoming the kinetic limitations. Moreover, CO_2 capture capacity was calculated as a function of pressure, temperature and reaction time; longer reaction times, as well as higher pressures and temperatures favoured the CO_2 capture. Siderite decomposition temperature was identified by mean of two methods, heating the sample in vacuum environment and plasma created in a dielectric barrier configuration; this regard proposes plasma as a novel method for decompose the siderite; plasma has been used mainly for CO_2 treatment, transforming it in gases less harmful for global warming. There were identified magnetite and hematite as decomposition products. Calcinated material could be used in a new carbonation calcination cycle according its mesoporous feature.

References

[1] Yu C, Huang H and Tan S 2012 *Aerosol and Air Quality Research* **12** 745-769

[2] Wiers B 2012 *Journal of the American Chemical Society* 7056-7069

[3] Rochelle G 2009 Amine scrubbing for CO_2 capture *Science* **325** 1652-1654

[4] Kummar S and Saxena S 2014 A comparative study of CO_2 sorption properties for different oxides *Materials for Renewable and Sustainable Energy* **3(30)** 1-15

[5] Zaman M and Hyung Lee J 2013 *Korean J. Chem. Eng,* **3** 1497-1526

[6] Ho M, Leamon G, Allinson W and Wiley D 2006 Economics of CO_2 and mixed gas geosequestration of flue gas using gas separation membranes *Journal of Membrane Science* **45** 2546-2552

[7] Zhao L, Riensche E, Menzer R, Blum L and Stolten D 2008 *Journal of Membrane Science* **325** 284-294

[8] Kummar S 2014 The effect of elevated pressure, temperature and particles morphology on the carbon dioxide capture using zinc oxide *Journal of CO2 Utilization* **8** 60-66

[9] Salvador C, Lu D, Anthony E and Abanades J 2003 Enhancement of CaO for CO2 capture in an FBC environment *Chem, Eng.* **96** 187-195

[10] Gotor F, Macias M, Ortega A and Criado J 2000 *Phys Chem.* **27** 495-503

[11] Hassanzadeh A and Abbasian J 2010 *Fuel* **89** 1287-1297

[12] Pannocchia G, Puccini M, Seggiani M and Vitolo S 2007 *Ind. Eng. Chem.* **46** 6696-6706

[13] Bhagiyalakshmi M, Jun Lee J and Tae Jang H 2010 *International Journal of Greenhouse Gas Control.* **4** 51-56

IOP Conf. Series: Journal of Physics: Conf. Series **935** (2017) 012045 doi:10.1088/1742-6596/935/1/012045

Compositional, thermal and microstructural characterization of the Nopal (*opuntia ficus indica*), for addition in commercial cement mixtures

C G Hernández Carrillo[1], **J A Gómez-Cuaspud**[1] **and C E Martínez Suarez**[1]
[1] Universidad Pedagógica y Tecnológica de Colombia, Tunja, Colombia

E-mail: carlosgabriel.hernandez@uptc.edu.co

Abstract. The Nopal (*opuntia ficus indica)* from remote times has contributed like food and additive product in prehispanic constructions; although it grows in all the Colombian territory is very little used and its contribution in mixtures of Colombian cement is unknown. In order to evaluate the hydration characteristics of Nopal, several Thermogravimetric Analysis (TGA) were performed to evaluate the optimal temperature of dehydration. Initially, the results show that around 175°C the weight loss is approximately 95%, this mass loss corresponds to the process of physical removal, suggesting that at least a remaining amount of 5% (w/w) has the ability to retain large amounts of water which is stored in the micro-structural deposits of Nopal. The evaluation by means Scanning Electron Microscopy (SEM), confirm that the whole cactus structure enables the water storage at cellular level. The results of infrared spectroscopy (FT-IR) and Energy Dispersive X-ray (EDX) analysis allowed the qualitative and semi-quantitative evaluation of the presence of functional groups and elemental chemical composition of Nopal respectively, mainly related with polysaccharide functional groups, which corresponds to 85% of the total composition. Other functional groups, are related with protein and mineral components. This found characteristics are relevant for the water retention in process that require the decrease of water consumption and the reinforcing of mechanical properties and durability, due to ability of Nopal mucilage to restore its hydration characteristics.

1. Introduction

The cement is one of the most durable and versatile building materials known to date, economical and the most used in the world, its production follows a trend of growth reaching 2.8×10^9 tons worldwide and an increase of 4×10^9 tons is expected by 2050 [1,2]. In Colombia, the production of cement in 2015 reached 12100 tons with a constant problem related with the excessive water consume, which increases the handling characteristic but leads to an increase in porosity, as consequence the final material exhibits a reduction in the mechanical resistance and a high permeability inducing to the decrease of durability [3,4]. However, if the water content is insufficient a decrease in workability appear, causing strong damage effects as excessive water content [5]. Although this phenomenon has been solved from pre-Columbian times due to the incorporation of derivatives of plant species as Cactus Genus *Opuntia,* also known as Nopal in cement mixtures, mainly concentrated in the region of Mexico with the Aztec culture in the construction process of the pre-Hispanic buildings [6,7]. This xerophytic plant characteristic of subtropical dry climate regions, originated in tropical and subtropical America, nowadays this vegetative specie is present in a great variety of agro-climatic conditions, where it could be cultivated. This cactus specie have a strong ability to withstand prolonged periods of dryness, due to the presence of a gelatinous

IOP Conf. Series: Journal of Physics: Conf. Series **935** (2017) 012045 doi:10.1088/1742-6596/935/1/012045

body called “mucilage” that form three-dimensional molecular networks capable of retaining large amounts of water in a stable cellular arrangement [8,9], reason that justify its use in cement mixtures, representing the objective of current work as one of the most relevant studies in this field, due to hydration effect that could be improved as well as another physicochemical properties. This vegetal product is used in form of a complex carbohydrate or gel with high molecular weight, which behaves like negatively charged polyelectrolytes, delaying the hydration of the cement because of the presence of hydroxyl and carboxyl groups along the chains of Nopal’s mucilage binding to water molecules by polar interactions specifically by hydrogen bonds [2,10], tending to form stable hydrocolloids causing greater cross linking between macromolecular bonds in a calcium rich system such as the cements, this increases thickening in this type of mixtures [11,12] , provide excellent results in the water absorption capacity; the Nopal has been used in the actual food industry as emulsifiers, water purifiers, comestible adhesives, plasticizers in mortars, among others [13,14]. To achieve this purpose, the work evaluates several percentage mixtures (w/w) and different setting times as been reported previously [6,15,16], contributing to provide a physicochemical characterization of native Nopal specie of Tunja city (Colombia), in order to evaluate its thermal behaviour, microstructural arrangement and its elemental composition, responsible for the improvement of properties in cement mixtures.

2. Materials and method

Initially, the Nopal specie was characterized with the help of the Botanical Garden of the Universidad Pedagógica y Tecnológica de Colombia, comparing and establishing qualitative similarities of the structural units of the plant such as stem, leaves, thorns and fruits, in order to determine its main characteristics related with the interest specie, because almost 300 species of Nopal are known, however, there are only 10 or 12 of the same are used as hydrating material [17,18]. For the preparation of the sample, was necessary to start from a sample of 2 years of age, which was treated to isolate the mucilage from the rest of Nopal plant in longitudinal sections with intervals of approximately 1cm, the outer cuticle was removed. The resultant material was dried in an electric oven at 40°C for 3 days, after which a sample was recollected to be grounded and sieved at 200 US standard before characterization process. Initially the dried mucilage was evaluated by infrared spectroscopy, using a Perkin-Elmer FTIR equipment between 4700-500cm^{-1}, using the ATR configuration. The TGA-DTA analysis was performed in a thermostatic balance SDT Q600 equipment, using 50mg of sample, at a heating rate of 25°Cmin^{-1} from room temperature to 500°C under air flow (50mLmin^{-1}). The percentage of weight as a function of temperature was analysed in terms of endothermic and exothermic processes. The microstructural properties were evaluated by means SEM and EDX on a *Zeiss-Evo MA 10* equipment. For the analysis, the samples were coated with platinum for a better resolution images in a sputtering *QUORUM Q150R ES* instrument.

3. Results and discussion

The botanical classification of Nopal is shown in Figure 1 and summarized in table 1; it allows to identify the best Nopal specie in the Tunja region for hydration process in cement mixtures, revealing that is a very common cactus without evidence of commercial exploitation with a high degree of proliferation in arid areas.

Figure 1. (a) Schematic images of the Nopal plant (Opuntia ficus indica). (b) Photographic images of the Nopal plant in Tunja, Colombia.

Table 1. Botanical classification of the Nopal (Opuntia ficus indica).

Structural unit of the plant	Description
Stem	Plants from shrub to arborescent, from 1.7 to 3m in height, with a primary stem, well defined. Dark chestnut, green or grey, cylindrical stem, 45cm long, 20cm in diameter.
Leaf	Cladodes usually elliptic, but also, oval, circular, oblong or rhombic 27 to 32cm wide and 44 to 63cm high when they reach 2 to 3 years of age.
Thorn	Thorns are usually absent, but sometimes there are few cladodes with a spine, usually acicular, sunken, and white.
Fruit	The fruit is usually turbid, sometimes spherical, cylindrical or elliptical, often bright yellow or red

The infrared spectroscopy results shown in Figure 2, allow to evaluate of qualitative form the presence of amide functional groups associated with polysaccharides, responsible for the higher water retention as has been established by the literature in which, this vegetative specie have been used as additive in cement mixtures [19]. The vibrating signals at approximately 1640 and 3280cm^{-1} could be identified with polysaccharide molecules in which the signals were compared with the corresponding SDBS database. The signals around 1600cm^{-1} identify the amide group corresponding with the amide vibration, composed by the coupled vibrations of the C=O, NH and CN bonds [20]. The presence of secondary amines could be identified by the signal around 1400cm^{-1}, corresponding with the N-H deformation and C-N deformation bonds [20]. Chandra [19] observed this behaviour in vegetative samples derived from cactus species with strong signals around 1400cm^{-1}, related with amines and amides. Other authors [20-22] associate these signals with the carboxylic acid group and the peptide bond linked with the C-H deformation vibrations of the polysaccharides. In other studies Bayar [18] determined that the area between 1200 and 950cm^{-1} is the region for carbohydrates fingerprint and it allows the identification of some functional groups of polysaccharides such as stretching C-O-C, bending O-H, and deforming (CH_3) vibrations. Finally, the band located at 2920cm^{-1} is associated with the vibrational bond C-H present in all organic samples with a strong signal around 3500cm^{-1} related with the hydroxyl group (O-H) [19,20,22], a characteristic band present in proteins, carbohydrates and polysaccharides in general.

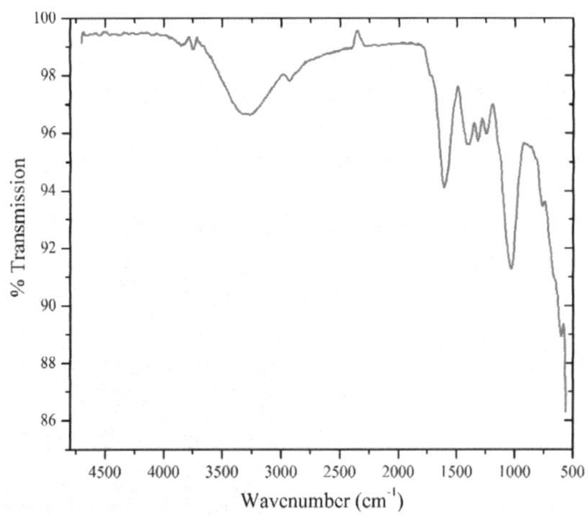

Figure 2. Infrared spectrum of dry mucilage extracted from Nopal specie.

The thermal analysis shown in Figure 3 evidence the progress of several exothermic reactions with weight loss around 90% of the initial mass, and related with the presence of water, which begins to evaporate around 50°C and progressively increase until a maximum value of 105°C and finish around 210°C in accordance with Pansu and Villegas [23,24]; this behaviour is common in organic samples and some research applied to cements has shown that the water retention capacity is slightly better when the mucilage is added dry at low doses and when is pre-mixed with the water in the place of mixing it as a dry powder in the other solid phases of the blends of cement as usual [11], Therefore by thermogravimetric analysis it was determined that the temperature of 50°C or less are optimal for a good dehydration of the Nopal in order to obtain better results when mixing it with cement.

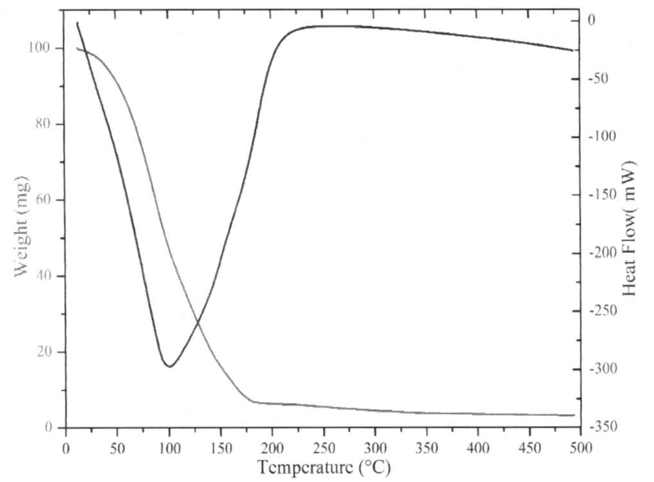

Figure 3. Thermal analysis (TGA-DTA) of the Nopal specie.

It is evident that the evaporation process is carried out in a single stage, clearly identified in which occur a weight loss around 95%, after the mass begins to stabilize. The corresponding heat flow is concordant with a process of physical removal of water retained in the material, after which exist a slight variation of mass related with the combustion of organic compounds (250-350°C) and the final loss of mass stabilizes around 350-500°C, implying that at least a remaining amount of 5% (w/w) has the capacity to retain large amounts of water for potential applications. The morphological evaluation of mucilage by SEM shown in Figure 4, confirm the morphological characteristics of the vegetative specie, determining that dry process produce an open cell morphology in which the plant stored the cellular water to fulfil its vital functions. The images reveal that all the lattice of the cactus focuses on water storage at the cellular level without evidence of another kind of another water storage mechanism in accordance with TGA results.

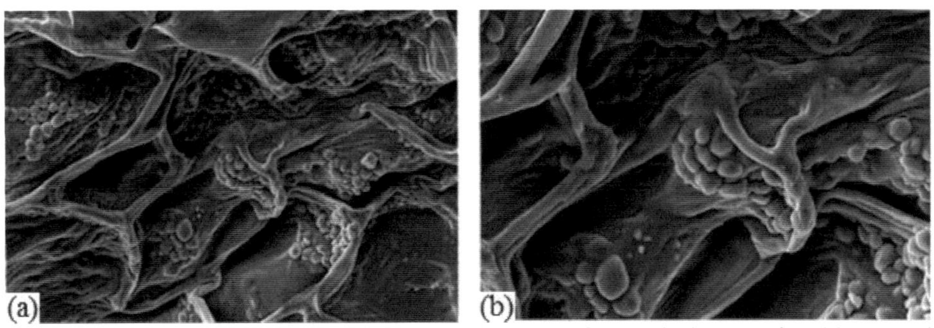

Figure 4. Scanning electron microscopy images of Nopal dry specie, (a), Nopal microstructure at 500x, (b) Nopal microstructure at 800x.

With the help of EDX analysis and energy dispersion, it was possible to provide an elemental composition of mucilage, determining that the majority of the components correspond to oxygen and carbon with at least 85% of the total composition and other elements such as sulphur, iron potassium and calcium in small amounts. There is no evidence of mineral elements that assist in the water retention process according to the results derived from FT-IR. It is therefore evident that in this case most of the carbon and oxygen contents are due to the presence of polysaccharide chains, as well as components such as sulphur in proteins and other organic compounds are responsible for the high water storage capacity In the Nopal plant, in this respect, the work done by Cardenas [8] determined that the Nopal has the capacity to withstand prolonged drought due to the presence of mucilage that form molecular networks capable of retaining large amounts of water, this property works like as an additive for good hydration in cement mixtures [25].

4. Conclusions

Based on the characterization by FT-IR and SEM-EDX, it was possible to determine that the morphology, molecular and elemental composition of Nopal are responsible for the retention of large amounts of water with small amounts of this plant's mass corroborated by TGA-DTA analysis; this is due to the fact that all the microstructure allows the storage of water at the cellular level supported by the presence of hydroxyl functional groups belonging to polysaccharide chains which, through chemical bonds, allow the retention and rheological changes at the molecular level in the water; this type of species could help in the obtaining of new cements and improve the process of hydration, durability and mechanical resistance in conventional cements with a low environmental impact.

References

[1] Montoya Y, Cadavid A and Gómez M 2009 Comportamiento mecánico y de fraguado de morteros de cemento pórtland gris tipo III con aditivos *Rev. Esc. de Ing. De Antioquia* **11** 39–49

[2] León F, Cano P, Lagunez L and Medina L 2014 Study of nopal mucilage and marine brown algae extract as viscosity enhancing admixtures for cement based materials *J. Const. and Building Mat.* **53** 109–202

[3] Schneider M, Romer M, Tschudin M and Bolio H 2011 Sustainable cement production present and future *J. Cement. and Conc. Research* **41** 642–50

[4] Perfetti del Corral M, Prada C and Freire F 2016 *Boletín Técnico: Estadísticas de cemento gris Diciembre 2015* (Bogotá: DANE) p 39

[5] Singh S, Munjal P, and Thammishetti N 2015. Role of water/cement ratio on strength development of cement mortar *J. Building Engineering* **4** 94–100

[6] Sáenz C, Sepúlveda E and Matsuhiro B 2004 Opuntia spp. mucilage's: A functional component with industrial perspectives *J. Arid Environments* **57** 275–90

[7] Hernández J, Coronado A, López T, Horta J 2008 Mejoramiento de morteros a través de Nopal *J. Professional Association for Cactus Development* **10** 126–32

[8] Cárdenas A, Arguelles M and Goycoolea 1998 On the possible role of *Opuntia ficus-indica* mucilage in lime mortar performance in the protection of historical buildings *J. Professional Association for Cactus Development* **3** 64–71

[9] Sáenz C 2007 *Boletín de servicios agrícolas: Utilización agroindustrial del nopal* vol 162 ed C Rosell (Roma: Food and Agriculture Organization)

[10] Medina L, Brito E, Torrestiana B and Katthainc R 2000 Rheological properties of the mucilage gum (*Opuntia ficus indica*) *J. Food Hydrocolloids* **14** 417–24

[11] Poinot T, Bartholin M, Govin A and Grosseau P 2015 Influence of the polysaccharide addition method on the properties of fresh mortars *J. cement and concrete research* **70** 50–59

[12] Lasheras M, Navarro I, Fernández J and Álvarez 2012 Effect of the addition of chitosan ethers on the fresh state properties of cement mortars *J. Cement Concrete Composites* **34** 964–73

[13] Hernández J, Coronado A, López T, Horta J 2008 Mejoramiento de Morteros a Través de nopal *J. Professional Association for Cactus Developmen* **10** 126–32

[14] Torres A and Cano J 2007 Las bondades del nopal *J. Construcción y Tecnología* **223** 44–50

[15] El-Etre A 2003 Inhibition of aluminum corrosion using Opuntia extract *J. Corrosion Science* **45** 2485–95

[16] Torres A, Martinez W and Alonso E 2012 State of the art on cactus additions in alkaline media as corrosion

inhibitors *J. International of Corrosion* **37** 835-41

[17] Kiesling R 1995 Origen, Domesticación y Distribución de Opuntia ficus-indica *J. Professional Association for Cactus Developmen* **22** 4747–56

[18] Felker P and Inglese P 2003 Short-term and long-term research needs for Opuntia ficus- indica (L.) Mill. Utilization in arid areas *J. of Professional Association for Cactus Development* **5** 131–52

[19] Chandra S, Eklund L and Villarreal R 1998 Use of Cactus in Mortars and Concrete *J. Cement. and Concrete Research* **28** 41–51

[20] Rodríguez S, Martínez H, Chávez C, Macías L, Zavala E, Garnica M and Chacõn L 2014 Extraction and characterization of mucilage from wild species of opuntia *J. Food Process Engineering* **37** 285–92

[21] Ramírez S 2008 Propiedades mecánicas y microestructura de concreto conteniendo mucílago de nopal como aditivo natural ed J Cano (Oaxaca: Instituto Politécnico Nacional) p 95

[22] Bayar N, Kriaa M and Kammoun R 2016 Extraction and characterization of three polysaccharides extracted from Opuntia ficus indica cladodes *J. International of Biological Macromolecules* **92** 441–50

[23] Pansu M and Gautheyrou J 2006 *Handbook of Soil Analysis, Mineralogical, Organic and Inorganic Methods* vol 13 ed Springer (Netherlands: Springer-Verlag Berlin Heidelberg) p 993

[24] Rodríguez E and Villegas E 2012 Caracterización de polímeros aplicando el método termogravimétrico *J. Métodos y Materiales* **2** 25–32

[25] Sánchez M, Martínez W, Alonso E, Guzmán E, Chávez H, García H, Torres A, Bernabé C, Arguello C, Hernandez S, Lara C, Gonzalez F, Velazquez J and Ruiz R 2015 Effect of the Addition of Dried Opuntia Cactus Fibers - Setting Times and Mechanical Resistance in Portland cement Mortar Base *2nd. Int. Conf. on Green Materials and Environmental Engineering (Phuket)* vol 1 (Amsterdam, Hong Kong, Paris: Atlantis Press) pp 121–125

IOP Conf. Series: Journal of Physics: Conf. Series **935** (2017) 012046 doi:10.1088/1742-6596/935/1/012046

Effects of CaO on the compaction and sintering by plasma of Powder-metallurgical iron

H G Ariza Suarez[1], A Sarmiento Santos[1] and C A Ortiz Otálora[1]
[1] Universidad Pedagógica y Tecnológica de Colombia, Tunja, Colombia

E-mail: geovannysy@hotmail.com

Abstract. This work the effect of the addition of Calcium Oxide (CaO) in the compaction and sintering of powder metallurgical iron Ancoor Steel 1000® is studied. Iron samples were made with proportions of: 0.5%, 1%, 1.5% and 2% by weight of CaO. The samples were sintered in a luminescent discharge furnace, in an atmosphere of H2+Ar at a temperature of 1150°C. XRD analysis was used to determine the formation of compounds, this analysis evidenced the formation of: hematite and magnetite, which were found both on the surface and in the volume. A characterization of the ability to protect against corrosive effects was carried out using the EIS electrochemical impedance spectroscopy method on the samples, in a solution of 1000ppm of chloride, with this procedure it was found that at a concentration of 0.5% and 1% CaO, the electrochemical impedance value is increased with values of 11.7MΩ, 2.2MΩ respectively.

1. Introduction

The Calcium oxide (CaO), is widely used since ancient times and has many applications in various sectors of industry. The CaO is used as catalyst for biodiesel production [1], is also used for the treatment of soils, and contaminated soils [2,3], and in the construction industry for the preparation of concrete [4,5]. The CaO is used in metallurgy as an oxide extracting agent in metals, it is also used to remove slag in the manufacture of steel and other metals [6,7]. The lubricating effect of CaO is mentioned in the wiredrawing process [8], however in the manufacture of parts via powder metallurgy it is not registered its use as lubricant. In this work, the effects of CaO in the powder metallurgy process were studied, the behaviour as lubricant during the production of samples in the green state was evaluated and its effect of oxide extraction in the sintering was observed. The oxides formed by the interaction of Fe and the oxygen present in the mixture are compounds such as hematite α-Fe_2O_3, magnetite Fe_3O_4, maghemite and goethite. The magnetite can form a protective layer that separates the metal from the aggressive environments [9,10], which improves the properties against corrosion.

2. Methodology

Four groups of test samples were made with the mixture of powder metallurgical CaO and Fe AncorSteel 1000®, The chemical composition of CaO is shown in Table 1. The samples were compacted at 700Mpa, in a cylindrical steel matrix and sintered in glow discharge plasma. The sintering conditions were: temperature of 1150°C for 30 minutes, atmosphere composed of H2+Ar, pressure of 6 Tor and heating rate of 100°C/min. The density of the green state samples and the sintered samples was calculated from their dimensions.

The compounds formed in the sintered samples were determined by X-ray diffraction test (XRD) on a PAN'alaltic X'pert PRO, with cobalt anode. The equipment configuration was: range from 25° to 104° for 2θ, a step of 0.0130° with permanence of 21.42 seconds, and a wavelength of 1.78901Å.

Table 1. Chemical composition CaO.

Specifications	Real value
Richness	95.8%
Chloride	<0.05%
Fe	<0.1%
Ni	<0.01%
Pb	<0.01%

The microhardness measurement was obtained using a Qualitest Vickers Microdurometer, the microdurometer was calibrated with 100gr of load for a time of 15 seconds. An analysis of electrochemical impedance spectroscopy EIS, was used to evaluate the behaviour of the test specimens in a corrosive environment, in a GAMRY PCI4G750. The EIS analysis which has been used to evaluate magnetite layers [11,12], The value of the electrochemical impedance is useful to show the interaction between the electrode and the electrolyte [13]. The solution used was sodium chloride of 1000ppm, the parameters used in the impedance analysis were: frequency scan of 10mHz to 100KHz, a cell composed of a working electrode with an exposed surface of 0.8cm^2, a Ag/AgCl reference electrode, and a platinum wire as a counter electrode.

3. Results

3.1. Compaction of the Fe-CaO mixture

The mixtures of the metallurgical powders were compacted at a pressure of 700Mpa, as seen in Figure 1. The shaped samples of 0.5%, 1%, 1.5%, and 2%, by weight of CaO was extracted from the compaction matrix without the use of any commercial lubricant for powder metallurgy. The compacting process shows that CaO has lubricating properties.

Figure 1. Compaction of the powder mixture Fe-CaO.

3.2. Density of samples with Fe-CaO mixture

The density values of the in green state samples and the sintered samples are shown in Table 2. The density is maintained around 6.9g/cm^3, for all groups of samples. The percentages of 0.5% and 1% by weight of CaO show an increase in density as a result of good sintering.

Table 2. Density of samples in green state and sintered

% of CaO	Density (g/cm^3)	
	Green state	Sintered 1150ºC
0.00%	6.88	6.92
0.50%	6.95	6.96
1.00%	6.92	6.94
1.50%	6.91	6.89
2.00%	6.93	6.91

A weak layer was formed on the surface of CaO samples during sintering, this layer increases the diameter and height dimensions affecting the density value.

3.3. Microhardness of sintered samples with Fe-CaO mixture

The microhardness values of the samples sintered at 1150°C are recorded in Table 3. The microhardness increases as a function of the CaO content, with a maximum value of 132.35 HV0.1 for a concentration of 0.5% by weight CaO and a minimum value of 127.74 HV0.1 at a concentration of 2% By weight CaO.

The highest microhardness values correspond to the samples with the highest density values.

Table 3. Microhardness of sintered samples to 1150ºC.

% of CaO	Microhardness HV0.1
0.00%	92.05
0.50%	132.35
1.00%	130.36
1.50%	129.05
2.00%	127.74

3.4. XRD analysis of the sintered samples

The analysis of XRD made to the specimens with 0.5%, 1%, 1.5%, and 2% by weight of CaO, showed in all the concentrations the characteristic peaks of the Iron-alpha, and the presence of the oxides of Iron. The iron oxides formed during sintering were magnetite and hematite. The results of the X-ray diffraction tests of the sintered samples are shown in Figure 2.

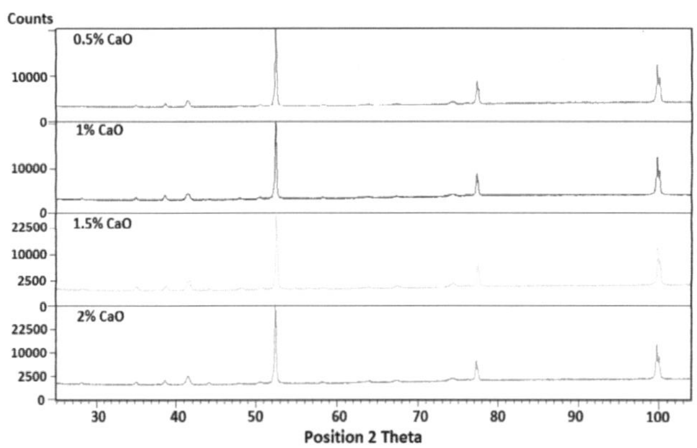

Figure 2. XRD analysis of samples whit 0.5, 1.0, 1.5 and 2% by wt of CaO

3.5. Electrochemical impedance spectroscopy analysis of the sintered samples

The electrochemical impedance analysis was performed on the sintered samples and their Bode diagrams are shown in Figure 3.

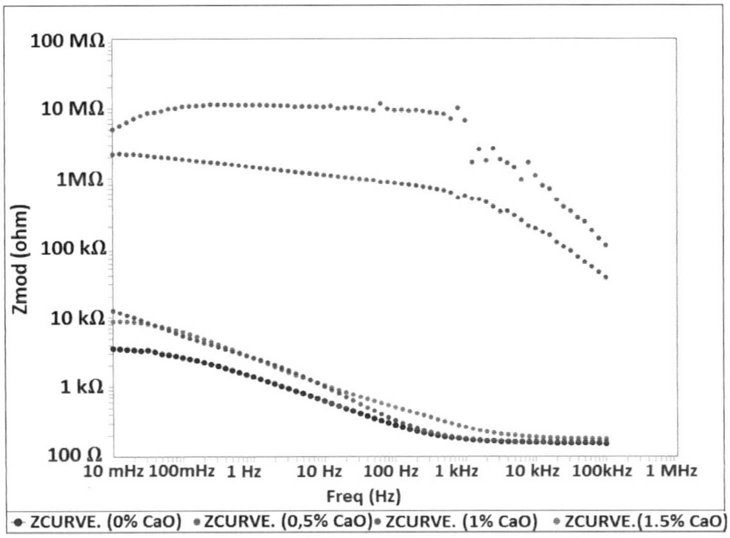

Figure 3. Bode diagrams of EIS spectra corresponding to mixture CaO-Fe sintered at 1150°C

The electrochemical impedance values for the different percentages of CaO in the test samples are shown in Table 4, the impedance value of the test sample without CaO is 3.56kΩ, and with an addition of 0.5% by weight of CaO the impedance value increases to a maximum of 11.7MΩ. Another significant value of 2.27MΩ is found with a content of 1% CaO. Test samples containing 1.5% and 2% by weight CaO do not have a significant impedance value.

Table 4. Bode diagrams of EIS Impedance

% of CaO	Impedance (Z) (Ohm)
0.00%	3.56×10^3
0.50%	11.70×10^6
1.00%	2.27×10^6
1.50%	8.87×10^3
2.00%	12.54×10^3

The high impedance values show that the samples have a more compact surface and there is less porosity, causing a lower transfer of charge and this is related to the density values obtained previously.

4. Conclusions

Samples were made via powder metallurgy of the CaO and Fe mixture, it was observed that CaO increases the microhardness and acts as a lubricant in the compacting step. The CaO reacts with the Fe during the sintering, formed Iron Oxides (hematite and magnetite) which were detected in the analysis of XRD. The impedance value of the EIS analysis is related to the density and porosity characteristics of the test specimens.

IOP Conf. Series: Journal of Physics: Conf. Series **935** (2017) 012046 doi:10.1088/1742-6596/935/1/012046

References

[1] Roschat W, Siritanon T, Yoosuk B and Promarak V 2016 *Energy Conversion and Management* **108** 459–467
[2] Robin V, Javadi A A, Cuisinier O and Masrouri F 2015 *Compu. and Geotechnics* **66** 189–202
[3] Yua S, Xua R, Houa L, Songa M and Liua J 2016 *Proce. Environmental Sciences* **31** 288–294
[4] Acharya P K and Patro S K 2016 *Journal of Cleaner Production* **131** 237-246
[5] Maheswaran S, Kalaiselvam S, Karthikeyan S K S, Kokila C and Palani G S 2016 *Cement and Concrete Composites* **66** 57-65
[6] Hao H, Zhang Y, Hao S, Zhang C, Jiang W and Cui P 2016 *Journal of Iron and Steel Research, International* **23** 884-890
[7] Wang Q, Wang D and Zhuang S 2017 *Construction and Building Materials* **151** 138–146
[8] ASM International Handbook Committee 2006 *Metalworking: Sheet Forming* vol 14B ed S L Semiatin (USA: ASM international)
[9] Machmudah S, Zulhijah R, Wahyudiono, Setyawan H, Kanda H and Goto M 2015 *Chemical Engineering Journal* **268** 76–85
[10] Montoya P, Martin C R, de Melo H G, Aok I V, Jaramillo F and Calderón J A 2014 *Electrochimica Acta* **124** 100–108
[11] Rodríguez-López A, Torres-Torres D, Mojica-Gómez J, Estrada-Arteaga C and Antaño-López R 2011 *Electrochimica Acta* **56** 8078–8084
[12] Bagherzadeh M, Haddadi H and Iranpour M 2016 *Progress in Organic Coatings* **101** 149–160
[13] Piratoba U and Olaya J J 2010 Electrochemical impedance - interpretation of typical diagrams with equivalent circuits *Dyna* **77(164)** 69-75

IOP Conf. Series: Journal of Physics: Conf. Series **935** (2017) 012047 doi:10.1088/1742-6596/935/1/012047

Mn-modification effects on Nb_2O_5 structural, optical and vibrational properties

A M Raba[1,2], **E Murillo**[1] **and M R Joya**[3]
[1] Universidad Francisco de Paula Santander, San José de Cúcuta, Colombia
[2] Universidad Pedagógica y Tecnológica de Colombia, Tunja, Colombia
[3] Universidad Nacional de Colombia, Bogotá, Colombia

E-mail: angelamercedesrp@ufps.edu.co

Abstract. In this study Mn-modification (at 1%, 2.5%, 5% and 10%) on Nb_2O_5 was carried out through the Pechini method; the annealing temperatures. 400°C, 500°C, 600°C and 700°C, were used to study the thermal stability of the system and the crystalline growth. The crystallization at high temperatures has been investigated because low annealing temperature cannot improve the crystallization and the effect of Mn-modification concentration is worthy of a further investigation. Before annealing the samples were analysed by TGA. The structural analysis through XRD is carried out to study a possible increase of the crystallite size (L) by increasing the Mn concentration; for the samples at 700°C and 1% and 2.5% concentrations, L is increased from 39.81 to 46.45nm while L changes from 32.67 to 35.63 with 5% and 10%, respectively. After Mn-modification the evolution of (100) and (180) peaks Nb_2O_5 was observed suggesting that they were well defined only until 2.5%. IR and Raman spectroscopies allowed to analyse vibrational modes present in the samples: the IR spectrums for the samples at 400°C, 500°C and 600°C showed a band around 1100cm^{-1} which can be associated to C-C(ν_{C-C}) and C-O(ν_{C-O}) vibrations; at 700°C the shoulder was distinguished around 750cm^{-1} characteristics of the Nb_2O_5 orthorhombic structure. Vibration modes associated with Nb_2O_5 were found with the Raman spectroscopy; these become less intense after Mn-modification. The band gap energy was obtained through UV-Vis spectrophotometry which revealed a slight increase at 700°C.

1. Introduction

Photocatalysis is a process based on the excitation of a semiconductor using radiation of adequate energy. The search for efficient highly processes in heterogeneous photocatalysis, which is a promising alternative of decontamination processes and it has directed the investigation to the optimization of existing photocatalysts and to the search of materials that may overcome the TiO_2 in its photocatalytics properties.

Niobium oxide is a p-type semiconductor with a wide band gap of about 3.4eV and has been proposed for a wide variety of applications, such as sensing materials [1-2], to support catalytic processes [3-6] and it has also attracted considerable interest as a raising material for electrochromic devices. All technological applications of oxide are determined by its structures and properties. In catalysis it can be correlated the catalytic activity with the crystalline phase; it has also been demonstrated that the Nb_2O_5 orthorhombic phase presents the greater catalytic activity. [7-8]

Choi *et. al.* [9] analysed the TiO_2 nanoparticles doping effect with 21 metals. There, in some cases, the TiO_2 oxidative capacity was improved of the and in others its reductive capacity. This effect leads to think in alternative materials that can overcome the TiO_2 properties when they are modified with transition metals.

Here, we report the synthesis of Mn-modification effects on Nb_2O_5 using a simple procedure involving the Pechini method. The value of Nb_2O_5 gap energy was reduced, which could increase its sensitivity introducing little manganese.

2. Materials and methods

$Nb_{2-x}Mn_xO_5$ (x=0.02, 0.05, 0.1, 0.2) samples were prepared by the polymeric precursor method - Pechini method. For this, Niobium Clorure ($NbCl_5$ 99% Sigma-Aldrich) and Manganese Nitrate ($Mn(NO_3)_2$•$4H_2O$ Merck) were used as starting precursors in a concentration previously established; distilled water the solvent employed. Hydrated Citric Acid (Panreac 99.5%) was used as a chelating and Ethylene glycol (Panreac 99.8%) as a polymerizing agent. The chelate was polyesterificated shaping a resin; the resin decomposition was done in a pre-calcination at 300°C for 4 hours. The material obtained was grounded in Agate mortar and to study the thermal stability, the powders obtained were subjected to thermal annealing at 400°C, 500°C, 600°C and 700°C for 2 hours.

Thermogravimetric Analysis (TGA) was used in order to detect the physical and chemical changes of the oxides depending on the temperature; for this, the DSC/TGA SDTQ600 equipment was used, applying a heating rate of 10°C/min. These spectres were analysed with the TA Instruments Universal Analysis 2000 analysis software. X ray diffraction (XRD) was done through an Xpert PANalytical Empyrean Series II Alpha 1 employing radiation Cu $K_{\alpha 2}$ (λ=1.54442Å) operating at 40 mA and varying 2θ between 5° and 80°. Infrared spectroscopy and Raman spectroscopy were used to analyse the vibrational behaviour of the samples; for this, a Prestige-21 Shimadzu IR spectrometer was used with a resolution of $8cm^{-1}$ and 8 shifted and a TG4000 Jobin Yvon micro-Raman commercial with line of Raman excitation λ=532nm. Shimadzu UV-Vis Cary 5000 spectrophotometer was used to obtain reflectance spectra of the samples.

3. Results and discussion

In Figure 1, the results of Thermogravimetric Analysis are presented showing an outstanding region of weight loss, between 320°C and 500°C, being associated to the temperatures of beginning and ending of decomposition of certain species in the samples, then the temperature chosen for the precalcination of the materials was adequate. The sample most thermally stable is the one synthesized at 1% Mn content and the least stable is the one synthesized at 10% Mn content.

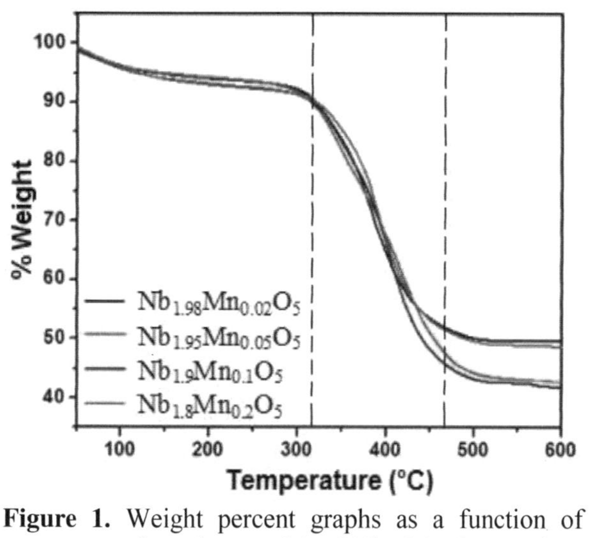

Figure 1. Weight percent graphs as a function of temperature for only precalcined $Nb_{2-x}Mn_xO_5$ samples.

X-ray diffraction patterns presented in Figure 2 show the presence of one phase for x=0.02 and x=0.05; structure orthorhombic with space group Pbam, corresponding to Niobium Pentoxide (this contrasts our results with the stored ones in Power Diffraction File). No reflections related to any secondary phase are detected in the XRD patterns, so for these x values the Niobium Pentoxide structure is not disturbed on substitution. Increasing the Mn concentration arises also peaks of low intensity. For x=0.1 and x=0.2 was found the structure orthorhombic [10] with space group Pbcn, corresponding to Manganese Niobium Oxide; it is believing that the Mn ions enter to Niobium pentoxide matrix occupying substitutional sites. Additional reflections consistent with Mn_3O_4 phase were observed, whose intensity depend on the annealing temperature. The small peak located around 36° for the sample modified with 10% of Mn is related to the cubic phase Mn_3O_4.

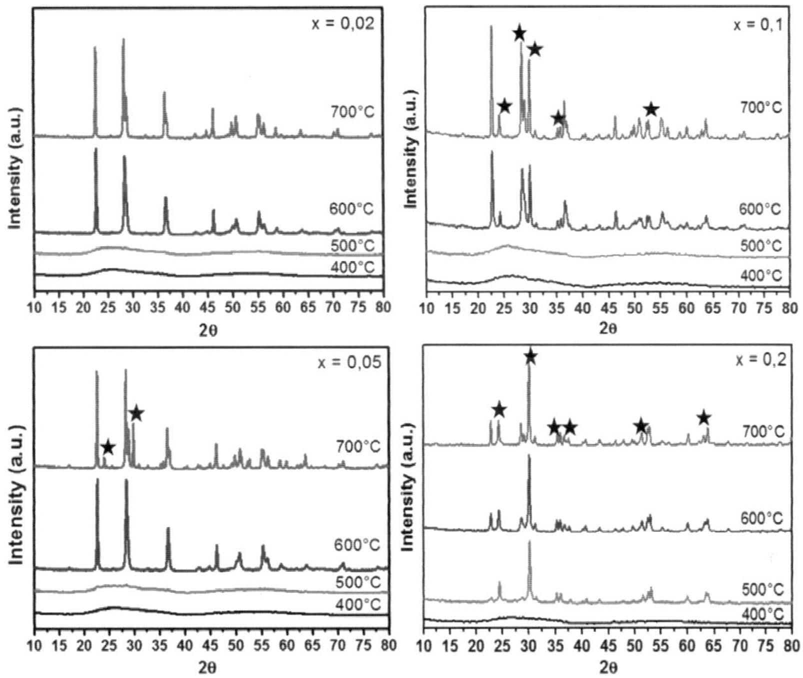

Figure 2. Comparison between XRD patterns of Mn-modified Nb_2O_5 at 1%, 2.5%, 5% and 10%. The star-type convention is used to highlight peaks that appear with the inclusion of metal.

Table 1 shows the crystallite size values (L) obtained through the Scherrer equation for the samples submitted at 700°C. It seems to have a slight tendency to increase as a concentration of the ion content increase too, only in the case when x pass from 0.02 to 0.05 or when x pass from 0.1 to 0.2, taking into account that two different phases were obtained. The width of the reflections increases with increasing Mn content, indicating decreasing in the particle size. In the same table, the respective lattice parameters are presented. An increase is expected in those as the Mn concentration is increased, due to the ionic radius of Mn^{2+} (0.80Å) is larger than the ionic radius of Nb^{+5} (0.75Å).

Figure 3(a) shows the evolution with the temperature of the infrared spectra of $Nb_{1.95}Mn_{0.05}O_5$ sample. Similar bands are found in the spectres rest at others Mn content. At 400°C, bands around 1600cm^{-1} are associated with the presence of residual organic components. At 500°C the spectra show a shoulder at about 350cm^{-1}, attributed to the amorphous phase of niobium pentoxide. A shoulder located around 1100cm^{-1} is distinguished for samples subjected to 400°C, 500°C and 600°C; this can correspond to the elongation type vibrations of the C-C(ν_{C-C}) and C-O (ν_{C-O}) groups, which are due to methoxy bridge species or may also be due to solvents, products and by-products of synthesis reaction of material. At 600°C, the small band at about 500cm^{-1} can be attributed to vibration ν_3 (Nb-O). At

700°C, the shoulder around 750cm^{-1} is characterized of the orthorhombic structure [11]. Figure 3(b) shows the Raman spectra varying Mn concentration for the samples submitted at 600°C. Raman modes for the calcination temperature of 400°C could not be observed. At 600°C and 700°C, bands correspond to vibrations of cations occupying centres of octahedrons and tetrahedrons are found in the 150-400cm^{-1} range. Bands in the 400-800cm^{-1} region are assigned to the symmetric and antisymmetric stretching mode of the Nb-O-Nb linkage [12].

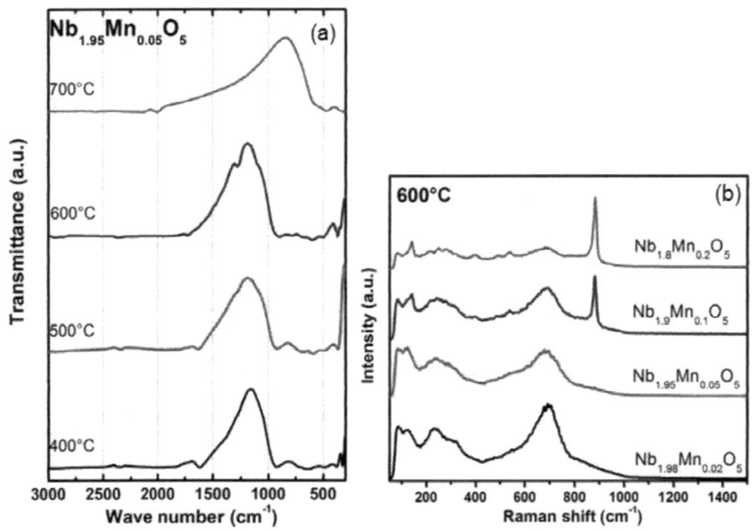

Figure 3. (a) Evolution with the temperature of the infrared spectra of $Nb_{1.95}Mn_{0.05}O_5$ sample. (b) Raman spectra varying Mn-content of the sample submitted at 600°C.

In Figure 4, the Diffuse Reflectance spectres are presented as a function of energy for the sample $Nb_{1.9}Mn_{0.1}O_5$, varying the temperature. Similar spectres were obtained for the samples rest. A decrease in the reflectance intensity is observed with increasing x; this is explained in terms of the lattice distortion due to the incorporation of the larger Mn^{2+} ions inside the Nb_2O_5 lattice.

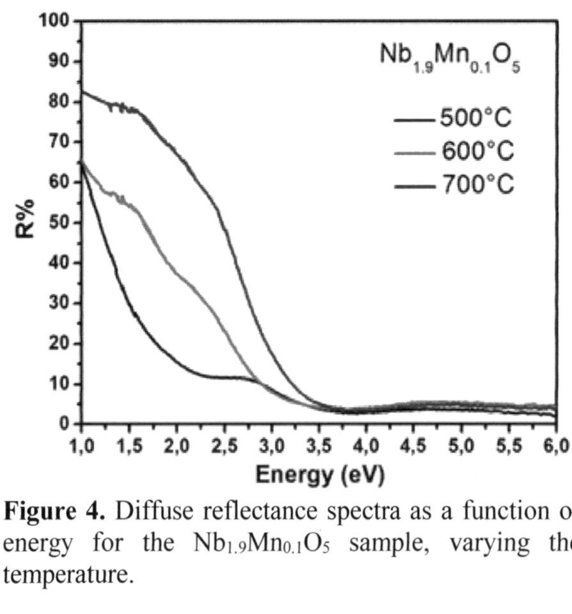

Figure 4. Diffuse reflectance spectra as a function of energy for the $Nb_{1.9}Mn_{0.1}O_5$ sample, varying the temperature.

The evolution of the gap energy values (E_g) with Mn content, at 700°C, is presented in the Table 1. In all cases the E_g value lightly increases at 700°C. In general, optical studies show a decrease in the band gap increasing Mn content, considering two cases separately: 1 and 2.5%, and 5 and 10%. Despite this behaviour, the E_g values obtained are suitable for photocatalytic purposes. The E_g values obtained can be attributed to the spin forbidden transition on Mn^{2+} in a tetrahedral environment.

Table 1. Crystallite size values, lattice parameters and gap energy values for the samples submitted at 700°C.

Sample	L(nm)[a]	JPDF cards No.	Lattice Parameters[b]	E_g (eV)
$Nb_{1.98}Mn_{0.02}O_5$	39.81	00-027-1003	a=6.1680Å b=29.3120Å c=3.9360Å	3.17
$Nb_{1.95}Mn_{0.05}O_5$	46.45			3.39
$Nb_{1.9}Mn_{0.1}O_5$	32.67	01-075-2159	a=14.4204Å b=5.7566Å c=5.0784Å	3.16
$Nb_{1.8}Mn_{0.2}O_5$	35.63	00-025-0543	a=5.0700Å b=5.7500Å c=14.4300Å	3.07

[a] The crystallite size values were estimated for the peak of maximum intensity.
[b] The lattice parameters were determined contrasting the results with the stored ones in Power Diffraction File.

4. Conclusions

$Nb_{2-x}Mn_xO_5$ (x=0.02, 0.05, 0.1, 0.2) nanoparticles have been synthesized using a simple Pechini method. With the increasing of Mn content, lattice parameters increase due to the larger ionic radius of Mn. Samples modified with the highest Mn content reveal the formation of small amounts of Mn_3O_4.

In general, the vibrational study shows that an increase of the synthesis temperature causes an increase in the material crystallinity because an increase in the synthesis temperature led to a sharp peak, this fact is verified with the values of the crystallite size.

Optical investigations through Diffuse Reflectance reveal that the presence of Mn^{2+} in tetrahedral coordination, possibly due to the substitution on Mn for Nb in the Nb_2O_5 lattice.

The presence of Mn^{2+} ions in the tetrahedral sites, the increased lattice parameters and the violet shift of the band gap, are some of the indicators of the incorporation on Mn ions inside the Nb_2O_5 lattice.

Acknowledgements

Authors acknowledge financial support from the Universidad Francisco de Paula Santander, San José de Cúcuta, Colombia, FINU Project No. 0.13-2016.

References

[1] Ficarro S B, Parikh J R, Blank N C and Marto J A 2008 *Anal. Chem.* **80** 4606
[2] Tamai T, Haneda M, Fujitani T and Hamada H 2007 *Catal. Commun.* **8** 885
[3] Cvelbar U and Mozetič M 2007 *J. Phys. D Appl. Phys.* **40** 2300
[4] Prado A G S, Bolzon L B, Pedroso C P, Moura A O and Costa L L 2008 *Appl. Catal. B* **82** 219
[5] Lin H Y, Yang H C and Wang W L 2011 *Catal. Today* **174** 106
[6] Sreethawong T, Ngamsinlapasathian S, Lim S H and Yoshikawa S 2013 *Chem. Eng. J.* **215** 322
[7] Lopes O F, Paris E C, Ribeiro C 2014 *Appl. Catal. B* **144** 800
[8] Raba A M, Falcony C, Rincón M 2016 *Respuestas* **21** 80
[9] Choi W, Termin A and Hoffmann M R 1994 *J. Phys. Chem.* **98** 13669
[10] Kato K and Tamura S 1975 *Acta Crystallogr. B* **31** 673
[11] Ristic M, et al. 2004 *Mater. Lett.* **58** 2658
[12] Raba A, Bautista-Ruíz J, Miryam R J 2016 *Mat. Res.* **19** 1381

IOP Conf. Series: Journal of Physics: Conf. Series **935** (2017) 012048 doi:10.1088/1742-6596/935/1/012048

Azimuthal polarization states for the optimization of a liquid crystal display Sony LCX038ARA in the pure-phase regime

C J Salamanca Vela[1], Y Torres Moreno[1], J Barrero Perez[1] and C Acevedo Caceres[1]
[1] Universidad Industrial de Santander, Bucaramanga, Colombia

E-mail: johanasalamanca29@gmail.com

Abstract. A liquid crystal display is an optoelectronic device contains molecules in an intermediate state between solid and liquid. In this work are showed numerical and experimental results of the azimuthal polarization states, which are used to optimize the pure-phase modulation of a Sony LCX038ARA LCD liquid crystal display for a linear polarized light beam with wavelength 632.8nm.

1. Introduction

A liquid crystal display is an optoelectronic device containing molecules that are in the intermediate state between solid and liquid [1]. Due the unique properties of these devices have been widely used as spatial light modulators (SLMs) [2]. Most of SLMs are of the nematic type are known as Twisted Nematic Liquid Crystal Displays (TN-LCDs). The TN-LCDs are widely used in applications in communications [3], medicine [4] and holography [5].

To characterize the complete behaviour of a TN-LCD is necessary to know the phase modulation of the LCD with applied electric field between their pixelated electrodes with respect to optical visible radiation [6]. In this work, we show the numerical and experimental results to generate states of equal azimuth i.e. [7] polarization states that allow to optimize the pure phase modulation of a Sony LCX038ARA liquid crystal LCD for a linear polarized light beam with a wavelength of 633.8nm using a system composed of a polarizer and a quarter-wave plate.

2. Methodology

The azimuth angle α and ellipticity angles ε can be used to describe polarization states over the sphere Poincaré [8]. The azimuthal polarization states represent southern curves over the sphere Poincaré with equal azimuth angle and variable ellipticity [9]. These states are generate used the LCD sandwiched between a polarizer and a quarter-wave plate appropriately oriented them [10]. To find the optimal orientations of the polarizer and the quarter-wave plate within the optoelectronic system we have generated approximately states of equal azimuth for the 26 grey levels using the schematic setup shown in Figure 1.

In addition, we have used the results for the intrinsic parameters total twist angle ($\emptyset$), angle of the molecular axis at the entrance face (Ψ_D) and maximum birefringence (β) of the TN-LCD with null applied voltage showed in the Table 1 with the purpose of find the rotor and retardation effect added on an arbitrary phase state [9].

IOP Conf. Series: Journal of Physics: Conf. Series **935** (2017) 012048 doi:10.1088/1742-6596/935/1/012048

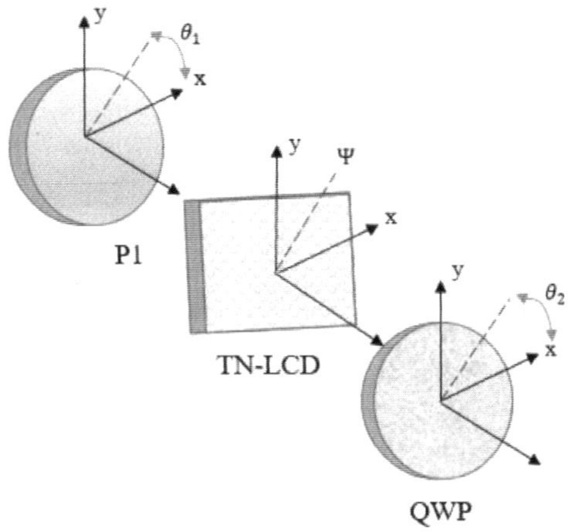

Figure 1. Schema of experimental setup used for measuring the optimal orientations of the polarizer (P) and the quarter-wave plate (QWP) in the azimuthal polarization state using linear polarized light to illuminate the LCD.

Table 1. Intrinsic parameters of the TNLCD [11].

Parameter	Value
Twister angle (∅)	+93.8°
Angle of the molecular axis (Ψ_D)	-14.9°
Birrefrigence (β)	2.126 rad

Using the experimental schema of Figure 1, it is possible obtain that the light transmitted by the optoelectronic system is given by [9]:

$$T = cos^2(\alpha_0 - \xi)cos^2\varepsilon_0 + sin^2(\alpha_0 - \xi)sin^2\varepsilon_{0,} \quad (1)$$

With,

$$\xi = \alpha_0 \pm \frac{\pi}{4} \quad (2)$$

The longitudinal coordinate over the Poincare sphere, $\varepsilon_{0,}$ is the elliptic angle and α_0 is the azimuth angle. Then, the output field can be writing as:

$$E_{out} = \frac{1}{\sqrt{2}}exp[j(\pm\varepsilon_0 + \delta)]\begin{pmatrix}1\\0\end{pmatrix} \quad (3)$$

2.1. Determination of angle theta 1(θ_1) y theta 2(θ_2)

To find the values of $\theta_1$and θ_2 that maintain constant the azimuth state for the 26 gray levels (g) we have used the least squares method with the purpose of determine numerically the angles of the polarizers and the quarter wave plate that minimized the standard deviation σ, which is defined as:

$$\sigma(\theta_1\theta_2) = \sqrt{\frac{1}{n-1}\Sigma(\alpha_g - \bar{\alpha})^2} \quad (4)$$

With the Stokes parameters given by [6],

$$s_0 = 1 \quad (5)$$

$$s_1 = \cos(2\theta_2)\left(cos^2\delta_{eq}\cos[2(\theta_1 - \theta_2 - \phi_{eq})]\right) + sin^2\delta_{eq}cos[2(\theta_1 - \theta_2 - \phi)] + \sin(2\theta_2)\sin(2\delta_{eq})\sin[2\theta_1 - \theta_2 - \phi_{eq}] \quad (6)$$

IOP Conf. Series: Journal of Physics: Conf. Series **935** (2017) 012048 doi:10.1088/1742-6596/935/1/012048

$$s_2 = sin(2\theta_2)\left(cos^2\delta_{eq}cos\left[2\left(\theta_1 - \theta_2 - \phi_{eq}\right)\right]\right) + sin^2\delta_{eq}cos[2(\theta_1 - \theta_2 - \phi)] + cos(2\theta_2)sin\left(2\delta_{eq}\right)sin\left[2\theta_1 - \theta_2 - \phi_{eq}\right] \quad (7)$$

$$s_3 = -sin^2\delta_{eq}sin[2(\theta_1 - \theta_2 - \phi)] + cos^2\delta_{eq}sin\left[2\left(\theta_1 - \theta_2 - \phi_{eq}\right)\right] \quad (8)$$

Where ϕ_{eq} and δ_{eq} are experimental equivalent rotor and retardation, respectively. We have used the Nelder-Mead method to minimize equation (6). The schema of the Figure 2 show the points used for the Nelder-Method. With this method we have calculate the optimal polarization states that generate a light beam with constant azimuth.

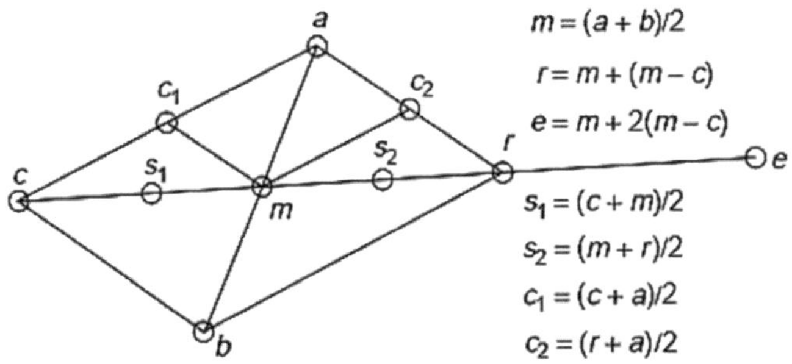

Figure 2. Notation used in the Nelder-Mead method [8].

3. Results

The Table 2 shows the optimal orientations of the polarizer and quarter wave plate that generate an azimuthal polarization state in the optoelectronic system. In this Table we also have calculate the standard deviation for these angles.

Table 2. Results of Nelder-Mead method for the TNLCD.

Parameter	Value
Angle Theta1 (θ_1)	78.18°
Angle Theta2 (θ_2)	61.48°
Minimum (σ)	4.44 rad

Using the numerical results of the Table 2, we have found that the azimuthal polarization state is obtained with the azimuth angle 17.31°. To corroborate the azimuth angle, we proceed to find the experimental value using the setup shown in left side of Figure 3. A random polarized light from a Helium-Neon laser (Spectra physics, Power 5mW) with wavelength 632.8nm in the transverse electromagnetic ground state mode TEM00 is filtered and collimated using the objective (O), the pinhole (PH) and the lens L1, respectively. The TN- LCD has been placed between two polarizers (P and P1) and an analyser (A). Finally, a power-meter (PM) is placed at the focal plane of the lens to detect efficiently intensity measurements of the TN-LCD. The experimental arrangement polarizers (P and P1), TN- LCD and an analyser (A), also includes one-quarter wave plates (QWP or λ/4), which has been adjusted to 78.18° with the purpose to obtain an equal-azimuth state.

The right side of Figure 3 shows the experimental results for a series of 26 images of grey levels ranging from g=0 to g=255 in steps of 10 displayed on the TN-LCD. In this Figure we can observe that the behaviour of the Stokes parameters S1 versus S2 is lineal, i.e., this state may be used like an equal-azimuthal polarization state with experimental azimuth of 17.31° [12] (put value). The discrepancy of the numerical and experimental result can be explained if we take into account that the quarter wave plate is the multiple order for 632.8nm.

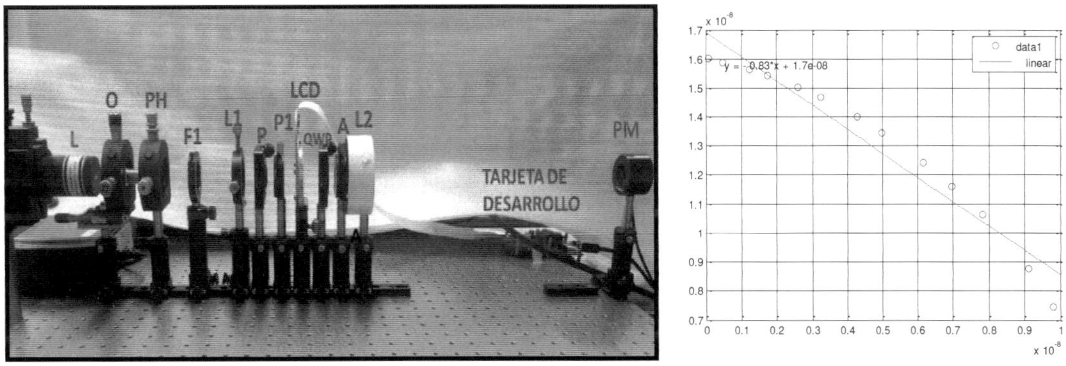

(a) (b)

Figure 3. (a) Experimental schema used to find the azimuth angle. (b) Experimental results for the Stokes parameter S1 and S2.

4. Conclusions

It was found that a state of equal azimuth in the optoelectronic system can be obtained for the $\theta_1 = 78.18°$ and the $\theta_1 = 61.48°$ orientations in the polarizer P and the quarter wave plate, respect to the molecular axis of the TN-LCD. This result is a cornerstone for the characterization of the matrix in pure-phase mode.

Acknowledgments

Research partially supported by Universidad Industrial de Santander with the projects 5708 and 1788. Thanks to Colciencias for financial support to project 110256934957: "Dispositivos ópticos para sistemas de alta dimensionalidad QKD (Quantum Key Distribution) basados en el momento angular orbital de la luz". C.H. Acevedo Cáceres acknowledges the support and sponsorship from Call 647-2015: National Doctoral program of Colciencias.

References

[1] M. G. Friedel 1922 The Mesomorphic states of matter *Annales de Physique* **18** 162-174
[2] Ehmke T, Heiko Nitzsche T, Knebl A. and Heisterkamp A 2014 Molecular orientation sensitive second harmonic microscopy by radially and azimuthally polarized light *Biomedical Optics Express* **5** 2231-2246
[3] Caihong K, Pei H, Fei S, Hongping Z, and Zhongyi G 2017 *IEEE Photo. Journ.* **9**
[4] Zhu W, Eckersorn N, Upadhya A, Li L, Rode A, and Lee W 2016 *Biomed. Opt. Expr.* **7** 2902
[5] Kelner y R, Rosen J 2015 *IEEE Trans. and Inform.* **12** 220
[6] E. Lueder 2010 *Liquid Crystal Displays: Addressing Schemes and Electro-Optical Effects 2nd edition* (Great Britain: John Wiley & Sons, Inc.) p 516
[7] Durán Bosch V 2008 *Optimización del funcionamiento de un modulador espacial de luz de cristal líquido mediante el modelo retardador-rotor. Aplicaciones en óptica adaptativa* (España: Universitat de Valencia)
[8] Chen R 2011 *Liquid crystal displays: fundamental physics and technology* (Great Britain: John Wiley & Sons, Inc)
[9] Duran V 2006 *J. of Appl. Phys.* **99** 113101
[10] S. Baldock, C. Reta-Tang, B. Beck, W. Gao, J. Cai and S. Hightower 2012 Prospecciones ortogonales de azimut amplio: adquisición y obtención de imágenes *First Break* **30(9)**
[11] Cely C, Acevedo C, and Moreno Y 2017 *J. of Phys.: Confer. Ser.* **786** 012009
[12] Yang W, Cao W, Chung S, and Morris J 2005 *Applied numerical methods using Matlab* (United States of America: John Wiley & Sons, Inc.)

IOP Conf. Series: Journal of Physics: Conf. Series **935** (2017) 012049 doi:10.1088/1742-6596/935/1/012049

Characterization using XRD of puzzolanic materials from residual sludge from water treatment

G Barón[1], A M Montaño[1] and C P González[2]
[1] Universidad Industrial de Santander, Bucaramanga, Colombia
[2] Universidad Pontificia Bolivariana, Bucaramanga, Colombia

E-mail: gcbarong@gmail.com , claudia.gonzalez@upb.edu.co

Abstract. The goal of this work is to do mechanical and chemical characterization of puzzolanic materials using compressive strength measurements and X-Ray Diffraction (XRD). These materials are composed of red clay and aluminous sludge produced by the treatment of potable water at Planta Algodonal, Ocaña, Norte de Santander, Colombia. Ceramic bricks were sintered to 1100°C and ten were characterized in their physically, mechanically and chemically properties. The results showed that the relationships with which the Colombian standards according to NTC 4017 (100KGF/cm2) for non-structural bricks are maintained for those containing 10% (105Kgf/cm^2) and 20% (102.9Kgf/cm^2) of sludge with respect to clay.

1. Introduction

The pozzolans are commonly siliceous materials or aluminosilicates which are capable of improving the properties of different products; this makes them useful in the formation of clay pieces which are subjected to a sintering process [1]. In Colombia, red and black clays are commonly used as raw materials for the formation of masonry products. In addition, sand is used as a complement. In recent years the study of materials has focused on the incorporation of industrial by-products or pozzolanic, considered waste [2], for the production of more environmentally friendly materials, one of these aluminosilicates is the sludge from the potable water treatment plants (PTAP).

In PTAPs during the water purification process, a quantity of coagulant, commonly aluminium sulphate, is added, thus generating the formation of sludge considered waste. These sludges are mostly composed of inorganic substances, feldspars, silts, fine sand [3] and are characterized by having a high concentration of silicon oxide and aluminium oxide. Approximately in PTAP, there are approximately 0.3% to 1% of these sludges compared to the total volume of treated water. Some researches around the world had found interesting properties in these sludges that allow them being part of glass-ceramics [4] and even used as coagulants [3]. In Colombia, there is no currently any regulation for the final disposal of sludge, which is why PTAPs discharge them further downstream from the watercourses from which they initially acquire their waters [5]. In Colombia the reuse of pozzolan materials mainly covers the construction sector, but in the case of sludge produced by water treatment, the reported information scarces, being thus used as additives for the production of Clinker, cement Portland, fertilizer in agricultural soils and as a raw material in recovery of aluminium [1-3,6-8]. For the manufacture of masonry units in Colombia, the Colombian Technical Standard (NTC) [9] must be taken into account, according to which the specimens created can be classified into units for structural and non-structural construction. In this study the compressive strength of non-structural type

IOP Conf. Series: Journal of Physics: Conf. Series **935** (2017) 012049 doi:10.1088/1742-6596/935/1/012049

pozzolanic materials was evaluated before and after chemical attacks with hydrochloric acid and sodium hydroxide, followed by X-Ray Diffraction (XRD) characterization of the crystalline phases present in the formed units.

2. Methods

The elaborate units are composed of 100: 0 90:10 80:20 70:30 60:40 50:50 40:60 30:70 20:80 10:90 and 0:100 red clay from the alluvial fan of the City of Bucaramanga and residual sludge from drinking water treatment plan in the PTP Algodonal, located on the road to the Batallon Santander in the municipality of Ocaña. The specimens manufactured were solid cubic units of 5cm×5cm×5cm, which had a sintering process following the heating rate described below: 0-105°C (120min) 105°C-600°C (120min) 600°C-1100°C (180min), finally allowed to descend to room temperature.

Once elaborated, the materials were characterized mechanically, chemically and structurally. In the first one it is necessary to evaluate the resistance to the compression presented by the units according to the criteria of the NTC 4017, chemically an evaluation of the corrosion that the units present, for that they were submitted to an acid medium and an alkaline means, using 20% hydrochloric acid and 20% sodium hydroxide respectively for one hour followed the evaluation of the mechanical strength presented by the material after the characterization of the clay, sludge and mixtures 90:10 and 80:20
By X-ray diffraction, these were taken on a BRUKER D8 ADVANCE powder diffractometer with DaVinci Geometry (40kV, 40mA, Cu Kα λ=1.5406Å) of "Laboratorio de Difracción de Rayos X" of the "Universidad Industrial de Santander, Parque Tecnológico Guatiguara".

3. Result and discussion

3.1. Mechanical characterization

Results from the compressive strength for the units of different clay: sludge mixtures, are presented in Table 1, Compressive strength test. It is observed that as the concentration of sludge in the unit increases, the mechanical resistance decreases considerably. So that the mixtures 90:10 and 80:10 meet the quality standards described by the standard for non-structural materials [9] with values of 105Kgf/cm^2 and 102Kgf/cm^2 respectively, being 100kgf/cm^2 the minimum value given by the norm. Units with relationships 70:30, 60:40 and 50:50 are too close to the minimum value of the standard, so, these could be used just as a non-structural material, together with aggregates that confer greater resistance to compression. Remaining ratios are well below the minimum value required by the units for which they are not suitable for the manufacture of non-structural materials.

Table 1. Compressive strength test.

Mixture	Area [cm^2]	Load of Break [Kgf]	Mechanical Strength [Kgf/cm^2]
100:0	25	2731.22	109.24
90:10	25	2625.10	105.00
80:20	25	2572.58	102.90
70:30	25	2433.15	97.32
60:40	25	2258.48	90.34
50:50	25	2057.31	82.29
40:60	25	1517.96	60.72
30:70	25	1335.25	53.41
20:80	25	1244.42	49.78
10:90	25	802.58	32.10
0:100	25	623.38	21.95

IOP Conf. Series: Journal of Physics: Conf. Series **935** (2017) 012049 doi:10.1088/1742-6596/935/1/012049

3.2. Chemical tests

Based on the data obtained above, the chemical resistance to the units of ratios was evaluated: 100:0 (used as reference), 90:10 and 80:20, these are the mixtures that meet the requirements of the NTC, the results obtained are observed in Table 2.

During the attack with hydrochloric acid, the immersed material initially absorbs a large amount of acid, due to its porosity. After 5 minutes, the absorption process becomes slow, once the immersion time and curing time elapses, the material acquires a dark tone with respect to the initial one. The size does not change considerably since the material is minimally corroded by the acid [10], thus generating slight deformations in the tips of the cube due to small parts detached. In the case of the attack with alkali, after the immersion the specimen is corrodes since a there is a high discoloration at the time of curing at room temperature; however, it does not generate a change in the size of the unit under test. Finally, the material after the curing time in the open air acquires mass, so that at a higher concentration of sludge in the specimen there is a higher degree of absorption of acid and base. The mechanical strength of the specimens was measured after 7 days of outdoor curing; For the 90:10 ratio it is observed that there is no significant change in the strength of the material while for the 80:20 ratio the change in mechanical strength is considerable. Therefore, this last mixture does not comply with the value stipulated by the standard.

Table 2. Chemical tests.

Mixture	Initial mass [g]	Final mass [g]	Mass difference [g]	Area [cm^2]	W [Kgf]	Mechanical Strength [Kgf/cm^2]
100:0 acid	174.01	186.41	12.40	25	2783.55	111.34
90:10 acid	157.93	180.82	22.89	25	2672.49	107.02
80:20 acid	155.02	193.48	38.46	25	1232.78	49.31
100:0 base	174.32	183.25	8.93	25	2742.98	109.72
90:10 base	163.34	180.58	17.18	25	2668.74	106.75
80:20 base	151.42	170.63	19.21	25	2161.62	86.47

3.3. Structural study

Mineralogical phases of the samples are analysed by XRD using de database PDF-2 [11] for clay (with and without sintering), sludge (with and without sintering) and mixtures of 90:10 and 80:20, Figures 1, 2 and 3 respectively. In the case of non-sintered clay, we observe the phases of Quartz (Q), Muscovite (M), Caolinite (C), Microcline (Mi), Albite (A), Cristobalite (Cr), Hematite (H) and Montmorillonite (Mo). Quartz, muscovite and kaolinite phases have the highest concentration in the sample, which is understandable since the silicon and aluminium oxides are the ones that are in the highest proportion.

Percentage in the sample of crystalline material is high: 83.1%. Clay after the sintering process presents the phases Quartz (Q), Hematite (H), Microcline (Mi), Illita (I), Mullite (Mu) and Albite (A). As can be observed due to the high temperature, there is a loss of the phases Muscovite (M), Cristobalite (C) and Montmorillonite (Mo), and at the same time the Illita (I) and Mullite (M) phases appear. All of the above causes the amount of crystalline material after the sintering process to decrease to 51.9%. Quartz should not be present in the sintered clay, since this phase at 870°C becomes tridimite and at approximately 1000°C becomes cristobalite. Kaolinite should be transformed to primary mullite at approximately 979°C [12], because it is a metastable phase and because of the heating rate used, this transformation was not carried out, thus taking form of quartz, which is corroborate in the quantitative analysis of XRD with an increase of concentration of this phase. The rest of the kaolinite, becomes amorphous. Expected transformations were not obtained even though the sintering process was carried out at temperatures reported in the literature. This could be due to the presence of microcline and albite in the clay, these two minerals are characterized by having a thermal effect of running, which causes that the phase transformations do not occur at the expected temperature but at higher temperatures. Sludge contains the Caolinite (C) phase and, to a lesser extent,

IOP Conf. Series: Journal of Physics: Conf. Series **935** (2017) 012049 doi:10.1088/1742-6596/935/1/012049

the phases Quartz (Q), Muscovite (M), Microcline (Mi), Albite (A), Magnesiohorn blende (Mg), Gibbsite (G) and Vermiculite V). In the sintering process phases such as magnesium blende, some phosphates and sulphates are formed, which result from the combination of atoms in the crystalline lattice. Finally, in the diffractogram of the mixtures, a combination of the minerals presents in the sludge and in the clay, is present, which is why there are small non-quantifiable quantities of the minerals Biotita, Paligorsquita, Bernalita, Chabasita.

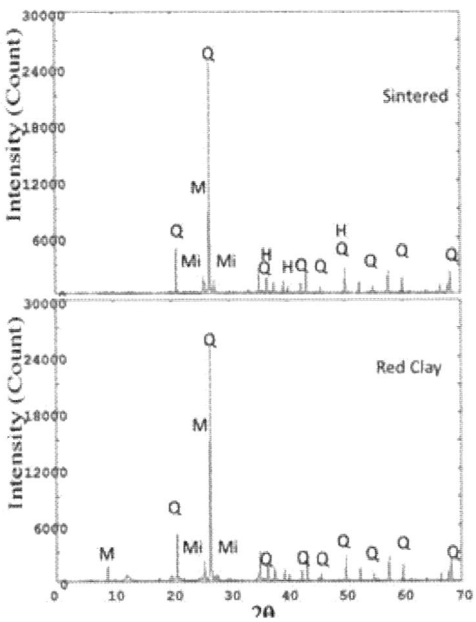

Figure 1. Clay diffractogram.

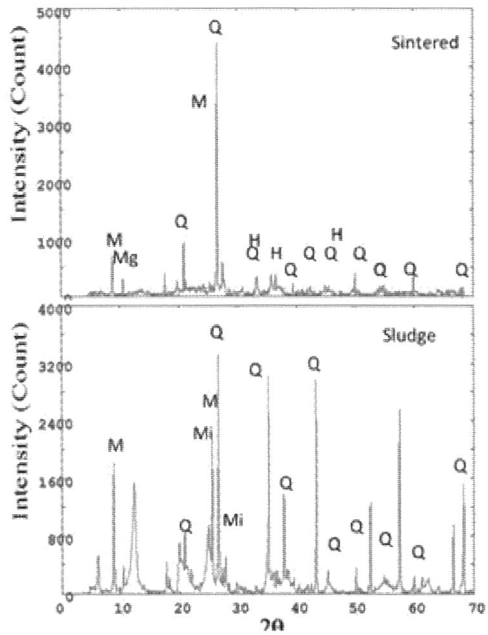

Figure 2. Sludge diffractogram.

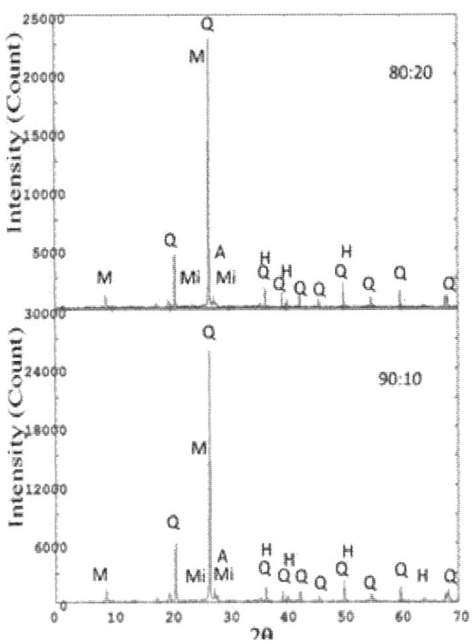

Figure 3. Mixtures Diffractograms

4. Conclusions

Sludge from the water purification process of the El Algodonal plant is useful as additive in non-structural building materials. The 90:10 and 80:20 mixtures comply with the parameters of mechanical resistance being that described by the Colombian Technical Standard of 100Kgf/cm^2 and 105Kgf/cm^2 and 102.9Kgf/cm^2 respectively. Absorption of acids and bases increases as the amount of pozzolan material increases in the sample, causing the material a decrease mechanical strength in 90:10 mixture after the attack with hydrochloric acid and sodium hydroxide, it complies with the mechanical resistance parameter described by the Colombian Technical Standard. The expected phase transformations were not metastable due to the presence of minerals such as microcline and albite. These minerals, when present, create a thermal phenomenon that raises the temperature at which a phase change occurs.

Acknowledgements

Authors are grateful to "Laboratorio de Difracción de Rayos X" and "Laboratorio de Ingeniería Civil" of the "Universidad Industrial de Santander (UIS)" and "Laboratorio de Ingeniería Civil" of the "Universidad Pontificia Bolivariana Seccional Bucaramanga".

References

[1] Torres P, Hernández D and Paredes D 2012 Uso productivo de lodos de plantas de tratamiento de agua potable en la fabricación de ladrillos cerámicos *Rev. Ing. Constr.* **27** 145–154

[2] Seco A, Ramirez F, Miqueleiz L, Urmeneta P, García B, Prieto E and Oroz V 2012 Types of waste for the production of pozzolanic materials – A Review *Review Industrial Waste* ed Kuan-Yeow Show and Xinxin Guo (Croatia: InTech) Chapter 7 pp 141–150

[3] Ahmad T, Ahmad H and Alan M 2016 Characterization of water treatment plant´s sludge and its safe disposal options *Procedia. Environ. Sci.* **35** 950-955

[4] Toya T, Nakamura A, Kameshima Y, Nakajima A and Okada K 2007 Glass-ceramics prepared from sludge generated by water purification plant *Ceram. Int.* **33** 573-577

[5] Grajales S, Monsalve J and Castaño J 2006 Programa de manejo integral de los lodos generados en la planta de tratamiento de aguas residuales de la universidad tecnológica de Pereira *Rev. Ing. Uni. Medellín* **31** 285-290

[6] Hernández D, Villegas J, Castaño J and Paredes D 2006 Aprovechamiento de lodos aluminosos generados en sistemas de potabilización, mediante su incorporación como agregado en materiales de construcción *Rev. Ing. Univ. Medellín* **5** 119–132

[7] Mantilla G 2015 Validación de uso de lodos generados en plantas de tratamiento de aguas residuales tipo UASB como insumo en la recuperación de suelos agrícolas *Esaica* **1** 18-23

[8] Gutiérrez-Rosero J A, Ramírez-Fajardo Á I, Rivas R, Linares B and Paredes D 2014 Tratamiento de lodos generados en el proceso convencional de potabilización de agua *Rev. Ing. Univ. Medellín* **13** 13–27

[9] ICONTEC NTC 4017 2005 *Métodos para muestreo y ensayos de unidades de mampostería y otros productos de arcilla* (Colombia: Icontec) pp 1–30

[10] Escoda L, Lledó M, Suñol J J, Roura P and Carda J 2003 Estudio sobre la resistencia quimica de baldosas ceramicas no-esmaltadas para pavimentos industriales *Bol. Soc. Esp. Cerám. Vidr.* **42** 85–88

[11] ICCD 2012 *PDF-2/inorganic+2016 (database)* ed S. Kabekkodu (USA: International Centre for Diffraction Data) (http://www.icdd.com/products/pdf2.htm)

[12] Macías López M A 2006 *Estudio de las transformaciones en estado solido de las arcillas de san jose de cucuta utilizadas en la fabricacion de baldosas mediante caracterizacion mineralogica por difraccion de rayos-x de muestras policristalinas* (Bucaramanga: Universidad Industrial de Santander)

IOP Conf. Series: Journal of Physics: Conf. Series **935** (2017) 012050 doi:10.1088/1742-6596/935/1/012050

High-temperature behaviour of ammonium dihydrogen phosphate

A Pardo[1], J Romero[1] and E Ortiz[1]
[1] Universidad del Atlántico, Barranquilla, Colombia

E-mail: adavidpardo@mail.uniatlantico.edu.co

Abstract. For over five decades, the high-temperature behaviour of ammonium dihydrogen phosphate, NH4H2PO4, has been a controversial subject; while initial works associate ionic conductivity increase around Tt=153°C to a physical transformation (structural phase transition), later research supports the chemical nature of the transformation. However, currently, the origin of the ionic conductivity increase is still not clear. To provide a possible interpretation for this phenomenon, a careful high-temperature thermal examination of this acid salt was conducted by means of Differential Scanning Calorimetry (DSC), Thermogravimetric Analysis (TGA), and Mass Spectroscopy (MS). The results show that when this acid salt is heated through Tt, a chemical decomposition into phosphoric acid (H_3PO_4) and ammonia (NH3) takes place. Considering that H_3PO_4 exhibits conductivity values around $10^{-3}S\cdot cm^{-1}$, our results suggest that the conductivity increase observed at around Tt is an exclusive consequence of the presence of this decomposition product.

1. Introduction

Ammonium dihydrogen phosphate, $NH_4H_2PO_4$ (ADP), has been widely studied because of its several technological applications. For instance, owing to its high efficiency of nonlinear conversion and wide optical transparency region, this compound has been used in short-wavelength laser technology, as well as in nonlinear and integrated optics [1,2]. Other applications include using ADP in electrochromic devices and long-term retardants [3-5]. Furthermore, it is known that ADP belongs to the KDP-family compounds, with general formula MH2PO4 (M=Na, K, Rb, Cs, NH4). These acid salts are dielectric crystals with hydrogen bonds. Under certain conditions, hydrogen ions can diffuse or hop by various mechanisms over the entire structure of these materials, due to excess proton positions [6]. Those compounds have been suggested for fuel cell applications because, upon heating, a superionic phase transition is observed [7-9]. However, the exact nature of the transformation leading to the "superprotonic phase" is still not clear and has been debated by many researchers [10-17].

The high-temperature behaviour of ADP is not free of controversy. Even though the conductivity increase around Tt has been observed by several authors [6, 18-23], some relate it to a structural phase transition [19-21], while others attribute it to a decomposition process [6, 18, 22-24]. Moreover, those who propose a chemical transformation disagree amongst themselves. For example, while Torijano *et. al.* [22] and Lim and Lee [24] propose the chemical decomposition into diammonium dihydrogen diphosphate, $(NH_4)2H_2P_2O_7$, and water (equation (1)), Abdel-Kader *et. al.* [18] and Cheng *et. al.* [23] suggest decomposition into ammonia, NH_3, and phosphoric acid, H_3PO_4 (equation (2)).

IOP Conf. Series: Journal of Physics: Conf. Series **935** (2017) 012050 doi:10.1088/1742-6596/935/1/012050

$$2NH_4H_2PO_4(s) \rightarrow (NH_4)2H_2P_2O_7(s) + H_2O(v) \quad (1)$$

$$NH_4H_2PO_4(s) \rightarrow NH_3(v) + H_3PO_4(l) \quad (2)$$

Here, s, v, and l denote solid, vapour and liquid state, respectively. Currently, the true nature of the transformation leading to the ADP conductivity increase, around Tt, remains unclear. This work seeks to shed light on the high-temperature behaviour of ADP and its relation with this conductivity jump at around this temperature.

2. Experimental details

Ammonium dihydrogen phosphate pure crystals (99.99%) were purchased from Sigma Aldrich. Differential scanning calorimetry measurements were carried out by using a TA Instruments, DSC 2920. The sample was heated from room temperature to 430°C, at a heating rate of 1°Cmin^{-1}. Ultra-pure argon was used as purge gas (50mLmin^{-1} flux). Temperature, baseline, and heat flow were carefully calibrated. A simultaneous TG-MS experiment was made by coupling a QMS BALZER ThermoStar™ equipment to a TA Instruments TGA 2920 apparatus. The purge gas (90mLmin^{-1} flux) and the heating rate were the same as used on the DSC experiment. The DSC and TG-MS experiments were controlled (and their respective data recorded) by using Universal Analysis and Quadstar software. Even though the TG signal was recorded up to 600°C, the MS data was limited to our temperature range of interest. All the MS parameters as well as the TG mass and temperature were also carefully calibrated. Platinum pans were used for all experiments.

3. Results and discussions

Figure 1 shows the thermal analysis results for two ADP samples. Upon heating, a DSC endothermic anomaly is registered with onset and peak temperatures of 153.8 and 188.9°C, respectively. This result is consistent with previous reports [18-20], where a sharp conductivity increase at 157°C is observed [21]. However, our DSC peak temperature does not coincide with that reported by Cheng *et. al.* [23] at 200°C. It should be taken into account that the heating rate used by these authors was higher (2.5°Cmin^{-1}) than that used in our study. This difference between the DSC peak temperatures of these measurements might be understood if the nature of the transformation is assumed as chemical (instead of physical). It is known that any chemical thermal decomposition is a time-dependent phenomenon. That is, a higher heating rate means higher onset decomposition temperature. Moreover, notice that before and after the DSC endothermic anomaly a sequence of very small anomalies is detected. Furthermore, the DSC anomaly does not correspond to a single well-defined peak, as typical for a physical transformation. Therefore, even though some researchers [19-21] attribute the conductivity increase to a structural phase transition, we support the hypothesis that when ADP is heated through T_t, a chemical transformation takes place.

To better understand this transformation, a simultaneous TG-MS experiment was carried out to evaluate the proposed hypothesis. To our knowledge, this is the first time that a simultaneous TG-MS experiment on ADP is reported. Figure 1 shows its respective TG-DTG results. Even though from the TG curve only two major weight loss events are observed, the corresponding DTG curve shows three peaks. The DTG onset and peak temperatures of the first peak are 153.9 and 188.1°C, respectively. The two additional DTG peak temperatures are 270.8 and 520.5°C, correspondingly. The TG thermogram also shows that weight is practically zero at 530°C, indicating that the sample totally decomposes into gaseous products at this temperature.

Thus, considering the coincidence between the DSC (153.8 and 188.9°C) and DTG (153.9 and 188.1°C) onset and peak temperatures of the first thermal event, we believe that the endothermic anomaly that starts around T_t=153°C is an exclusive consequence of a chemical thermal decomposition process, rather than a structural phase transition.

IOP Conf. Series: Journal of Physics: Conf. Series **935** (2017) 012050 doi:10.1088/1742-6596/935/1/012050

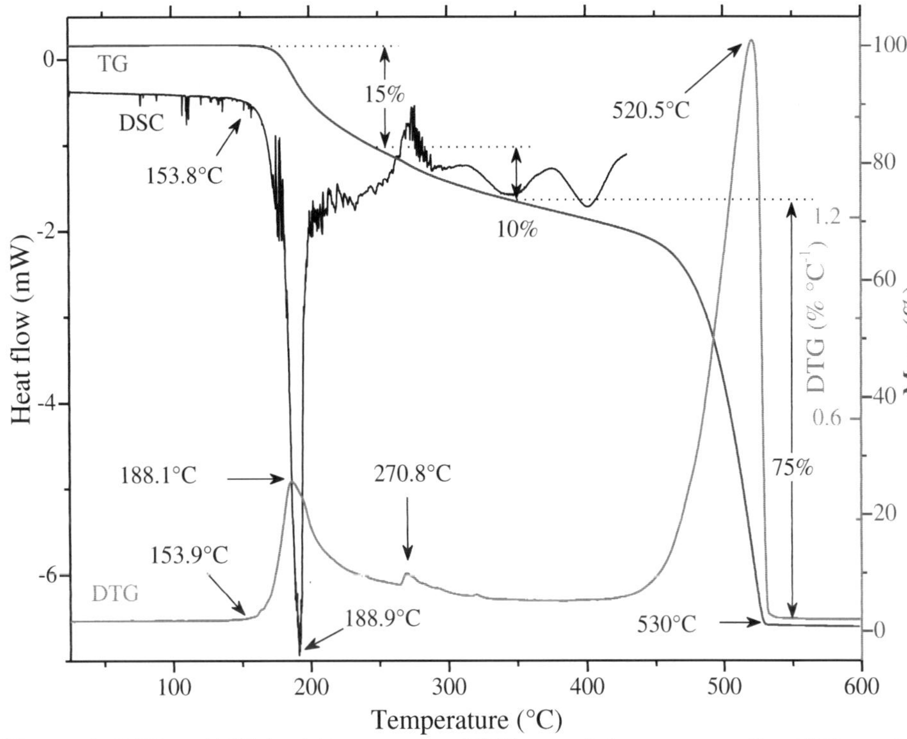

Figure 1. DSC and TG (and its respective DTG curve) thermograms for ADP crystals. The high DSC temperature limit was fixed at 430°C because our main interest is the thermal anomaly associated to the conductivity increase. The heating rate was fixed at 1°C min^{-1}.

To examine which of the proposed thermal decomposition equations takes place, we need to know which molecule evolves when the weight loss occurs. According to the TG thermogram in Figure 1, the first DTG peak offset temperature is 250°C. Here, the weight loss corresponds to 15%. Consistent with the stoichiometry of equation (1), the total water evolution requires a mass loss of 7.8%, which corresponds to almost half of the mass loss measured in our experiment. On the other hand, for the respective stoichiometry of equation (2), the total ammonia evolution requires a mass loss of 14.8%. This value matches better with our TG measurement than that expected by equation (1). Therefore, the TG result suggests that ammonia (instead of water) evolves from the sample, by following equation (2). To confirm this last statement, our TG-MS results (for the 100-250°C temperature range) are shown in Figure 2. For comparison purposes and Figure clarity, only the DTG curve is shown together with the MS results. As seen from the m/z=14 (N^+), 15 (NH^+), 16 (NH_2^+), and 17 (NH_3^+) ionic current signals, peaks were detected at 188.4, 188.4, 188.6, and 188.7°C, respectively. Although the m/z=15 signal seems to have no peak, upon zooming (as shown in the insert) a well-defined peak is observed. From this Figure, it is evident the coincidence among all the MS peak temperatures. Also, those peaks coincide with the DTG peak temperature. Besides, the relative intensity of the MS peaks matches correctly with the mass spectrum of ammonia. Therefore, we conclude that when ADP is heated through T_i=153°C, a decomposition process starts, reaching the maximum decomposition rate at 188.1°C, according to equation (2).

Although Harris and Vela [6] and Cheng *et. al.* [23] also observed ammonia evolution when ADP is heated through 157°C, they did not use the TG-MS technique, which offers accurate results. In addition, through optical observation on the DSC cell, we found that at temperatures above 160°C, a liquid film coats the sample, as reported by Cheng *et. al.* [23]. If this transformation were a structural phase transition, there is no reason to see this liquid film. Moreover, it is known that if a sample undergoes a solid-solid phase transition, its mass must remain constant, but this is not the present case. Considering that phosphoric acid exhibits high conductivity values, around 10^{-3}S·cm^{-1} [25], we believe that the ADP conductivity jump observed at 157°C [21] is exclusively related to the phosphoric acid production.

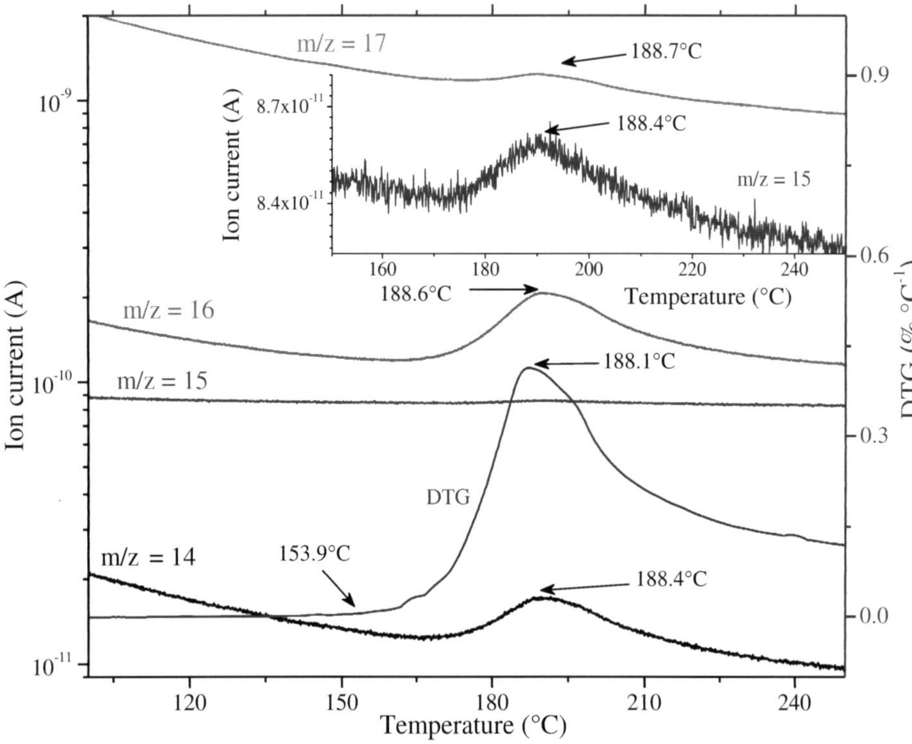

Figure 2. Simultaneous TG-MS (specifically DTG-MS) thermogram of an ADP crystal, at the temperature range 100-250°C. The insert magnifies the m/z=15 signal. The heating rate was fixed at 1°C min^{-1}.

It has been proposed that all the KDP-family compounds follow a similar decomposition kinetic process [13, 24]. According to this statement, upon heating, these acid salts decompose into water and a solid polymeric material, equation (3):

$$nMH_2RO_4(s) \rightarrow MnH_2RnO_3n + 1(s) + (n-1)H_2O(v), n \geq 1 \quad (3)$$

where M=Na, K, Rb, Cs, NH_4; *s, v* mean solid and vapour states; and n stands for the number of reactant molecules. However, according to our results, ADP is the KDP-family compound exception because instead of water and a polymeric product, ammonia and a liquid acid are the decomposition products. Consequently, the charge carrier in ADP does not correspond to the H_3O^+ ion, as proposed by Ortiz *et. al.* [12,17] for the KH_2PO_4 and CsH_2PO_4 cases. Rather, it is explained in terms of the dissociation of phosphoric acid [25]. Despite this difference, upon heating, all the KDP-family compounds undergo a chemical transformation at the same temperature in which a physical transformation has been reported. For instance, Ortiz *et. al.* [12,17] presented evidence that the tetragonal→monoclinic structural phase transition in KH_2PO_4, and the low proton conduction→superprotonic conducting phase transition in CsH_2PO_4 observed at 180 and 230°C, respectively, do not exist; instead, they propose that the thermal and conductivity behaviours observed are exclusive consequence of a chemical transformation.

Returning to the TG thermogram in Figure 1, the second and third DTG peaks register offset temperatures of 350 and 530°C, respectively. Their corresponding partial weight losses are 10% and 75%. According to stoichiometric calculations, we believe that these weight loss events are related to the chemical decomposition of phosphoric acid into pyrophosphoric acid and water; and then pyrophosphoric acid into methaphosphoric acid and water, respectively, as suggested by Abdel-Kader *et. al.* [18].

4. Conclusions

The ADP high-temperature behaviour was carefully studied by means of DSC and simultaneous TG-MS. Upon heating, an endothermic anomaly was found with onset and peak temperatures of 153.8 and 188.9°C, respectively. This thermal event is related to weight loss. Certainly, not only the DTG onset (153.9°C) and peak (188.1°C) temperatures match the respective DSC onset and peak temperatures, but also the DTG and DSC peak temperatures fit well with the 14 (N^+), 15 (NH^+), 16 (NH_2^+), and 17 (NH_3^+) ionic current signal peak temperatures at 188.4, 188.4, 188.6, and 188.7°C, respectively. These peaks, which fit the ammonia mass spectrum, indicate that at Tt=153°C ADP decomposes into phosphoric acid and ammonia, equation (2). We conclude that the conductivity increase observed is an exclusive consequence of the presence of phosphoric acid in the sample and not due to a structural phase transition [19-21] or to a chemical transformation leading to a polymeric material and water [22,24], as proposed.

Acknowledgements

This work was supported by the Research Department of Universidad del Atlántico (research project No. CB40-FGI2016, Grant No. 000721-10-05-2017).

References

[1] Rashkovich L 1991 *KDP-Family single crystals* (Bristol: Adam Hilger)
[2] De Vries S A *et al.* 1999 *J. Cryst. Growth* **205** 202–214
[3] Pu H and Huang P 2006 *Mater. Lett.* **60** 1724-1727
[4] Tayabee R and Ghadamgahi M 2017 *Arab. J. Chem.* **10** S757-S5764
[5] Liodakis S, Bakirtzis D and Dimitrikopoulos A P 2003 *Thermochim. Acta* **399** 31-42
[6] Harris L B and Vella G J 1971 *J. Chem. Phys.* **58(10)** 4550-4557
[7] Baranov A I, Khiznichenko V P, Sandler V A and Shuvalov L A 1988 *Ferroelec.* **81** 183-186
[8] Baranov A I, Khiznichenko V P and Shuvalov L A 1989 *Ferroelectrics* **100** 135-141
[9] Chandra S and Kumar A 1990 *Solid State Ionics* **40-41** 863-868
[10] Goñi-Urtiaga A, Presvytes D and Scott K 2012 *Int. J. Hydrog. Energy* **37** 3358–3572
[11] Bronowska W and Pietraszko A 1990 *Solid State Commun.* **76** 293-298
[12] Ortiz E, Piñeres I and Leon C 2016 *J. Therm. Anal. Calorim.* **126(2)** 407-419
[13] Lee K S 1996 *J. Phys. Chem. Solids* **57(3)** 333-342
[14] Haile S M, Chisholm C, Sasaki K, Boysen D A and Uda T 2007 *Faraday Discuss.* **134** 17-39
[15] Haile S M, Boysen D A, Chisholm C and Merle R 2001 *Nature* **410** 910-913
[16] Botez C, Tacket R, Hermosillo J, Zhang J, Zhao Y and Wang L 2012 *Solid State Commun.* **213** 58-62
[17] Piñeres I, Ortiz E, De la Hoz C, Trochez J and León C 2016 *Ionics* **23(5)** 1187-1195
[18] Abdel-Kader A, Ammar A A and Saleh S I 1991 *Thermochim. Acta* **176** 293-304
[19] Subhadra V K, Syamaprasad U and Vallabhan G P 1983 *J. Appl. Phys.* **54** 2593-2596
[20] Sharon M and Kalia A K 1977 *J. Solid State Chem.* **21** 171-183
[21] Viswanath R S and Miller P J 1979 *Solid State Commun.* **32** 703-706
[22] Torijano E, Vargas R, Diosa J and Mellander B E 2000 *Phys. Stat. Sol.* **220** 659-662
[23] Cheng R H, Yen C and Shern C S 2005 *J. Appl. Phys.* **98** 044104
[24] Lim A R and Lee K S 2013 *Solid State Sciences* **21** 54-58
[25] Whitehead G 1961 *J. Appl. Chem.* **11** 136-410

IOP Conf. Series: Journal of Physics: Conf. Series **935** (2017) 012051 doi:10.1088/1742-6596/935/1/012051

Evaluation of corrosion products formed by sulfidation as inhibitors of the naphthenic corrosion of AISI-316 steel

J A Sanabria-Cala[1], N D Montañez[1], D Laverde Cataño[1], D Y Peña Ballesteros[1] and C A Mejía[1]

[1] Universidad Industrial de Santander, Bucaramanga, Colombia

E-mail: sanabriacalaj@gmail.com

Abstract. Naphthenic acids present in oil from most regions worldwide currently stand as the main responsible for the naphthenic corrosion problems, affecting the oil-refining industry. The phenomenon of sulfidation, accompanying corrosion processes brought about by naphthenic acids in high-temperature refining plant applications, takes place when the combination of sulfidic acid (H_2S) with Fe forms layers of iron sulphide (FeS) on the material surface, layers with the potential to protect the material from attack by other corrosive species like naphthenic acids. This work assessed corrosion products formed by sulfidation as inhibitors of naphthenic corrosion rate in AISI-316 steel exposed to processing conditions of simulated crude oil in a dynamic autoclave. Calculation of the sulfidation and naphthenic corrosion rates were determined by gravimetry. The surfaces of the AISI-316 gravimetric coupons exposed to acid systems; were characterized morphologically by X-Ray Diffraction (XRD) and X-ray Fluorescence by Energy Dispersive Spectroscopy (EDS) combined with Scanning Electron Microscopy (SEM). One of the results obtained was the determination of an inhibiting effect of corrosion products at 250 and 300°C, where lower corrosion rate levels were detected. For the temperature of 350°C, naphthenic corrosion rates increased due to deposition of naphthenic acids on the areas where corrosion products formed by sulfidation have lower homogeneity and stability on the surface, thus accelerating the destruction of AISI-316 steel. The above provides an initial contribution to oil industry in search of new alternatives to corrosion control by the attack of naphthenic acids, from the formation of FeS layers on exposed materials in the processing of heavy crude oils with high sulphur content.

1. Introduction

Currently the world reserves of light crude oil have decreased significantly because their low concentrations of sulphur and naphthenic acids facilitate their refining, transport and storage, positioning them as the first choice of oil industry to supply the growing energy demand in national and international market. At this time, new alternatives are being explored for the processing of heavy crude oil that will position them as a strategic energy resource, which contribute to generate economic growth for the country and to sustain the current demand of fuels and other petroleum products in Colombia [1-3]. Heavy crude oils possess physicochemical characteristics such as: 1) high viscosity, 2) low API gravity and 3) high sulphur content, heavy metals and naphthenic acids. These characteristics condition the refining processes due to the great deterioration caused on the materials exposed to naphthenic corrosion and sulfidation, whence the processing of heavy crude oil is currently becoming an industrial challenge for the petrochemical sector [4].

The damages caused by naphthenic corrosion and sulfidation in the alloys of primary distillation units and their transfer lines is aggravated by the high pressure and temperature conditions that they are constantly subjected. In the refining of heavy crude oil with high sulphur content >2%w/w, it becomes a highly corrosive agent for materials exposed to temperatures >200°C [5]. Under these conditions, the thermal breakdown of sulphur generates the formation of H_2S, which is directly responsible for the sulfidation phenomenon. According to several investigations, the sulfidation of alloys leads to the formation of FeS as corrosion products on the material surface, among which is possible to find: 1) pyrite, 2) troilite and 3) mackinawite [5-6]. These products can reduce the corrosion rate caused by the attack of other corrosive species in the system, such as naphthenic acids, because they are formed on the surface and act as a physical barrier protecting the material [7]. Likewise, minimal changes in temperature and exposure time of petrochemical processes can lead to different results in the morphology of the formed FeS layers, and consequently increase the corrosion rate due to the formation of a galvanic couple on the exposed steel surface [7-8].

The formation of homogeneous and thermodynamically stable FeS layers depends on variables such as: 1) sulphur concentration in the heavy crude oil, 2) temperature, 3) fluid velocity and 4) exposure time of the material [2]. Therefore, in this study, the influences of temperature and exposure time on the formation of stable corrosion products from the sulfidation of AISI-316 steel were determined, exposed to the processing of a heavy crude oil with high sulphur content. Moreover, the anti-corrosive capacity of formed corrosion products was evaluated as a possible inhibitor of the naphthenic corrosion rate of AISI-316 steel, exposed to refining conditions of synthetic crude with a total acid number (TAN) equivalent to 1mg KOH/g of crude oil. The gravimetric and microscopic results determined an inhibitory effect of the AISI-316 steel naphthenic corrosion rate, at temperatures between 250 and 300°C and exposure times of 144 and 192 hours, due to an iron sulphide layer formed as a corrosion product, whereby the naphthenic corrosion rates of the material presented a decreasing behaviour at these conditions in the system [9]. The above provides an initial contribution in the constant search of new alternatives directed towards the control of corrosion by the attack from naphthenic acids in oil industry, using a pollutant such as sulphur, present in high concentrations in the Colombian heavy crude oil.

2. Experimental

The coupons used for gravimetric tests were machined from AISI-316 steel, obtaining a rectangular shape with dimensions of: 37mm long, 12mm high and 2mm wide. The surface of the gravimetric coupons was prepared metallographically using sandpaper, initially of silicon carbide No. 400 until reaching a cloth of 0.05μm. To carry out the gravimetric tests, a type Parr model 4848 batch dynamic autoclave was used. All tests were performed in triplicate at a single stirring speed of 150rpm. The corrosion rate, in units of mpy (mils per year), was calculated according to the guidelines of ASTM Standard G1-03 (Standard practice for preparing, cleaning and evaluating corrosion test specimens) [10]. For the formation of corrosion products by the sulfidation of AISI-316 steel, tests are carried out at temperatures of 250, 300 and 350°C, and exposure times of 96, 144 and 192 hours.

During each test, the corresponding gravimetric coupons are placed in the reactor, and immersed in 1500mL of heavy crude oil with a sulphur weight percent of 2.5% w/w; API gravity of 12.2 and a density of 0.9848g/mL. A constant flow of high purity analytical nitrogen is supplied to the autoclave for one hour prior to the gravimetric tests in order to reduce the oxygen content dissolved in the crude. Once the sulfidation corrosion products are formed on the AISI-316 steel coupons surface, the inhibitory potential of naphthenic corrosion is evaluated by immersing the coupons with corrosion products in a solution of mineral oil and naphthenic acid with a TAN equivalent to 1 mg KOH/g crude oil; temperatures of 250, 300 and 350°C, and exposure times of 48, 96, 144 and 192 hours. After the tests are finished, the gravimetric coupons are cleaned for 1 hour with acetone in an ultrasonic bath and oven-dried at a temperature of 30°C for 5 minutes. Then, the coupons were weighed in an analytical balance with an accuracy of 0.0001g, labeled and stored in desiccator for gravimetric and superficial analysis. The corrosion products formed were characterized by: Scanning Electron Microscopy (SEM), in a device model Quanta Feg 650 equipped with System of Energy Dispersive of X-ray Spectroscopy (EDS); and

X-Day Diffraction (XRD), using a brand BRUKER model D8 DISCOVER diffractometer with DaVinci Geometry in beam mode.

3. Results and discussion

Preliminary studies have determined that at temperatures between 250 and 350°C and exposure times less than 96 hours, the H_2S present in the system exhibits a corrosive behaviour towards exposed AISI-316 steel. In addition, the SEM-EDS characterization did not evidence the formation of FeS layers on the material surface because the formation of FeS layers is governed by diffusion phenomena in solid state that depend on long exposure times, making each system unique [8]. Table 1 presents the corrosion rates by sulfidation of AISI-316 steel, where it is possible to observe that the corrosion rate increases as a function of temperature and exposure times. This behaviour is attributed to thermal breakdown of sulphur at high temperatures in hydrogen sulphide, which generates a difference of pressure in the system and favours its chemical reaction on the material surface, resulting in the formation by chemisorption of FeS-type corrosion products on the AISI-316 steel surface.

Table 1. Corrosion rates by sulfidation of AISI-316 steel (mpy).

Temperature (°C)	Corrosion Rate ± Standard Deviation (mpy)		
	96(h)	144(h)	192(h)
250	0.0381±0.0002	0.0508±0.0003	0.0762±0.0003
300	0.0762±0.0003	0.0762±0.0002	0.0953±0.0002
350	0.1524±0.0001	0.1778±0.0002	0.2096±0.0003

The morphological analysis of the corrosion products formed on the AISI-316 steel surface was carried out using SEM. The selection of the coupons was made based on the results obtained from the gravimetric analysis; which indicated that at operating conditions of 350°C and an exposure time of 192 hours, the highest corrosion rate by sulfidation of AISI-316 steel was obtained. In the micrograph of Figure 1(A), the surface of the AISI-316 steel before being subjected to the sulfidation phenomenon is detailed. Furthermore, in the micrograph of Figure 1(B), is observed that the FeS grains formed on the surface of AISI-316 steel have a compact irregular crystal structure; finally, a homogeneous and stable layer whose grain size varies in a range from 396nm to 566nm is shown in the micrograph of Figure 1(C). On the other hand, the diffractograms shown in Figure 2 evince that the crystalline phase, present with greater intensity on the AISI-316 steel surface, exposed for 192 hours to a sulfidating atmosphere at 350°C of temperature, corresponds to an iron sulphide layer.

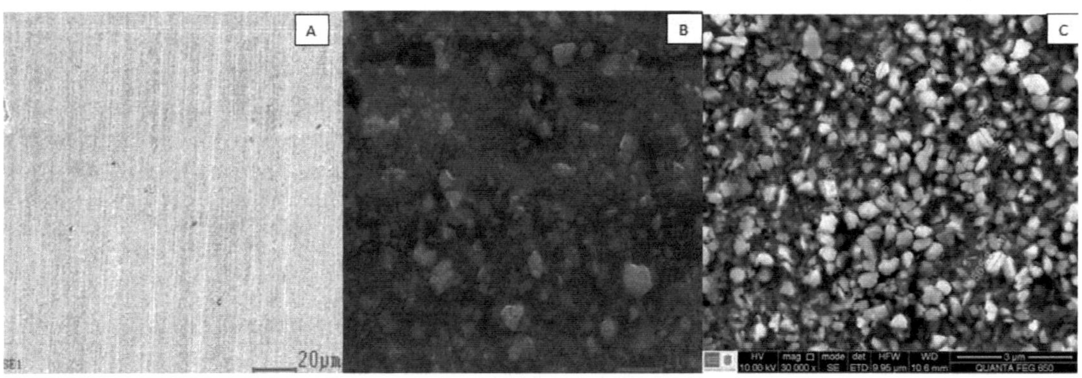

Figure 1. SEM images of AISI-316 coupon surface at 192h and 350°C.

Figure 3 relates the naphthenic corrosion rates of AISI-316 steel exposed to the processing of a crude simulated with naphthenic acid of TAN=1 at temperatures of 250, 300 and 350°C, and exposure times of 48, 96, 144 and 192 hours. Figure 3(A) shows that the naphthenic corrosion rate of AISI-316 steel behaves in a linearly increasing manner due to the increase of all exposure times and temperatures in the system for coupons that do not have corrosion products formed by sulfidation on their surface. The highest deterioration of AISI-316 steel occurs at a temperature of 350°C and exposure time of 192 hours, reaching values of naphthenic corrosion rate of 1.7mpy. This is because a higher temperature increases the reaction kinetics of naphthenic acids with the steel surface, resulting in higher corrosion rates in the material [7-9]. On the other hand, zone 1 of Figure 3(B) shows that the naphthenic corrosion rates of AISI-316 steel tend to stabilize at exposure times between 96 and 192 hours for temperatures of 250 and 300°C, which is due to the inhibitory effect generated by the iron sulphide layer formed on the material surface. On the contrary, zone 2 of Figure 3(B) shows that the naphthenic corrosion rates had a greater increase, and therefore, generated a great deterioration in the material compared to the corrosion rates of the coupons without iron sulphide on their surface, at a temperature of 350°C and times between 48 and 144 hours. This occurs due to the deposition of corrosive species, such as naphthenic acids, over areas where the iron sulphide layer is less stable, which causes their destruction and consequently an increase in the naphthenic localizes corrosion rate of the steel [6]. According to the obtained results, the iron sulphide layer formed by sulfidation on AISI-316 steel surface has influence on the naphthenic corrosion rates of the material, achieving values up to 20% lower than operating temperatures between 250 and 300°C and exposure times between 144 and 192 hours [8].

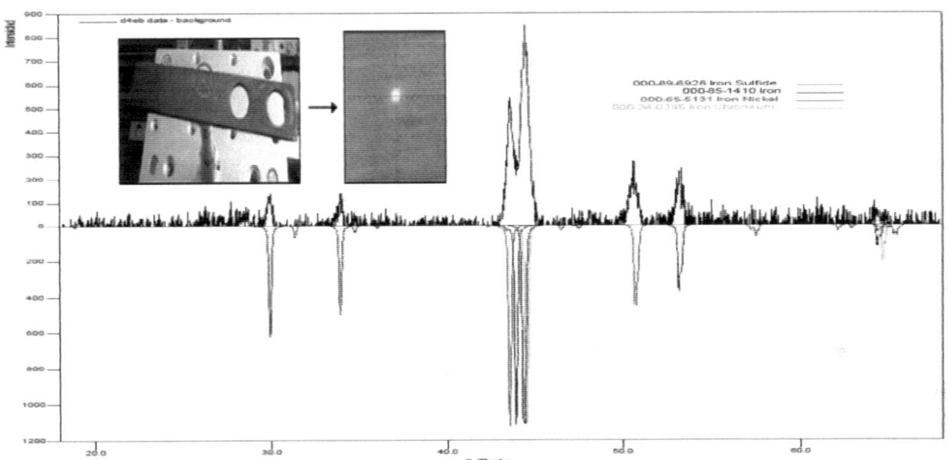

Figure 2. XRD of coupon surface at 192h and 350°C.

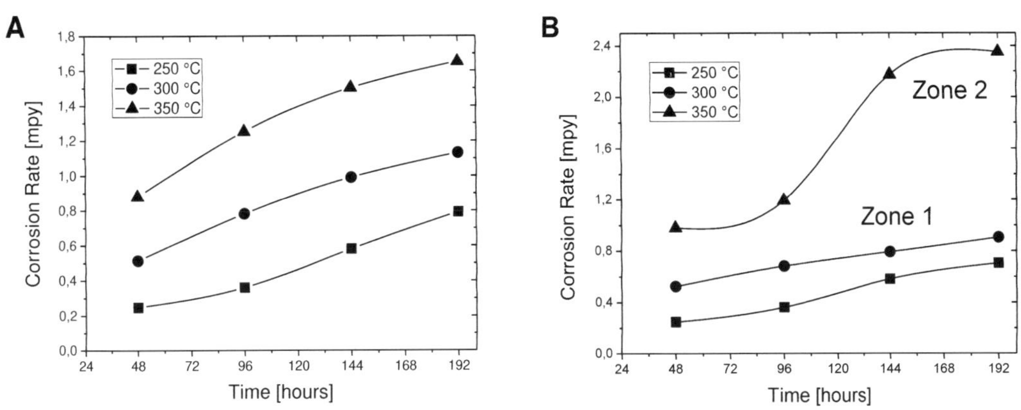

Figure 3. Corrosion rate of AISI-316 steel (A) Coupons without FeS and (B) With FeS.

The EDS analysis of the micrograph of Figure 4(A) is related in Table 2. There is a decrease from 9.97% to 6.77% in the sulphur content of the iron sulphide layer formed on AISI-316 steel surface after being exposed to naphthenic corrosion at a temperature of 300 C and an exposure time of 192 hours, conditions which the iron sulphide layer protects the material by reducing its naphthenic corrosion rate. Therefore, the EDS analysis of the micrograph of Figure 4(B), presented in Table 2, indicates that at exposure times of 192 hours and a temperature of 350°C, the sulphur content on the AISI-316 surface decreases from 9.97% to 4.78%, evidencing an aggressive attack to the iron sulphide layer by naphthenic acids. This damage in the iron sulphide layer favours the generation of a galvanic couple on the material surface, which can accelerate the deterioration of AISI-316 steel.

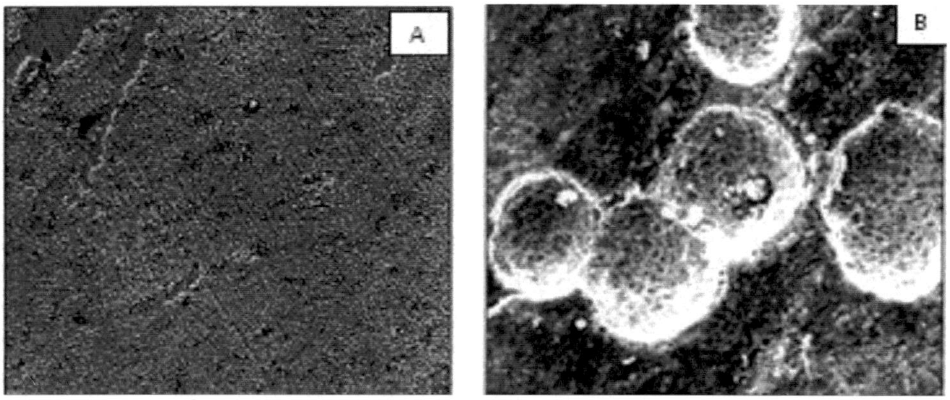

Figure 4. SEM of AISI-316 steel coupon surface.

Table 2. EDS of FeS layer exposed 192 hours to a simulated crude with TAN=1

Element	Wt% (300°C)	Wt% (350°C)
C	2.75	3.01
Si	0.76	0.72
Mo	2.33	2.13
S	6.77	4.78
Cr	15.70	16.05
Ni	8.52	8.22
Fe	55.24	49.24

4. Conclusions

The developed gravimetric tests determined the great influence of exposure time and temperature as fundamental variables in the formation of a stable and homogeneous corrosion product of iron sulphide on the AISI-316 steel surface, at the refining conditions of a heavy crude oil with high sulphur content. A reduction of the naphthenic corrosion rate of AISI-316 steel, exposed to a simulated crude with a TAN=1, was determined at temperatures between 250 and 300°C, and exposure times between 144 and 192 hours. This behaviour is due to the anti-corrosion effect generated by the formation of a homogeneous and compact layer of iron sulphide, as corrosion product of the steel sulfidation. Morphological analysis using surface-characterization techniques such as SEM-EDS showed the damage caused in the iron sulphide layer by the attack of naphthenic acids at a temperature of 350°C and an exposure time of 192 hours. The characterization by XRD allows determining the formation of an iron sulphide layer as a corrosion product, which acted as a physical barrier between the metal and the naphthenic acids, generating a protection towards the material; hence, it is possible to increase the useful life of the materials exposed to the processing of crude with high acidity content.

Acknowledgments
The authors want to thank "Universidad Industrial de Santander, Vicerrectoría de Investigación y Extensión" and "Grupo de Investigaciones en Corrosión"; for research funding of this project under internal code 1857.

References

[1] Miranda-Mejia L, Laverde D and Molina D 2015 *Energy & Fueks* **29** 7595-7600
[2] Sanabria-Cala J, Laverde D, Vásquez C, Blanco C and Quiroga H 2014 *Revista ION* **27** 35-41
[3] Sanabria-Cala J, Laverde D, Peña Ballesteros D and Merchan Arenas D 2017 *Chemical Engineering Transactions* **57**
[4] Alizadeh M and Bordbar S 2013 *Corrosion Science* **70** 170–179
[5] Zheng Y, Ning J, Brown B, Young D and Nešić 2015 Mechanistic study of the effect of iron sulfide layers on hydrogen sulfide corrosion of carbon steel *CORROSION 2015 (Salt Lake City)* (Houston: NACE International) 5933
[6] Bai P., Zheng S, Zhao H, Ding Y, Wu J and Chen C 2014 *Corrosion Science* **87** 397–406
[7] Jun L, Yuichi K, Takayuki N, Yasuchika H and Fushimi K 2016 *Electrochim Act* **220** 304-311
[8] Sun H, Fang H and Hudgins D 2012 Investigation of effects of iron sulfide on corrosion and inhibition of carbon steel in h2s containing conditions *CORROSION 2012 (Dallas)* (Houston: NACE International) 1651
[9] Kaneko M and Isaac H 2002 *Corrosion Science* **44** 1825-1834
[10] ASTM G1-03 2011 *Annual book of ASTM standards* vol 03.02 (USA: ASTM International)

IOP Conf. Series: Journal of Physics: Conf. Series **935** (2017) 012052 doi:10.1088/1742-6596/935/1/012052

The effect of zinc thickness on corrosion film breakdown of Colombian galvanized steel

A Sandoval-Amador[1], J E Torres Ramirez[2], P A Cabrales-Villamizar[1], D Laverde Cataño[1] and D Y Peña-Ballesteros[1]
[1] Universidad Industrial de Santander, Bucaramanga, Colombia
[2] Universidad Nacional de San Martin, Argentina

E-mail: anderson84f@outlook.com

Abstract. This work studies the corrosion behaviour of Colombian galvanized steel in solutions of chloride and sulphate ions. The effect of the thickness and exposure time on the film's breakdown susceptibility and protectiveness of the corrosion products were studied using potentiodynamic polarization curves and electrochemical impedance spectroscopy. The corrosion products were analysed using SEM-EDS and XRD. The samples with a higher thickness level in the zinc film (Z180) have the lowest corrosion rate. In this case, one of the products that was formed by the chemical reactions that occurred was Zinc hydroxide, which exhibits a passive behaviour as observed in the Pourbaix curves of the obtained potentials and in how the different Ph levels of the solutions worked. The sheets with the highest thickness (Z180) had the best performance, since at the end of the study they showed the least amount of damage on the surface of the zinc layer. This is because the thickness of the zinc layer favours the formation of simonkolleite, which is the corrosion product that protects the material under the conditions of the study.

1. Introduction

The search for steel with corrosion resistance optimized for use in the construction, the automotive and the manufacture of household appliances industry has motivated numerous research work [1-7]. In this sense, galvanized steels have played a very important role in the improvement of the corrosion resistance of steels used in these areas [8-11]. These steels with zinc-based coatings act as a barrier layer and as cathodic protection. The effect of the barrier layer is produced by the insulation of the steel from the environment through two different layers formed during the different stages of the corrosion process. The first barrier is the zinc coating itself, which is progressively replaced in its insulation function by the formation of a zinc corrosion products layer. Cathodic protection means that even if the steel substrate is exposed, the zinc layer will still corrode [12-14]. The zinc-based coating can be obtained by electrodeposition, thermal spraying or immersion in liquid metal baths (hot dip) at 450°C [5-16]. The most industrially popular zinc coatings in Colombia are the galvanized ones (Zn-0.2% p/p de Al) [2].

Recently, Ramirez *et. al.* [1] discovered that the size of the zinc crystal is a determinant factor in the corrosion resistance of this material; according to their results a smaller size of the zinc crystal favours the corrosion resistance of the sheet of galvanized Steel. On the other hand, Sandoval et al. [2] have found that the corrosion potential and the corrosion rates of the galvanized steel sheets increase with the immersion time and observed chopped corrosion in sheets that were studied after 30 days of immersion.

For that reason, the effect of galvanizing the thickness and the exposure time on a solution with chlorides and sulphates on the corrosion resistance of a galvanized steel in Colombia, was evaluated using potentiodynamic polarization curves and electrochemical impedance spectroscopy.

2. Methodology

In order to evaluate the breakdown of the protective layer of galvanized Steel in this study, the following variables were taken into account: time (0, 7 and 14 days of exposure), thickness of the galvanized layer (90, 120, and 180g/m^2 of Zinc) and effect of Cl^- ion (15.15g/L NaCl) y SO_4^+ (16.08g/L Na_2SO_4). The samples used in this work were named Z90, Z120 and Z180, these names were given considering the thickness of the galvanized layer. Two sample groups were cut, the first group's dimensions were 2.0cm×2.0cm. These samples were used to conduct a saline exposure test and to characterize the corrosion products formed after 14 days of immersion. The second group of samples was cut with dimensions of 6.0cm×10.0cm. These were used to carry out the electrochemical tests on the galvanized Steel. In both cases, the area of exposure of the samples was 1cm^2. For the electrochemical characterization, a flat three–electrode cell was used (see Figure 1) and the open circuit potential was recorded for 1 hour before each electrochemical test. The electrochemical impedance spectroscopy was obtained by applying 10mV AC, performing a frequency sweep of 0.01 to 100000Hz and recording 10 points per decade. Potentiodynamic polarization curves were performed within a potential range of –500mV to 2000mV, with a scanning speed of 1mV/s. Finally, the morphology of galvanized Steel sheets and corrosion products was observed by Scanning Electron Microscopy (SEM); the elemental composition of the corrosion products was identified by EDS and the crystalline phases were identified by XRD.

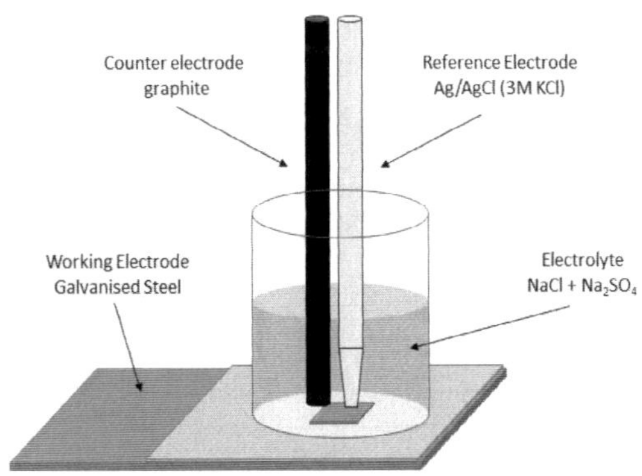

Figure 1. Electrochemical cell schematic.

3. Results and analysis

Figure 2 shows the Nyquist diagrams obtained under the study conditions and the equivalent circuit models used to make the impedance spectra adjustments. In Figure 2(a), an inductive lasso is show in sheet Z90, which is associated with the adsorption of species on the surface. These species generally result in the formation of corrosion products on the surface of the galvanized Steel and they can be the reason for the good performance shown by this sheet in the initial test. Figure 2(b) shows the effect of the sulphate and chloride ions, it can be observed that the sulphate ion causes the greatest deterioration of the galvanized sheets. This happens because this ion forms compounds that are soluble in the solution and are reabsorbed by the surface continuously. The intermediate behaviour that can be observed in sheet Z120 is due to the fact that the corrosion products created by sulphate protect the surface of the sheet, but since they are corrosion products with a porous morphology, these allow the passage of chloride ions, which cause the breakdown of the galvanized Steel sheet Z180. This is due to the thickness of the galvanization and to the formation of passivating layers on its surface.

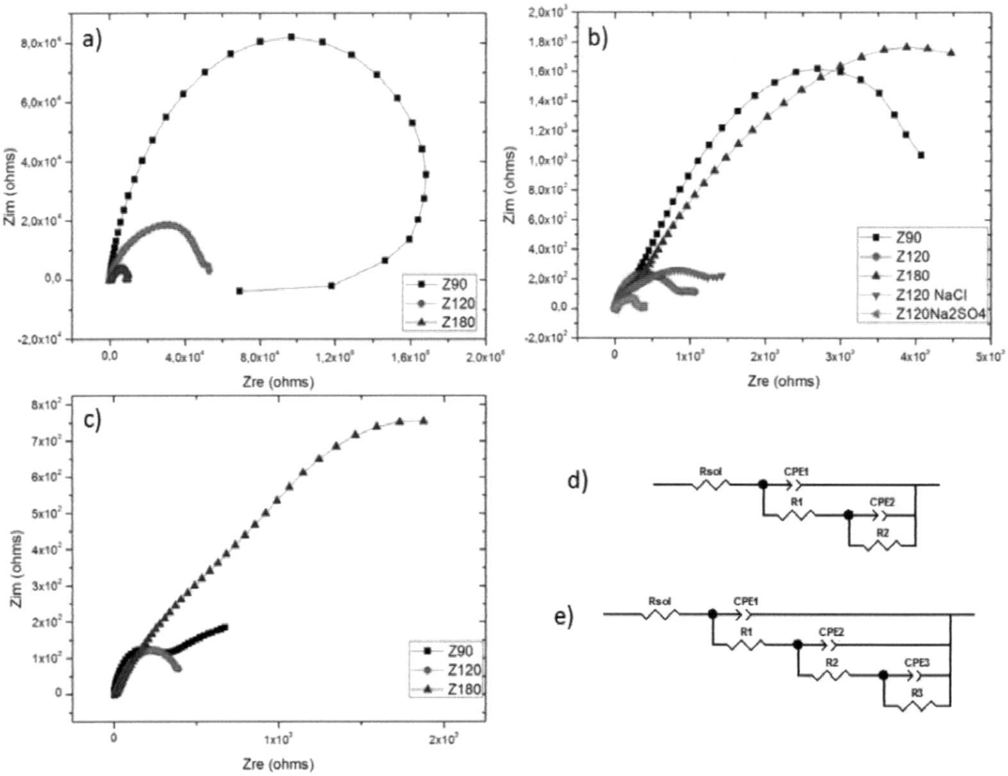

Figure 2. Nyquist diagrams of galvanized steel exposed at different times, (a) 0 days, (b) 7 days, (c) 14 days, (d) and (e) Equivalent circuits used to adjustment EIS data.

Figure 2(e) shows the equivalent circuit with which the impedance spectrum exposed sheet Z120 was adjusted for 14 days, while the circuit of Figure 2(d) was used to adjust the other spectra. Table 1 shows the parameters of the equivalent circuits. The parameter Rp refers to the polarization resistance in different sheets and this was determined by the sum of contributions made by the resistance of the galvanized layer and interface metal coating. It can be observed that the Z180 galvanized Steel sheet was the one with the highest Rp after the 15 days of immersion. This high value in the parameter Rp is associated with a good corrosion resistance.

Table 1. Polarization resistance obtained from the adjustment to equivalent circuits of the EIS data.

Time	0 days			7 days					14 days		
Sample	Z90	Z120	Z180	Z90	Z120	Z120 NaCl	Z 120 Na_2SO_4	Z180	Z90	Z120	Z180
Rp ($k\Omega/cm^2$)	205	52.7		5.62	1.01	1.81	0.35	7.4	1.47	0.47	3.54

Figure 3 shows a very low slope at the beginning of the anode zone, which is associated with a rapid increase in current. This suggests that the anode processes are present in the cathodic zone since the reactions are associated to processes controlled by concentration. This happens because zinc tends to form corrosion products spontaneously on its surface under the study conditions. On the other hand, it can be seen that the Cl^- and SO_4^+ ions acting separately can generate shifts in the corrosion potential towards more positive values. In addition, the presence of the sulfate ion is able to decrease the density

IOP Conf. Series: Journal of Physics: Conf. Series **935** (2017) 012052 doi:10.1088/1742-6596/935/1/012052

of the passivation current, which is attributed to the formation of corrosion products that protect the sheet of galvanized steel.

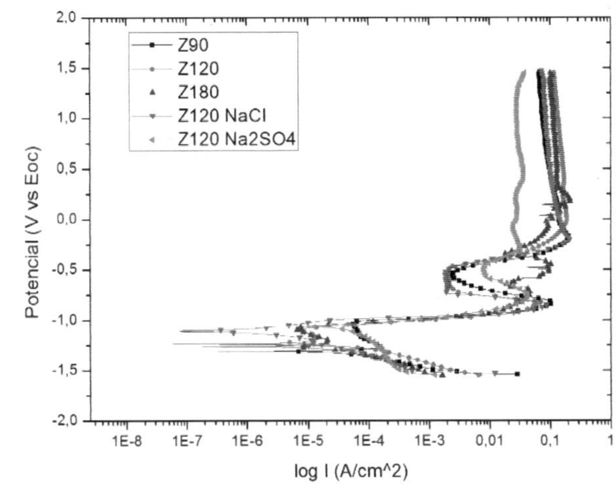

Figure 3. Potentiodynamic polarization of galvanized steel at 7 days.

In Figure 4, it can be observed that after 15 days of exposure, corrosion with red colour products formed on the Z90 sheet and white corrosion, which shows that a total breakdown of the galvanizing layer has been achieved, was also formed. The white corrosion is attributed to the formation of oxide compounds and the precipitation of salt that protects the material from corrosion. Additionally, it can be observed that as the exposure time increases, the corrosion rate is higher except for the sheet Z180 at 14 days, where there is a decrease in the corrosion rate attributed to the formation of simonkolleite as shown in EDS and DRX results.

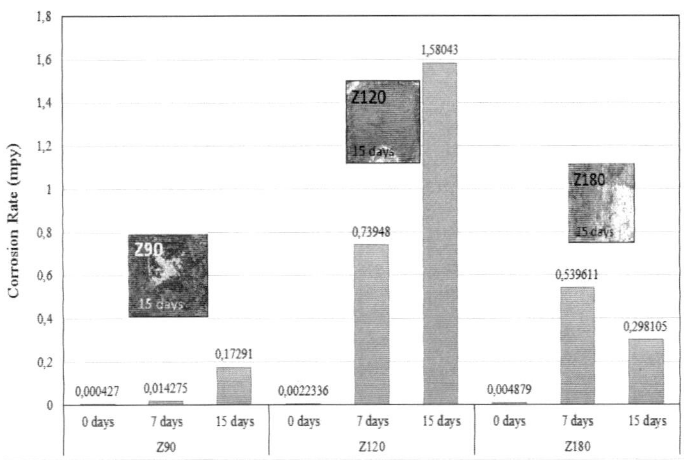

Figure 4. Corrosion rates of galvanized steel sheets at different exposure times.

The elemental composition maps of Figure 5 show that in the Z90 sheet there are dark areas in the zinc map that corroborate that the zinc layer has been destroyed. Images corresponding to sheet Z120 show the formation of a salt agglomerate which can be observed on the surface, where zinc is displaced by oxygen, sulphur, chlorine and sodium. This indicates that the zinc layer has been destroyed and corrosion products have been formed. On the other hand, it is possible to observe the formation of a homogeneous layer of salt on the surface, which is a characteristic of white corrosion. It is also possible to observe that the distribution of zinc on the surface is uniform and that the base material has not been attacked yet. Through the XRD analysis, it was found that corrosion products formed on the galvanized Steel sheet Z90 (Gordaite, Zinc and sodium hydroxide) and Z120 (Gordaite, Zincite, Sodium Chlorate and Ammonium Sulphate) are corrosion products that demonstrate the attack and deterioration of the protective zinc layer. The products formed on sheet Z180 (simonkolleite, zincite, zinc hydrogen sulphate, halite, sodium, chlorate, sodium sulphate and gordaite) showed that sheet Z180 had the best performance against corrosion and less damage in the galvanized layer, which is mainly attributed to the thickness of the layer and to the formation of simonkolleite on the surface of the sheet.

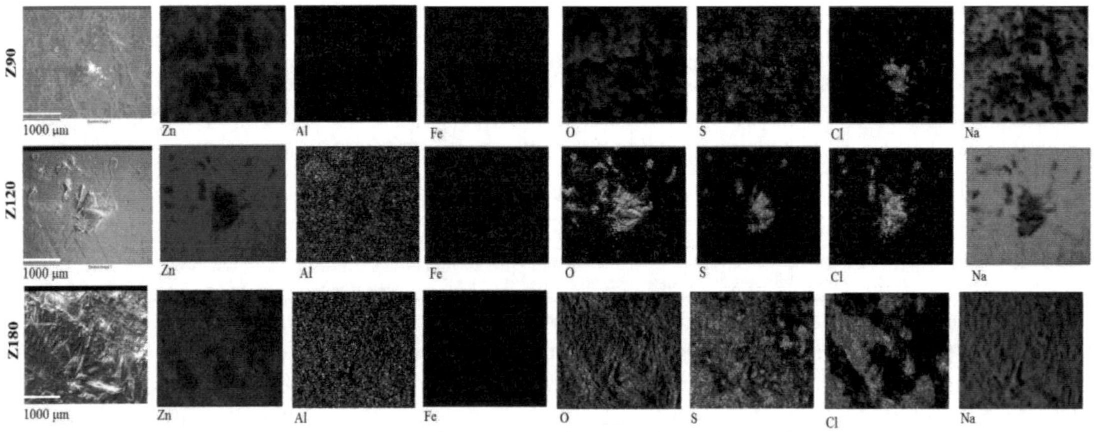

Figure 5. Chemical composition maps of exposed galvanized steel at 14 days exposure.

4. Conclusions

The exposure time of samples to chlorides and sulphates and grain size are factors that negatively affect the corrosion behaviour of galvanized Steel sheets. The thickness of the galvanization improves the corrosion properties of galvanized steel sheets, therefore, the penetration of chloride ions into the substrate will be more difficult. Although the Z90 sheet had the best performance against corrosion due to its small grain size, its coating was completely corroded with red corrosion on its surface, which is a characteristic of the iron oxides, indicating that the substrate was corroded.

References

[1] Ramirez-Jerez J A, Valero Alvarado A, Peña Ballesteros D Y and Torres Ramirez J E 2015 *Revista Facultad de Ingeniería,* **24** 99
[2] Peña Ballesteros D Y, Sandoval Amador A, Estupiñan Duran H A, Sierra Fajardo D, and Quintero Ascanio L. 2015 *Informador Técnico* **79** 127
[3] Azaroual M A, Harrak E E, Touir R, Rochdi A, and Touhami M E. 2016 *J. Mol. Liq.* **220** 549
[4] Chen Z Y, Persson D and Leygraf C 2008 *Corr. Sci.* **50** 111
[5] Jaramillo B, Calderón J A, Castaño J G. 2006 *Revista Facultad de Ingeniería.* **37** 200
[6] Padilla V, Alfantazi A 2014 *Constr. Build Mater.* **66** 447
[7] Osorio W R, Freire C M, Garcia A 2005 *J. Alloy Compd.* **397**, 179
[8] Almeida E and Morcillo M 2000 *Surf. Coat. Technol.* **124** 169
[9] Almeida E and Morcillo M 2000 *Surf. Coat. Technol.* **124** 180
[10] Almeida E, Morcillo M 2000 *Surf. Coat. Technol.* **124** 44
[11] Yasakau K, Kallip S, Lisenkov A, Ferreira m G S, Zheludkevich M L 2016 *Electrochim. Acta* **211** 126
[12] Soriano C and Alfantazi A 2016 *Constr. Build Mater.* **102** 904
[14] Amirudin A and Thierry D 1996 *Prog. Org. Coat.* **28** 59
[15] Neufeld A. K, Cole I S, Bond A M and Furman S A 2002 *Corros. Sci.* **44** 555
[16] Ploypech S, Jearanaisilawong P and Boonyongmaneerat Y 2013 *Surf. Coat. Technol.* **223** 1

IOP Conf. Series: Journal of Physics: Conf. Series **935** (2017) 012053 doi:10.1088/1742-6596/935/1/012053

Effect of the temperature and the CO_2 concentration on the behaviour of the citric acid as a scale inhibitor of $CaCO_3$

K Blanco[1], H Aponte[1] and E Vera[1]
[1] Universidad Pedagógica y Tecnológica de Colombia, Tunja Colombia

E-mail: katherineblancobuitrago@gmail.com

Abstract. For all Industrial sector is important to extend the useful life of the materials that they use in their process, the scales of $CaCO_3$ are common in situation where fluids are handled with high concentration of ions and besides this temperatures and CO_2 concentration dissolved, that scale generates large annual losses because there is a reduction in the process efficiency or corrosion damage under deposit, among other. In order to find new alternatives to this problem, the citric acid was evaluated as scale of calcium carbonate inhibition in critical condition of temperature and concentration of CO_2 dissolved. Once the results are obtained it was carried out the statistical evaluation in order to generate an equation that allow to see that behaviour, giving as result, a good efficiency of inhibition to the conditions evaluated the scales of products obtained were characterized through scanning electron microscopy.

1. Introduction

The scales generation associated with the production of formation and injection water, can lead to oilfield production problems which are often costly to remedy. The calcium carbonate ($CaCO_3$) is one of the most common scale that is present in production wells and surface installations. The formation of carbonate scale can affect the production because the pipe and the flow lines lock, equipment fouling and corrosion under deposit [1].

Scale formation from calcium carbonate deposition is a major drawback in many industrial facilities, including oil and gas industry, leading to obstruction problems then causing productivity reduction and economic loss [2]. To solve this phenomenon different alternatives have been proposed, among the scales inhibitor. [3]. One of the methods of control of scales more used in refineries and chemical plants is the use of chemicals called incrustation inhibitors [3-7].

Carboxylic Acids have been used as a chemical specie to formulate different scale inhibitors as Polyphosphino carboxylic Acid, carboxymethyl inulin [8]. In the present work, the results of evaluation of citric acid as a scale inhibitor, dosed at different concentrations (15, 25 and 40ppm), are presented under the influence of the critical variables for the formation of $CaCO_3$ crystals. In order to determine their effectiveness to specific conditions of temperature, flow rate and CO_2 concentration.

2. Experimental procedure

2.1. System used for the tests performed

The experiments carried out for the evaluation of the inhibitor of incrustation composed by acid. Were carried out in the laboratory by means of a closed circuit assembly. The designed system consists of a saturated solution of calcium carbonate that lies in a glass vessel and has agitation as well as constant

IOP Conf. Series: Journal of Physics: Conf. Series **935** (2017) 012053 doi:10.1088/1742-6596/935/1/012053

CO_2 flow is driven by means of hoses through a pump at a constant speed towards the test capillary found in a heating bath by means of which the desired temperature is achieved, the solution passes through the Capillary and is redirected to the container containing the solution. For the development of the system used, it was taken as reference the one found in the research. The kinetics of carbonate scaling-application for the prediction of downhole carbonate scaling (See Figure 1) [3].

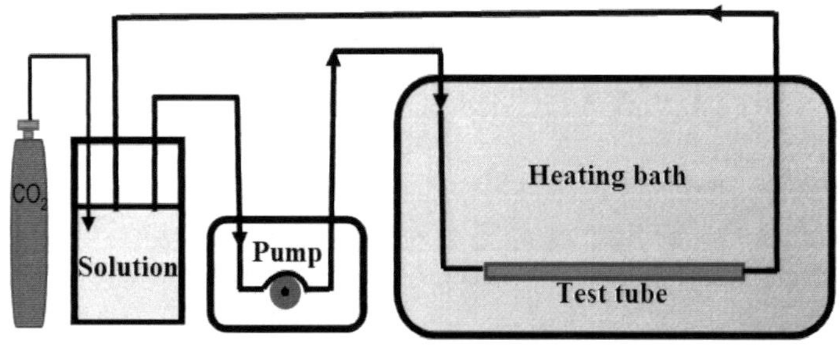

Figure 1. System used for the tests performed.

2.2. Material
The material used for the development of the tests corresponds to an AISI 316L stainless steel capillary (AISI 316L stainless steel), with an external diameter of 6.36mm and a wall thickness of 0.89mm and 150mm in length.

2.3. Conditions
The tests performed lasted 36 hours, at temperatures of 70°C and 80°C, CO_2 dosages of 1py3/h and 3py3/hr handling a flow rate in a laminar flow regime (Re=127.46), Once the solution is saturated by the gas it is determined that the pH is 6.80 this is obtained with a pH meter HANNA.

2.4. Brine
A supersaturated solution of calcium carbonate, prepared by weighing 10g of calcium carbonate and 33g of sodium chloride, was used, which were taken to a Beaker and dissolved in water. Then this solution is taken to a 1L balloon to be titrated. This solution is prepared for each test performed [9].

2.5. Calculation of inhibition efficiency
To determine the efficiency of citric acid as a calcium carbonate scale inhibitor, the weight in grams of the crystal layer deposited in the test capillary was taken into account. The calculation of the inhibition efficiency was carried out by the following equation:

$$Efficiency = \frac{W_{white} - W_{inhibitor}}{W_{white}} \quad (1)$$

Where, W_{white}=scale gained without inhibitor dosage in grams. $W_{inhibitor}$=scale gained with inhibitor dosage in grams.

3. Results and discussion

3.1. Efficiencies obtained and effect of variables
Citric acid very well counteracts the formation of calcium carbonate scale, since it is an organic tricarboxylic acid, these carboxyl groups can lose a proton when they are in solution, producing the citrate ion, said citrate ion, forms salts with metal ions within which calcium is found. For the case under study, calcium carbonate reacts with citric acid to form calcium citrate, CO_2 and water as follows in the Figure 2.

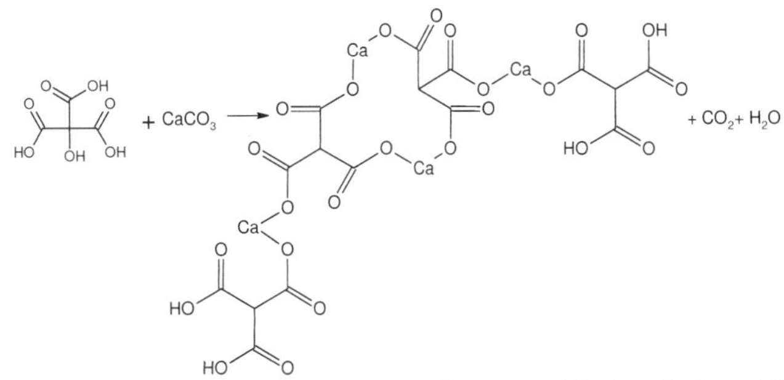

Figure 2. Reaction of calcium carbonate with the citric acid.

According to the study, citric acid shows good behaviour as a calcium carbonate scale inhibitor, in Table 1, the results obtained are presented, it can be observed that the efficiencies obtained for most cases are found on 98% efficiency.

Table 1. Results of incrustation efficiencies obtained.

Evaluation Conditions	Dosing of inhibitor (ppm)	Efficiency (%)
	15	99.56
70°C-1ft^3/CO_2	25	99.34
	40	99.34
	15	99.36
70°C–5ft^3/CO_2	25	99.08
	40	99.15
	15	98.08
80°C-1ft^3/CO_2	25	94.40
	40	98.73
	15	63.73
80°C–5ft^3/CO_2	25	76.21
	40	98.63

From the analysis performed, it is determined that as the temperature and the CO_2 concentration in the system increases, the efficiency of the inhibitor is affected, presenting a decrease in the same, thus having a greater influence on the CO_2 dosage, since the stability Of the complex that form the chemical species chelating with the metal ion, in this case $Ca^{2+,}$ are affected by the changes of pH, which in this case are generated with the addition of CO_2, affecting the way in which the complexes can react. A statistical analysis was performed, which correlates all the studied variables, thus obtaining a mathematical equation that allows predicting the inhibition behaviour of the acid, with which the behaviour previously analysed is corroborated.

$$\%Eficiencia = 18{,}07405 + 1{,}1936T + 96{,}9002C + 2{,}0037D + 0{,}002901D^2 - 1{,}3850TC - 0{,}03105TD - 2{,}4106CD + 0{,}03443TCD \quad (2)$$

Where T=Temperature, C=Addition of CO_2 and D=inhibitor dosage.

3.2. Morphology and characterization of crystals obtained

Figure 3 and 4, the crystals formed using citric acid inhibitor are shown, they are significantly smaller than those found in tests carried out without the use of inhibitor, in addition there are holes in the embedded material and the Grains formed have a geometry different from that of white tests, which in this case, corresponds to conic forms mostly hollow, all this, is due to the action generated by the

IOP Conf. Series: Journal of Physics: Conf. Series **935** (2017) 012053 doi:10.1088/1742-6596/935/1/012053

inhibitor in the formation of the incrustation. Calcium carbonate is polymorphic, i.e. it may be formed or existing in more than one crystalline structure. The polymorphs of calcium carbonate are three; Aragonite, vaterite (unstable phases), and calcite (more stable phase). [10-14]. Based on the shape of the crystals obtained, it is determined that the phase that is formed conforms to one of the unstable phases of the calcium carbonate, possibly aragonite. [12]. The morphology shows that, with this inhibitor, the formation of the most stable phase of calcium carbonate is not achieved, as well, it is observed that the formed crystals are significantly smaller than those obtained with the tests without inhibitor use, presenting sizes of 899.9nm and 978.3nm.

Figure 3. Scale layer formed citric acid inhibitor. Crystals obtained at 5000X increase without inhibitor dosing.

Figure 4. Scale layer formed citric acid inhibitor. Crystals obtained at 10840X. T=70°C, 3260mL/min CO_2 and 15ppm.

3.3. Characterization realized by XRD

The XRD analysis was carried out in order to obtain more information about the chemical composition of the formed incrustations and to identify the present classes. Figure 5 shows the results obtained from XRD analysis on the scale obtained with the use of citric acid inhibitor. In the spectrum, the presence of calcium carbonate, which in this case is presented as aragonite, is present This result confirms the evidence obtained by scanning electron microscopy, where an elongated and weak crystal morphology is observed, with very thin layers and very small crystals.

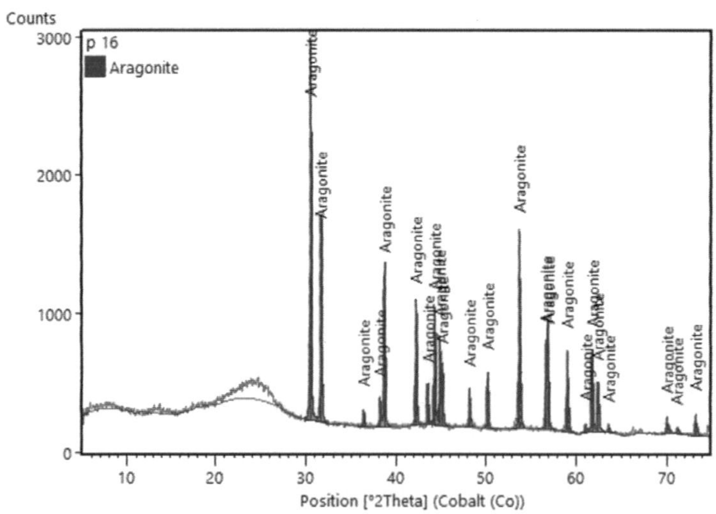

Figure 5. Characterization made to the incrustation products obtained.

IOP Conf. Series: Journal of Physics: Conf. Series **935** (2017) 012053 doi:10.1088/1742-6596/935/1/012053

4. Conclusions

Evaluation efficiencies using citric acid as a $CaCO_3$ scale inhibitor are found to be above 98% in most cases, this means that the inhibitor exhibits good behaviour against scale. It was found that with this inhibitor, the efficiencies are excellent even when dosed at 15ppm which is the lowest amount used, so in this case this would be the optimal dosage of inhibitor to use.

It was determined that the increase of CO_2 to the system decreases the efficiency of inhibition, this is possibly because the behaviour of the molecules as inhibitors depends on the changes pH; The CO_2, causes pH in the system to change, thus generating instability in the mechanism of inhibition.

With the use of inhibitor, a remarkable change in the morphology of the crystals is observed, as well as the decrease in the size of the crystals, indicating the positive effect of the inhibitor against the decrease of $CaCO_3$ salt formation.

References

[1] Zhang Y, Shaw H, Farquhar R and Dawe R 2001 *Journal of Petroleum Science and Engineering* **29** 85
[2] Sousa M, Signorelli F, Bertra C 2016 *Journal of Petroleum Science and Engineering* **147** 468
[3] Yang Q, et al. 2001 *Journal of Colloid and Interface Science* **240** 608
[4] Hasson D, Bramson D, Limoni B and Semiat R 1996 *Desalination* **108** 67
[5] Khormali A, Petrakov D and Javad M 2016 *Journal of Petroleum Science and Engineering* **147** 843
[6] Tantayakom V, Sreethawong T, Scott H, Morales F, Chavadej S 2005 *Journal of Colloid and Interface Science* **284** 57
[7] Chauan K, Kumar R, Kumar M, Sharma Pand Chanshyam S 2012 *Desalination* **305** 31
[8] Allan F and Neville A 2001 *Desalination* **281** 340
[9] NACE TM 0374 2007 *Laboratory screening test to determine the ability of scale inhibitors to prevent the precipitation of calcium sulfate and calcium carbonate from solution (for oil and gas production systems)* (USA: NACE International)
[10] Hung H, Han L and Hsieh C 2016 *Powder Technology* **302** 160
[11] Katsifaras A and Spanos Nikos 1999 *Journal of Crystal Growth* **204** 183
[12] Prasad S and Subramanian G 2012 *Desalination* **297** 38
[13] Zhang Y and Dawe R 2000 *Chemical Geology* **163** 129
[14] Muryantos S, Bayuseno A, Ma`mun M, Usamah, and Jotho 2014 *Procedia Chemistry* **9** 69

IOP Conf. Series: Journal of Physics: Conf. Series **935** (2017) 012054 doi:10.1088/1742-6596/935/1/012054

Electrochemical techniques implementation for corrosion rate measurement in function of humidity level in grounding systems (copper and stainless steel) in soil samples from Tunja (Colombia)

Y Salas[1], L Guerrero[2], J Blanco[3], C Jimenez[4], S P Vera-Monroy[5] and A Mejía-Camacho[5]

[1] Universidad INCCA, Bogotá, Colombia
[2] Sigemec SAS, Tunja, Colombia
[3] Universidad Pedagógica y Tecnológica de Colombia, Tunja, Colombia
[4] Pontificia Universidad Javeriana, Bogotá, Colombia
[5] Universidad Manuela Beltrán, Bogotá, Colombia

E-mail: yecid.salas@unincca.edu.co

Abstract. In this work, DC electrochemical techniques were used to determine the corrosion rate of copper and stainless-steel electrodes used in grounding, varying the level of humidity, in sandy loam and clay loam soils. The maximum corrosion potentials were: for copper -211 and -236mV and for stainless steel of -252 and -281mV, in sandy loam and clay loam respectively, showing that in sandy loam the values are higher, about 30mV. The mechanism by which steel controls corrosion is by diffusion, whereas in copper it is carried out by transfer of mass and charge, which affects the rate of corrosion, which in copper reached a maximum value of 5mm/yr and in Steel 0.8mm/yr, determined by Tafel approximations. The behaviour of the corrosion rate was mathematically adjusted to an asymptotic model that faithfully explains the C.R. as a function of humidity, however, it is necessary to define the relation between the factor □ established in the model and the precise characteristics of the soil, such as the permeability or quantity of ions present.

1. Introduction

Among the electrical protection systems that exist against overload events, buried electrodes are a simple, economical and mass-use system in numerous electrical distribution systems both industrial and residential level [1-4]. In previous studies, the behaviour against corrosion of copper and stainless steel electrodes against sandy loam and clay loam was established, but it was only carried out with a moisture level [5]. The moisture level is perhaps the most relevant factor in the ionic conduction of corrosive agents such as sulphides and chlorides [6-8] the soils present considerable variation of this the level of humidity during its period of operation [9]. Therefore, it is necessary to determine the effect of moisture on the corrosion rate.

In a parallel work [10] to this one is standardized the electrochemical tests with synthetic soils designed to accurately simulate the behaviour of a determined soil. In contrast, in this work, the original soils extracted were used and their level of humidity varied, a procedure that is described in detail in the methodology.

It seeks to know in detail the problem of corrosion in grounding systems, which are differentiated by their requirements in terms of electrical performance, apart from their acceptable behaviour against aggressive agents of the terrain.

The purpose of this work is to analyse the influence of soil characteristics and humidity on the morphology and properties defined for grounding electrodes. The decomposition rates were established by electrochemical techniques using potentiodynamic curves and linear polarization resistance measurements [16].

2. Methodology

For this work two types of soils characteristic of the city of Tunja, Colombia were referenced, as Guerrero *et. al.* [5], where a single level of humidity was used. The matrix of the experimental design was defined with two types of soil: sandy and clay loam, two electrode materials and 7 different levels of humidity, to make an adequate sweep in each experimental treatment.

The soil samples were taken to the respective analysis laboratory at the "Universidad Pedagógica y Tecnológica de Colombia (UPTC)" to obtain data such as: texture, pH, percentage of organic material, phosphorus, aluminium, potassium, calcium, magnesium, cation exchange capacity and characterization of electrical conductivity for these two types of soil [5]. The electrolyte used in the cell was obtained from soil samples prepared with the 7 moisture levels, from 10 to 40%.

According to the respective table consulted in the Electrical Installations Regiment RETIE [1], IEEE Standars [11, 13-14] and ICONTEC Guide [15], the test electrodes were chosen: copper and stainless steel. These electrodes are most commonly used in earthing systems [12].

The experimental matrix considers three factors (Table 1): soil type, with two levels: sandy loam and clay loam; Electrode material, with two levels: copper and stainless steel, and humidity, with seven levels: from 10 to 40% humidity.

Table 1. Notation of the experimental matrix with the three factors considered.

Soil Factor	Electrodes	Humidity Factor	FH
Sandy Loam	Copper	10%	10
Clay Loam	Stainless Steel	15%	15
		20%	20
		25%	25
		30%	30
		35%	35
		40%	40

Potentiodynamic curves were used to obtain an accurate estimate of the corrosion rate of the metal in the solution with the soil. The current is measured in the cell with a potential sweep and is presented in logarithmic scale. The potential sweep was made from about 500mV to its open circuit potential

3. Results and discussion

3.1. Potentiodynamic curves

The graphs obtained for each electrode system and terrain are detailed in the following Figures 1 and 2.

If the Tafel approximation is used, for the copper electrode in the experimental arrangements with the two types of soil, it is observed that as the moisture increases the corrosion rate shifts to the right less than a decade. For moisture values greater than 30% no significant variations in rate and corrosion potential are observed.

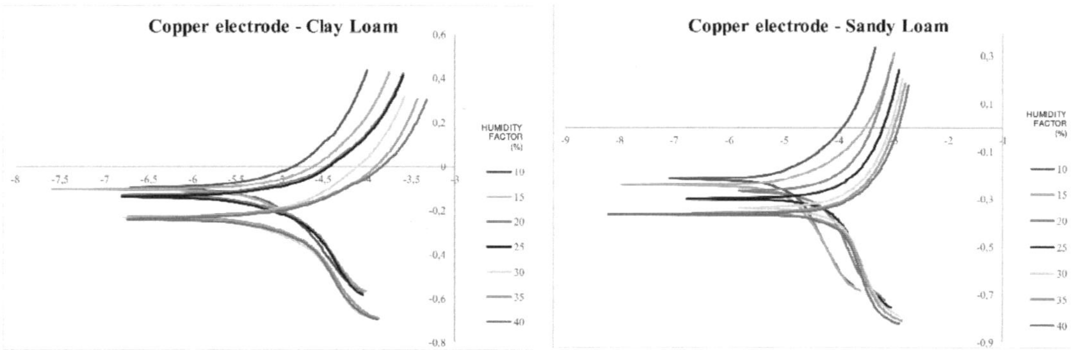

Figure 1. Potentiodynamic curves for Copper electrode in clay and sandy loam soil with variation in moisture.

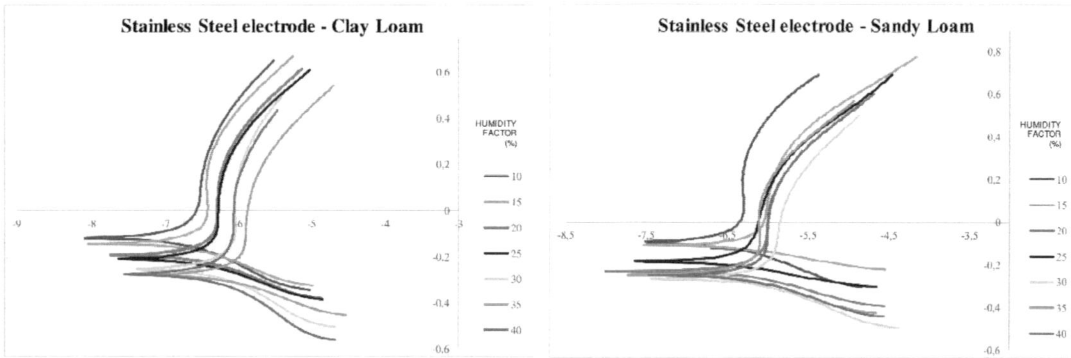

Figure 2. Potentiodynamic curves for Stainless Steel electrode in clay and sandy loam soil with variation in moisture.

The potentiodynamic curves for stainless steel present the typical asymmetry between the anodic branch and the cathodic branch, because the anodic branch presents an infinite slope. The Tafel approximation will be calculated with a few points close to the corrosion potential. The behaviour of the steel for the two arrangements, visually, presents little variation.

3.2. Summary of results

The results obtained after the Tafel approximation for each treatment can be synthesized in Tables 2 and 3.

Table 2. Synthesis of results of electrochemical tests in the copper.

Humidity %	Copper Clay Loam			Copper Sandy Loam		
	E corr (mV)	I corr (µA/cm^2)	CR (mm/yr)	E corr (mV)	I corr (µA/cm^2)	CR (mm/yr)
10	-104.4	31.27	0.3622	-63.63	55.62	0.6443
15	-113.3	49.86	0.5776	-97.77	56.45	0.6548
20	-121.2	60.79	0.7042	-118.4	215.4	2.495
25	-130.4	60.66	0.7026	-136.0	331.2	3.837
30	-235.9	62.51	0.7241	-179.9	428.1	4.960
35	-236.2	62.69	0.7261	-201.1	458.7	5.313
40	-236.3	62.82	0.7277	-211.3	460.2	5.331

Table 3. Synthesis of results of electrochemical tests in the stainless.

Humidity %	Stainless Clay Loam			Stainless steel - Sandy Loam		
	E corr (mV)	I corr (μA/cm²)	CR (mm/yr)	E corr (mV)	I corr (μA/cm²)	CR (mm/yr)
10	-119.8	10.50	0.1217	-98.34	17.72	0.2052
15	-142.6	10.98	0.1272	-123.4	27.91	0.3233
20	-181.3	13.48	0.1561	-177.5	33.38	0.3867
25	-216.7	15.94	0.1846	-202.1	33.24	0.3850
30	-250.3	28.06	0.3250	-266.0	36.08	0.4179
35	-274.8	30.17	0.3495	-247.7	37.70	0.4368
40	-281.2	32.11	0.3739	-251.7	39.50	0.4576

3.3. Discussion

In the copper electrodes, when comparing the Tafel slopes, a control is verified by transfer of load and mass, fact that is evidenced more clearly when registering rates of corrosion always greater than the stainless steel. Although the remarkable increase in the corrosion rate for the copper - free arrangement must present an additional mechanism.

In Figure 2, potentiodynamic curves of the stainless steel for the different arrangements, present a zone in the anodic branch where the current becomes constant, evidence of the control mechanism of corrosion by diffusion.

Figure 3 show the corrosion rates for copper and stainless steel in the two experimental areas. As expected, the rates for copper are comparatively higher than those of stainless steel in the two soils, which is explained by their conductivity that facilitates ionic mobility.

The corrosion rate values for a copper electrode in the sandy loam soil are significantly higher than the other soil-electrode combinations. There is no concrete explanation for this phenomenon, because errors were ruled out in the experimental procedure following the replicate protocol in the treatments.

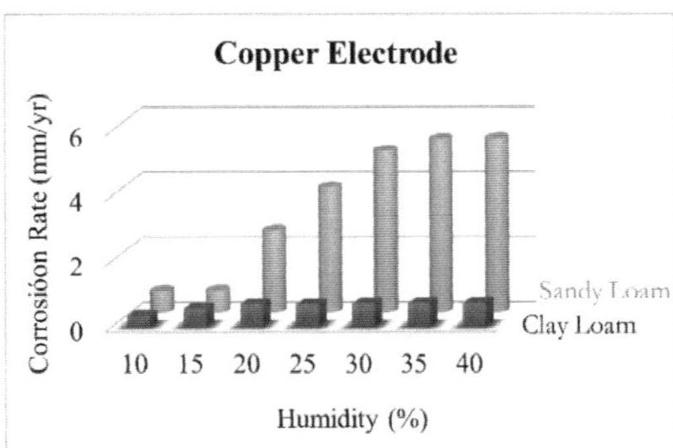

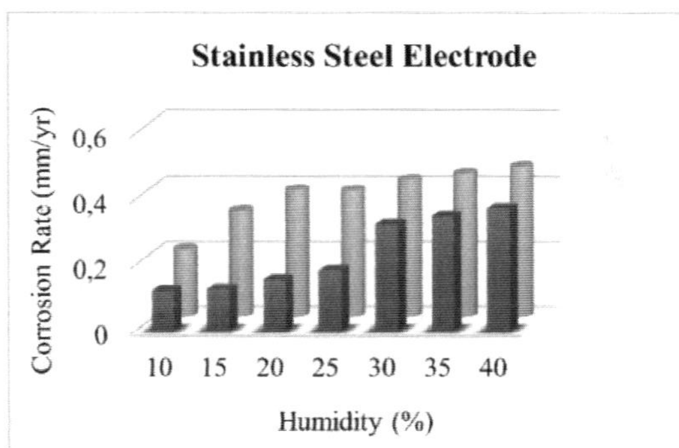

Figure 3. Comparison of corrosion rate for experimental samples.

The behaviour for any soil regardless of the electrode used is similar qualitatively. The rate of corrosion undergoes large increases in moisture ranging from 10 to 20 or 25%, while for moistures over 30% tends to stabilize. This suggests that there is a limiting corrosion rate for a particular type of soil with the highest possible moisture.

3.4. Mathematical model

The behaviour of the corrosion rate as a function of humidity has a noticeable tendency in the data obtained in the laboratory for the potentiodynamic curves. The moisture is deduced as a fundamental factor in the phenomenon of metals corrosion buried in the soil because it determines the base environment of the electrolyte for the conduction of dissociated ions in the water. At higher humidity, a tendency of increase in the speed of corrosion is obvious. However, this trend is not linear. A final

IOP Conf. Series: Journal of Physics: Conf. Series **935** (2017) 012054 doi:10.1088/1742-6596/935/1/012054

corrosion value is observed at the asymptotically tended by this rat as the moisture is increased. It is induced that the ion exchange between the electrolyte and dry soil should be zero. Therefore, we are looking for a model that replicates this functional dependency.

The solution of the first-order differential equation deduced for this phenomenon would be:

$$C.R. = C.R.F.\left(1 - e^{-\frac{h}{\rho}}\right) \quad (1)$$

That determines the behaviour of C.R. (corrosion rate) In terms of h (soil moisture %). C.R.F. is called final corrosion rate. And ρ is a factor arising from proportionality with the same units as the humidity (percentage) and that would involve soil properties and instant humidity. In a more complete model this value must include factors such as soil permeability to moisture and electrode reactive ions and that are present in the soil.

The regression that can be implemented for the experiments with variation of humidity arise from the clearance of h as a function of C.R (equation 2). The behaviour of C.R. Depending on the factor is detailed in Figure 4.

$$h = -\rho \cdot \ln\left(1 - \frac{C.R}{C.R.F.}\right) \quad (2)$$

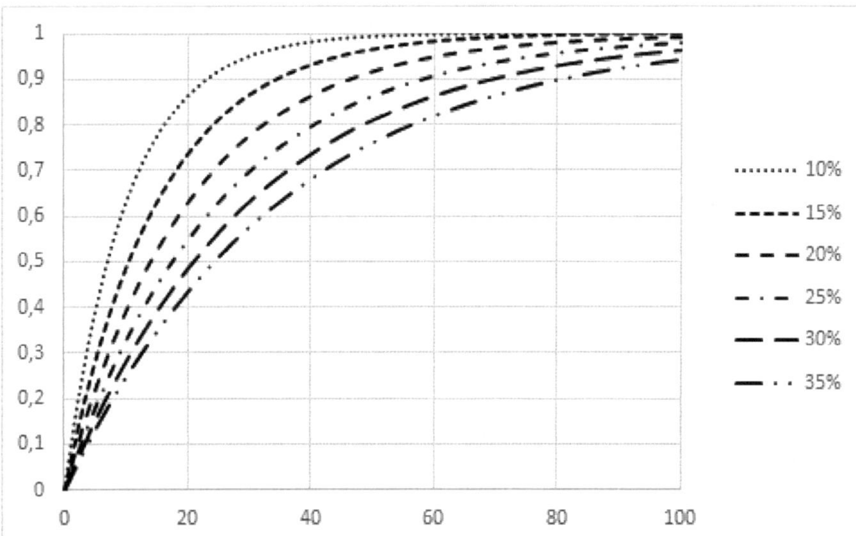

Figure 4. Unit C.R. versus humidity level for several values of factor ρ.

4. Conclusions

Corrosion rates were established for the different experimental arrangements, the behaviour of the copper electrode standing out when interacting with the sandy loam soil, experimental samples that yielded very different results to the other arrangements.

A marked difference between the rate of corrosion when measured for copper or for steel was observed. The mechanisms in which these elements control corrosion is very different: while in steel there is control by diffusion, in copper it is made by transfer of mass and charge. The Franco-sandy floor was the most aggressive against metal electrodes.

The level of humidity plays a leading role in the phenomenon of corrosion of buried structures. However, the variability of terrain when exposed to the weather did not allow precise and rapid measurements of the corrosion rate, a problem that solves the electrochemical techniques allowing the generation of rapid and controlled responses to influential factors.

The generated mathematical model faithfully explains the phenomenon of change of C.R. as a function of moisture. However, it is still necessary to find the relation between the factor defined in the model and the precise characteristics of the soil, such as the permeability or the quantity of ions present in the soil.

References

[1] Ministerio de Minas y Energía 2013 Reglamento tecnico de normas electricas (RETIE) *Anexo general del RETIE Resolución 90708 de 2013* (Colombia: Ministerio de Minas y Energía) (https://www.minminas.gov.co/retie)
[2] Montaña J 2011 Recommendations for grounding systems in lightning protection systems *Ing Invest* **31(suppl 2)** 5-10
[3] Anggoroa B Yutadhia R 2013 *Proc. Tech.* **11** 1156–1162
[4] Ghavamian A Maghami M Dehghan S Gomes C 2015 *Eng. Fail. Anal.* **57** 434-443
[5] Guerrero L Salas Y Blanco J 2016 *J, Phys.: Conf. Ser.* **687** 1
[6] Acosta L Castañeda A Corvo F Berardo L 2010 *Cenic. Ciencias Quimicas* **41** 1-11
[7] Yahaya N LimK Noor N Othman S Abdullah A 2011 *Malays J. Civ. Eng.* **23(1)** 24-32
[8] SicaI Y Kenny E Portella K Campos D 2007 *J. Brazil Chem. Soc.* **18(1)** 153-166
[9] Ehteram A Aisha H 2014 *Arab. J. Sci. Eng.* **39** 5421-5435
[10] Matamala G E, Carranza J A, Jara S E 2001 Corrosión de conductores de cobre en suelos dopados con aditivos semiconductores y salinos, para el mejoramiento de los sistemas de puesta a tierra Jornadas *SAM CONAMET (Pasadas)* (Argentina: Sociedad Argentina de Materiales) 707-14
[11] IEEE 2012 Guide for measuring earth resistivity, ground impedance, and earth surface potentials of a grounding system *IEEE Std 81-2012* 1-86
[12] Yu L Chow M Bowen J 1998 *IEEE Ind. Appl. Mag.* **4(2)** 32-36.
[13] IEEE 2000 *IEEE* Guide for safety in ac substation grounding *IEEE Std 80-2000* 1-192
[14] IEEE 1983 IEEE Guide for measuring earth resistivity, ground impedance, and earth surface potentials of a ground system part 1: Normal measurements *IEEE Std 81-1983* 1-44
[15] NTC 2050 1998 *Código eléctrico colombiano NTC 2050* (Colombia: ICONTEC)
[16] Fouda A Abo-Shohba R Husien W Ahmed E 2015 *Global Journals of Research in Engineering C: Chemical Engineering* **15(3)** 9-24

IOP Conf. Series: Journal of Physics: Conf. Series **935** (2017) 012055 doi:10.1088/1742-6596/935/1/012055

Electrochemical characterization of corrosion in materials of grounding systems, simulating conditions of synthetic soils with characteristics of local soils

Y Salas[1], L Guerrero[2], S P Vera-Monroy[3], J Blanco[4] and C Jimenez[5]

[1] Universidad INCCA, Bogotá, Colombia
[2] Sigemec SAS, Tunja, Colombia
[3] Universidad Manuela Beltrán, Bogotá, Colombia
[4] Universidad Pedagógica y Tecnológica de Colombia, Tunja, Colombia
[5] Pontificia Universidad Javeriana, Bogotá, Colombia

E-mail: yecid.salas@unincca.edu.co

Abstract. The integrity of structures buried in earthing becomes relevant when analysing maintenance and replacement costs of these systems, as the deterioration is mainly due to two factors, namely: the failures caused in the electrical systems, which are due to the system. Failure in earthing due to corrosion at the interface cause an alteration in the structure of the component material and generates an undesirable resistivity that cause malfunction in this type of protection systems. Two local soils were chosen that were categorized as sandy loam and clay loam type, whose chemical characteristics were simulated by means of an electrolyte corresponding to the amount of ions present determined by a soil characterization based on the CICE (effective cation exchange coefficient), which allows us to deduce the percentage of chloride and sulphate ions present for the different levels established in the experimental matrix. The interaction of these soils with grounding electrodes is a complex problem involving many factors to consider. In this study, the rates and corrosion currents of the different soils on two types of electrodes, one copper and the other AISI 304 stainless steel, were approximated by electrochemical techniques such as potentiodynamic curves and electrochemical impedance spectra. Considerably higher speeds were determined for copper-type electrodes when compared to those based on steel. However, from the Nyquist diagrams, it was noted that copper electrodes have better electrical performance than steel ones. The soil with the highest ionic activity turned out to be the sandy loam. The clay loam soil presents a tendency to water retention and this may be the reason for the different behaviour with respect to ionic mobility. The diffusion control in the steel seems to alter the ionic mobility because its corrosion rates proved to be very similar regardless of the type of soil chemistry. In general, corrosion rates fell since tenths of a millimetre every year to the hundredths of a millimetre in the case of the steel-based electrode, which are relatively small corrosion rates.

1. Introduction

Any electrical installation that follow the recommendations of RETIE [1] and IEEE Standards [6-15] has a grounding system, to prevent people near, electrical or electronic devices, can be subjected to high voltages of contact or transferred that surpass the thresholds of tolerance affecting the human body or damage electrical compounds when a fault occurs. The requirement of earthing for electrical installations covers the electrical system and give supports to metal structures that in the event of a temporary overvoltage, can trigger a permanent fault between the earthing structure and the network. The objectives of a grounding system are: the safety of people, the protection of installations and electromagnetic compatibility [2-5].

In the procedure of earthing is necessary investigate the characteristics of the soil especially the resistivity. Within the tests that must be done as part of the inspection and as a requirement of evidence through records, must be recollected at least the following information: General conditions of the system conductors, corrosion level, state of the connections and conductors, electrical components and resistance values [6-7].

This work addresses the problem of comparing the degradation of the grounding electrode material from an approach chemical differentiation of elements present in each soil. It simulates soil chemistry through a very methodical approach designed specifically for this work. use salts that's will separation in anions and cations, using a study of typical soils that determine the present cations, the possibility of having soils based on sulphates or based on chlorides is deduced to consider two types of corrosion operationally different.

2. Methodology

2.1. Electrodes composition
The copper composition was determined by X-ray florescence (MINIPAL 2), according to ASTM B 115. The stainless-steel composition was made by Optical Emission Spectrometry according to ASTM E 415 (Q8 MAGELLAN BRUKER QUANTRON).

2.2. Soils characterization
The corrosion evaluation was carried out by simulating the saline composition of two local soils, sandy loam and clay loam, supplementing them with chlorides and sulphates in adequate proportions. The characterization was performed by evaluating: texture, pH, hardness, cation exchange, electrical conductivity, percentage of organic matter, phosphorus, potassium, calcium and magnesium, in the UPTC soil laboratory. The resistivity of the soils was determined using the Wenner method.

2.3. Metallographic analysis of electrodes
It was performed using the DACE TECHNOLOGIES NANO 200 DT polisher, with grit # 80 to # 1200, the micrographs were evaluated on the LEICA DVM 2500.

2.4. Electrodes electrochemical test
Potentiodynamic evaluation was performed using a rotating ring electrode coupled to a galvanostat potentiostat (Gamry 750) [16], with a sweep of -250 to 250mV and a speed of 2mV/s.

3. Results and discussion

3.1. Soil resistivity

Table 1. Resistivity of clay loam and sandy load soils with the Wenner method.

Clay Loam Soli				Sandy Loam Soil			
Spacing (m)	Profile 1 ρ(1) (Ω-m)	Profile 1 ρ(2) (Ω-m)	Average Resistivity ρ(m)(Ω-m)	Spacing (m)	Profile 1 ρ(1) (Ω-m)	Profile 1 ρ(2) (Ω-m)	Average Resistivity ρ(m) (Ω-m)
1	700.87	710.89	701.38	1	301.36	301.79	301.58
3	805.93	705.95	755.94	3	329.49	287.51	308.50
5	830.86	690.68	760.77	5	349.96	291.08	320.52
7	790.75	705.89	748.32	7	330.11	294.47	312.29
10	830.76	740.85	785.86	10	354.91	316.19	335.55

The average value of the resistivity measure for the clay loam soil was 750.45Ω m (Table 1), Generating a resistance of 45.22Ω calculated by EIS method, in all cases the value was around 50Ω. For the sandy loam soil, the resistivity was 315.69Ω m and the resistance of 19.62Ω, with the IES method the value oscillates between 18 and 40Ω. These data indicate that electrolytic simulation of soils was acceptable and data that presented the greatest deviation, regarding the expected results was the stainless steel in loamy loam soil with base sulphate electrolyte.

3.2. Morphological properties of electrodes

The morphological evaluation showed that in the two soils with chlorides based electrolyte, corrosion by stinging was observed, being that it is more aggressive in the copper electrodes (Figures 1(a) and (b)) than in stainless-steel electrodes (Figures 1(C) and (d)).

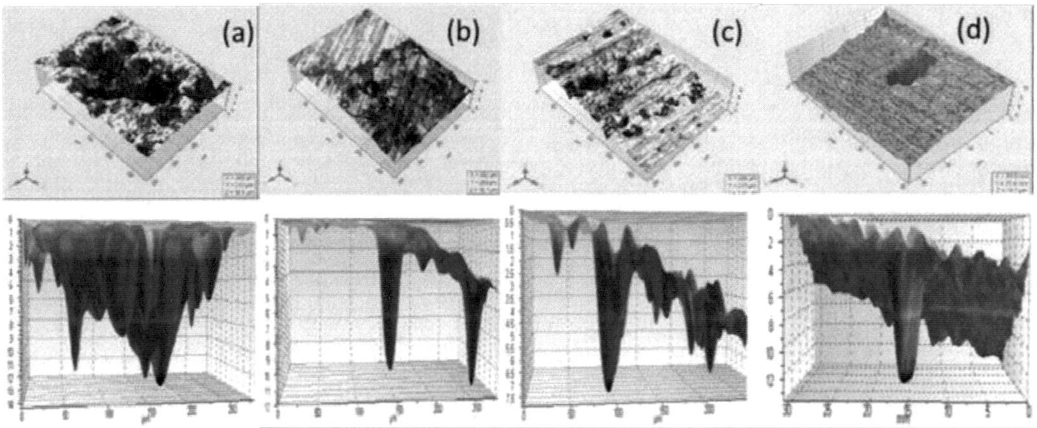

Figure 1. Electrode morphology. Micrographs and mappings with chlorides based electrolyte (a) sandy loam - copper (b) clay loam - copper (c) sandy loam - stainless steel and (d) clay loam - stainless steel.

3.3. Potentiodynamic curves

The Figures 2 show that sandy loam soils present more negative potentials. The difference between corrosion potentials is around of 50mV and there are appreciable differences in the current densities of corrosion for the copper electrodes. The stainless-steel electrode shows a change in the potential corrosion greater than the copper electrodes. The most negative potentials correspond to the electrodes evaluated in soils with basic sulphate electrolyte.

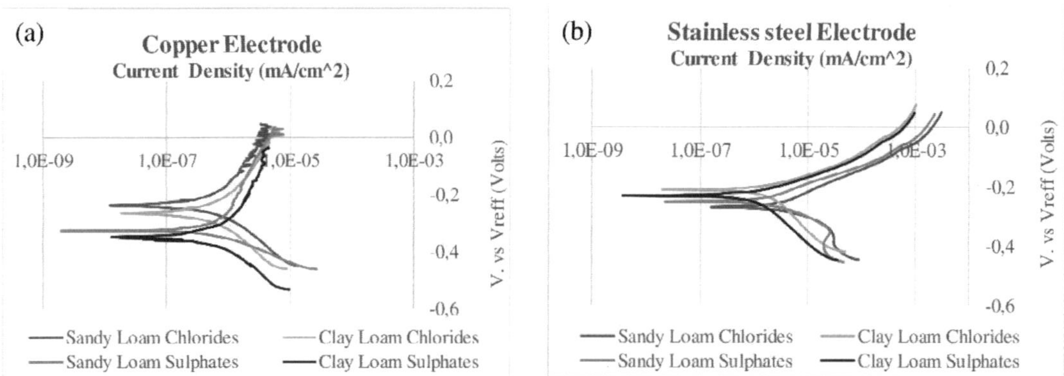

Figure 2. Potentiodynamic curves (a) copper electrode and (b) stainless steel electrode.

3.4. Electrochemical impedance spectroscopy

The Bode graphs show the capacitive behaviour of the stainless steel (Figure 3(b)) that shows the presence of its passivation layer in contrast to the copper (Figure 3(a)) that presents minor variations in the phase angle. In Nyquist graphs the value of resistance of the solution is observed which are corresponded with measurement of resistivity in the field.

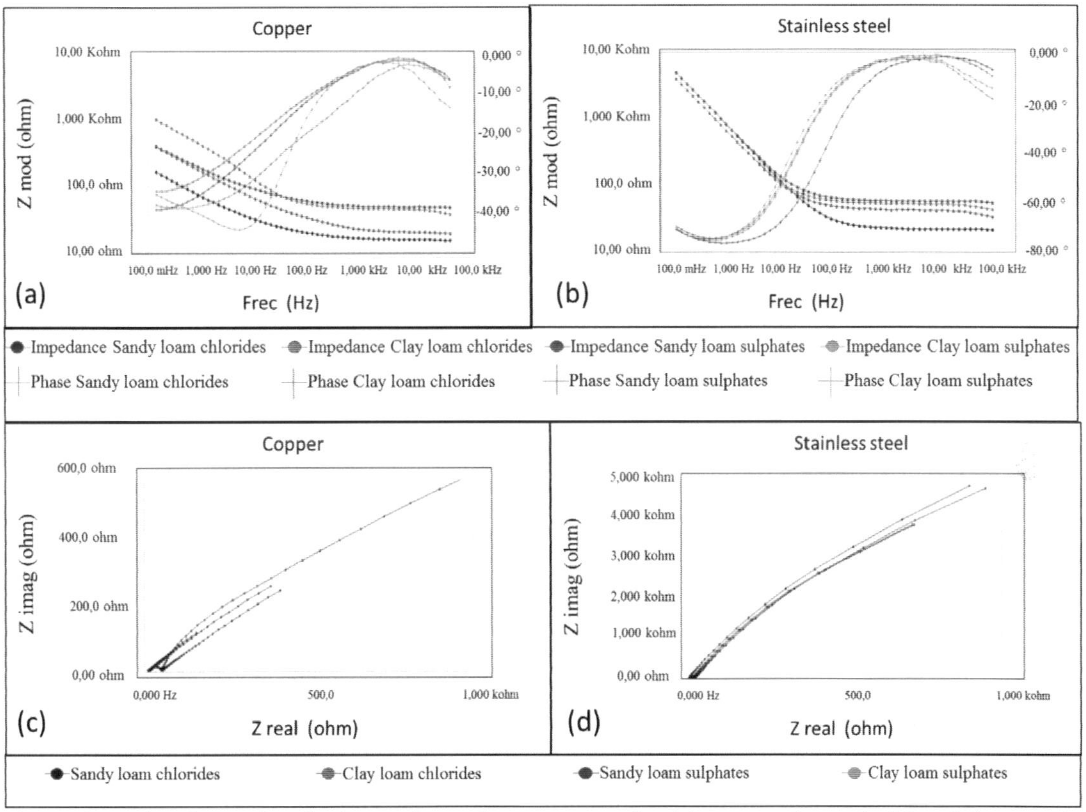

Figure 3. Bode Electrochemical Impedance Spectroscopy for (a) copper electrode (b) stainless steel electrode and Nyquist for (c) copper electrode (d) stainless steel electrode.

3.5. Circuits

The physical significance derives from the simple corrosion phenomenon that must be presented by the electrodes immersed in the solution that simulates the soil and therefore the CPE and Randless circuits (Figure 4) were chosen. The second circuit, chosen with diffusion, was tried to test due to the 304 steel electrodes, which when being stainless presented an anodic branch with considerable slope in the Tafel approximation, which was carried out on the potenciodynamic curves.

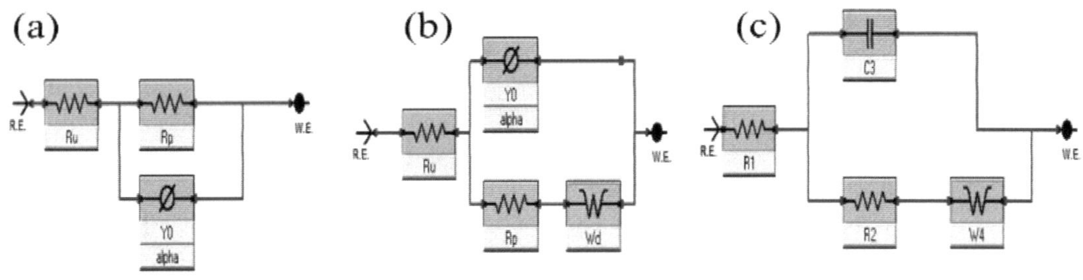

Figure 4. Equivalent circuits chosen. (a) CPE, (b) CPE with diffusion and (c) Randles.

IOP Conf. Series: Journal of Physics: Conf. Series **935** (2017) 012055 doi:10.1088/1742-6596/935/1/012055

Using the Gamry software, the parameters were deduced for each equivalent circuit, taking into account that Rp is the resistance value used in the calculation of corrosion rate in each case.

Table 2. compares the results for the different methods in terms of corrosion rate. The most similarity between the Tafel approximation of the potenciodynamic curves and the modelling made on the basis of the Randless circuit is observed in both the order of magnitude and its qualitative behaviour.

Table 2. Comparative synthesis of the calculated corrosion rates for each experimental treatment.

	Tafel		LPR		EIS CPE modelling		EIS CPE diffusion modelling		EIS Randless modelling	
Soil	Corrosion rate		Corrosion rate		Corrosion rate		Corrosion rate		Corrosion rate	
	mpy	mm/yr	mpy	mm/yr	mpy	mm/yr	mpy	mm/yr	mpy	mm/yr
SLC	0.875	0.0222	7.64	0.194	0.332	0.00843	2.93E-17	7.45E-19	44.6	1.13
CLC	0.819	0.0208	4.16	0.106	1.143	0.02903	1.16E-17	2.94E-19	20.0	0.507
SLS	0.740	0.0188	3.49	0.089	0.162	0.00410	5.05E-18	1.28E-19	22.8	0.580
CLS	0.485	0.0123	2.02	0.051	2.770	0.07036	2.75E-18	7.00E-20	8.21	0.209
SLC	0.243	0.0062	0.20	0.005	0.131	0.00333	3.25E-18	8.26E-20	0.480	0.012
CLC	0.263	0.0067	0.21	0.005	0.127	0.00323	1.70E-18	4.32E-20	0.480	0.012
SLS	0.227	0.0058	0.22	0.006	0.102	0.00259	3.78E-19	9.59E-21	0.564	0.014
CLS	0.299	0.0076	0.22	0.006	0.117	0.00298	1.49E-19	3.77E-21	0.566	0.014
CG	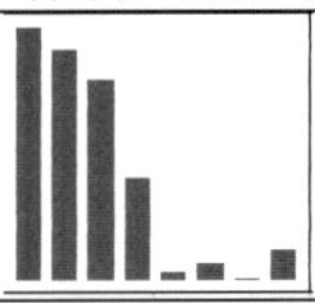		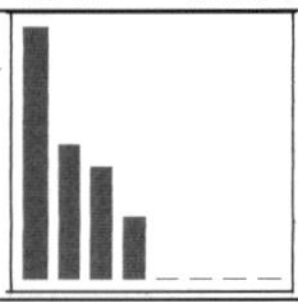		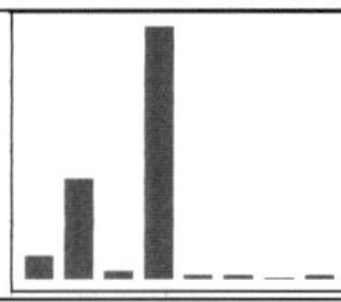		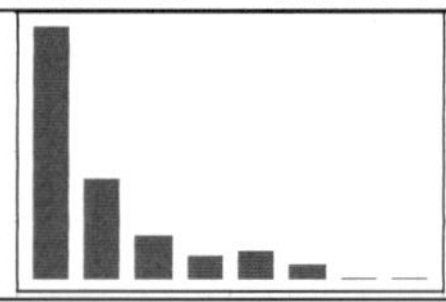		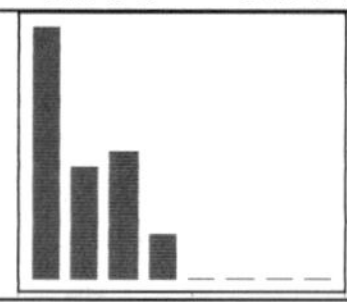	

SLC-Sandy Loam Chlrorides, CLC-Clay Loam Chlorides, SLS-Sandy Loam Sulphates, CLS-Clay Loam Sulphates, CG – Comparative Graphics (No scale)

The corrosion rate, determined by Tafel approximations, shows lower values than those calculated with the Linear Polarization method, by an order of magnitude, although they have the same tendency. Steel electrodes have high anodic slopes suggesting that diffusion control is present. The corrosion rates of the stainless-steel electrode are always lower than those of copper.

4. Conclusions

The interface parameters of the materials in the simulated soils were evaluated using the EIS, LPR and Tafel approximation techniques for potentiodynamic curves. Resistivity measurements showed good simulation of the soils studied.

The electrochemical parameters were correlated with the factors used in the experimental matrix that defined the greater susceptibility to copper corrosion and the greater aggressiveness of soils in chloride electrolyte. The corrosion rate was determined using the electrochemical parameters obtained for the best models. The Tafel approximation presented lower values than those calculated by the LPR method, although with the same trend. The values obtained with the CPE approximation had the same order of magnitude as the LPR but with different behaviour. The results showed that sandy loam soil and the copper electrode present higher values of corrosion rate, chloride electrolytes are more reactive than sulphates and the stainless-steel electrode is more stable than copper electrode.

Copper exhibits poor corrosion performance but good electrical performance if RETIE standard is followed. The load transfer resistance value in stainless steel determines poor electrical performance if this type of electrodes is used in grounding systems.

IOP Conf. Series: Journal of Physics: Conf. Series **935** (2017) 012055 doi:10.1088/1742-6596/935/1/012055

References

[1] Ministerio de Minas y Energía 2013 Reglamento técnico de normas eléctricas (RETIE) *Anexo general del RETIE Resolución 90708 de 2013* (Colombia: Ministerio de Minas y Energía) (https://www.minminas.gov.co/retie)

[2] Rodríguez L Acevedo R Saxton G 2009 *UCT* **13(52)** 259-264

[3] Astorga J Iriarte Y Peralta L *Ing. Energética* **36(2)** 146-154

[4] Jeff Roberts, Hector J Altuve, and Daqing Hou 2001 *Review of Ground Fault Protection Methods for Grounded, Ungrounded, and Compensated Distribution Systems* (USA: Schweitzer Engineering Laboratories)

[5] Montaña J 2011 Recommendations for grounding systems in lightning protection systems *Ing Invest* **31(suppl 2)** 5-10

[6] G Vijayaraghavan, Mark Brown, Malcolm Barnes 2004 Introduction and Basics *Practical Grounding, Bonding, Shielding and Surge Protection* (Oxford: Elsevier Ltd.) chapter 1 1-12

[7] Sanz J Pérez L Ocampo F 2007 *Scientia Et Technica* **13(34)** 121-126

[8] Parise G Parise L Martirano L 2015 *IEEE T. Ind. App.* **51(6)** 5017-5022

[9] Maynard K Mitolo M O'Reilly D 2015 *IEEE T. Ind. App.* **51(6)** 5385-5389

[10] Lim S Gomes C Ab Kadir M 2013 *Int. J. Elec. Power* **47** 117-128

[11] Prabhakar C Deshpande R 2014 Evaluation of soil resistivity and design of grounding system for hydroelectric generating station in a hilly terrain — A case study *International Conference on Advances in Energy Conversion Technologies (ICAECT)* (USA: IEEE) 104-107

[12] Luke Y Chow M Bowen 1998 *IEEE Ind Appl Mag* **4**.2 32-36.

[13] IEEE 2000 *IEEE* Guide for safety in ac substation grounding *IEEE Std 80-2000* 1-192

[14] Kam S Oh S Lee G Kim H Boo C Jeong Y 2014 Possibility of groundwater contamination of copper grounding materials by soil corrosion and influx of various aqueous solutions in Jeju island *International Conference on Lightning Protection (ICLP)* (USA: IEEE) 671-75

[15] IEEE 1983 IEEE Guide for measuring earth resistivity, ground impedance, and earth surface potentials of a ground system part 1: Normal measurements *IEEE Std 81-1983* 1-44

[16] Fouda A Abo-Shohba R Husien W Ahmed E 2015 *Global Journals of Research in Engineering C: Chemical Engineering* **15(3)** 9-24

IOP Conf. Series: Journal of Physics: Conf. Series **935** (2017) 012056 doi:10.1088/1742-6596/935/1/012056

Effects of oxidation-nitridation in the presence of water vapor on ASTM A335 P92 steel using SEM-EDS and XPS characterization techniques

J C Orozco[1], V Kafarov[1], D Y Peña[1] and A Alviz[1]
[1] Universidad Industrial de Santander, Colombia

E-mail: ingjorozcoa@gmail.com

Abstract. This research studies the physical and chemical changes in steel ASTM A335 P92, produced from a typical refinery corrosion environment. The environment evaluated was oxidation-nitridation with the presence of water vapor. In this study five (5) exposure times were selected: 1, 20, 50, 100 and 200 hours; As well as two (2) analysis temperatures: 450 and 550°C. The working pressure used was one (1) atmosphere. Bearing in mind the kinetic study, the behaviour shown in ASTM A335 P92 steel describes an accelerated growth until 50 hours, after this time growth is less. For the tests carried out at 450°C, the kinetic constant was $2x10\text{-}8g^2mm^{-4}h^{-1}$, as well as for 550°C the calculated kinetic constant was $3.1\times 10\text{-}7g^2mm^{-4}h^{-1}$; through the SEM-EDS characterization techniques, it was possible to appreciate a good adhesion and homogeneity of the layers formed on the metal matrix until a time of exposure of 100 hours at 450 and 550°C, different from that evidenced to 200 hours of exposure where the layer formed near to the substrate showed detachment, this is attributed to the formation of hydroxides product of water vapor. Among the results obtained are the elemental composition, the presence of nitrides such as Si_3N_4, also $NSiO_2$ and NSi_2O, molybdenum oxides: MoO_2 and MoO_3 and iron oxides: FeO and Fe_2O_3 can be evidenced.

1. Introduction

The permanent need to maintain and improve efficiency in industrial production processes promotes the use of high temperatures, which favours the presence of more severe corrosive environments on the materials used by the different equipment. Combustion gases and water vapor represent a constant problem in equipment operating in high temperature processes, mainly due to the use of mixtures of natural gas and other streams obtained during the refining process commonly called refinery gases; a very common example is found in boiler tubes, steam generators and furnaces in the petrochemical industry [1]. When ferritic steels are subjected to high temperatures and long periods of time, failures in materials and structures begin to appear [2-5]. According to Gil, of all operational failures in industry, the most important is corrosion with 33% of cases [6]. On the other hand, it has been determined that the presence of water vapor stimulates the formation of volatile species in the form of hydroxides, stimulating the increase in the speed of corrosion of the material. One of the main reasons because the steels with nine percent of chromium (9% Cr) present such high corrosion rates in the presence of combustion gases, is due to the presence of water vapor [7-10]. In these steels, the negative effect of water vapor on oxidation is greater, as during the dry oxidation a thin layer of chromite (Cr_2O_3) is produced, which is continuous and adherent, which is a protective layer. However, in atmospheres with water vapor does not occur the formation of a layer with the same characteristics, since the water reduces the stability of the film of chromite, because it allows the diffusion of the atoms of Iron (Fe) through the oxide layer increasing the formation of iron oxides.

Among the alloys used in combustion equipment is steel ASTM A335 P92 (9Cr-1Mo modified), which owes its use to the aggregates of alloying elements such as tungsten (W) and boron (B) which contribute to improving mechanical properties such as creep resistance and corrosion resistance at high temperatures [11]. High temperature corrosion includes studies of the different effects that may occur according to the composition of the gas mixture: oxidation, nitridation, water vapor, among others. However, on this alloy there are few reported studies that contemplate all the aforementioned corrosive effects, and which are typical of combustion atmospheres typical of actual refining processes. For this reason, this study intends to evaluate the performance of ASTM A335 P92 steel in an atmosphere similar to those generated in the hydrocarbon combustion and transformation of a refinery furnace (oxidation-nitridation with water vapor), with temperatures on the surface between 450°C and 550°C, in order to understand the corrosive phenomena which, cause damages in this steel used in tubes for combustion processes in boilers.

2. Methodology

ASTM A335 P92 steel coupons with dimensions of 20mm×15mm×2mm were used. The coupons were machined and sanded with SiC paper from sandpaper No. 180 to 600, then cleaned in an acetone bath in ultrasound for a time of 10 minutes [11]. The tests were carried out in a tubular horizontal oven at temperatures of 450 and 550°C in an atmosphere of 79.66% N_2-1.92% O_2-18.40% H_2O (% mol), atmospheric pressure and a volumetric flow of 212.3cm^3/min N_2-O_2 and 0.0358cm^3/min H_2O. The selection of the study atmosphere was selected bearing in mind the study carried out at the Barrancabermeja´s refinery, in which four (4) characteristic mixtures were determined from the historical chromatographs of different refinery units [12]. The times used for each test were 1, 20, 50, 100 and 200 hours of exposure. Before and after each test the coupons were weighed on a high precision digital scale.

To understand the corrosive effects on the material, characterizations such as optical microscopy, hardness and microhardness were carried out before exposure, as well as after exposure. Scanning Electron Microscopy (SEM), a technique that allowed us to study the morphology of the layers and phases formed in the material. An elementary collective profile (Line) was also determined, as well as the punctual elemental composition in each of the phases, by means of the energy-dispersive X-Ray Spectroscopy (EDS). For the X-ray Photoelectron Spectroscopy analysis (XPS), the samples were placed on a carbon conductive belt in metal samples. In the measurements, a monochromatized Al Kα (FOCUS 500) X-ray source was used at 200W. The passing energy of the hemispherical analyser was set at 100eV for the general and 60eV spectra for high resolution spectra. The surface charge compensation was monitored with a Flood Gun (device FG 15/40-PS FG500) operated at 58μA and 2eV. Finally, the hardness tests were performed with a Rockwell A hardness tester and microhardness with the digital microdurometer in Vickers respectively.

3. Results

Initially, a metallographic characterization of ASTM A335 P92 steel was performed to determine microstructural changes. In Figure 1(a), the initial microstructure composed of a ferritic matrix with a martensitic structure is observed. Figures 1(b) and (c) show the microstructures after exposure to the atmosphere at 450 and 550°C respectively, where the steel presented atomic diffusion by heating generating precipitation of carbides of tungsten, chromium, molybdenum and manganese among others. It was possible to show the precipitation of carbides and a martensite structure

In the Table 1 shows the chemical composition of the study material, this analysis carried out in SMT Laboratorio Prove Tecnologiche ed Analisi (Tenaris).

IOP Conf. Series: Journal of Physics: Conf. Series **935** (2017) 012056 doi:10.1088/1742-6596/935/1/012056

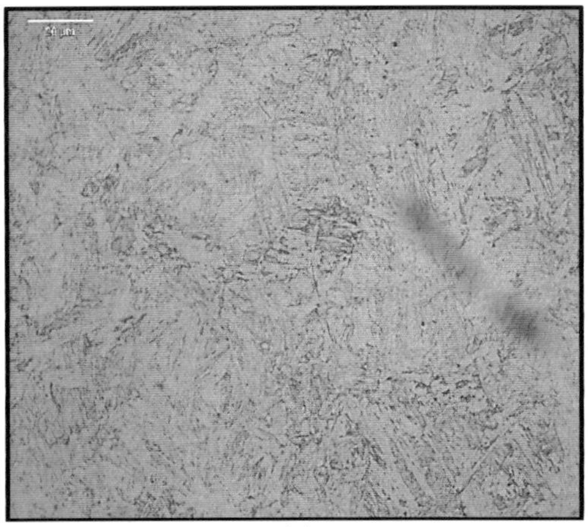

(a)

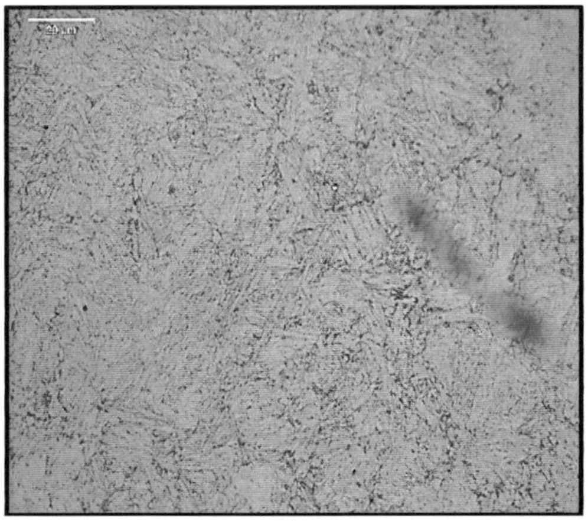

(b)

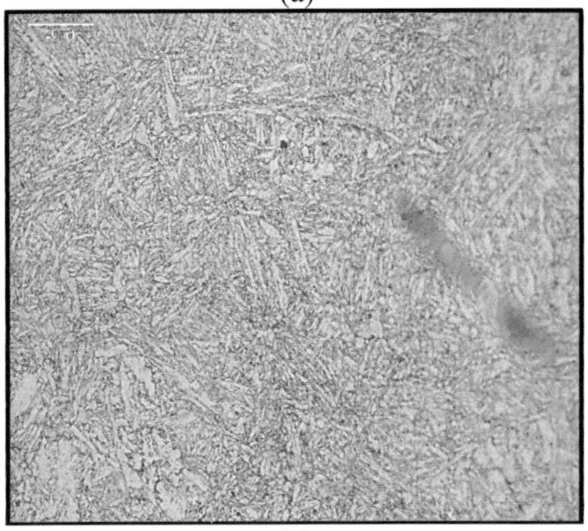

(c)

Figure 1. (a) Metallography P92 steel before exposure (b) Metallography P92 steel after exposure to the environment O_2-N_2-H_2O for 100h at 450°C (c) Metallography P92 steel after exposure to the environment O2-N2-H2O for 100h at 550°C.

Table 1. Chemical composition of steel ASTM A335 P92 (% weight).

Element	C	Mn	Si	P	S	Cr	Mo
Percentage	0.115	0.454	0.220	0.013	0.0033	9.14	0.4
Element	Ni	Al	V	Nb	W	N	B
Percentage	0.119	0.011	0.165	0.055	1.979	0.039	0.0022

Source: SMT Laboratorio Prove Tecnologiche ed Analisi (Tenaris)

By means of the SEM technique, the formation of three homogeneous layers on the surface of the steel P92 after a time of exposure to 100 hours in the atmosphere O2-N2-H2O at 450 and 550°C could be evidenced. According to what was reported in the literature, in addition to the XPS technique, the outer layer corresponded to hematite (Fe_2O_3), the intermediate layer to Magnetite (Fe_3O_4), and the inner layer corresponded to a spinel of chromium and iron [13]. It was observed that the inner layer has a homogeneous structure for both working temperatures. The presence of voids in the interface of the outer layer and the intermediate layer is observed, allowing signs of detachment of the layers from one to another. The hydroxides generally have greater volatility than the corresponding oxide, and this leads to loss in the protection of the material. Thus, the rapid diffusion of water into oxides is possible because by variation of protons located in the oxide ions that move by transferring one oxygen to another. Because the concentration of OH^- ions is high, allow the formation of compounds with cationic vacancies, and this in part is responsible for the observed increase in corrosion rates (Figure 2).

On the other hand, using the EDS analysis the determination of the elemental composition of three points in each layer formed after a time of exposure to 100 hours at 450°C was carried out. Through of this technique it was possible to identify the elements present in the layers formed on the surface (Figure 3).

Through the images of the outer layer in contact with the atmosphere and the intermediate layer, it is possible to observe the high percentage in the iron and oxygen elements, characteristic of the oxides referred to above according to the SEM characterization technique; in the inner layer in contact with the base metal it is possible to observe a higher concentration of the chromium element in comparison to the other two layers.

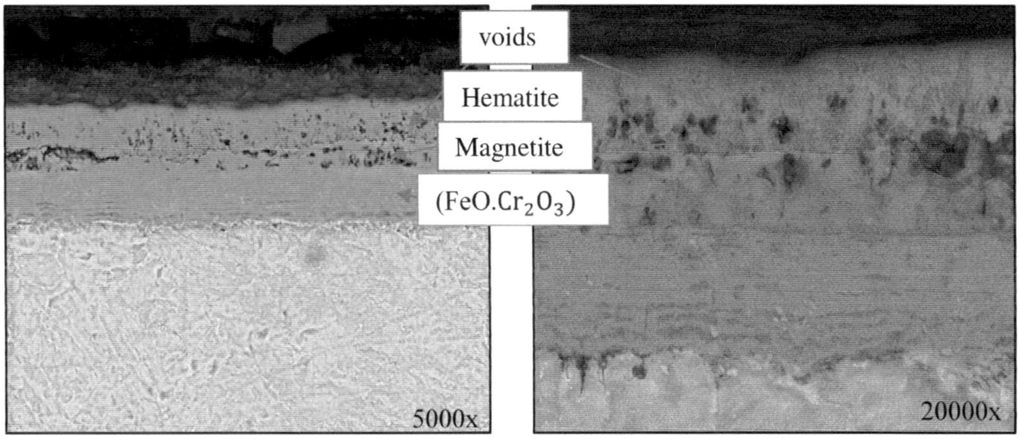

Figure 2. Layers formed on the surface of P92 after exposure to the O_2-N_2-H_2O environment for 100 hours (a) 450°C and (b) 550°C.

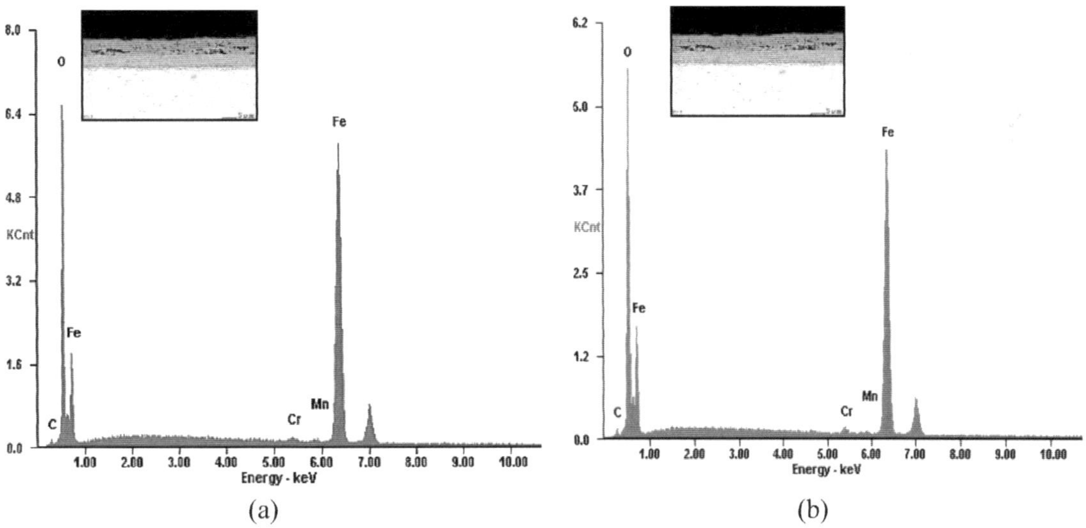

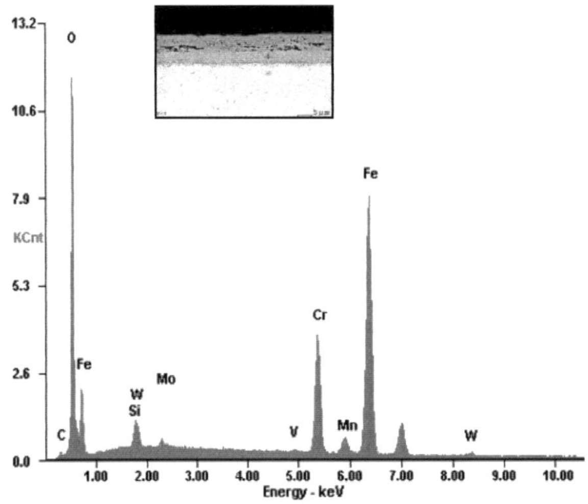

(c)

Figure 3. EDS of P92 steel at 100 hours and 450°C. (a) outer layer (b) intermediate layer and (c) inner layer in contact with the base metal.

With X-ray Photoelectron Spectroscopy (XPS) technique, it was possible to determine the elemental surface composition and the products formed, from the chemical properties of the elements. The analysed elements were chosen bearing in mind the composition of the steel and the corrosive environment. Table 2 shows the percentages by weight of the elements present.

Table 2. Elemental percentage at 100 hours, 450 and 550°C.

Element	450°C (%w)	550°C (%w)
O	25.446	22.799
Cr	2.4670	4.7805
N	0.6038	0.05850
Mo	1.4999	4.4601
Fe	32.666	30.599

It is verified the presence in greater proportion of elements like oxygen, chromium, and iron. On the other hand, bearing in mind that the steel studied contains 0.039% by weight of the initial elemental composition, the increase in the weight percentage of the nitrogen content in the surface is corroborated.

Regarding the electronic information and the products formed after the exposure to the corrosive environment, the following results were presented bearing in mind the XPS Handbook [14] and the program (simplified) containing the equipment software: Oxygen: It was found with atomic level 1s and according to the binding energy was determined the presence of metal oxides, SiO_2, and Al_2O_3. On the other hand, from the metal oxides, the binding energy of O_1 of many compounds and species is within a very narrow range. Thus, O_1 peaks tend to be wide, with multiple components superimposed and components due to water and/or organic contamination overlap directly with SiO_2, for example. Nitrogen: The atomic level is 1s and according to the binding energy indicates the presence of metal nitride Si_3N_4 and compounds like NSi_2O and $NSiO_2$. Chromo: The atomic level is 2p and strongly overlaps with Auger of Zn LMM peaks, which is very frequent in samples of steels. In the presence of high concentrations of zinc, it may be difficult to assign the chemical energy of the chromium using the Cr 2p region, due to the strong overlap with the Zn LMM Auger region. The Cr 2p peak has spin-orbit components. Thus, the presence of metallic chromium could be determined, but not the oxides present in P92 steel. Molybdenum: The atomic level is 2s and the presence of metal Mo, MoO_2 and MoO_3 is evidenced. Iron: The atomic level is 2p and tends to overlap strongly with Auger Ni LMM line peaks. Iron metal, FeO and Fe_2O_3 are present.

Also, the kinetic study was carried out by means of the analysis of the curves obtained of the variation of the mass as a function of the time of exposure to the working temperatures. The data recorded to determine the corrosion rate were expressed as mass gain per unit area (g/mm^2). The curves show a clear parabolic tendency up to 50 hours of exposure for both temperatures, after this time the steel has a mainly protective characteristic, attributed to the stabilization of solids by means of the alloying elements such as tungsten, vanadium, among others (Figure 4). Based on non-linear regressions, it was possible to establish the speed laws for the two working temperatures. These can be seen in Table 3.

The kinetic constants obtained are lower than those reported by Ehlers at 650°C, as expected, for a similar oxidizing environment, composed of N_2-1% volO_2- (2-7% vol) H_2O [15].

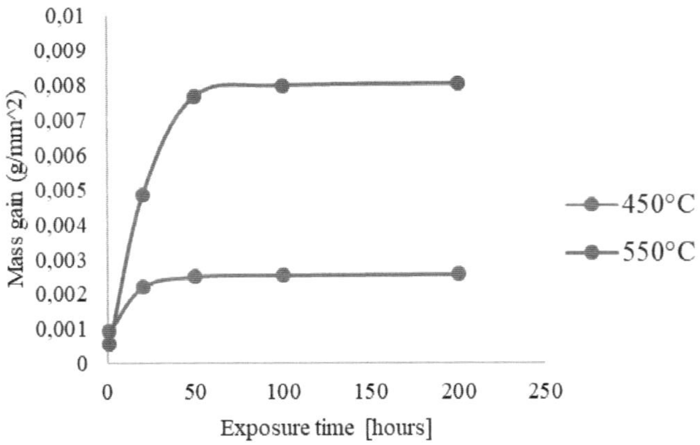

Figure 4. Exposure time vs mass gain of P92 steel at 450 and 550°C.

Table 3. Kinetic constants and speed laws for 450 and 550°C.

Work temperature (°C)	Speed law	Kinetic constants ($g^2\ mm^{-4}\ h^{-1}$)	Percentage error
450	$x^2 = 2.037x10^{-8}t + 3x10^{-6}$	$2.037x10^{-08}$	$2.6380x10^{-06}$
550	$x^2 = 2.77x10^{-07}t + 2x10^{-05}$	$2.77x10^{-07}$	$2.806x10^{-05}$

4. Conclusions

Metallographic analyses showed changes in the steel microstructure, indicating the presence of carbide precipitation at the grain boundaries and a martensitic structure with thinner needles. On the other hand, the formation of protective oxide layers was evidenced on the surface of the material, which maintains in good condition the properties in service of this steel at these temperatures and working time. The results obtained from SEM-EDS and XPS characterization techniques showed congruence between the corrosion products identified in the P92 steel, among which are nitrides, and oxides, finally, the kinetics of the oxidation-nitriding process showed a behaviour with an accelerated growth until 50 hours, then the variation was less; also, was possible evidence the direct relationship between the corrosion rate and the temperature in the two kinetic constants found for 450 and 550 °C.

Acknowledgments

The authors would like to express their gratitude to the Colombian network of knowledge on energy efficiency (RECIEE) for all the support provided for the achievement of this work. We would also like to thank "Colciencias" through the program Young Researchers Convention No. 761 of 2016; to the (Universidad Industrial de Santander" for all the efforts that it makes for the accomplishment of studies in this scientific area, especially to your laboratories of SEM and XPS.

IOP Conf. Series: Journal of Physics: Conf. Series **935** (2017) 012056 doi:10.1088/1742-6596/935/1/012056

References

[1] Alvarez E and Infazon D 2009 *Oxidación isotérmica y cíclica del acero ASTM A335 P92 a alta temperatura* (Bucaramanga: Universidad Indusrial de Santander)

[2] Van Wortel J C, Etienne C F and Arav F 1992 Application of modified 9chromium steels in power generation components *Conference Proceedings The manufacture and properties of steel 91 for the power plant and process industries: ECSC Information Day, (Dusseldorf)* (Germany: TIB) paper 4.2

[3] Fujita T 1992 Current progress in advanced high Cr steel for high temperature applications *ISIJ International* **32(2)** 175.

[4] Birchenall G.E 1981 A brief history of the study of oxidation of metals and alloys *High Temperature Corrosion Proceedings* (USA: Nace) p 3

[5] Jones D A 1996 *Principles and prevention of corrosion second edition* (USA: Prentice Hall)

[6] Gil L 2013 Corrosión en la industria del petróleo y el uso de tecnologías de recubrimientos como alternativa de protección *Memorias del Segundo Encuentro Internacional de Investigadores en Materiales y Tecnología del Plasma* (Bucaramanga: Universidad Pontificia Bolivariana) p 24

[7] Khanna A S, Rodriguez P and Gnanamoorthy J B 1986 *Oxid. Met.* **26** 171-200

[8] Fujii C T, Meussner R A and J. Electrochem 1963 *Soc.* **110** 1195-1204

[9] Pujilaksono B, Jonsson T, Halvarsson M, Svensson J and Johansson L 2010 *Corros. Sci.* **52** 1560–1569

[10] Birks N and Arnold E 1983 *Introduction to high temperature oxidation of metals* (London: Edward Arnold Publishers Ltd)

[11] ASTM G1-03 2011 *Metals Standard practice for preparing, cleaning, and evaluating corrosion test specimens* (USA: ASTM International)

[12] Cala O M, Merriño L, Kafarov V and Saavedra J 2013 Efecto de la composición del gas de refinería sobre las características del proceso de combustión *Rev. Ingenierías Universidad de Medellín* **12(23)** 101-112

[13] Serna J 2003 *Oxidación, carburación y sulfidación de aleaciones ferríticas Fe-9Cr-1Mo modificadas en ambientes con hidrocarburos a temperaturas entre 550 y 750°C* (Bucaramanga: Universidad Industrial de Santander)

[14] Moulder J 1992 *Handbook of X-ray photoelectron spectroscopy: A reference book of standard spectra for identification and interpretation of XPS data* ed Jill Chastain, Roger C. King (USA: Physical Electronics) p 43-49

[15] Ehlers J *et al* 2006 Enhanced oxidation of the 9%Cr steel P91 in water vapour containing environments *Corros. Sci.* **48(11)** 3428–3454.

Study of the continuous corrosion in an oxidation environment derived from the theoretical combustion products in a refinery. Case study: Ferritic steel ASTM A335 P91

A Alviz Meza[1], V Kafarov[1] and D Y Peña Ballesteros[1]
[1] Universidad Industrial de Santander, Bucaramanga, Colombia

E-mail: anibalalvizm@hotmail.com

Abstract. Corrosion studies are key element that ensure the correct functioning of equipment in the industrial sector. The oxidation phenomena were evaluated, taking as a case study steel ASTM A335 P91 (P91), a steel of typical use in equipment that work at high temperatures. Five (5) exposure times were selected for the experimental development: 1, 20, 50, 100 and 200h; as well as four (4) analysis temperatures: 450, 550, 650 and 750°C. Through the metallographic analysis was possible to evidence the presence of multiple carbide precipitates and a ferritic structure, after all the temperatures tested. On the other hand, the analysis of hardness and microhardness showed an increase for all the evaluated temperatures. These increases were mainly related to the precipitation of carbides in the coupons of P91. Regarding to the chemical analysis, it was possible to conclude that after 200h of experimentation in each temperature, a layer of duplex oxide, composed mainly of hematite, magnetite and spinel iron-chromium, was formed in the O_2/H_2O atmosphere. Finally, the kinetic study demonstrated that the oxide layer formed on each coupon of P91 was of protective character.

1. Introduction

The most important reaction in the processes developed at high temperatures is oxidation, being that when the oxide layers formed are protective, other types of corrosion can be prevented, as is the case of internal corrosion. However, not all oxide layers formed are protective; in the absence of water vapour a protective chromium-rich spinel scale is developed, but upon introduction of water vapor breakaway oxidation occurs [1,2].

The oxidation behaviour of Cr-Mo steels in water vapor, supercritical water, air and pure oxygen has been extensively studied [3–11]. When there is presence of water vapor, it has been extensively reported that the oxide layer formed is constituted by a duplex structure; composed of Fe and Cr oxides (Fe_2O_3, Fe_3O_4 and $Fe_{3-x}Cr_xO_4$, Cr_2O_3, FeO). The outer layer is predominantly in iron oxides, while the innermost layer contains a significant amount of chromium. In addition, the effect of the water vapor has a disruptive effect on the oxide layers, generating the formation of volatile oxides such as: Cr(g), CrO_2(g), CrOOH(g), $CrO_2(OH)_2$(g), $Cr(OH)_6$(g), FeOH(g), $Fe_{0.947}O$(g), Fe_2O_3(g), which generate voids, pores, cracks and detachment of oxide layers, facilitating the damage by other forms of corrosion [12].

However, although that the oxidation process on ASTM A335 P91 (P91) steel has been extensively studied, there is little information on how this phenomenon occurs under simulated combustion conditions in a refinery.

2. Methodology

The methodology selected to evaluate the HTC on P91 steel was divided into a simulation part and an experimental part, as follows in the next sections.

2.1. Simulation methodology

The simulation in this study required the selection of a mixture model of RG from a furnace of Barrancabermeja's refinery; it was done through the statistical study of the most representative compounds found in the historical chromatography's records. With the RG mixture were obtained the theoretical combustion products (TCP) in Aspen HYSYS 8.6, which were used in HSC chemistry 5.11 to obtain the theoretical corrosion products (TCOP). The procedure is shown in the Figure 1.

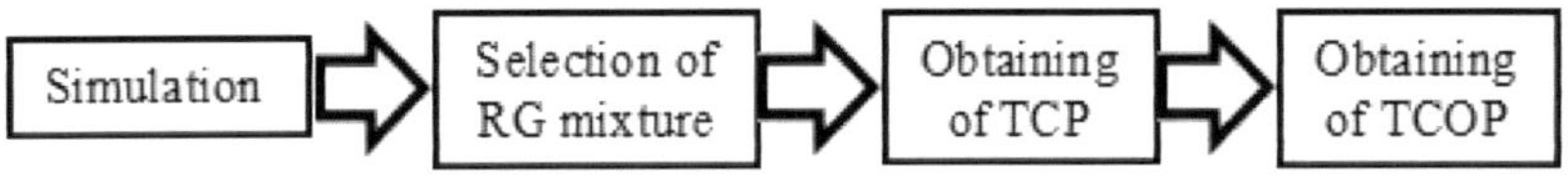

Figure 1. Methodology for the simulation of theoretical corrosion products.

2.2. Experimental methodology

For the experimental development, the installation of corrosion coupons of P91 steel in the furnaces from the experimental scheme shown in Figure 2 was realized. The dimensions selected for each coupon were 15×10×2mm (height×length×Width), with enough area to get appreciable mass gain.

As is shown in Figure 2, to complete the required mixture of O_2/H_2O, the humidification of the O_2 gas was performed through the use of a bubbler. This bubbler was initially calibrated at the appropriate temperature, in order to obtain the simulated molar composition of the TCP in the furnaces.

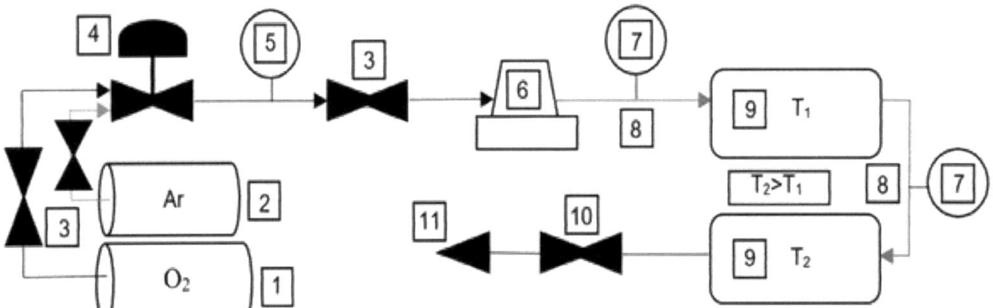

Figure 2. Experimental setup, where: (1) Cylinder with O_2, (2) Cylinder with Air, (3) Flow valve, (4) Mass flow meter, (5) Flow regulator, (6) Humidifier, (7) Pressure gauge, (8) Heating cord, (9) Horizontal tubular electric furnace, (10) Relief valve, (11) Gas and steam outlet.

With regard to the preparation of the coupons; the coupons were sanded before testing with silicon carbide paper, until reaching the sandpaper 600 on all the sides. Each coupon was subjected to ultrasonic cleaning in acetone, dried and marked, as is recommended by standard ASTM G1. The coupons were characterized by the following techniques: SEM-EDS, XRD, hardness, microhardness and optical microscopy. In addition, the kinetic study was carried out taking into account the dimensions of the coupons, as well as the mass gained by each one in the different times of exposure. The complete experimental methodology is shown in the Figure 3.

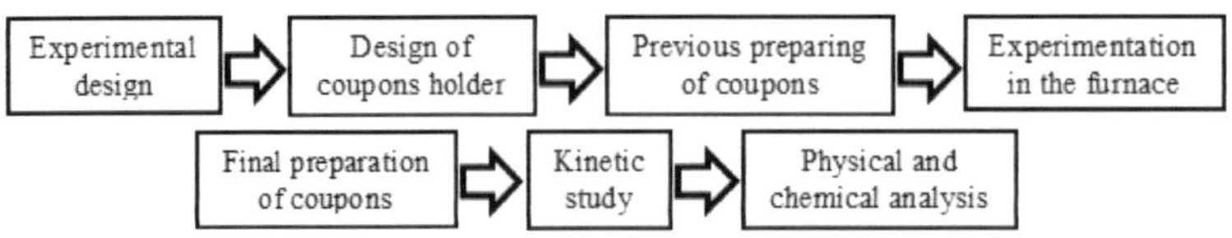

Figure 3. Methodology for the experimental development.

3. Results and discussion

It is important to mention that the temperature in this study was selected according to a thermography taken from the radiation zone of a furnace from Barrancabermeja's refinery. So, the temperature in this work was selected between 450 and 750°C.

3.1. Determination of combustion products and gas mixture

The selection of the RG model mixture was made based on the average of the compounds of higher occurrence in a chromatographic study of a refinery furnace [13]. The Table 1 shows the composition.

Table 1. Composition of the mixture of RG.

Compounds	CH_4	C_2H_6	C_4H_{10}	C_2H_4	C_3H_6	H_2S	H_2
%Molar	55	10	4	5	2	4	20

The TCP were determined for a 10% of air in excess in an adiabatic combustion furnace simulated in Aspen HYSYS 8.6. The TCP are shown in Table 2.

Table 2. Composition of the combustion gases.

Compounds	CO_2	O_2	N_2	H_2S	H_2O
% Molar	9.10	1.74	72.15	0.34	16.67

Starting from the molar ratio of the TCP, the molar composition of the O_2/H_2O sub-mixture was obtained (Table 3).

Table 3. Composition of gas mixture used.

Compounds	O_2	H_2O
% Molar	9.47	90.53

Finally, to obtain the flows used, a mass flow of 333.4mg/min was assumed for the complete mixture of the TCP. Thus, with the physical properties of the gases at 298.15K, a volumetric flow of 5.01cm^3/min for the O_2 gas and 35.83mg/min for the mass flow of water vapor in the humidifier were calculated.

3.2. Theoretical combustion products

By using the HSC Chemistry program and the phase stability diagram, theoretical potential corrosion products were established between 450 and 550°C, for one atmosphere of pressure. The molar ratio used between steel and gas was 1/1000 as is recommended in other studies [14]. The Table 4 shows the composition of the P91 steel.

Table 4. Composition of P91 steel.

%Mo	%C	%Si	%Mn	%P	%S	%Ni	%Cr	%V	%Nb	%Al	%N	%Fe
0.989	0.106	0.768	0.316	0.013	0.003	0.271	8.439	0.024	0.008	0.006	0.015	88.807

Among the corrosion products found on the P91 steel for all the temperatures tested, mainly oxides were obtained: hematite (Fe_2O_3), magnetite (Fe_3O_4), wplustita ($Fe_{1-x}O$), chromium oxide (Cr_2O_3), silicon oxide (SiO_2), molybdenum oxide (MoO_3), manganese oxide (MnO_2), CrO_2, FeOOH and spinels of Cr and Mn (FeCr2O4 and Fe_2MnO^4). On the other hand, using the phase stability diagram between 450 and 750°C, it was showed that is thermodynamically feasible the stabilization of iron oxide (Fe^2O^3) and chromium oxide (Cr_2O_3) on P91 steel in an atmosphere of O_2/H_2O, as shows the dot in Figure 4.

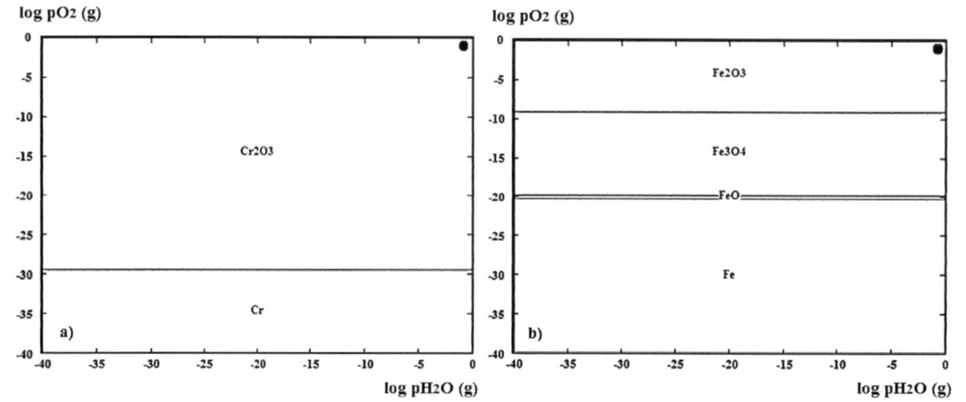

Figure 4. Phase stability diagram between 450 and 750°C: (a) Cr-N-O and (b) Fe-N-O.

4. Experimental results

The results of the physical and chemical analysis performed, in addition to the kinetic study, are presented as follows.

4.1. Metallographic analysis

In Figure 5(a), it is possible to observe fine grained, martensitic structure, with homogeneous distribution of precipitates in the matrix, characteristic behaviour of normalized and tempered P91 steels. Additionally, the increase of elongated and rounded carbides could be noted with the increase in temperature (Figure 5(b)).

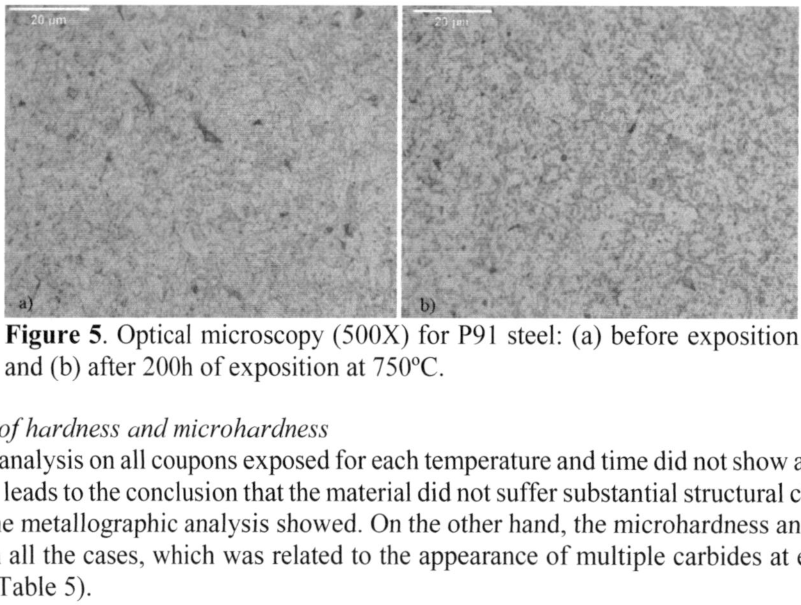

Figure 5. Optical microscopy (500X) for P91 steel: (a) before exposition and (b) after 200h of exposition at 750°C.

4.2. Analysis of hardness and microhardness

The hardness analysis on all coupons exposed for each temperature and time did not show any significant changes. This leads to the conclusion that the material did not suffer substantial structural changes during the tests, as the metallographic analysis showed. On the other hand, the microhardness analysis showed an increase in all the cases, which was related to the appearance of multiple carbides at each exposure temperature (Table 5).

Table 5. Analysis of hardness and microhardness after 200 h of test in all the temperatures evaluated. The force applied by each analysis was respectively 588.40 and 0.49N.

	Initial state	450°C	550°C	650°C	750°C
Hardness (Rockwell A)	47.00	47.65	47.20	49.64	49.80
Microhardness (Vickers HV)	229.40	244.05	243.48	288.40	255.40

4.3. Scanning electron microscopy with x ray microanalysis

Through this study, it was possible to observe the morphology of the oxide layers deposited on P91 steel. In addition, from the EDS analysis it was identified the elemental profile of the different oxide layers. In Figure 6, two layers of oxides can be observed for all the temperatures studied, after 200h of exposure to the O_2/H_2O atmosphere. As has been reported extensively in other oxidation studies on P91 steel, these two oxide layers correspond to: an outer layer of hematite and magnetite and an internal, which is the thickest, composed mainly by an iron-chromium spinel. This latter only of protective behaviour at 450 and 550°C. Additionally, the presence of cracks, pores and voids in the outermost layers can be observed, so it is also suggested the formation of volatile oxides, a characteristic behaviour during the phenomenon of breakaway oxidation on P91 steel. Finally, a depletion of the alloying elements in the metal interface is also observed, an effect that became more severe with the increase in temperature.

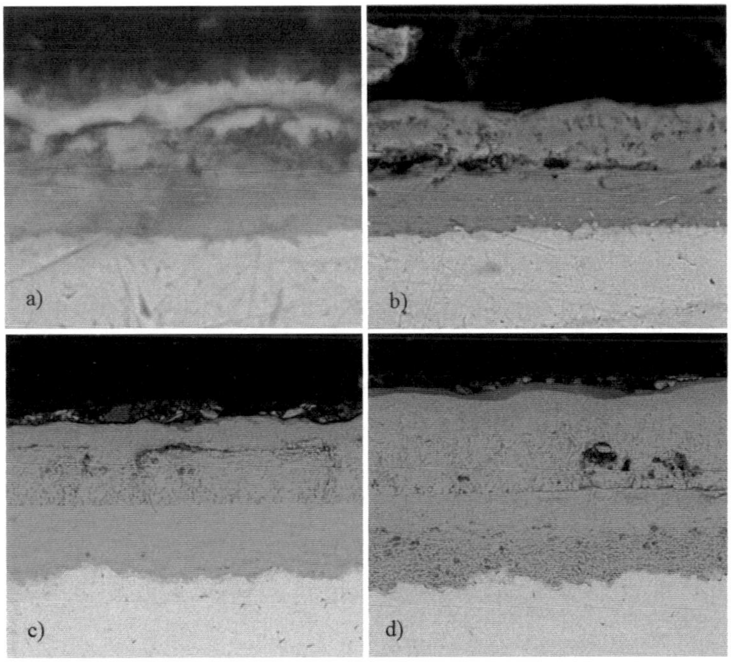

Figure 6. SEM for coupons of P91 exposed to 200h at: (a) 450°C (5000X), (b) 550°C (2000X), (c) 650°C (500X), (d) 750°C (250X).

Regarding the oxidation mechanism, it is possible to take into account the explanation given by Quadakkers *et al.* [3]: The breakdown of the protective spinel scale is accompanied by formation of a rapidly growing magnetite layer and an inner scale consisting of Cr_2O_3 precipitates in a "FeO-matrix". Then the separation of the oxide layers is due to the appearance of multiple vacancies in the spinel layer, which promotes the formation of the magnetite layer. Further the corrosion mechanism depends of the "O_2/H_2O bridges", which are in charge of the transport processes between the voids, as is suggested by Rahmel and Tobolski [15]. Finally, due to the formation of large voids, diffusion of the Fe atoms between the layers is difficult, resulting in a layer of hematite, which is an oxide in equilibrium with the surrounding environment. From that point, the growth of the oxide layers is largely temperature dependent.

Otherwise, in Figure 7 the presence of the three oxide layers formed on P91 coupons, are evident when the fluctuations between chromium, iron and oxygen are monitored. In addition, Mn, Si and Mo are present mostly in the innermost oxide layer.

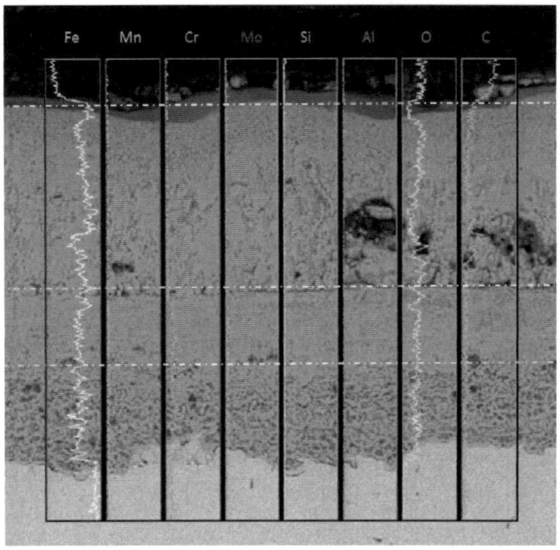

Figure 7. SEM-EDS for coupons of P91 exposed to 200h at 750°C (250X). The elemental concentration increases from left to right in each box.

4.4. X ray diffraction analysis
Through the conventional X ray diffraction analysis, it was identified the compounds formed on the surface of P91 steel. The presence of magnetite (Fe_3O_4) and hematite (Fe_2O_3) were confirmed. However, considering the results of EDS, the formation of chromium-rich spinels in the inner oxide layers is not ruled out, nor is the presence of other oxides.

4.5. Kinetic study
The behavior of the graphs in Figure 8 is close to parabolic, which suggests the presence of layers of protective oxides, where the diffusive processes decrease with time. The results obtained through the analysis of SEM support the presence of a protective internal oxide layer, composed mainly of iron-chromium spinel.

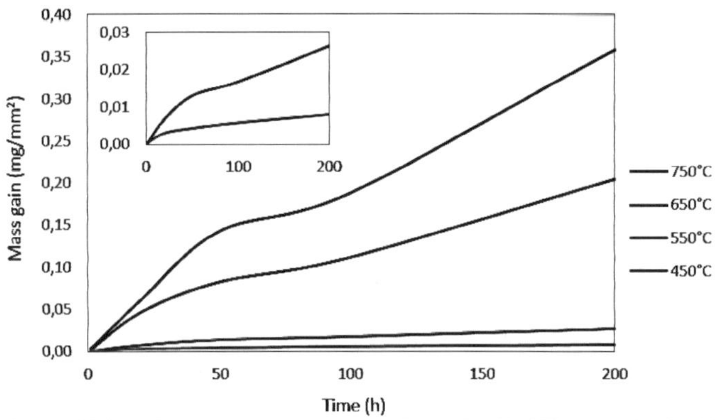

Figure 8. Thickness of the oxide layer against time for each study temperature.

The kinetic constants showed a clear increase in the oxidation rate at 650 and 750°C, compared to 450 and 550°C (Figure 6). Finally, is possible to say that the relationship between the oxidation kinetics and the temperatures used, was not linear but rather quadratic for P91 steel (Table 6).

Table 6. Kinetic constants for temperatures tested.

	450°C	550°C	650°C	750°C
Kinetic constants ($mg^2 * mm^{-4} * h^{-1}$)	$3.30x10^{-7}$	$5.12x10^{-6}$	$2.07x10^{-4}$	$6.41x10^{-4}$

5. Conclusions

Coupons of P91 were exposed to an oxidation environment obtained from a model refinery mixture and the molar ratios of its theoretical combustion products. It was found that there was congruence between the experimental and simulations results; where the compounds reported by EDS and XDR analysis showed the presence of external oxide layers rich in iron and inner layers rich in chromium. Also, the phenomenon of breakaway oxidation between the inner and outer layers was presented, revealing the weak protective character of iron oxide layers in wet environments. These results are similar to those reported by several researchers in wet environments, information very well compiled by Saunders *et. al*. [7]. These conclusions, were supported by kinetic analysis, where parabolic behavior was evidenced in the mass gain graphs; noting that the oxidation kinetic constants presented a quadratic relation with the increase in temperature. further, the hardness and microhardness analysis showed respectively no substantial changes and a slight increase. This behavior suggests small changes in the structure of the material and the presence of multiple precipitates of carbides, as was corroborated in the ferritic structure of P91 steel using the optical microscopy.

Acknowledgments

The authors want to give thanks to "Universidad Industrial de Santander, Decanatura de la Facultad de Ingeniería Fisicoquímicas, Centro de Investigación para el Desarrollo Sostenible en Industria y Energía and Grupo de Investigaciones en Corrosión".

References

[1] Young D J 2008 *High Temperature Oxidation and Corrosion of Metals* vol 2 (Amsterdam: Elsevier) pp 539-589
[2] Lai G Y 2007 *High Temperature Corrosion and Materials Applications* vol 1 (USA: ASM International) pp 5-62
[3] Ennis P J and Quadakkers W J 2008 *Creep-Resistant Steels* **19** 519–535
[4] Ehlers J, Young D Y, Smaardijk E J, Tyagi A K, Penkalla H L, Singheiser L and Quadakkers W J 2006 *Corrosion Science* **48** 3428–3454
[5] Laverde D, Gómez-Acebo T and Castro F 2004 *Corrosion Science* **46** 613–631
[6] Swaminathan S, Mallika C, Krishna N G, Thinaharan C, Jayakumar T and Mudali U K *Corrosion Science* **79** 59–68
[7] Saunders S R J, Monteiro M and Rizzo F 2008 *Prog. Mater. Sci.* **53** pp 775–837
[8] Ostwald C and Grabke H J 2004 *Corros. Sci* **46** 1113–1127
[9] Zurek J, Wessel E, Niewolak L, Schmitz F, Kern T U, Singheiser L and Quadakkers W J 2004 *Corros. Sci.* **46** 2301–2317
[10] Castañeda S I, Bolivar F J and Perez F J 2010 *Oxid. Met.* **74** 61–78
[11] Martinelli L, Desgranges C, Rouillard F, Ginestar K, Tabarant M and Rousseau K 2015 *Corrosion Sciencie* **100** 1–29
[12] Perez F J and Castañeda S I 2006 *Oxid. Met.* **66** 231–251
[13] Cala O M, Kafarov V and Saavedra J 2013 *Rev. Ing. Universidad de Medellin* **12(23)** 101–111
[14] John R 2010 Sulfidation and mixed gas corrosion of alloys *Shreir's Corrosion* vol 1, ed Tony J.A. Richardson (Amsterdam: Elsevier B. V.) p 247
[15] Rahmel A and Tobolski J 1965 *Corros. Sci.* **5(5)** 333–346

IOP Conf. Series: Journal of Physics: Conf. Series **935** (2017) 012058 doi:10.1088/1742-6596/935/1/012058

Correlation between the thermal performance and the microstructure of the material used in medium and high voltage transformer terminals

Y J Salas[1], S P Vera-Monroy[2], A Mejia-Camacho[2] and W Rivera[3]

[1] Universidad Incca, Bogotá, Colombia
[2] Universidad Manuela Beltrán, Bogotá, Colombia
[3] Universidad Antonio Nariño, Tunja, Colombia

E-mail: yecid.salas@unincca.edu.co

Abstract. In Colombia, energy companies neglect the distribution that represents the main and most valuable process, presenting shortcomings in prevention and forecasting programs, using contractors who perform corrective maintenance of the components without guaranteeing the quality and performance of the materials. Within the process, the terminals determine the effective connection between the voltage line and the transformer, which have faults that are evidenced by the thermal deterioration of the material. In this work, a diagnosis of the thermal performance of these components was carried out and it was correlated with the microstructure, observing variations of the working temperature, with a thermography camera, for three types of terminals, which were classified by X-ray fluorescence in brass Z20, Z40 and Z60, and for two types of connection, copper and aluminium. The microstructure results showed that copper is the conductor that degrades the terminals faster, evidencing cracking of the material; on the other hand, the Z40 brass was the most stable with the lowest temperature variation regardless of the conductor diameter; however, in all cases the behaviour of higher temperature to lower calibre is satisfied.

1. Introduction

The secondary output terminals are electrical devices of distribution transformers, responsible for connecting the transformer to user connections, these terminals suffer wear by meteorological agents that affect their integrity and compromise their operation [1-3], causing power interruptions, generating economic losses for the company [4-5], who must take urgent intervention and compensation measures, on the other hand, there is discomfort in users because it affects daily lives, the food rots or because the services offered remain inoperative [6-7].

The physicochemical deterioration of terminals affects transformer performance, reduction of effective conduction area and increase in resistance due to the oxides formation, common anomalies that cannot be completely stopped, for that reason, to extend the lifetime of these devices and to develop effective strategies to reduce their impact, are problems that the scientific community strives to solve [8-12].

The thermal monitoring of electrical substations is a routine and effective exercise to quickly perform measurements, evaluating the operation of different electrical parts, easily identified by the relative increase of temperature that they present before failure, this increase of temperature accelerates the processes of degradation that affect in a unlike way to the connections, because they are manufactured in different materials [13-15].

In this paper, the correlation between the deterioration presented in secondary output terminals of distribution transformers for Energy Company from Boyacá, is determined by means of a thermal, of composition and microstructure study.

2. Methodology

Analysis of terminals started whit measurement of work temperatures, they were obtained in peak hours, performing 4 replicates using a thermographic camera, collecting the samples showing higher temperature, to which they were made: firstly, composition analysis by X-Ray Fluorescence (XRF) and secondly, analysis of microstructure, on the parts of smaller diameter that suffer greater wear by Joule effect, carefully following the methodologies described in Table 1.

Table 1. Ratio of ASTM standards used in microstructural analysis.

Analysis	Standards ASTM	Analysis	Standards ASTM
Metallographic preparation	E 03–01	Micrographs	E 407–99
Grain size	E 112–96	Inclusions	E 45–97

The experimental matrix was constructed based on three factors (Table 2). (a) type of connection, copper base and aluminium base; (b) manufacturing material of terminals, brasses Z20, Z40 and Z60 and (c) calibre of the connectors, thicknesses: 1/0, 2/0 and 4/0.

Table 2. Experimental matrix.

Connection	Material of terminals	Connection	Material of terminals
	Brass Z20		Brass Z20
Copper	Brass Z40	Aluminium	Brass Z40
	Brass Z60		Brass Z60

3. Results and discussion

The series of terminals used (Figure 1) in physical appearance are of varied materials, to establish the composition of each of them, metallographic and microstructure tests were carried out, including the grain size.

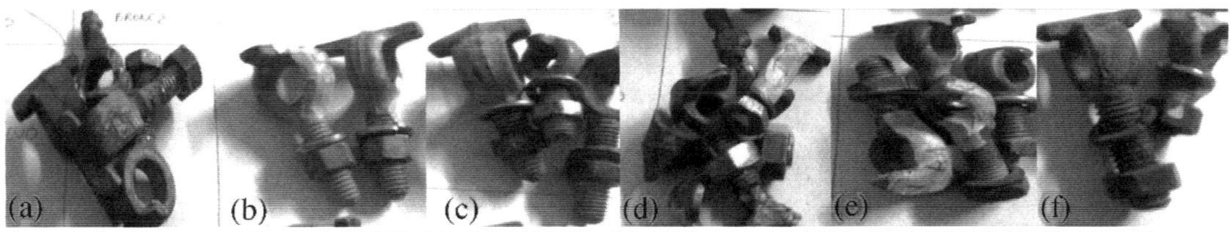

Figure 1. Brass terminals (a) Z20 Al, (b) Z20 Cu (c) Z40 AM (d) Z40 Cu (e) Z60 Al (f) Z60 Cu.

3.1. Composition analysis using FRX

The composition of the samples (Table 3) in the polished inner section (core) was typical of a muntz brass (60-40) and a bearing brass (80-20) with few traces of iron inherent in the manufacturing process. The external section of the unpolished samples, they had chlorine contamination, which affects the integrity of the material, as it forms cuprous and cupric chlorides, generating an unstable blue-green patina, corroding the piece, these salts reacts with the environment to produce hydrochloric acid, modifying the corrosion rate of the base material.

Table 3. Percentage composition of the terminals, obtained by FRX.

Sample	Section	Composition (%)			
		Fe	Cu	Zn	Cl
Z40	Core	0.38	59.2	40.4	-
Z20	Core	0.68	79.6	20.0	-
Z20	External	0.50	39.2	56.3	4.10
Z60	External	0.78	0.6	0.10	98.5

3.2. Counting inclusions

The micrographs of Figure 2, show that material of analysed pieces is homogeneous, with no evidence of inclusions. During scanning, cracks were observed in the Z60-Cu and Z60-Al, (Figure 3(a)) samples, possibly due to residual stresses confined within material generated by sudden temperature changes, a pore is also observed (Figure 3(b)), as a result from process of solidification, which is common in melt-formed pieces.

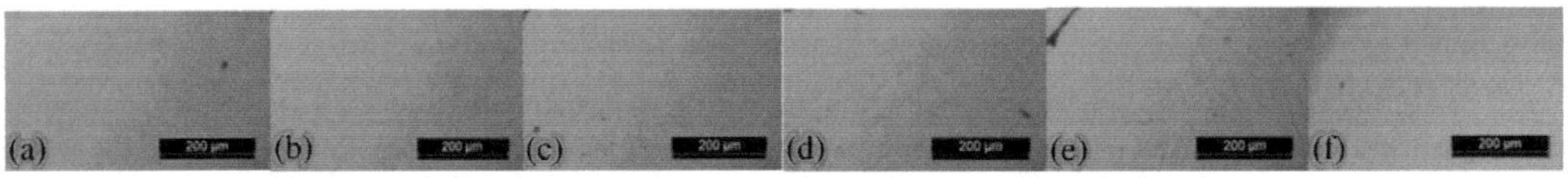

Figure 2. Micrographs (a) Z20 Al, (b) Z20 Cu (c) Z40 AM (d) Z40 Cu (e) Z60 Al (f) Z60 Cu.

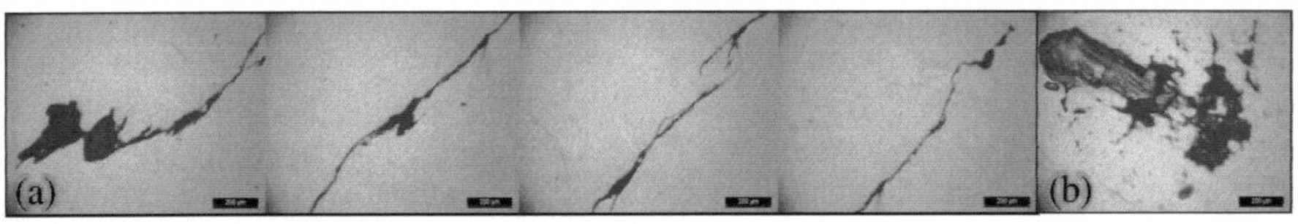

Figure 3. (a) cracks present in sample Z60 (b) pore present in sample Z60-Copper.

3.3. Determination of grain size

Figure 4(a) shows the typical microstructure of a brass moulded in sand, where it is observed a single-phase present due to the Zn content, the black spots correspond to small pores and fissures due to stresses created in the process of cooling. Figure 4(b) shows a change in the grain elongation, which is like obtained in a brass 60-40 of phase β, that when they cooled initiates precipitation of α phase, which has higher concentration of copper. Figure 4(c) presents the regular grain of a brass with low Zn content, its granulometry is 4 sizes, when compared to template. Figure 4(d) shows a microstructure similar to that presented in a Brass 60-40 after being initially subjected to heating to 600°C and then to an abrupt cooling, for this reason, α phase is precipitated and the excess copper is retained, the microstructure also has an elongation of grain which hinders determine its size. Figure 4(e) and (f) show a clear grain growth, such as that found in muntz brass after being subjected to heating to 800°C and later cooled abruptly, this thermal modification favours that the grains evolve to more equiaxial and great forms.

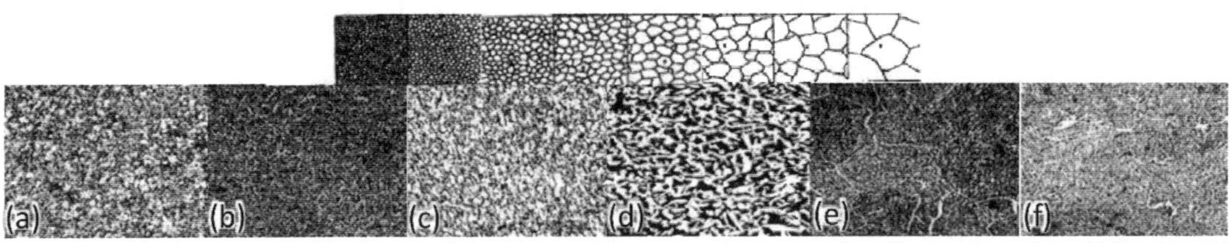

Figure 4. Grain size–100X. At top of the image the comparison template is displayed (a) Z60–Al, (b) Z40–Al, (c) Z20–Al, (d) Z60–Cu, (e) Z40–Cu y (f) Z20–Cu.

3.4. Operating temperature
As a result of operating temperature averages (Figure 5) it was found that the fact of making copper connections to any terminal always has higher operating temperature when compared with aluminium, also it was established that in the two connections the manufacturing material with the highest operating temperatures is the Z60 brass, this fact is more evident in copper connection, as far as the conductor calibre, which is inversely proportional, compared to the operating temperature.

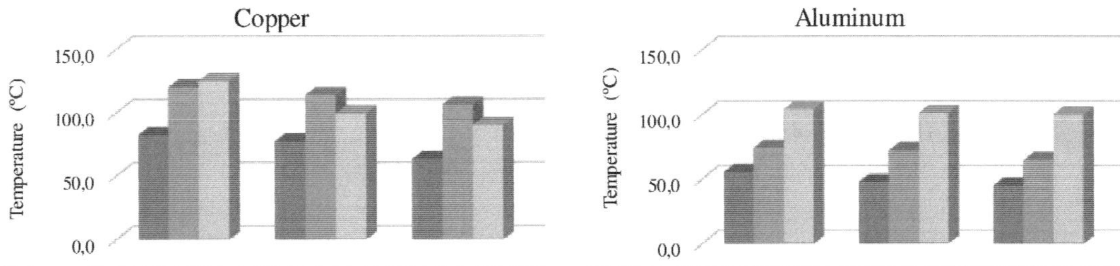

Copper	Caliber 1/0	Caliber 2/0	Caliber 4/0
Brass Z20	82.6	77.4	63.5
Brass Z40	120.4	114.0	106.7
Brass Z60	125.2	99.8	90.3

Aluminum	Caliber 1/0	Caliber 2/0	Caliber 4/0
Brass Z20	55.5	47.8	45.3
Brass Z40	74.1	72.3	64.8
Brass Z60	104.3	101.4	100.3

Figure 5. Average operating temperatures for terminals connected to copper and aluminium.

4. Conclusions
The reduction of the calibre increases the working temperature, which is higher in the copper connections, degrading the material of the terminals more quickly.

The high operating temperature in terminals, such as the one presented in sample Z60, accelerate the degradation processes, increasing the number and size of the cracks, as evidenced in the micrographs taken to this material.

The Z40 brass presented high operating temperatures, however, no significant degradation was found in obtained micrographs, showing that its composition presents a greater thermal stability.

Most terminals that are continuously exposed to sudden changes in temperature since their manufacture, show a weakening in their granulated structure, thus facilitating their deformation continuous and subsequent failure.

References
[1] Salas B Schorr M Carrillo M Zlatev R Stoycheva M Lopez G Vargas L Terrazas G Ocampo G 2012 *Air Quality - New Perspective* ed G Lopez Badilla, B Valdez and M Schorr, (Croatia: InTech) chapter 13
[2] Cárdenas E Garcia S Fernández F Dzul A 2012 *Rev. Fac. Ing. Univ. Antioquia* **64** 57
[3] Cárdenas C Pardo O Barajas E Santos A 2016 *Tecciencia* **11(20)** 27-33
[4] Davis J 2000 The Effects and Economic Impact of Corrosion *Corrosion: Understanding the basics* ed Davis & Associates (Ohio: ASTM International) chapter 1 pp 1-17
[5] Cortés E 2013 *Andamios* **10(21)** 345-369
[6] Navas D Ramírez H Echeverry I 2013 *Revista Esc. Ing. Antioq.* **19** 23-31
[7] Gomez V Peña R Hernández C 2012 *Inf. y Tec.* **23(2)** 109-116
[8] Murugan R Ramasamy R 2015 *Eng. Fail. Anal* **55** 182-192
[9] Kusumadevi G Gurumurthy G 2015 *J. Elec. Sys. Inf. Tech.* **2(2)** 172-177
[10] Shayan T Jan T Afzal R Khan A 2015 *International Conference Data Mining, Civil and Mechanical Engineering (Bali)* (Indonesia: International Institute of Engineers) (http://iieng.org/allproceedings.php/29) pp 49-52
[11] Martin Y Corvo F Castaleda A Valdés González Pérez J Portilla C 2007 *Cenic Ciencias Quimicas* **38(1)** 219-225
[12] Song J Wang L Zibart A Koch C 2012 *Metals.* **2** 450-477
[13] Nazmul A Taib S 2013 *App. Therm. Eng.* **61(2)** 220-227
[14] Mora E, Venegas E 2011 *Rev. Met. UTO* **30** 57-64
[15] Adbin Z Kumar A 2014 *Int. J. Elec. Power. Energ. Syst.* **3** 753-59

IOP Conf. Series: Journal of Physics: Conf. Series **935** (2017) 012059 doi:10.1088/1742-6596/935/1/012059

Assessing of the effect of hydrodynamics in the growth of scale ($CaCO_3$) in rotating cylinder electrode

M García[1], E J Espinosa Cardenal[1] and E Vera[1]
[1] Universidad Pedagógica y Tecnológica de Colombia, Tunja, Colombia

E-mail: marceladelpilar.garcia@uptc.edu.co

Abstract. Fouling is a common problem related to different industries in which the solubility of dissolved minerals in water changes leading to the formation of deposits on the surface of the equipment, for instance, in oil production the scale can be formed on the wellbore, on the well tubing or superficial facilities. Over the last years, several studies have been carried out in order to determine the effect of operational conditions in the scale growth but there are no conclusive results. This paper presents the results of calcium carbonate deposition on the surface of stainless steel AISI 316 at 71°C, and the influence of rotating velocity on the quantity and morphology of the generated scale. The tests were carried out by using a rotating cylinder electrode and X-Ray Diffraction (DRX), and Scanning Electron Microscopy (SEM) of the scale obtained was analysed. The results show that the amount of scale formed changes regarding the velocity of the fluid, as well as the morphology of the deposited crystals.

1. Introduction

Scale occurs due to the deposition of unwanted minerals, originally dissolved in processes of fluids, on equipment surfaces [1]. In general, the solubility limit of the liquid must be exceeded at the operating conditions for precipitation to occur [2]; For example, when a solution has a salt of inverse solubility like the $CaCO_3$ among its components, the supersaturation will appear when increasing the temperature of the solution. After the solution is supersaturated, nucleation occurs and very small crystals are formed. There are two types of nucleation: the homogeneous one in which the crystals form spontaneously in the solution and the heterogeneous one in which they are created on surfaces or on foreign bodies [3] (metallic surfaces or mineral in contact with the liquid, particulate material, and so on), these are two different processes and each one of them has its own mechanism and kinetics [4].

Formation of scales is a common problem in the equipment used throughout different industries such as water cooling systems, evaporators, oil extraction and production equipment, desalination plants using evaporation techniques, among others [5]. In the case of the oil industry the growth of deposits is known as the biggest problem presented, and this can appear on surface and well bottom equipment either in production or injection wells. The scale contributes to the wear of the equipment due to the corrosion presented under the deposits and the restrictions in the flow, provide a decrease in production rate [6].

Due to problems caused by scales, several researches have been carried out to determine the effect of variables that affect the growth and inhibition of scale, one of them is the flow rate of the solution [7], which can interfere with the amount, resistance and the adhesion of the obtained deposit, some studies have shown that at higher flow rates the deposition rate is increased due to the decrease in diffusion resistance and this eases the growth of more tenacious crystals when working in laminar flows, but when turbulent flow is handled there is a decrease in the thickness of the scale due to the increase in the removal rate and greater shear forces between the deposit and the liquid [8].

Among the deposits formed the most common is calcium carbonate ($CaCO_3$) which has three anhydrous polymorphs: calcite, aragonite and vaterite, presenting a trigonal, orthorhombic and a hexagonal crystalline system respectively, which can be observed in Figure 1[9,10]. The properties of the compound vary because each structure has different characteristics, one of these factors is the thermodynamic stability, in the case of the calcite, which is the most stable form, whereas the aragonite and the vaterite presents a stability decrease. The vaterite is the least stable phase of all, therefore it is more soluble and easier to remove [11]; so, the study of the morphology and characterization of the scales is an important aim.

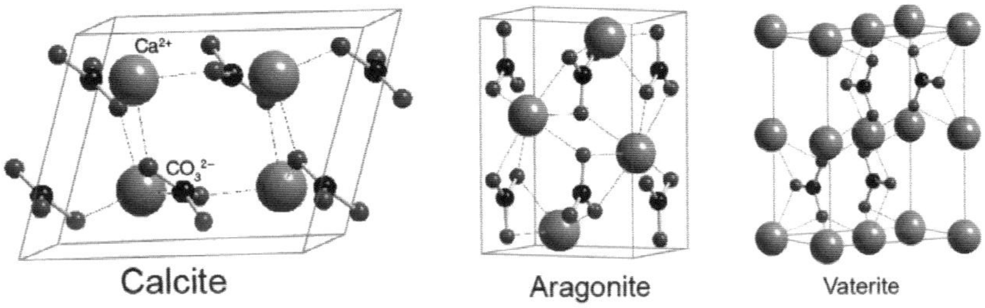

Figure 1. System crystalline of polymorphs of $CaCO_3$ [12].

In this paper we present the results of using a Rotating Cylinder Electrode (RCE) to study the formation of calcium carbonate scale on steel AISI 316 and we also determine the influence of flow rate on the amount and morphology of generated deposit.

2. Experimental equipment and procedure

The surface in which the scale was deposited was a cylindrical sample of stainless steel AISI 316 of 1cm in length and 1.2cm of external diameter. Before starting the test, it was polished with sandpaper ranging in number from 600 to 1500, and after that, it was washed with acetone and dried in a stream of air. The experiments were performed using a Rotary Cylinder Electrode (RCE) in which the substrate was placed and immersed in the saturated solution heated by means of a heating plate and the suspension solution was maintained by using a magnetic stirrer, as shown in Figure 2.

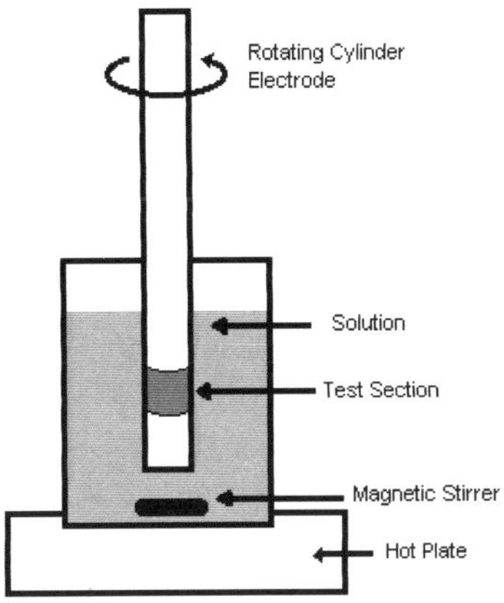

Figure 2. Experimental assembly for the precipitation of calcium carbonate.

The tests were performed for a period of 24 hours at ambient pressure and the temperature was controlled at 70°C by means of an immersion thermometer, the rotating speed ranged from 1000 to 1900rpm. A supersaturated solution consisting of $CaCl_2.2H_2O$ and $NaHCO_3$ dissolved in 1L of

IOP Conf. Series: Journal of Physics: Conf. Series **935** (2017) 012059 doi:10.1088/1742-6596/935/1/012059

distilled water was used and the analytical grade reagents were weighed on a balance VIBRA HT224R to 4 decimal precision.

Before and after each test the sample was weighed to determine the amount of deposit formed by means of the difference in mass measurements. The scale was characterized by X-ray diffraction using an equipment PANalitycal X'Pert-PRO with cobalt anode and a sweep scanning between 5 and 90°, driving a step time of 0.5 seconds, with which the crystalline system was established. SEM was used an equipment CARL ZEISS RA-ZEI-001 where micrographs were obtained using the secondary electron detector and backscattered electron detector.

3. Results

3.1. Rotating speed

The tests were performed at different rotational speeds of the sample, which were found within a turbulent regime v>0.6m/s [13] approximately 955rpm for the test conditions used, thus the flow velocity is able to generate cut forces between the solution and the obtained deposit, generating possible detachments of the scale. It can also exist more quantity of ions in the proximity of the substrate at a greater speed causing an increase in the amount of the generated scale. The results are presented in Table 1, where an increase in the rate of deposition regarding the flow velocity is shown.

Table 1. Amount of scale for each of the tests.

P.No[a]	Rotating Speed (rpm)[b]	Area (mm^2)	Deposition Rate (mg/m^2h)	Amount (mg)
1	1000	376.55	712.6	6.4
2	1250	376.99	733.8	6.6
3	1500	377.01	791.3	7.1
4	1750	376.82	1629.1	14.7
5	1900	377.30	2197.6	19.9

[a] Number Test
[b] Rotating Speed in revolutions per minute

3.2. Morphology

The micrographs at 2000X using the Scanning Electron Microscope for each of the tests are shown in the Figure 3.

Figure 3 is observed that at a lower speed few crystals are formed and deposits in the form of leaflets cover a large part of the test surface. By increasing the speed, a greater coverage of the surface can be observed but no increase in the amount of crystals is evidenced. The following tests (Figure 3(c)) show the formation of calcium carbonate crystals in large clusters surrounded by material without apparent crystallinity covering almost the entire surface until the formation of a compact crust on the sample.

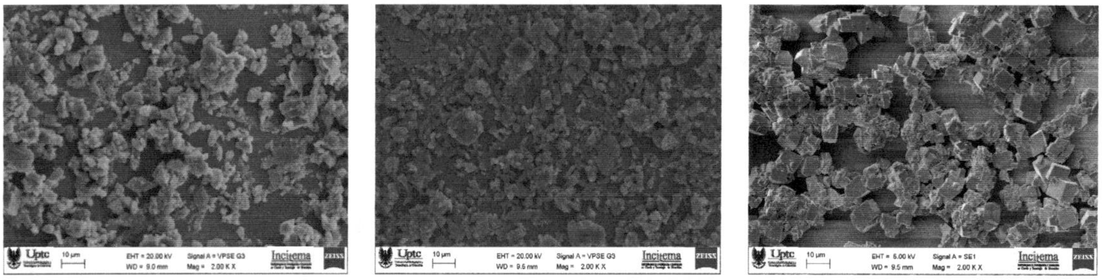

(a) (b) (c)

(d) (e)

Figure 3. Micrographics (a), (b), (c) (d), and (e) show morphology of $CaCO_3$ for test to rotating speed 1000, 1250, 1500, 1750 and 1900rpm respectively.

When the deposits were analysed by means of XRD, it was established that the deposit obtained consists of calcium carbonate in the form of calcite and sodium chloride in a percentage of 89 and 11 respectively.

4. Conclusions

At higher rotational speed an increase in the amount of deposit formed can be observed, in addition to presenting a greater coverage of the sample surface and a high degree of crystallinity of the scale generated.

As for the conditions to which the tests were carried out, it is observed that the deposit is mainly compound of calcite, as indicated in the micrographs and in the results provided by the XRD.

References

[1] Zhao X and Chen X D 2013 *Heat Transfer Engineering* **34** 719-32
[2] Bott T R 1997 *Experimental Thermal and Fluid Science* **14** 356-60
[3] Hoang T A 2015 *Mineral Scales and Deposits* ed Z Amjad and K Demadis (Amsterdam: Elsevier B.V) pp 47-83
[4] Al Nasser W N, Shaikh A, Morriss C, Hounslow M J and Salman A D 2008 *Chemical Engineering Science* **63** 1381-1389
[5] Tlili M M, Rousseau P, Ben Amor M and Gabrielli C 2008 *Chemical Engineering Science* **63** 559-566
[6] Bin Merdhah A B and Yassin A M 2008 *Emirates Journal for Engineering Research* **13** 69-79
[7] Muryanto S, Bayuseno A P, Ma'mun H, Usamah M and Jotho 2014 *Procedia Chemistry* **9** 69-76
[8] Hoang T A 2015 *Mineral Scales and Deposits* ed Z Amjad and K Demadis (Amsterdam: Elsevier B.V) pp 47-83
[9] Mullin J W 2001 *Crystallization* (Oxford: Butterworth-Heinemann) pp 27
[10] Pancorbo F J 2011 *Corrosión, degradación y envejecimiento de los materiales empleados en la edificación* (Barcelona: Marcombo) p 234
[11] MacAdam J and Parsons S A 2004 *Rev Environ Sci Biotechnol* **3** 159-169
[12] Shriver D, Weller M, Overton T, Rourke J and Armstrong F 2014 *Inorganic Chemistry* (Great Britain: Oxford University Press) p 875
[13] Larrea Idiarte A E 1998 *Estudio de problemas de incrustaciones evaluación técnica y estimación de costos de químicos anti-incrustantes para diseños de tratamientos tipo "squeeze" a las areniscas u y t del campo Shushufindi* (Guayaquil: Escuela Superior Politécnica del Litoral) p 20

IOP Conf. Series: Journal of Physics: Conf. Series **935** (2017) 012060 doi:10.1088/1742-6596/935/1/012060

Evaluation of corrosion inhibitor simulating conditions of operation

O Gómez[1], H Aponte[1], E Vera[1] and Y Pineda[1]
[1] Universidad Pedagógica y Tecnológica de Colombia, Tunja, Colombia

E-mail: gomezoscarjavier@hotmail.com

Abstract. Operating conditions at the head of oil wells are critical, in addition to injecting water to increase the pressure while maintaining production cause deterioration in the metallic structures that transport fluids. One way to maintain integrity is the injection of inhibitors which plays an important role in protecting the pipes. In this study a molecule N-PHENYL NITRONE was obtained, which was evaluated by electrochemical tests (LPR) and Tafel Polarization Curves in a reactor controlling environments containing different dosages, pressure values, temperatures and flow velocity using working electrodes tubing API N 80, the reactor was connected to a potentiostat to determine corrosion rates, allowing the analysis of the influence of each variable on the protective behaviour of the inhibitor, and its efficiency against the decrease of the deterioration of the pipes. Corrosion products are analysed by X-Ray Diffraction (XRD). Photographic records of the surface verify the formation of iron carbonate ($FeCO_3$). In addition, a mathematical analysis of the independent variables is performed to evaluate the effect that it has on the efficiency of corrosion inhibition.

1. Introduction

The analysis of residual corrosion inhibitor concentration in production waters has been used for years to monitor the corrosion protection of oilfield systems. [1] A corrosion inhibitor is a material that fixes or covers the metal surface, providing a protective film that stops the corrosive reaction [2]. The corrosion inhibitors are added in small concentrations to the corrosive medium, acting at the metal-solution interface, significantly reducing the corrosion rate of the metal material, exposed to it [3]. Corrosion inhibitors generally control corrosion by forming thin films that modify the medium on the metal surface [4]. The use of corrosion inhibitors is one of the most common methods to control corrosion [5], due to their excellent performance in "sweet" and "acid" environment.

Dry CO_2 gas is not itself corrosive at the temperatures encountered within oil and gas production systems but is so when dissolved in an aqueous phase through which it can promote an electrochemical reaction between steel and the contacting aqueous phase. CO_2 will mix with the water, forming carbonic acid making the fluid acidic [6,7] and is by far the most prevalent form of attack encountered in oil and gas production [8].

These inhibitors are mainly constituted of organic nitrogen compounds, being the most common, which have as part of their structure long chains of hydrocarbons (usually C18). Today these inhibitors are constituted by long aliphatic chains (diamine) or long carbon chains (imidazolines) [9]. Polymers are widely used as effective corrosion inhibitors [10]. Corrosion inhibition by these organic compounds is mostly attributed to their adsorption ability on metal surface and the subsequent formation of a protective layer, which can block the active sites on metal surface and thereby suppress the corrosion attack [11]. The adsorption process of organic molecules on metal surface depends mainly on the surface charge of metal, the chemical structure of organic molecule and the type of aggression medium [12].

2. Experimental procedure

Figure 1 shows the reactor used to perform the tests, showing the parts of the system, it is connected to a potentiostat to perform electrochemical tests at the moment of obtaining the values of each of the variables at work.

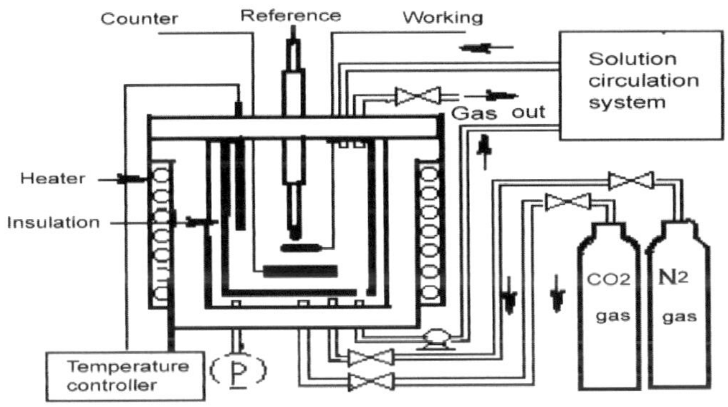

Figure 1. Autoclave for measurement of corrosion kinetics, high pressure and high temperature.

The test solution is introduced into the reactor vessel and closed with the equipment lid, then the safety clamps are located and the clamping screws are tightened. Because the oil well systems have very low or no oxygen concentrations, the N_2 inlet valve is opened to the reactor, the gas passage is maintained for 1 hour to remove the oxygen present in the vessel. After the removal of oxygen, CO_2 injection is performed to the value required by the test, because of this the equipment has an analogy pressure gauge and a pressure sensor. Additional variables such as temperature and flow rate were monitored at the control panel.

At the end of the test, the system is depressurized and the gas is sent out of the laboratory by means of a gas extraction line. The jaws are removed and the lid is removed from the equipment. The electrochemical specimen is disassembled and cleaned with acetone before being labelled and placed in a re-sealable bag with silica gel inside and finally stored in a desiccator for further analysis.

2.1. Electrochemical evaluation

A potentiostat is connected to each of the electrodes to form an electrochemical cell, after 45 minutes of potential stabilization the measurement of resistance to linear polarization and the Tafel polarization curve is performed, then, the Corrosion rate is analysed and determined. Inhibitors usually adsorb on the metal surface, protecting, forming films, reducing corrosive processes by these actions:

- Increasing the anodic or cathodic polarization (Tafel Curves).
- Reducing the movement and diffusion of ions to the metal surface.
- Increasing the electrical resistance of the metal surface [9].

2.2. Efficiency of a corrosion inhibitor

The evaluation of corrosion inhibitors is carried out according to the NACE G 170 standard. The efficiency of a corrosion inhibitor is defined by equation (1) [13]:

$$\text{Efficiency (\%)} = \frac{Vcorr(blanco) - Vcorr(Inh)}{Vcorr(blanco)} * 100 \tag{1}$$

2.3. Materials and methods

The basic assembly of a test consists of the preparation of the solution, with distilled water and the addition of nitric acid or sodium hydroxide, according to the required value. The experimental units are cylindrical API N80 steel pieces, with an area of 2.2cm^2.

2.4. Variables chosen and their values to perform the tests

Tests will be carried out to complete the parametric control, taking into account the chosen levels, in a random way by submitting the treatments with each level of variable shown in Table 1.

Table 1. Variables and levels.

Variables	Pressure CO_2 (psi)	Temp. (°C)	pH	Flow rate (m/s)	Dosage (ppm)
Levels	300	40	4	0	20
	500	80	8	0,5	50

2.5. Mathematical model

The regression model is a fractional factorial design, which predicts the effectiveness of the inhibitor for the established ranges. A final number of 32 tests was obtained, however 16 tests were taken; Measuring the effect of each of the variables, using coded data, for high levels it is tabulated as 1 and for low levels as -1.

3. Results and discussion

Thirty-two tests were performed in which the total combinations were completed, and tests were performed without inhibitor or blank solution to determine dosing efficiency. According to Table 2. for the first 8 tests the parameters were fixed (TEMP 40°C, PH 4); For tests 9-6 (TEMP 80, PH 4); 17 to 24 (TEMP 40°C, PH 8) and from test 25 to 32 (TEMP 80, PH 8).

Table 2. Results.

No.	a	b	c	d	e	No.	a	b	c	D	e
1	300	20	0	69.83	70.59	17	300	20	0	56.04	56.13
2	300	20	0.5	69.52	70.86	18	300	20	0.5	55.93	56.40
3	300	50	0	70.67	68.95	19	300	50	0	57.6	56.74
4	300	50	0.5	70.11	69.23	20	300	50	0.5	57.12	57.02
5	500	20	0	65.31	65.90	21	500	20	0	54.29	51.44
6	500	20	0.5	63.57	66.17	22	500	20	0.5	52.73	51.71
7	500	50	0	68.5	67.67	23	500	50	0	54.42	55.46
8	500	50	0.5	68.16	67.95	24	500	50	0.5	54.28	55.74
9	300	20	0	85.18	85.11	25	300	20	0	61.83	62.11
10	300	20	0.5	84.78	82.60	26	300	20	0.5	60.96	59.60
11	300	50	0	87.02	83.47	27	300	50	0	62.29	62.73
12	300	50	0.5	85.9	80.97	28	300	50	0.5	60.54	60.22
13	500	20	0	79.86	80.42	29	500	20	0	56.73	57.42
14	500	20	0.5	78.17	77.91	30	500	20	0.5	56.39	54.91
15	500	50	0	83.25	82.19	31	500	50	0	58.8	61.45
16	500	50	0.5	79.58	79.69	32	500	50	0.5	58.71	58.94

a: Pressure CO2 (psi)
b: Dosage (ppm)
c: Flow rate (m/s)
d: Efficiency obtained experimentally (%)
e: Efficiency obtained in the mathematical model (%)

3.1. Characterization of corrosion products

The corrosion products found after each test without inhibitor have an opaque and black-grey colour. Specimens that were tested with inhibitor did not evidence adherence of corrosion products. The final appearance of the specimens is presented in Figure 2.

Figure 2. Final appearance of the specimen. (a) Test without inhibitor and (b) Test with inhibitor.

The absence of orange or reddish tones indicates that there is no significant formation of oxides, indicating that CO_2 and N_2 injection adequately displaced the oxygen from the test solution.

3.2. X-ray diffraction analysis

X-ray diffraction was applied with flush film technique to obtain information on the chemical composition of the corrosion products. Table 3, presents the products and elements identified.

Table 3. XRD Products.

Element or Compound	Chemical formula
Iron carbonate (Siderite)	$FeCO_3$
Magnetit	Fe_3O_4
Iron	Fe

The result obtained in the X-ray diffraction was the presence of corresponding peaks with siderite, which is an iron carbonate type ($FeCO_3$). The detected iron carbonate is the characteristic product of corrosion in aqueous environments with CO_2 at the pH, temperature and pressure levels used. There is presence of magnetite (Fe_3O_4).

3.3. Mathematical model

According to the obtained results, a model is established by means of fractional factorial design, that allows to predict the efficiency of inhibitor N-Phenyl Nitrone, to the conditions evaluated. Each of the variables studied has a direct influence on the protective behaviour of the inhibitor on the metal surface, and the model obtained with each of its coefficients is defined by equation 2.

$$\%Efficiency = 74.4795 - 0.0348166 \times P_{CO2} - 0.185835 \times T + 5.76374 \times pH - 0.30025 Dosis + 6.105 \times Flow\ rate + 0.0005683 \times P_{CO2} \times Dosis - 0.053343 \times T \times pH + 0.0018791 \times T \times Dosis - 1.39 \times pH \times flow\ rate \quad (2)$$

3.4. Effect of each variable

Table 4 shows the analysis performed on each of the independent variables, and with each of their combinations that have an effect on the protective behaviour of the inhibitor.

The model allows defining that for this type of inhibitor, the variable that has a significant effect on the inhibitor behaviour is the temperature and the combination between pH and temperature is more influential, for the positive data have a directly proportional effect is shown and in the negative data its effect is inversely proportional to the efficiency.

IOP Conf. Series: Journal of Physics: Conf. Series **935** (2017) 012060 doi:10.1088/1742-6596/935/1/012060

Table 4. Effect results of the variables. Value of each variable and effect of combinations between variables.

Effect	Value	Combination Effect	Value	Combination Effect	Value
CO_2	-3.0	CO_2 - Temperature	-0.5	Temperature - Dosage	0.3
Temperature	-17.6	CO_2 - pH	0.2	Temperature - Flow rate	1.1
pH	8.9	CO_2 – Dosage	1.7	pH – Dosage	-0.5
Dosage	1.2	CO_2 - Flow rate	0.6	pH – Flow rate	-1.4
Flow rate	-1.1	Temperature – pH	-4.3	Dosage – Flow rate	-0.6

4. Conclusions

The N-PHENYL-NITRONE molecule has a good protective behaviour in the evaluated conditions, however a reduction of the barrier characteristic is evidenced by the presence of more aggressive conditions.

Iron oxide corrosion products were generated, characteristic of environments rich in CO_2 in the aqueous phase, in basic pH the siderite film is stronger.

The molecule can be applied in high temperature systems. Although it shows a degradation, a protective characteristic is evidenced. Under static conditions the inhibitor tends to behave better, because there is no dynamic agent that can affect adhesion.

The obtained model presents a reliability of 97%, for the ranges studied.

References

[1] E. Ojeda, R Hing 2011 *Evaluación de inhibidores de la corrosión obtenidos a partir de la vinaza residual de las destilerías de alcohol etílico con aditivos* (Santiago de Cuba: Universidad de Oriente)

[2] R. Álvarez, J. Mendoza, S. Castillo and J. Marín 2013 Studies on the development of new efficient corrosion inhibitors for crude oil pipelines: Electrochemical impedance spectroscopy results for 1,8-Naphthyridines *Journal of the Mexican Chemical Society*

[3] Sandianes V, Uruchurtu C, Calderón S, Calderón S 2012 Estudio de un inhibidor de corrosión en medio ácido *Tecnología Química* **XXXII(2)** 112-120

[4] D C Pacheco 2017 *Estudio de la velocidad de corrosión de tres aceros comerciales usados en la construcción, en un dispositivo experimental* (Quito: Universidad Central del Ecuador)

[5] E.S.M. Sherif 2013 Electrochemical investigations on the corrosion inhibition of aluminum by 3-amino-1,2,4-triazole-5-thiol in naturally aerated stagnant seawater *J. Ind. Eng. Chem.* **19(6)** 1884-1889

[6] Ambrish Singh, Yuanhua Lin, I.B. Obot, E.E. Ebenso, K.R. Ansari, M.A. Quraishi 2015 *Appl. Surf. Sci.* **356** 341-347

[7] K.S. George, S. Nesic 2007 *Corrosion* **63** 178–186

[8] Singh, A., Lin, Y., Ansari, K. R., Quraishi, M. a., Ebenso, E. E., Chen, S., & Liu, W 2015 Electrochemical and surface studies of some Porphines as corrosion inhibitor for J55 steel in sweet corrosion environment *Applied Surface Science* **359** 331–339

[9] Marianela G. 2011 Métodos de control de corrosión. (Venezuela: Universidad del Zulia)

[10] M.A. Migahed, S.M. Rashwan, M.M. Kamel, R.E. Habib 2016 Synthesis, characterization of polyaspartic acid-glycine adduct and evaluation of their performance as scale and corrosion inhibitor in desalination water plants *J. Mol. Liq.* **224** 849–858

[11] H.M. Abd El-Lateef, M.A. Abo-Riya, A.H. Tantawy 2016 Empirical and quantum chemical studies on the corrosion inhibition performance of some novel synthesized cationic gemini surfactants on carbon steel pipelines in acid pickling processes *Corrosion Science* **108** 94–110

[12] D. Daoud, T. Douadi, H. Hamani, S. Chafaa, M. Al-Noaimi 2015 Corrosion inhibition of mild steel by two new S-heterocyclic compounds in 1 M HCl: experimental and computational study *Corrosion Science* **(94)** 21–37

[13] ASTM G170-06 2012 *Standard guide for evaluating and qualifying oilfield and refinery corrosion inhibitors in the laboratory* (USA: ASTM International)

IOP Conf. Series: Journal of Physics: Conf. Series **935** (2017) 012061 doi:10.1088/1742-6596/935/1/012061

Comparison of two temperature control techniques in a forced water heater solar system

E Hernández[1], R E Guzmán[1], A Santos[1] and E Cordoba[1]
[1] Universidad Pontificia Bolivariana, Floridablanca, Colombia

E-mail: emil.hernandez@upb.edu.co

Abstract. a study on the performance of a forced solar heating system in which a comparative analysis of two control strategies, including the classic *on- off* control and *PID* control is presented. From the experimental results it was found that the two control strategies show a similar behaviour in the solar heating system forced an approximate settling time of 60 min and over- elongation 2°C for the two control strategies. Furthermore, the maximum temperature in the storage tank was 46°C and the maximum efficiency of flat plate collector was 76.7% given that this efficiency is the ratio of the energy of the radiation on the collector and the energy used to heat water. The efficiency obtained is a fact well accepted because the business efficiencies of flat plate collectors are approximately 70%.

1. Introduction

Automatic control has played an important role in the advancement of engineering and science. Given the development in control systems, for space vehicles, guided missiles and robots, this has become an integral part of modern industrial processes and manufacturing. For example, automatic control is fundamental for the control of pressure, temperature, humidity, viscosity, level and flow in the process control industries, to improve productivity and simplify the work of many repetitive and routine manual operations, making it attractive to achieve optimum performance of the systems [1-2]. On the other hand, solar collectors are devices that are used to heat water, for heating systems and swimming pools, and can be used both at home and industrial level. Because they are equipment of easy acquisition thanks to its technical - economic relation.

It is composed of a solar radiation absorbing plate that is responsible for transferring the thermal energy to the circulating fluid through a coil, a transparent cover and a metal box isolated on the sides and the bottom, where the elements Previously described [3,4].

It is composed of a solar radiation absorbing plate that is responsible for transferring the thermal energy to the circulating fluid through a coil, a transparent cover and a metal box isolated on the sides and the bottom, where the elements Previously described [3,4].

In the following work we use a forced solar heating system with a flat plate collector, as shown in Figure 1, Figure 2 and Figure 3, in order to find the curves of radiation vs efficiency, according to the solar rays that effect the collector surface, the parameters obtained were compared with similar studies carried out by the company CENSOLAR (Solar Energy Studies Center) [5-8].

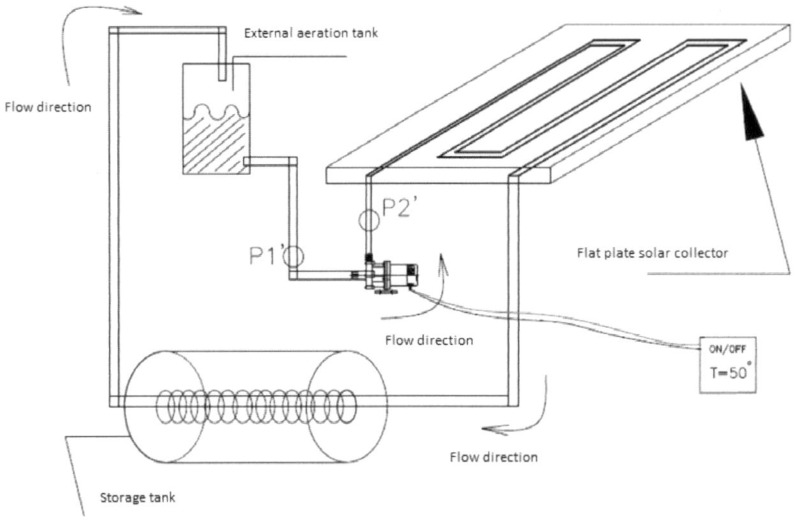

Figure 1. Schematic of the forced solar heating system.

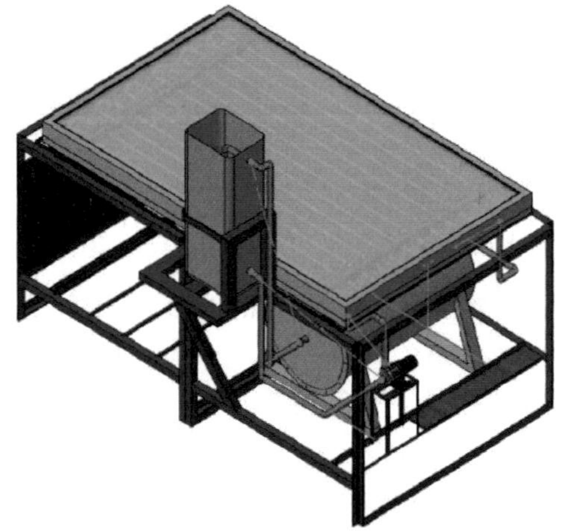

Figure 2. Forced solar heating built system.

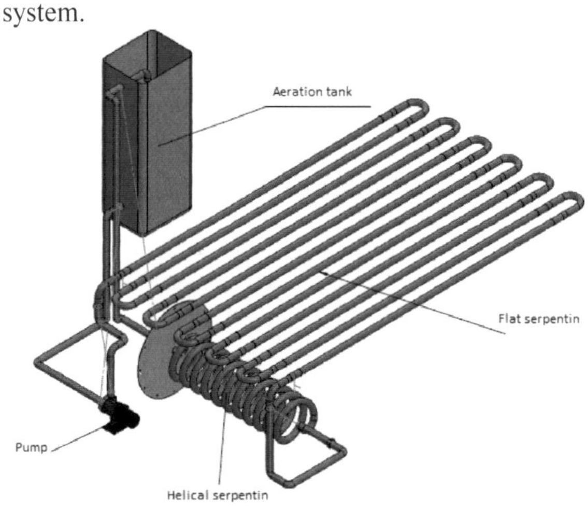

Figure 3. Forced solar heater pipe system.

2. Theory

Numerous investigations have been carried out on thermosiphon solar water heaters with flat plate collectors. Runsheng Tang *et al.* [9], carried out a theoretical and experimental study on a solar water heater thermosyphon and concluded that for a given temperature in the tank, the outlet temperature in the system with a vertical cylindrical tank was slightly larger than a horizontal cylindrical tank.

Soteris A. Kalogirou [10], showed, through a simulation, that the best performance system is obtained with small pipe diameters in the collector with a range of 15mm-35mm with similar efficiencies, therefore the decision depends on the costs. It also determined that the optimum slope of the collectors is the latitude plus 10°, although a smaller slope does not significantly affect the performance, and the optimum distance between the top of the collector and the bottom of the tank is 15cm. The useful energy that is obtained in a solar collector, can be determined with equation (1).

$$Q_{útil} = mC_p(t_s - t_e) \tag{1}$$

Where m=mass flow, C_p=Specific water heat, t_s=Output temperature, t_e=Inlet temperature. The performance of the solar collector depends on the useful energy and the solar radiation, according to equation (2).

$$\eta = Q_{útil}/(S \times I) \tag{2}$$

Where: I=Intensity of solar radiation in W/m^2 and S=catchment area in m^2. The useful energy can be calculated as the difference between the energy that is captured and that which is lost for thermal reasons. The energy absorbed can be expressed according to equation (3).

$$Q_{abs} = a \times I \times S \tag{3}$$

Where a=Coefficient of losses. Thermal losses are proportional to the collector surface and the temperature difference between the absorber and the environment, according to equation (4).

$$Q_{per} = b(t_m - t_a)S \tag{4}$$

$$t_m = (t_e + t_s)/2 \tag{5}$$

The performance is expressed according to equation (6).

$$\eta = Q_{útil}/(I \times S) = \left(Q_{abs} - Q_{per}\right)/(I \times S) = a - b(t_m - t_a)/I \tag{6}$$

$$\eta = 0.83 - 6.8(t_m - t_a)/I \tag{7}$$

The values of the constants for flat plate collectors can be taken as [11-12]:

$$a = 0.83 \text{ and } b = 6.8 \tag{8}$$

3. Methodology

The forced system consists of a solar collector, an accumulator tank with heat exchanger, with a capacity of 150 litres, a centrifugal pump of magnetic drive and an aeration tank. All of this is coupled, as shown in Figure 2, in which the installation is presented. The dimensions of the tank and the manifold were made according to the recommendations of the company Censolar for forced systems installations.

The thermal performance test of the collector was carried out considering equation (9) for which FR τα= constant and FR U=m=constant can be assumed [3]. The equation of this characteristic curve of the collector can be assimilated quite accurately to that of a line in which the variable on the abscissa axis is $(t_m - t_a)/I$ and whose slope is FRU. For a c.p.p. Of average type can be approximated a=0.83 and m=6.8, obtaining the equation (9).

$$\eta = 0.83 - 6.8(t_m - t_a)/I \tag{9}$$

In equation (9), if $x = (t_m - t_a)/I$ is obtained

$$\eta = 0.83 - 6.8x \tag{10}$$

The value of I(w/m2) radiation was measured with a Kipp & Zonen CM3 pyranometer and the temperatures with thermocouples type k located at the ends of the collector, the results are shown in Figure 4, where the maximum efficiency obtained was 76.7% [13].

Figure 5 shows the behaviour of the two control strategies in which for the on-off case a set point of 30°C and a hysteresis of 1.5°C are used, for the PID case the values of the parameters with the auto-tuning function of the controller P=2.9, I=342 and D=71. An approximate settling time of 60 minutes and over-elongation of 2°C can be observed for the two control strategies [14].

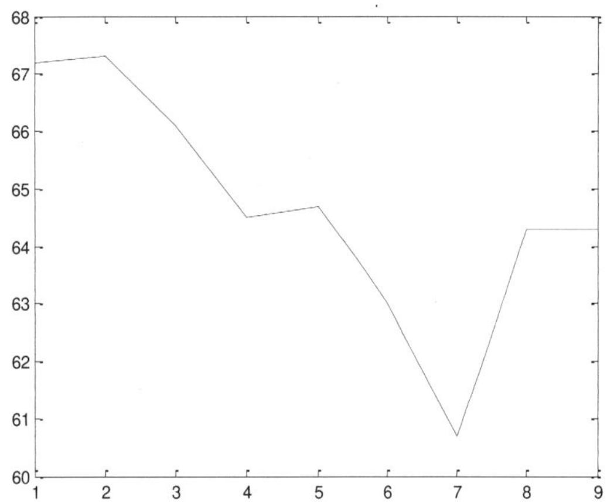

Figure 4. Efficiency of forced system collector. Is observe the behaviour the efficiency percentage in function of time measurement each 30min.

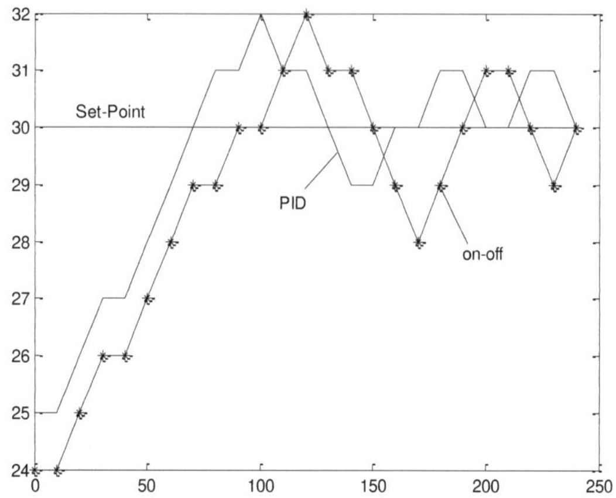

Figure 5. PID and ON-OFF Control in a forced solar heater. Is observe the behaviour the temperature in degree Celsius in function of time in minutes.

4. Conclusions

A forced solar water heating system is designed, constructed and evaluated with a flat plate manifold. This equipment is operated by a centrifugal pump of 12V, has a storage capacity of water of 150 litres, which can be used to supply hot water.

According to the experimental results, it is observed that the collector reaches a maximum efficiency of 76.7% and a maximum temperature of 46°C in the storage tank.

It was found that the forced flow system allows the storage tank to be positioned below the collector level, which is an advantage when compared to the thermosyphon system where the storage tank must be above the collector.

With an average flow rate of 0.3gpm the pump provides sufficient mass flow to the water to extract the heat from the manifold and deliver it to the coil quickly and steadily, raising the water temperature to a maximum of 50°C.

References

[1] K Ogata 2003 *Ingeniería de control moderna, 4ta edición* (España: Pearson Educación)

[2] R S Burns 2001 *Advanced Control Engineering, 1st edition* (Oxford: Butterworth-Heinemann)

[3] V J Roldan 2012 *Necesidades energéticas y propuestas de instalaciones solares* (España: Editorial Paraninfo)

[4] Norton B and Probert S D 1986 Thermosiphon Solar Energy *Water Heater* **3** 125

[5] Ramon *et al.* 2006 Diseño de la automatización de una planta generadora de bio-gas *Revista de Tecnologías de Avanzada* **2(8)** 48-54

[6] Burbano C, Restrepo H and Sabogal J 2006 *Sciencia et Technica* **31** 85

[7] B Zabalza 2007 *El ahorro energético en el nuevo código técnico de la edificación* (España: Editorial FC)

[8] B Valdez Salas 2006 *Tecnología en la UABC* (México: Miguel Ángel Porrúa)

[9] Runsheng *et al.* 2010 Experimental and modeling studies on thermosiphon domestic solar water heaters with flat-plate collectors at clear nights *Energy Conversion and Management* **51** 2548-2556

[10] Soteris A. Kalogirou 2014 Flat-plate collector construction and system configuration to optimize the thermosiphonic effect *Renewable Energy* **67** 202-206

[11] Gupta Cl and Garg HP 1968 *Solar Energy* **12** 163

[12] Méndez Muñiz J M, Cuervo García R 2008 *Energía Solar Térmica* (España: FC Editorial)

[13] M Rodríguez 2009 Desarrollo de la energía solar en Colombia y sus perspectivas *Revista de Ingeniería* **28** 84-86

[14] E Hernández 2010 Construcción y análisis de una válvula de control en lazo abierto en un sistema de refrigeración *Revista de Tecnologías de Avanzada* **1(15)** 25-29

IOP Conf. Series: Journal of Physics: Conf. Series **935** (2017) 012062 doi:10.1088/1742-6596/935/1/012062

Comparative study of the energy potential of cyanide waters using two osmotic membrane modules under dead-end flow

Y García-Díaz[1], E Quiñones-Bolaños[1], C Bustos-Blanco[1], L Vives-Pérez[1], C Bustillo-Lecompte[1] and M Saba[1]
[1] Universidad de Cartagena, Cartagena, Colombia

E-mail: ygarciad1@unicartagena.edu.co

Abstract. The energy potential of the osmotic pressure gradient of cyanide waters is evaluated using two membrane modules, horizontal and vertical, operated under dead-end flow. The membrane was characterized using Scanning Electron Microscopy (SEM) with Energy Dispersive X-ray Spectroscopy (EDS). The membrane is mainly composed of carbon, oxygen, and sulphur. The properties of the membrane were unchanged and had no pore clogging after exposure to the cyanide waters. Potentials of 1.78×10^{-4} and $6.36\times10^{-5}Wm^{-2}$ were found for the horizontal and vertical modules, respectively, using the Van't Hoff equation. Likewise, the permeability coefficient of the membrane was higher in the vertical module. Although the energy potential is low under the studied conditions the vertical configuration has a greater potential due to the action of gravity and the homogenous contact of the fluid with the membrane.

1. Introduction

Power generation by osmotic pressure due to the salinity gradient consists in the use of a semipermeable membrane, which divides two solutions with different salinity and chemical potential [1]. The transport of water from the solution with a low concentration towards the concentrated solution generates a pressure on the water of low concentration. The pressure generated could be exploited with the use of a hydraulic turbine and an electric generator [2].

The current installed capacity of this type of energy is merely 0.05MW worldwide [3], and the characteristics of the membranes and configuration must be improved to make the system competitive [4]. On the other hand, this technology has outstanding benefits, such as having the lowest emission of greenhouse gases (GHGs) [5], it does not consume either water or salt in the process, and requires low flow to generate enough energy to supply remote populations [6]. It can also be used in hybrid applications, including its combination with other renewable energies, wastewater treatment plants [7,8], and drinking water systems [9].

Cyanide waters from tailings in the mining industry present an enormous potential because of the high concentrations of NaCN salts in the range of 700–800$mg\cdot L^{-1}$. These NaCN concentrations have a negative impact on the environment, reducing growth and swimming performance in fishes [10]. NaCN is also a potent inhibitor of respiration, acting on mitochondrial cytochrome oxidase, blocking electron transport, which results in decreased oxidative metabolism and oxygen utilization causing lactic acidosis, which can be fatal [11].

It has been appraised that the global energy potential by salinity gradient is between 1600 and 2000TW per year [12] and can generate $1MW\cdot m^{-3}s^{-1}$ of water passing through the membrane [13]. In the case of Colombia research has been carried out on the potential of the salinity gradient [14] but it should be noted that its scope has been only theoretical.

The objective of this study is to determine the energy potential of cyanide waters using the osmotic pressure and salinity gradient. Membrane systems fed with cyanide waters were used with two types of osmotic modules with a vertical and horizontal configuration as a precedent for process optimization and scale-up.

2. Theoretical calculations

The energy density in an osmotic pressure system is defined as the power (W) per unit area of membrane (A). A high W value determines the requirements between the area of the membrane and the size of the plant. The ideal output force is obtained by equation (1):

$$W = A(\Delta\pi - P)\Delta P \tag{1}$$

where P is the initial pressure, $\Delta\pi$ (bar) is the pressure difference, and ΔP is the effective hydraulic pressure. For diluted solutions, the osmotic pressure ($\Delta\pi$) is found by approximation using the *Van't Hoff* equation:

$$\pi = V_i C_i RT \tag{2}$$

where V_i is the number of ions of solute formed in the solution (V_i=2 for NaCN), C_i is the molar concentration of solute, R is the gas constant, and T the absolute temperature (K), V_iC_i is the total concentration of ions (kmol·m^{-3}).

3. Experimental procedure

Thirty tests were performed in each module configuration (vertical and horizontal) to determine the energy potential. A General Electric (GE) Osmonics membrane was used for each module under six concentrations of NaCN (25, 50, 100, 200, 500, and 700mg·L^{-1}). The experimental procedure was developed in three phases described below.

3.1. Polymeric membrane characterization

The membranes were subjected to a preparation and drying test in which four 3cm×1cm membrane samples were immersed in cyanide water with four concentrations of NaCN (25, 50, 100 and 200mg·L^{-1}) for 24 hours. After the immersion period, the samples were taken to a vacuum desiccator with a volume of 1.5L for 24 hours. Then, dried membrane samples were passed to a Scanning Electron Microscope (JEOL JSM-6380LV, Tokyo, Japan) where images were taken on both sides of the membrane samples. The approximations were 100×, 1000×, 10000× and 100000×. Finally, the elemental composition of the membranes was determined by Energy Dispersive X-ray Spectroscopy (EDS) for each membrane sample.

3.2. Experimental runs

As seen in Figure 1, each membrane module (vertical and horizontal) had a stainless-steel and a perforated Teflon sheet as support media. Neoprene gaskets were used to seal the points of attachment of the two containers. The volume of each module was 8L. Cyanide water and fresh water were fed into each container side by hydrostatic pressure. The only difference between the modules was the orientation of the membrane, performing on an active layer facing draw solution (AL-DS) orientation.

3.3. Energy potential calculation

The permeability coefficient of the membrane and a correlation between the hydraulic and osmotic pressure differences ($\Delta\pi$–ΔP) were established to determine the energy potential. The time at which the pressure change has started was also measured. Then, having the permeability coefficient and pressure differences for each concentration, the potential of energy potential could be calculated using equations (1) and (2).

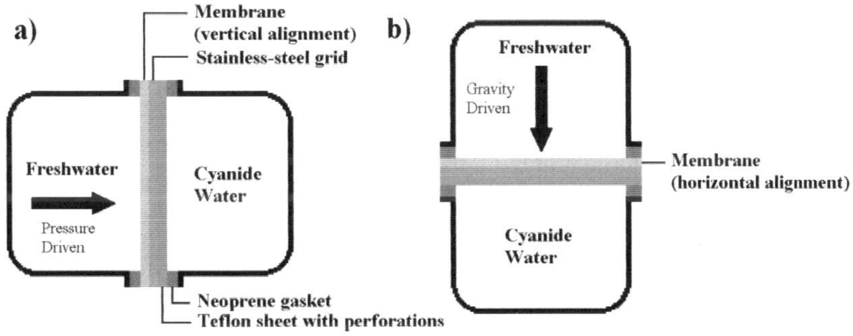

Figure 1. Schematic diagram of the osmotic modules (a) horizontal (b) vertical.

4. Results

4.1. Polymeric membrane characterization

The spectrum analysis by EDS (Figure 2) revealed contents of carbon (C) and oxygen (O) due to the presence of polyamide (polymer). Low percentages of sulphur (S) were also found, which is added to improve the thermal stability and other characteristics of the polymers in percentages between 5 and 10% [15].

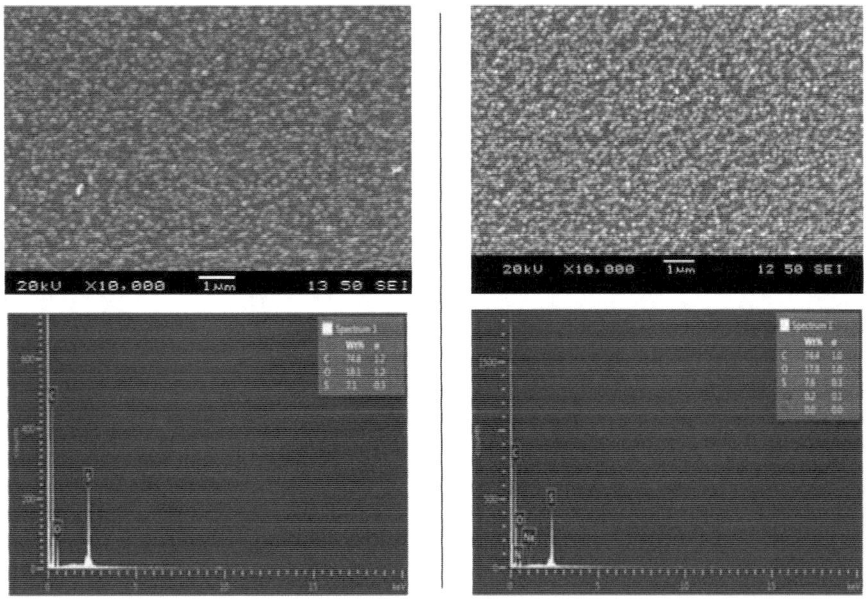

Figure 2. SEM and XRD images of (a) dry membrane sample and (b) membrane samples after being submerged into 200mg NaCN L^{-1}.

The surface of the membrane did not suffer any changes, such as variation or adhesion of material on the surface, and had no pore obstruction after being exposed to cyanide waters. However, it is observed trace values of Na and N in the EDS as expected due to the exposure of the membrane to the cyanide water.

4.2. Energy Density

The slope values correspond to the permeability coefficient which was $3.26\times10^{-3} m\cdot s^{-1}\cdot bar^{-1}$ for the vertical module and $7.51\times10^{-4} m\cdot s^{-1} bar^{-1}$ for the horizontal module (Figure 3). Results demonstrate the flow rate that passes through the vertical module is about 23% higher than that of the horizontal configuration.

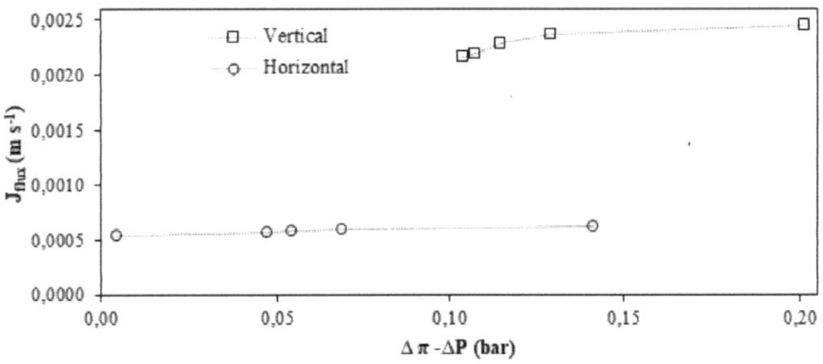

Figure 3. Pressure difference with respect to permeate flow for permeability coefficient determination.

Table 1 revealed the maximum pressure reached in the horizontal module was about 25% lower than that of the vertical configuration. The exposure time suggests the vertical module doubled the response time of the horizontal one. Finally, the energy density per membrane area was determined using equation (1).

Table 1. Pressure variation, osmotic starting time and power in experimental set-up.

	Concentration ($mg\cdot L^{-1}$ of NaCN)	Pressure (cm)	Time (s)	Power density (Wm^{-2})
Horizontal module	25.0	30.87	570	5.51×10^{-6}
	50.0	32.53	568	5.97×10^{-6}
	100	33.73	578	6.88×10^{-6}
	200	34.83	580	8.71×10^{-6}
	700	35.77	578	1.78×10^{-5}
Vertical module	25.0	40.30	186	3.27×10^{-5}
	50.0	41.57	190	3.39×10^{-5}
	100	42.13	184	3.62×10^{-5}
	200	44.47	188	4.07×10^{-5}
	700	45.53	186	6.36×10^{-5}

On the other hand, an exponential relationship between the electric power and the cyanide concentration is shown in Figure 4, where the vertical module exhibits a more favourable performance with a maximum value of $6.36\times10^{-5} Wm^{-2}$ for the concentration of $700 mg\cdot L^{-1}$ of NaCN.

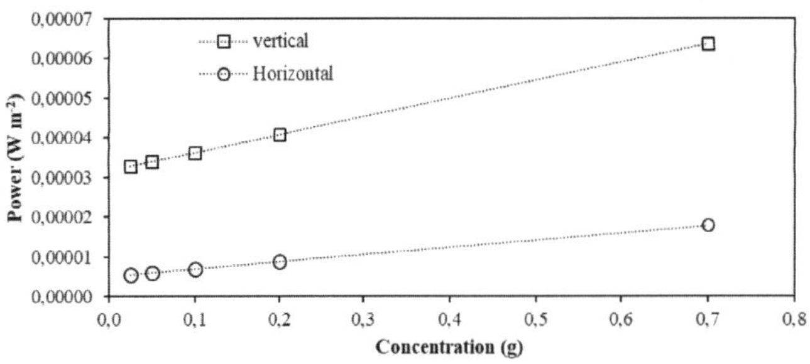

Figure 4. Potential relationship between power and NaCN concentration.

IOP Conf. Series: Journal of Physics: Conf. Series **935** (2017) 012062 doi:10.1088/1742-6596/935/1/012062

5. Conclusions

The energy density of vertical and horizontal configurations of an osmotic module with dead-end flow using cyanide waters as the draw solution was evaluated. The results show that the horizontal module has an energy potential of $1.78\times10^{-5}Wm^{-2}$; whereas, the vertical module achieved $6.36\times10^{-5}Wm^{-2}$ for a maximum NaCN concentration of 700 $mg\cdot L^{-1}$. The membrane permeability coefficient was $7.51\times10^{-4}m\cdot s^{-1}bar^{-1}$ and $3.26\times10^{-3}m\cdot s^{-1}bar^{-1}$ for the horizontal and vertical modules, respectively, which means an increase in permeability of 23% in the vertical module. On the other hand, the start time of the osmotic pressure in the vertical module was over 50% faster than that the horizontal configuration. The characterization of the membrane confirmed its properties did not change or presented pore obstruction due to contact with cyanide waters. Thus, it is recommended to optimize the process by testing diverse types of membrane and different configurations for future scale-up.

Acknowledgments

The authors gratefully acknowledge "Universidad de Cartagena", and "Colciencias" (Colombia) for financial support Osmotic power generation through the financing of this project (Project Code 1107-715-52479).

References

[1] Hussain A, Arif SM, and Aslam M 2017 Emerging renewable and sustainable energy technologies: state of the art *Renew. Sust. Energ. Rev.* **71** 12–28

[2] Cipollina A and Micale G 2016 *Sustainable Energy from Salinity Gradients* (London: Elsevier) pp 1-350

[3] OES 2016 *Ocean energy systems annual report* Ocean Energy Systems (OES) (Lisbon: The Executive Committee of Ocean Energy Systems) pp 1–188

[4] Kempener R and Neumann F 2014 *Salinity gradient energy technology brief* (Abu Dhabi: Irena) pp 1–28

[5] Loeb S, Van Hessen F, and Shahaf D 1976 Production of energy from concentrated brines by pressure retarded osmosis. II. Experimental results and projected energy costs *J. Membr. Sci.* **1** 249–269

[6] Maisonneuve J, Pillay P and Laflamme C 2015 Osmotic power potential in remote regions of Quebec *Renew. Energy* **81** 62–70

[7] Kim D, Kim J, Shon H, and Hong S 2015 Pressure retarded osmosis (PRO) for integrating seawater desalination and wastewater reclamation: energy consumption and fouling *J. Membr. Sci.* **483** 34–41

[8] Wan CF and Chung TS 2015 Osmotic power generation by pressure retarded osmosis using seawater brine as the draw solution and wastewater retentate as the feed *J. Membr. Sci.* **479** 148–58

[9] Husnil YA, Harvianto GR, Andika R, Chaniago YD, and Lee M 2017 Conceptual designs of integrated process for simultaneous production of potable water, electricity, and salt *Desalination* **409** 96–107

[10] Ramos Avilez HV 2012 *Remoción de cianuro en relaves auríferos utilizando biochar producido a partir de tallos de gliricida sepium* (Cartagena: Universidad de Cartagena) pp 1–116

[11] Harris MA 2016 Cold-water starch dissolution bioreactors: role in treating heap-leached wastewater from precious metal and REE ores *Geobiotechnological Solutions to Anthropogenic Disturbances* (Gewerbestrasse: Springer) pp 393–406

[12] Naghiloo A, Abbaspour M, Mohammadi-Ivatloo B, and Backhtari 2015 Modeling and design of a 25 MW osmotic power plant (PRO) on Bahmanshir river of Iran *Renew. Energy* **78** 51–59

[13] Gerstandt K, Peinemann KV, Skilhagen SE, Thorsen T, and Holt T 2008 Membrane processes in energy supply for an osmotic power plant *Desalinization* **224** 64–70

[14] Osorio AF, Ortega S, and Arango-Aramburo S 2016 Assessment of the marine power potential in Colombia *Renew. Sust. Energ. Rev.* **53** 966–977

[15] Le Sueur C T 1947 Sulfur-containing polyamides Patent US2430859 A (USA: United States Patents) pp 1–3

IOP Conf. Series: Journal of Physics: Conf. Series **935** (2017) 012063 doi:10.1088/1742-6596/935/1/012063

Design of a residential microgrid in Lagos del Cacique, Bucaramanga, Colombia

D Bellon[1], O A González Estrada[2] and A Martínez[1,3]

[1] Universidad de Santander, Bucaramanga, Colombia
[2] Universidad Industrial de Santander, Bucaramanga, Colombia
[3] Foundation of Researchers in Science and Technology of Materials, Colombia

E-mail: alejandrom@udes.edu.co , danielabm@outlook.com

Abstract. In this paper is presented a model that analyses the options to provide energy to an interconnected house in Lagos del Cacique, Bucaramanga, Colombia. Three power supplies were considered: photovoltaic, 1 kW wind turbine, and a 2.6kW gasoline generator, as well as a battery for energy storage. The variables considered for the sensitivity analysis correspond to the price of gasoline and the variation in loads. The simulation results suggest an optimal configuration of microgrids in generator -photovoltaic panel-battery. Sensitivity variables were specified in order to evaluate the effect of uncertainty. The simulation was done through the Homer software and the results of the combinations of sources are suggestions of the same.

1. Introduction

The fast growth of global energy demand has led to an increase in pollution and global warming. Alternative energies represent a clean source of generation and therefore their development is crucial to minimize the effects of climate change; likewise, its use has implied the concept of distributed generation, which associates the technical and environmental elements that must be considered with special attention [1-4]. Thus, it is of interest to estimate the correct type of energy for a given project, according to the specific conditions of the area, seeking the execution of the optimized system, which in a first phase must be carried out under computational techniques [5-7].

The modelling and planning of microgrids represent the crucial stages and critics in the process of design and execution [8-10], given the large number of components to integrate and the characteristics of each one From the Homer software, a simulation of a microgrids was done for a house without interconnection in Lagos del Cacique, Bucaramanga, Colombia, on which the elements are integrated according to the specific characteristics of the zone. The work details the costs of investment, operation and maintenance, and the configuration with better environmental and economic benefits.

2. Planning parameters

The scheme of the hybrid power system is shown in Figure 1(a). The components include wind turbines, photovoltaic panels, a generator, the low primary load for a residential stratum 5 and converters (AC to DC). Figure 1(b) shows the load profile used in the present work.

The generator selected is from Kipor brand, with a capacity of 2.6kW, with single-cylinder OHV engine and output regulation system that allows for stable voltage. The investment and replacement costs are assumed as US $1282.84/kW. In reference to the costs of maintenance and operation of the generator, were assumed as $0.05USD/hr.

IOP Conf. Series: Journal of Physics: Conf. Series **935** (2017) 012063 doi:10.1088/1742-6596/935/1/012063

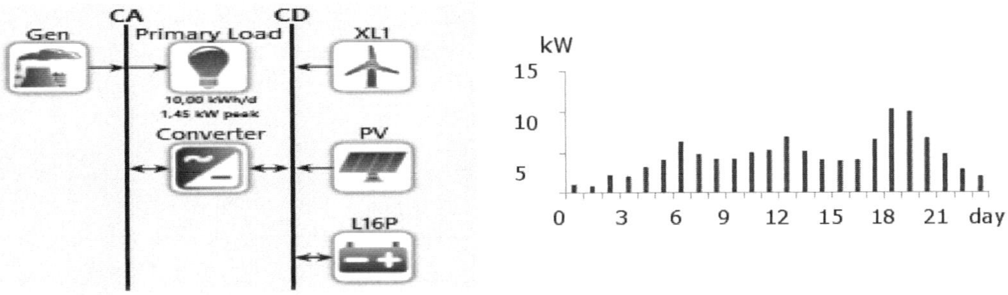

(a) (b)

Figure 1. (a) Scheme of the system, (b) Load profile of the area under study.

The cost of fuel for the generator is assumed as \$0.69USD/l. The wind turbine selected corresponds to a Bornay AE-BOR-B15.48, of 1.5kW and 48v, two blades, with digital regulator. The cost of maintenance and operation were assumed as \$1,000 USD/year. The estimated life of the equipment is 20 years. Investment and replacement costs were assumed as \$4,967.89, according to market prices.

The photovoltaic panel is Eco-Worthy of 1kW. The useful life is assumed of 25 years, with a reduction factor of 90%. The investment cost of the panel is \$1,447USD, with an equal replacement cost, and an operation and maintenance cost were assumed of \$72USD/year.

The inverter corresponds to a "Horizontal Grid Inverter", with a capacity of 1kW, and works in full wave, what improves the system. The efficiency of the inverter is assumed as 90%, with a useful life of 15 years. Investment and replacement costs are estimated at \$270.11USD in the market, the cost of operation and maintenance are disregarded.

The last of the components corresponds to the battery, was selected a Trojan L16P 420Ah, 6V, of deep cycle. His cost of investment and replacement are \$319USD and his cost of maintenance and operation are \$20USD/year. For the present study was located 8 batteries in series, with initial and final charge of 100% and 30% respectively. The performance of the battery life is 1,075kWh.

On the other hand, according to the Solar Radiation Atlas of Colombia, the city of Bucaramanga presents an average daily radiation between 4.5 and 5.5kWh/m^2, with stable behaviour throughout the year, which coincides with the values predicted by HOMER, according to NASA's database [11], which establishes an average radiation of 5.16 kWh/m^2/day. Figures 2(a) and 2(b) show the solar resource profile during the 12 months of the year. Wind statistics are obtained according to Homer, by NASA's database; was found that the average wind speed in the selected area is approximately 2.15m/s. The Figure 2(b) shows the profile of the wind resource.

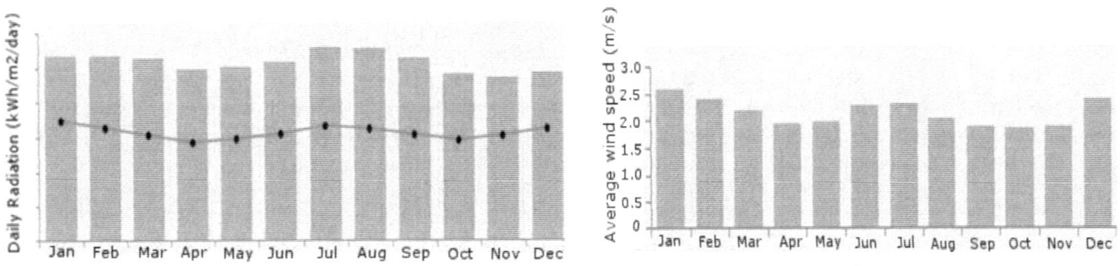

(a) (b)

Figure 2. (a) Solar resource of the study area; (b) Wind resource of the study area.

The variables considered for the sensitivity analysis correspond to the values of load and the price of gasoline. The discount rate and inflation are assumed as 8% and 2% respectively. Was assumed the lack of annual capacity of 2% and the project life cycle of 25 years.

3. Analysis and discussion of results

Economic aspects play an important role in any simulation and Homer is no exception. The Net Annual Cost (CAN), include the initial cost, replacement components, maintenance and others cost, within the lifetime of the project. Table 1 shows the first 8 grid configurations, with better cost-benefit according to the CAN.

Table 1. Cost-Effective Grid Configurations.

Cost/CDE (USD)	Cost/CAN (USD)	Cost/ Operating Cost (USD)	Cost/ Initial investment (USD)
0.926525	43718.37	2952.34	5551.947
1.119128	52806.38	3767.269	4104.947
1.27527	60.174	3840.967	10519.84
1.467517	69245.27	4654.601	9072.838
1.601744	75578.79	5747.117	1282.838
1.636236	77206.3	5740.186	2999.948
1.992439	94013.87	6767.969	6520.837
2.005599	94634.8	6704.069	7967.837
0.926525	43718.37	2952.34	5551.947

Two initial microgrid configurations stand out, as they represent the highest profitability. Thus, the first configuration of HOMER suggests a hybrid system of a solar panel, a gasoline generator of 2.6kW, 8 batteries and an inverter. For this case, the average cost per kWh of useful electrical energy produced by the system is \$0.927USD, the lowest of all other configurations. The second configuration suggests the generator power output of 2.6kW, 8 batteries and an inverter. This configuration presents an average cost per kWh of energy produced by the system of \$1.12USD.

The best possible system architecture in terms of profitability and operation evidences a total CAN of \$34,375USD, with 0% excess electricity and a maximum renewable penetration value of 227.7. Figure 3 shows the monthly average of electricity production with this configuration.

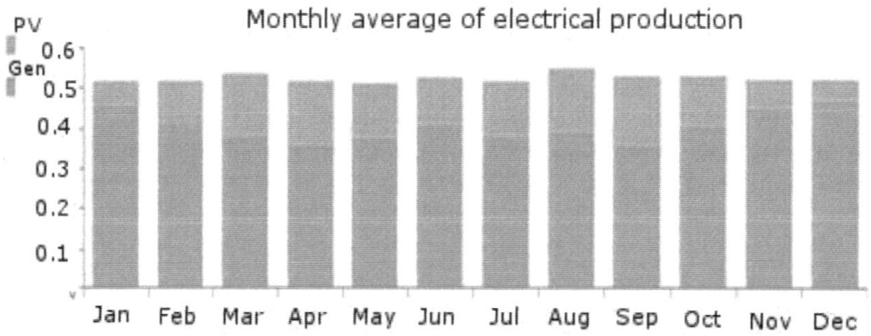

Figure 3. Monthly average of production with more optimal configuration.

The sensitivity analysis seeks to evaluate the way in which the fluctuations of certain variables affect the optimal configuration of the system [12]. In the present study, the variables fuel cost (gasoline) and load values were evaluated.

For the initial study, an approximate gasoline cost of \$0.69/l was estimated. In order to evaluate the effect of the variations of this cost on the system, prices increased by US \$0.05/l. For the second variable, an annual scaled average of 10kWh/d is established, with an increase of 1kWh/d.

According to the result of the sensitivity analysis for wide ranges of gasoline prices and charge prices, was observed that, with any of the fluctuating values, regardless of the change of the variables, the system generator- photovoltaic panel- battery is the best option.

Based on the results of the simulation it is possible to determine what type of distributed generators to use in a wide range of changes of the established variables. Figure 4 shows the costs of different systems and the different configurations suggested by Homer, finding that the lowest cost is $42,582.28USD.

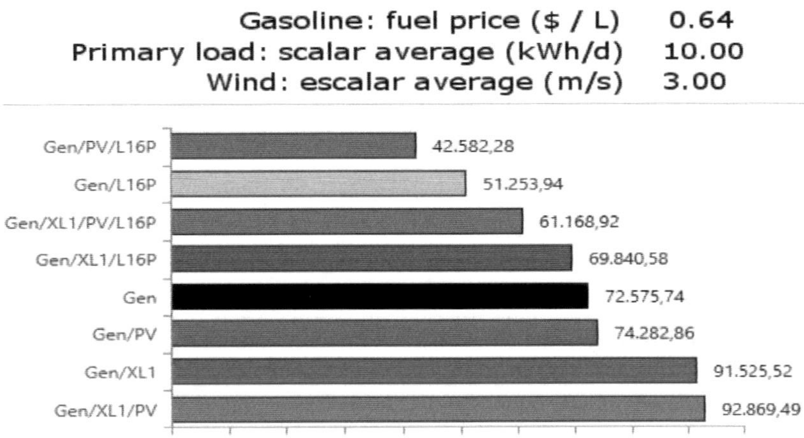

Figure 4. Costs according to configuration.

4. Conclusions

It was possible to establish, by Homer simulation, the optimum microgrid configuration, which provides the greatest technical and financial benefits. This led to the determination that the configuration that integrates generator, photovoltaic panels and battery, with a final cost of $42,582.28USD, represents the microgrid of greater viability according to the established parameters, for future projects.

References

[1] Su W, Yuan Z, Chow M 2010 Microgrid planning and operation: solar energy and wind energy. *Power and Energy Society General Meeting, IEEE* **11571203**

[2] Huan Jaiyi, Jian Chuanwen, Xu Rong 2008 A review on distributed energy resources and microGrid *Renewable and Sustainable Energy Reviews* **12(9)** 2472-2483.

[3] Stern N, Peters S, Bakhshi V, Bowen A, Cameron C, et al 2006 *Stern review: the economics of climate change* (London: HM Treasury) p 679

[4] S. Solomon, G.K. Plattner, R. Knutti, P. Friedlingstein 2009 Irreversible climate change due to carbon dioxide emissions *Proc. Natl. Acad. Sci.* **106(6)** (2009) 1704–1709

[5] Omar Hafez, Kankar 2012 Optimal planning and design of a renewable energy based supply system for microgrids *Renewable Energy* **45** 7-15

[6] Masters C L 2002 Voltage rise: the big issue when connecting embedded generation to long 11kV overhead lines *Power Eng. J.* **16(1)** 5–12

[7] Quezada VH Méndez, Abbad J Rivier, San Roman T Gomez 2006 Assessment of energy distribution losses for increasing penetration of distributed generation *IEEE Trans. Power Syst.* **21(2)** 533–40

[8] Dario Gandini, Anibal T. de Almeida 2017 Direct current microgrids based on solar power systems and storage optimization, as a tool for cost-effective rural electrification *Renewable Energy* **111** 275-283

[9] Manur, A. *et al.* 2017 Simple electric utility platform: A hardware/software solution for operating emergent microgrids *Applied Energy* (In Press)

[10] Peças Lopes JA, *et al.* 2007 Integrating distributed generation into electric power systems: A review of drivers, challenges and opportunities *Electric Power System Res.* **77(9)** 1189-1203

[11] NASA 2008 Surface meteorology and Solar Energy *Global horizontal radiation, monthly average values over 22-year period (July 1983- June 2005)* (https://eosweb.larc.nasa.gov/sse/)

[12] Connolly D, *et al.* 2010 A review of computer tools for analysing the integration of renewable energy into various energy systems *Applied Energy Apr.* **87(4)** 1059-1082

IOP Conf. Series: Journal of Physics: Conf. Series **935** (2017) 012064 doi:10.1088/1742-6596/935/1/012064

Construction, study and mathematical modelling of the electrical behavior of CZTiS photovoltaic material as function of time and synthesis temperature

M Patarroyo[1], **H Pérez**[1], **J Gómez**[1] **and E Vera**[1]

[1]Universidad Pedagógica y Tecnológica de Colombia, Tunja, Colombia

E-mail: miguel.patarroyo@uptc.edu.co

Abstract. This work reports the obtaining, study and mathematical modeling of a photovoltaic material type CZTiS (Cu_2ZnTiS_4), in function of two thermodynamic variables such as temperature and hydrothermal synthesis time. The obtention the materials had to be validated through the implementation of characterization techniques such as X-Ray Diffraction (XRD) and solid state Impedance Spectroscopy (IS). The results of the characterization, allowed to confirm in all cases the obtention of the materials, with a crystalline structure concordant with a tetragonal geometry space group of *I-42m*, a preferential crystalline orientation in the plane (1 1 2), with crystal sizes in the nanometric order (5-6nm). The electrical characterization, showed a semiconductor behavior of the solids concordant with values of conductivity that can be modeled by the analysis of variance and verified through the Kramers-kronig transform. The information obtained confirms that the temperature variable is the most strong influence in the electrical behavior of the CZTiS material, which is in agreement with similar works evaluated with alternative techniques [1].

1. Introduction

The type CZTiS (copper, zinc, titanium, sulfur) made in the Uptc and obtained by a synthesis process using a hydrothermal pathway is a low cost alternative of semiconductor that has been evaluated with the possibility of using as material for construction of photovoltaic cells. The optical and electrical properties with kesterite structure achieved in this research show that CZTiS has an acceptable electric behavior. The CZTiS has been manufactured using this technique, analyzed and validated using statistical and mathematical methods such as variance and the Kramers-Kronig equations in order to find a cost-effective process and where possible replace other high-cost techniques such as those mentioned in [2]. The temperatures and times used in this work (250, 275, 300°C and 24, 48 and 72 hours) served to mathematically model the electrical behavior of the studied material, as well as to compare the results of the characterization with works that used temperatures of 473K a 773K, with 100K intervals and investigations with temperatures of 500 to 575°C at 25°C intervals and different techniques as can be seen in [2, 3]. The mathematical model found allowed to interpolate and predict the conductivity of the material at different temperatures.

The X-Ray Diffraction (XRD) was used to characterize the manufactured product, from which it was possible to obtain information that corroborated the type of material sought CZTiS Cu_2ZnTiS_4), on the other hand, by means of The electrical Impedance Spectroscopy (SI), a technique used to collect experimental data, which consists of applying a sinusoidal electrical disturbance (voltage or current) to the study material, obtained numerical information used to perform the analysis of the variance and for the treatment of the data by means of the Kramers-Kronig transforms, in this stage several impedance tests were carried out since due to the applied voltage and to strange factors there can be some type of noise or error that influences the measurements [4–6].

2. Methodology

The preparation of the CZTiS material started from the respective precursors in the form of Cu and Zn nitrates, titanium isopropoxide and thiourea, which were stoichiometrically dosed in a teflon vessel with magnetic stirring and deposited in an steel autoclave . The synthesis temperatures were established according to the thermodynamic criteria to form the respective kesterite phase (250, 275 and 300°C). Likewise, the time values were established at 24, 48 and 72 hours, being these variables the main factors to be evaluated in the present work. At the end of each process the materials were obtained as a black precipitate, which was washed with absolute ethanol repeatedly to reduce the effect of impurities generated in the process. Subsequently, the materials were dried for 80 minutes in an electric oven at a temperature of 100°C, obtaining the respective materials in each case in the form of a compact powder. The solids were characterized by XRD on PANalytical X′pert pro MPD equipment, equipped with an X′celerator detector in Bragg-Bretano configuration, using the Cu Kα (λ=1.54Å). The diffractograms were taken between 10 and 90°and the results were analyzed using the X′Pert High Score software. The solid state impedance spectroscopy (IS) analyzes were performed on a GAMRY potentiostat-galvanostat between 1 and 10MHz obtaining numerical information from the set of samples, which were uniaxially compacted at 10Mpa with a diameter of 10mm and a thickness of 1.0mm. Four measurements were performed on each sample to verify the reproducibility of the data and each analysis was corrected by a reference cell to avoid noise in the determinations.

3. Analysis and results

Initially, the CZTiS materials were characterized by XRD, obtaining the presence of characteristic crystalline structures and associated with the pure tetragonal phase of kesterite and identified with PDF *card* 00-026-0575, with space group *I-42m* (1 2 1) and cell parameters a=5.427 angstrom and b=10.848 angstrom, as shown in Figure 1. In all cases, there is no evidence of relevant structural variations related to this crystalline phase.

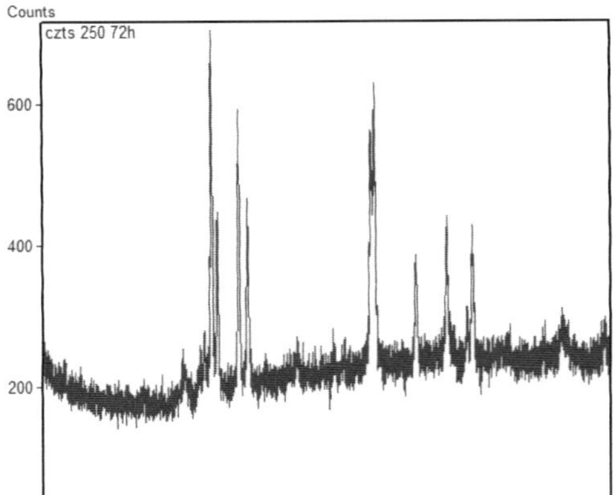

Figure 1. Structural characterization by XRD of CZTiS material.

IOP Conf. Series: Journal of Physics: Conf. Series **935** (2017) 012064 doi:10.1088/1742-6596/935/1/012064

The results of the electrical characterization of the materials, in the form of Nyquist and Bode diagrams, show a favorable semiconductor behavior in agreement with preliminary works [7] And as shown in Figure 2. The data of these diagrams were collected and analyzed by means of the Kramers-Kronig equations, with the final purpose of validate the proposed method related to the analysis of the impedances as a function of thermodynamic parameters of temperature and time of synthesis. The data were analyzed taking into account the proposed variables, recording the maximum values of the impedance modules in each test, as indicated in Table 1.

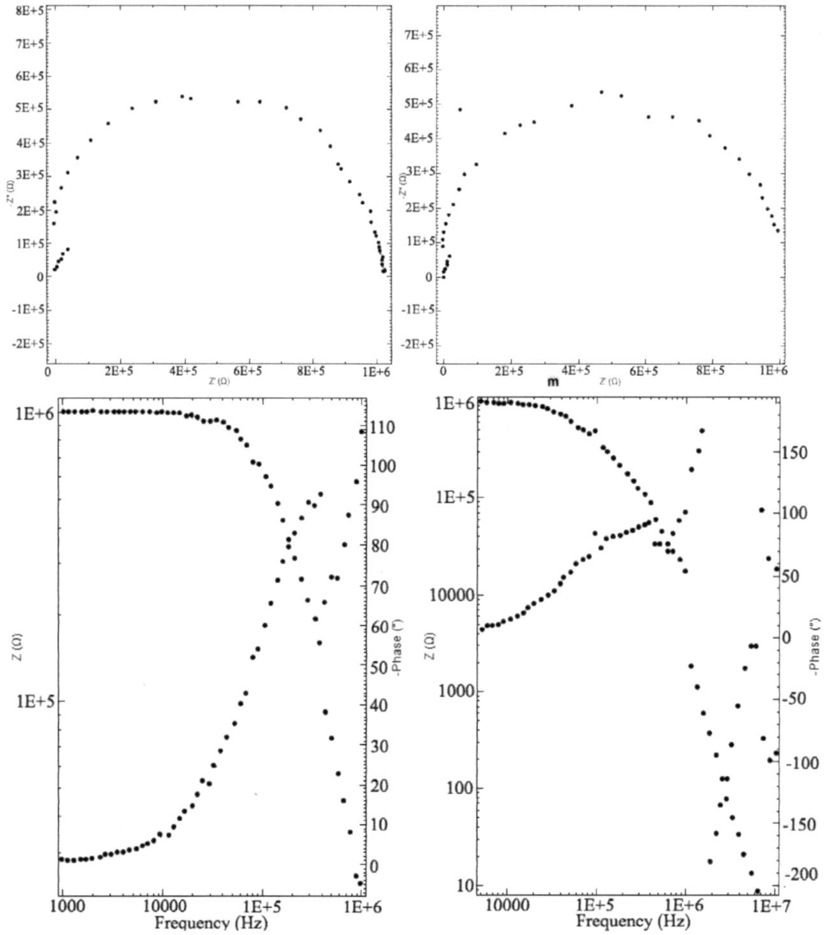

Figure 2. Nyquist and Bode diagrams for 24-48h and 275-300°C.

To find the conductivity of the material, was used the inverse of the maximum values of the impedance modules of each test, as indicated in Table 2. These data were statistically analyzed by the variance obtaining a correlation factor for temperature and time, as well as the interaction of these two variables, in order to check which of these is the factor that most influences the conductivity of the material as is indicated in the Table 3.

As Fo>1, of the treatments provided (time and temperature), which has the greatest influence on the conductivity of the material is the temperature, it is also observed that the temperature correlation factor is greater than F, that is, $Fo_{(T)} > F$, so with this variable the experiment has a 97% of success. Although the correlation factor of the time variable is close to 1, it has little influence on the conductivity of the material.

IOP Conf. Series: Journal of Physics: Conf. Series **935** (2017) 012064 doi:10.1088/1742-6596/935/1/012064

Table 1. Impedance spectroscopy (CZTiS) to different times and temperatures.

T (°C)	Registry 1	Registry 2	Registry 3	Registry 4	Time in h
300	769.74	3787.87	530.85	3562.81	24
300	3024.25	2643.64	6017.47	2154.54	48
300	3938.44	500.37	8765.01	6431.26	72
275	8412.06	8944.22	8294.3	15535.48	24
275	9475.23	7151.24	12359.13	10023.38	48
275	1785.34	8421.64	11432.14	8941.36	72
250	16463.65	6275.14	13612.17	10172.35	24
250	18688.88	8761.5	14672.51	16707.95	48
250	13036.76	10594.106	19903.7	18164.13	72

Table 2. CZTiS conductivity material at different times and temperatures.

T (°C)	Registry 1	Registry 2	Registry 3	Registry 4	Time in h
300	1.29×10^{-3}	2.64×10^{-4}	1.88×10^{-3}	2.80×10^{-4}	24
300	3.30×10^{-4}	3.78×10^{-4}	1.66×10^{-4}	4.46×10^{-4}	48
300	2.53×10^{-4}	1.99×10^{-3}	1.14×10^{-4}	1.55×10^{-4}	72
275	1.19×10^{-4}	1.12×10^{-4}	1.20×10^{-4}	6.43×10^{-5}	24
275	1.055×10^{-4}	1.398×10^{-4}	8.09×10^{-5}	9.97×10^{-5}	48
275	5.60×10^{-4}	1.187×10^{-4}	8.74×10^{-5}	1.118×10^{-4}	72
250	6.07×10^{-5}	1.59×10^{-4}	7.34×10^{-5}	9.83×10^{-5}	24
250	5.35×10^{-5}	1.14×10^{-4}	6.81×10^{-5}	5.98×10^{-5}	48
250	7.67×10^{-5}	9.43×10^{-5}	5.02×10^{-5}	5.50×10^{-5}	72

Table 3. Variance analysis of the electrical behavior of the material CZTiS.

Source of variation	Sum of squares	Degrees of freedom	mean squares	F_0	Value P
Synthesis time	2.63×10^{-7}	2	1.315×10^{-7}	0.766	< 0.01
Synthesis temperature	2.19×10^{-6}	2	1.095×10^{-6}	6.387	< 0.01
Interaction	4.86×10^{-7}	4	1.215×10^{-7}	0.709	< 0.01
Fault	4.63×10^{-6}	27	1.714×10^{-7}	$F_{0.05:4:27}=2.73$ $F_{0.05:2:33}=3.27$	
Total	7.57×10^{-6}	35			

4. Application of Kramers Kronig transforms

These equations relate the real and imaginary part of a complex analytic function [8]. Now well, as the impedance is formed by a real part (resistance) and an imaginary part (reactance), these equations (KK) can be used to calculate and verify the experimental data of the impedance spectra of the CZTiS material [9,10]. Therefore, making use of the KK equations, it is possible to transform the complex part of the impedance in the imaginary part and reciprocally [11,12].

4.1. How to verify the reliability of the data

The literature evidences several ways to validate the experimental data of a system; one is adjusting several or a polynomial of Z' or Z'' with ω and replacing them in the corresponding equations of KK and integrating numerically. Another way is making use of the models of measurement, these are used by means of general equivalent circuits (Maxwell or Voigt circuits) and always comply with the KK. The component equations of the impedance for a given frequency in a parallel circuit are: [13].

$$Z'(\omega_i) = R_1 + \sum_{k=2}^{M} \frac{R_k}{1+(\omega_i \tau_k)^2} \tag{1}$$

$$Z''(\omega_i) = -\sum_{k=1}^{M} \frac{\omega_i R_k \tau_k}{1+(\omega_i \tau_k)^2} \quad \text{con} \quad t_k = (\omega_k)^{-1} \tag{2}$$

The easier way to calculate the transformed components is: 1. Find the imaginary components from the real ones, for this the real experimental values Z', the values of the angular frequency and the time constant τ_k are replaced for each frequency in the equation (1), this allows to establish a system of linear equations in R_k, 2. Once this system is established, it is solved to obtain the values corresponding to R_k, 3. These values are replaced in the equation (2) obtaining of this way the transformed imaginary components, which will be noticed $Z''(trans)$. 4. The $Z''(trans)$ values are compared with the imaginary experimental data (Z'') by imaginary residuals (Δ'') for each frequency, as shown in the following equation.

$$\%(\Delta'')_i = \frac{Z''(trans)_i - Z''_i}{|Z_i|} * 100 \tag{3}$$

5. Mathematical model for data conductivity of the CZTiS material in function of the time and temperature of synthesis

Using the data from the Table 2, it was possible to fit a mathematical model capable of describing the electrical behavior of the CZTiS material as a function of time and temperature of synthesis, the model was adjusted using the method of least squares [14], the obtained surface verifies that the experimental data remains closely or are on the graph, as evidenced in the Figure 3.

$$C(t,T) = 3.74 \times 10^{-4} - 1.158 \times 10^{-6} t - 5.85 \times 10^{-8} T - 1.30 \times 10^{-9} t * T \tag{4}$$

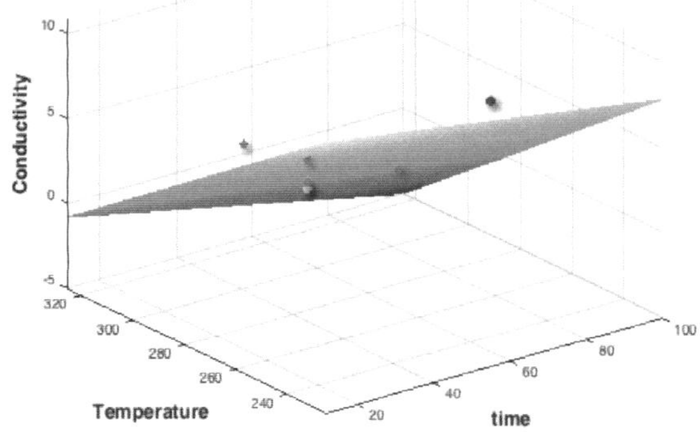

Figure 3. Curve fitted to the conductivity data of CZTiS material at 250, 275, 300°C and times of 24, 48, 72h.

IOP Conf. Series: Journal of Physics: Conf. Series **935** (2017) 012064 doi:10.1088/1742-6596/935/1/012064

6. Conclusions

The characterization of the properties of the material, using X-Ray Diffraction (XRD) technique, allowed to verify the obtention of a material type CZTiS and the quality of the same. The shape of the Nyquist diagrams showed good electrical behavior of the material, since the material is not as resistive. With the inverse of the maximum impedance modulus, an appreciable value of the conductivity could be obtained, as can be seen in the Table 2. The model fitted to the conductivity data was made possible by the different impedance spectra of the material, this model was adjusted by the method of least squares. The statistical analysis of the variance allowed to detect that the factor that predominates in the manufacture of this type of materials is the synthesis temperature. It is observed that the conductivity of the CZTiS material manufactured with the characteristics and process described above is linked to the synthesis temperature and very little to the reaction time of the material. he Kramers-Kronig equations are of a lot importance to validate the impedance data, with this it is possible to ensure the validity of the process. The construction of a model and the adjustment of the conductivity data obtained through the impedance spectra allow us to interpolate and to know point approximations of the electrical behavior of the CZTiS material.

References

[1] Suryawanshi M P, Shin S W, Agawane G L, Gurav K V, Ghorpade U V, Hong C W, Gaikwad M A 2015 A Promising Modified SILAR Sequence for the Synthesis of Photoelectrochemically Active Cu2ZnSnS4 (CZTS) Thin Films *Israel Journal of Chemistry* **55** 1-6

[2] Nuttee Khottoommee, et al. 2015 Structural and optical properties of Cu_2ZnSnS_4 films deposition by screen coating technique of sol-gel process *SNRU Journal of Science and Technology* **7(2)** 100-105

[3] Shin S W, et al. 2015 A chemical approach for synthesis of photoelectrochemically active Cu_2ZnSnS_4 (CZTS) thin films *Solar Energy* **110** 221-230

[4] Becerra R A 2016 *Síntesis y caracterización de películas delgadas de* Cu_2ZnSnS_4 *depositadas por CBD, asistida con membrana de difusión* (Colombia: Universidad Nacional de Colombia)

[5] Boukamp B A 1993 Practical application of the Kramers-Kronig transformation on impedance measurements in solid state electrochemistry *Solid State Ionics* **62** 131-141

[6] Antaño L Q R 1997 *Aplicación de un algoritmo basado en un modelo de medición para la detección de errores en las medidas experimentales de impedancia* (México: Universidad Autónoma Metropolitana-Iztapalapa)

[7] Patarroyo M, Vera E, Pineda Y, Gómez J 2017 Impedance spectroscopy in photovoltaic materials of Cu_2ZnSnS_4 (CZTS) and use of the KK transform *Journal of Physics: Conf. Series* **786** 012014

[8] Urquidi-Macdonald M Real S, Macdonald D D 1986 Application of Kramers-Kronig transforms in the analysis of electrochemical impedance data II transformations in the complex plane *Journal of the Electrochemical Society* **133(10)** 2019-2024

[9] Matthew E, Orazem M 1991 On the Application of the Kramers-Kronig relations to evaluate the consistency of electrochemical impedance Data *The Electrochemical Society Inc* **138(1)** 67-76

[10] Dambrowski J 2013 Validation of impedance-data and of impedance-based modeling approach of electrochemical cells by means of mathematical system theory *The 39Th Annual Conference Of The IEEE Industrial Electronics Society (Vienna)* (USA: IEEE) 1-7

[11] Bayer M H, Schneider I A 2013 Application of the Kramers Kronig relations to locally resolved impedance data of polymer electrolyte fuel cells *Journal of Electroanalytical Chemistry* **689** 42-45

[12] León C, Martín J M, Santamaría J, Skarp J, Gónzalez D G, Sánchez Q F 1996 *Use of Kramers-Kronig transforms for the treatment of admittance spectroscopy data of p-n junctions containing traps* (USA: American Institute of Physics) 7830-7837

[13] Boukamp B A 1995 A linear Kronig-Kramers transform test for immittance data validation *The Electrochemical Society Inc.* **142** 6

[14] Montgomery D C 2004 *Diseño y análisis de experimentos* (USA: Limusa Wiley)

Hydrothermal synthesis and characterization of the semiconductor material Cu_2ZnTiS_4

H G Pérez Bustos[1], M Patarroyo-Mesa[1], J A Gómez-Cuaspud[1], Y Pineda-Triana[1] and E Vera-López[1]
[1] Universidad Pedagógica y Tecnológica de Colombia, Tunja, Colombia

E-mail: hellem.perez@uptc.edu.co

Abstract. This paper describes the results of the synthesis and characterization of a quaternary semiconductor based on Cu_2ZnTiS_4 (abbreviated CZTiS), using a hydrothermal technique. The results confirm that time (24, 48 and 72 hours) and temperature (250, 275, and 300°C) factors, used in synthesis process, regulate different levels of electrical conductivity in these materials. The results of ultraviolet spectroscopy (UV) analysis, confirm the production of semiconductor solids with Band-gap values around 1.4eV, being coherent with a strong absorption in the ultraviolet region. The X-Ray Diffraction analysis (XRD), confirm that there is an opposite and proportional relationship between the crystal sizes, the reaction times and the synthesis temperature. In all cases, the particle sizes were 50-100nm. The results derived from Scanning Electron Microscopy (SEM), confirm the obtaining of homogenous materials with optimal morphological properties in accordance with synthesis method. Similarly, the composition values derived from the Energy-Dispersion X-ray Spectroscopy analysis (EDS), shown that obtained compositions are in agree with expected values. Finally, the results of electric characterization, confirm that used synthesis parameters show a strong dependence on the conductive behaviour of solids being the most relevant result the shown by the solid obtained at 300°C and 72 hours of reaction in accordance with preliminary works.

1. Introduction

The world's largest source of energy throughout history has been through the exploitation of fossil resources (oil and coal), which have allowed high levels of energy production with respect to the amount of matter used and the amount of energy obtained [1]. However, despite their good efficiency in responding to global energy needs, this type of energy generates two main problems: one, non-renewable energy and two, generate toxic greenhouse gases to the detriment of the quality of life of many people and the environment in general. At present, the percentage of fossil resources available for its exploitation is exhausted, which means that in a few decades it will not be possible to respond to the global energy demand through this type of resources, in addition to the restrictions that present the problem environmental. It is for this reason that the search for new sources of clean and self-sustaining energy generation is especially important, especially those related to the use and use of solar energy. In this context, the use and use of solar energy through the use of solar panels, which basically use the photoelectric effect as a mechanism of electrical generation through materials of a semiconductor nature, have shown an important evolutionary process since its inception as panels of Silicon to alternative materials of low cost and toxicity, among which are the Kesterite type based on quaternary systems type Cu_2ZnSnS_4 (Abbreviated CZTS), which have allowed good conversion levels and a notable reduction in production costs. Although these materials have achieved good results,

IOP Conf. Series: Journal of Physics: Conf. Series **935** (2017) 012065 doi:10.1088/1742-6596/935/1/012065

efficiency processes remain low compared to traditional silicon materials, which is why the possibility of incorporating new elements into the basic structure such as titanium, that allow to reduce the values of band-gap, maintaining the efficiency of the system. One of these materials is called CZTiS (Cu_2ZnTiS_4) which consists of a semiconductor material in which the traditional tin has been replaced by titanium in the same stoichiometric proportions. This type of material, therefore, became a very promising material for the development of new photovoltaic devices, because it reduces the band gap values of the traditional CZTS up to 5% around 1.5eV. Based on the above, the present work proposes the synthesis and characterization of this type of material (CZTiS), by a route of hydrothermal synthesis that allows obtaining better experimental conditions for the obtaining of a pure phase, that will be reflected in the improvement in Conductivity and reaction conditions, taking into account the variables time and temperature [7-9].

2. Experimental

For the synthesis of the photovoltaic material of Cu_2ZnTiS_4 through a hydrothermal route, was started from the corresponding precursors of the material in stoichiometric quantities in the form of Copper nitrate ($Cu(NO_3)_2$), Zinc acetate ($Zn(CH_3COO)_2$), Titanium butoxide ($C_{16}H_{36}O_4Ti$), Thiourea (CH_4N_2S), all from Sigma-Aldrich commercial house. The precursors were metered stoichiometrically in aqueous medium to obtain 1.0g of material. The mixture of the components was made in a Teflon vessel and taken to a steel autoclave, where the temperature conditions were adjusted in each case (24, 48 y 72 hours). The systems were maintained under specific reaction conditions, simultaneously varying both time and temperature of synthesis, obtaining a total of nine samples as indicated in Table 1. After the synthesis process, each solid was repeatedly washed with absolute ethanol to remove possible impurities generated in the synthesis process. At the end, the solids were dried in an oven at 110°C for 2 hours.

Table 1. Parameters established for the synthesis of photovoltaic materials type CZTiS.

Material	Reaction temperature (°C)	Time of synthesis (h)
CZTiS	250	24
CZTiS	250	24
CZTiS	250	24
CZTiS	275	48
CZTiS	275	48
CZTiS	275	48
CZTiS	300	72
CZTiS	300	72
CZTiS	300	72

In order to analyse the morphological, structural and electrical properties of the materials obtained, different characterization techniques were used, such as visible ultraviolet spectroscopy (UV-VIS) were used in a scanning range of 190-1000nm in a spectrophotometer UV-VIS Mapada Instruments 8000 PC. Structural analyses were made by X-Ray Diffraction (XRD), in an equipment PANALytical X´pert PRO MPD, using radiation Cu Kα (λ=1.54186Å) between 20 and 90° 2θ with steps of 0.02°. The diffraction patterns were analysed using X'Pert software High Score. Analyses of infrared spectroscopy were made in a Perkin-Elmer FTIR equipment between 4700 and 500cm^{-1}. Scanning Electron Microscopy analysis (SEM) were developed in an JEOL 2100 equipment endowed with a LaB_6 thermionic barrel, an acceleration voltage of 200kV and a CCD camera that allows to obtain images of high resolution. X-ray Photoelectron Spectroscopy (XPS) analysis, were performed on a surface characterization platform XPS/ISS/UPS-

IOP Conf. Series: Journal of Physics: Conf. Series **935** (2017) 012065 doi:10.1088/1742-6596/935/1/012065

ACenteno SPECS. The platform provided with an energy analyser PHOIBOS 150 2D-DLD and a lightning source X Al Kα monochrome (FOCUS 500) was operated at 200W. The passing energy of the hemispherical analyser was set at 100 eV for the general and 60eV spectra for high resolution spectra. The surface load compensation was controlled with a flood pistol (dispositive FG 15/40-PS FG500) operated at 58μA and 1.5eV. The spectra were analysed with the program CasaXPS (Casa Software Ltd) using the CasaSSI GL 15 library for values R.S.F. a Shyrley baseline was used for this purpose. (BE) was corrected using the *C1s* at 284eV. Finally, the electrical analyses were developed in a potentiostat-galvanostat AUTOLAB between 1 and 10Mhz using Tablets of 10mm and 1mm of thickness obtained at 5.0Mpa of pressure. The data were adjusted and corrected by a reference cell to avoid noise in the acquisition of the data.

3. Results and discussion

The formation of Cu_2ZnTiS_4 materials can be described under the hydrothermal reaction conditions by the following half-reactions describing the three-step process identified in equations (1), (2) and (3).

$$2\ Cu_{(l)} + \frac{1}{2}\ S_{2(s)} \rightarrow Cu_2S_{(s)} \tag{1}$$

$$Cu_2S_{(s)} + TiS_{2(s)} \rightarrow Cu_2TiS_{3(s)} \tag{2}$$

$$Cu_2TiS_{3(s)} + ZnS_{(s)} \rightarrow Cu_2ZnTiS_{4(s)} \tag{3}$$

After the synthesis process under controlled temperature and time conditions, the obtained materials were repeatedly washed with absolute ethanol and dried in an oven at 110°C for 2 hours before the physicochemical characterization processes were performed.

The results of the UV-VIS spectrophotometry analysis on the different samples suspended in ethylene glycol between 190-1000 allowed determining the corresponding inflection points for each system to determine the band gap values using equation 4. In all cases the value Average was 1.40eV in agreement with the obtaining of semiconductor materials. In spite of this, it is clear that there is a proportionality relationship between the synthesis temperature and the band gap value, with the highest values being for the solid of the highest reaction temperature [11].

$$E = \frac{h\,c}{\lambda} \tag{4}$$

These results are in agreement with those obtained by T. E. Manjulavalli, *et. al.* [2], who has reported prohibited band gap for these same systems between 1.37 and 1.6eV, with efficiency values of around 0.16%. The results of infrared spectroscopy in all the samples allowed to identify that the materials show absence of organic phases related to free and derived sulphur species that could affect the performance of the systems. Signals identified around 800-1000cm^{-1}, Correspond to typical signals associated with the characteristic M-S and M-O vibrations present by the potential formation of Cu and Zn metal sulphides [3].

The X-ray results of the samples synthesized and shown in Figure 1, make it possible to determine that by varying the reaction conditions, more defined signals are generated for the highest reaction time and temperature (300°C-72h), this being due to the remarkable improvement the crystallinity of the samples. The detailed analyses using the X'Pert® High Score program and the comparison with the ICDD database, allowed to establish that the materials present a crystalline system *I-42m* (1 2 1) with cell parameters *a*=0.5427, *b*=0.5427, *c*=1.0848nm with preferential crystalline orientation along the plane (1 1 2) in all the Cases as shown in Figure 2. Additionally, the presence of secondary

IOP Conf. Series: Journal of Physics: Conf. Series **935** (2017) 012065 doi:10.1088/1742-6596/935/1/012065

phases such as ZnS, Cu_2S and Cu_2TiS_3, which decrease the percentage of synthesis efficiency of $Cu_2ZnTiS_4$and are made more noticeable at low temperatures and reaction times, is identified.

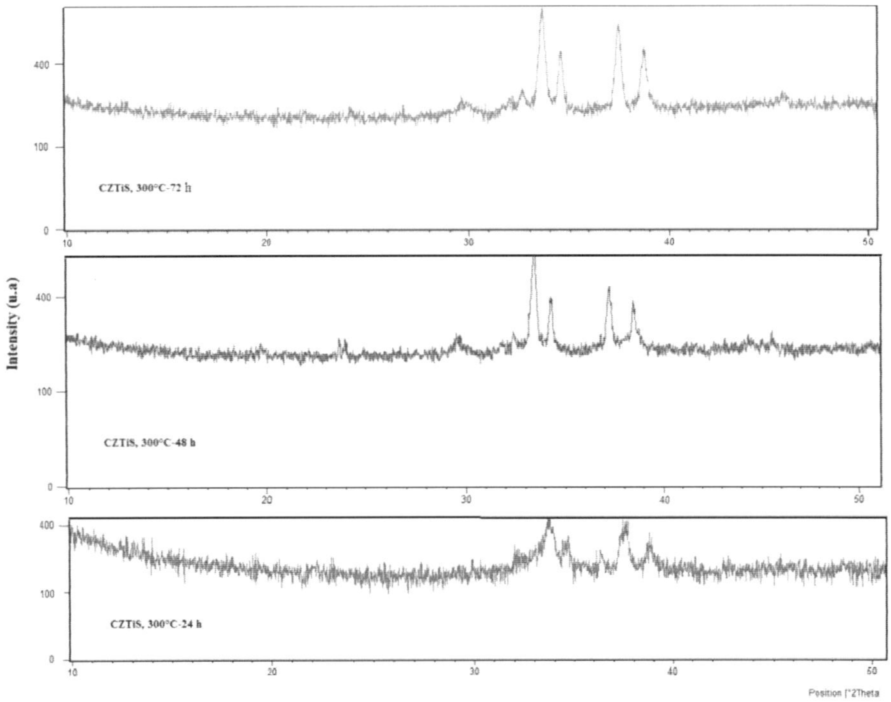

Figure 1. X-ray diffraction pattern for CZTiS systems under different reaction times and the maximum temperature studied-300°C.

The results of the analysis by scanning electron microscopy (SEM) allowed us to validate the identified morphological results and the surface characteristics of the synthesized materials as can be seen in the images of Figure 2. The results shown for the most representative materials of each series determine that the proposed synthesis method provides a series of surface features of interest with a heterogeneous grain distribution. According to the work of *Wang et. al.,* [3] the morphology of the materials is improved with increases in the temperatures of synthesis, since the formation of amorphous agglomerates is dramatically reduced reason why the obtained results correspond with the expected results. [4]

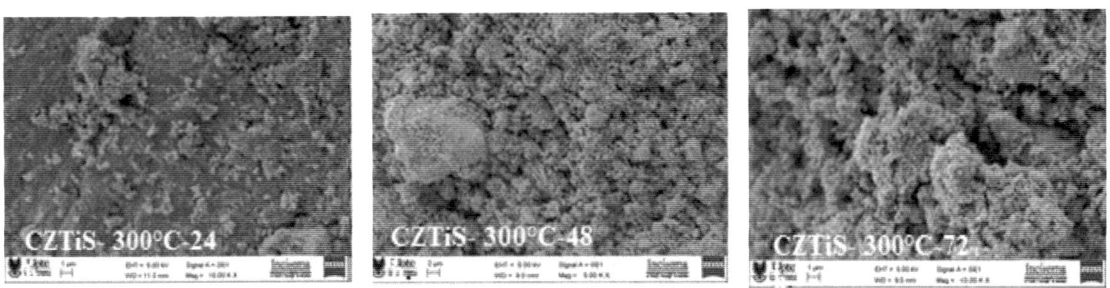

Figure 2. Scanning electron micrographs for CZTiS materials obtained at 300°C under different reaction time conditions.

The results show that the different characteristic elements of the material (Cu, Zn, Ti, S) exhibit signals. The results of XPS results on CZTiS type materials with reaction conditions of 300°C and 72 hours of reaction Characteristics, which allow analysing the oxidation states in which the constituent elements are found as indicated in Figure 3. The copper signals, confirm a state of oxidation Cu^{2+}, while those of zinc are attributed to the presence of Zn^{2+}, with a characteristic bonding energy at 1046 and 1023eV. For its part, titanium shows a typical bonding signal at 459 and 468eV related to the presence of an oxidation state 4^{+}, while in the case of sulphur the presence of two signals at 163 and 168 eV determining their presence in form of metal sulphide with a defined oxidation state and in the form of S^{2-}, which allows to validate the preliminary characterization results and to confirm the presence of the crystalline structure sought. In the case of sulphur results, it should be noted that secondary phases such as sulphides (mainly zinc or copper) may be present, which explain the slight modifications in bond energy compared to preliminary works [5,6,10].

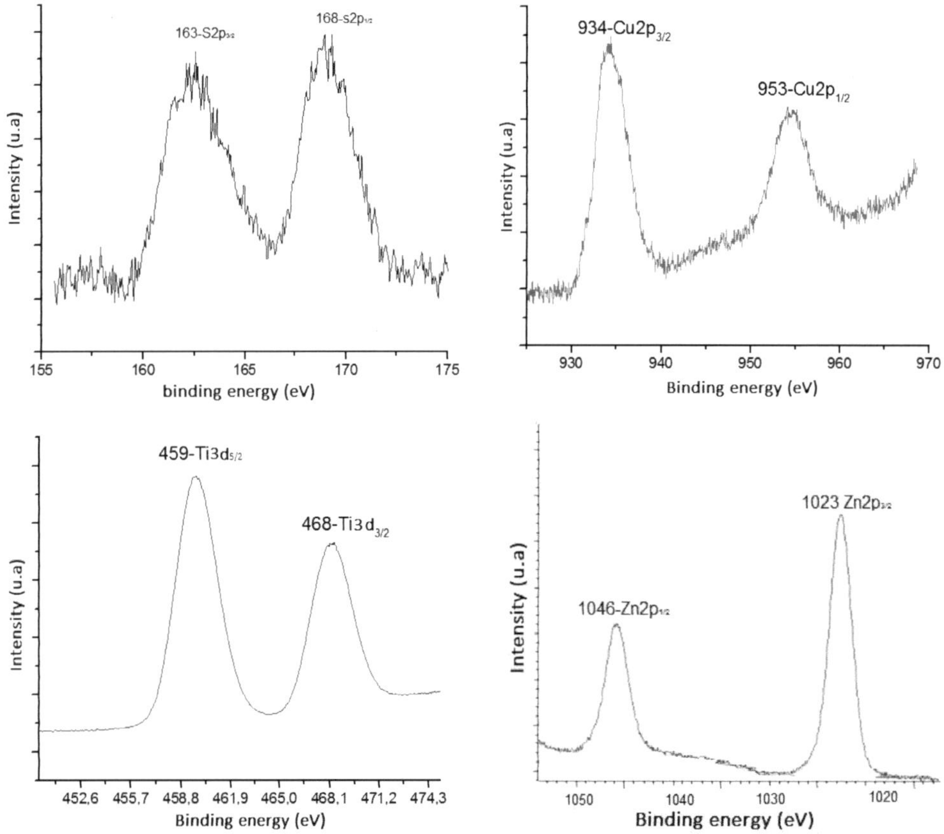

Figure 3. XPS spectra for the material synthesized under the highest reaction conditions.

Finally, the electrical characterization by solid state impedance spectroscopy allowed the determination of a semiconductor behaviour along the series of materials obtained at a higher temperature, which allowed the determination of values of conductivity of 1.29×10^{-3}, 7.98×10^{-3} and $4.46 \times 10^{-4} S \cdot cm^{-1}$ for the samples obtained at 24, 48 and 72 hours respectively. The results were validated by mathematical refinement processes and correspond to the expected conductive behaviour in this type of solids, in agreement with the work developed by L. Wang *et. al.* [5], according to which the temperature is the most critical factor and not the reaction time the most Influences the conductivity values [12].

IOP Conf. Series: Journal of Physics: Conf. Series **935** (2017) 012065 doi:10.1088/1742-6596/935/1/012065

4. Conclusions

The synthesis methodology used in the production of CZTiS photovoltaic materials was successful in that it allowed obtaining materials with relevant properties in terms of composition, structure and electrical performance, being the results of the samples obtained at higher times and temperatures of Synthesis (300°C 72 hours). The results of infrared and ultraviolet spectroscopy confirmed the achievement of semiconductor materials with a band gap around 1.5eV for potential applications with little evidence on the presence of sulphates in the materials obtained. The structural results evaluated by X-ray diffraction confirm the presence of a crystalline phase consistent with the CZTiS material with evidence of secondary phases associated with sulphides, which can be corrected by the development of subsequent thermal stages in agreement with preliminary works. Finally, the results of X-ray Photoelectron Spectroscopy (XPS) and electric characterization allow us to infer that the semiconductor response of materials, especially those obtained under the most extreme reaction conditions, is based on the expected chemical composition for these solids. Although the analyses show that the best composition results are obtained using the highest temperature and the highest reaction time due to the consolidation of a better kesterite type phase, it is clear that all can be improved to the extent that Subsequent heat treatment processes are implemented that reduce the presence of the secondary crystalline phases for their potential use as a photoabsorbent layer in photovoltaic devices.

References

[1] N. Asim, K. Sopian, S. Ahmadi, Kasra Saeedfar, M. A. Alghoul, Omidreza Saadatian, Saleem H. Zaidi 2012 *Renewable and Sustainable Energy Reviews* **16(8)** 5834–5847
[2] A. Kannan, T. Manjulavalli and J. Chandrasekaran 2016 *Procedia Engineering* **141** 15– 22
[3] J. Wang, P Zhang, X Songa, L Gao 2015 *RSC Adv.* **5** 1220-1226
[4] W. Huang, C. Tseng, S. Chang Solvothermal 2012 *Langmuir* **288** 496–501
[5] S. M. Camara, L.Wang and X. Zhang 2013 *Nanotechnology* **24** 495401
[6] J. M. Skelton, A. J. Jackson, M. Dimitrievska, S.K. Wallace, and A. Walsh 2015 *APL Materials* **3(4)** 041102
[7] D. Pardo and L. Bailón 2006 *Fundamentos de Electrónica Digital* (España: Universidad de Salamanca)
[8] Todorov 2008 *Preparación y estudio de películas delgadas con aplicaciones fotovoltáicas* (España: Universitat JAUME I) pp 20-29
[9] H. Dana, C. Hurlbut and C. Klein 2006 *Manual of Mineralogy* (New York: John Wiley & Sons, Inc) pp 73-87
[10] S. Chen, X. Gong, A.Walsh 2009 *Applied Physics Letters* **94** 041903
[11] R. Marti Valls, E. Barrachina, D. Fraga, I. Calvet, T. Stoyanova and J. Bautista 2015 Práctica de laboratorio *Síntesis y depósito de capas fotoabsorbentes tipo CIGS para dispositivos fotovoltáicos* (España: Universitat Jaume I) pp 37-40
[12] J. Balenzategui Manzanares 2008 Módulo: Gestión Eficiente de la Energía. *Fundamentos de la conversión fotovoltáica: la célula solar* (España: Escuela de Organización Industrial) pp7-11

Chemical obtaining of $LiMO_2$ and LiM_2O_4 (M=Co, Mn) oxides, for cathodic applications in Li-ion batteries

A Y Neira-Guio[1], J A Gómez Cuaspud[1], E Vera López[1], Y Pineda Triana[1]
[1] Universidad Pedagógica y Tecnológica de Colombia, Tunja, Colombia

E-mail: ariatna.neira@uptc.edu.co

Abstract. This paper describes the synthesis and characterization of two spinel and olivine-type multicomponent oxides based on $LiMO_2$ and LiM_2O_4 systems (M=Co and Mn), which represent the current state of the art in the development of cathodes for Li-ion batteries. A simple combustion synthesis process was employed to obtain the nanometric oxides in powder form (crystal sizes around 5-8nm), with a number of improved surface characteristics. The characterization by X-Ray Diffraction (XRD), Scanning and Transmission Electron Microscopy (SEM, TEM) and X-Ray Fluorescence (XRF), allowed to evaluate the morphology and the stoichiometric compositions of solids, obtaining a concordant pure crystalline phase of $LiCoO_2$ and $LiMn_2O_4$ oxides identified in a rhombohedral and cubic phase with punctual group R-3m (1 6 6) and Fm-3m (2 2 5) respectively. The electrical characterization of materials developed by impedance spectroscopy solid state, allowed to determine a p-type semiconducting behaviour with conductivity values of 6.2×10^{-3} and 2.7×10^{-7} S for $LiCoO_2$ and $LiMn_2O_4$ systems, consistent with the state of the art for such materials.

1. Introduction

The lithium ion batteries are defined as storage systems for the electrochemical energy conversion through the exchange of lithium ions (Li^+) between an anode (graphite) and a cathode of oxide nature (usually olivines) [1]. Since its first commercialization in 1991 [2], the Li-ion batteries play an outstanding position in the market with a production of billions of units per year, for powering portable electronic devices and currently used as generation systems in hybrid (HEV), electric (EV) and plug-in electric (PHEV) vehicles [3]. From point of view of functional applications, these devices operate by the interaction of an anode an electrolyte and a cathode, being the latter a critical component that can be influenced by aspects like crystalline phase, purity and particle size, which determines the efficiency of the battery. Among the most advantaged materials for these applications are the olivines ($LiCoO_2$-$LiFePO_4$) and spinels ($LiMn_2O_4$), whose electrochemical properties are strongly depending on aspects as homogeneity and microstructure, which is related with the used synthesis method [4].

This situation has promoted the research in the synthesis of ceramic systems to improve the conduction phenomena, preferably to development of mixed electronic ionic conductors (MIEC´s), which represents the opportunity to improve the quality of the Li-ion batteries [5]. Among the most used synthesis methods in the preparation of cathodic materials are the so-called solid-state reaction, co-precipitation, sol-gel, hydrothermal and Pechinni technique, among others [6], which has led to the obtaining of highly effective materials for a wide form of applications. However, the need to improve the performance of these electrochemical ceramics in the cyclic process of electric load, require the preservation of the morphologic and surface properties in materials, which cannot always be achieved by implementing of traditional synthesis methods [7]. It is for this reason that the present work tries to find from the perspective of the materials chemistry, a route based on combustion process that allow to obtain ceramic oxides based on an olivine ($LiCoO_2$) and spinel ($LiMn_2O_4$) configuration; which permit to improve the current performance of the lithium-ion batteries cathodes.

2. Experimental

For the synthesis of lithium oxides, corresponding nitrates of $LiNO_3$, $Co(NO_3)_2.6H_2O$ and $Mn(NO_3)_3.9H_2O$ were adjusted to a 1.0mol·L^{-1} concentration. The amount of each precursor in each system was adjusted according to the corresponding hydrolysis constants with the subsequent addition of a solution of citric acid 1.0mol·L^{-1} in a molar ratio 0.5:1, with respect to the total cation content as has been established elsewhere [8]. Each system remained at reflux for 2 hours, after which the resulting sol was heated at 120°C to evaporate the solvent and obtain the corresponding solid precursor, which were calcined at 850°C for 15 hours to guarantee the consolidation of crystalline phases. The calcined solids were characterized by X-Ray Diffraction (XRD), on a PANalytical X'per PRO MPD equipment provided with an ultra-fast X'Celerator detector in Bragg-Brentano configuration, using the Cu K radiation (λ=1.54Å). The diffractograms were taken between 10 and 90° 2θ and the results were analysed using the X'Pert High Score software [9].

The morphological evaluation of solid was performed by Scanning Electron Microscopy (SEM), in a Leica-Zeiss LEO 440 electron gun equipment with an accelerating voltage of 30kV. The analysis by Transmission Electron Microscopy (TEM) was done on a JEOL 2100 equipment provided with a thermionic LaB_6 gun with an accelerating voltage of 200kV. The composition of the solids was evaluated by X-ray fluorescence (XRF) on a Bruker S8 Tiger system, while the electrical and conductivity properties of the materials were evaluate using the solid state impedance spectroscopy technique (IS), in a GAMRY potentiostat-galvanostat instrument between 1 and 10MHz [10].

3. Results and discussion

The structural characterization of the oxides by XRD using the High Score X'Pert software and the ICDD database, led to a classification according to the reference materials $LiCoO_2$ ICSD code: 00-016-0427 with rhombohedral geometry and space group R-3m (166) and $LiMn_2O_4$ ICSD code: 00-018-0736, with cubic geometry and space group Fm-3m (225), according to preliminary works [11-12]. The crystallite size was obtained by means of the Debye-Scherrer equation, resulting in the presence of nanometric crystallites in the range of 5-8nm, using the most intense diffraction signals in each case. The Figure 1 shows the diffraction patterns obtained by Rietveld refinement, confirming the effectiveness of the synthesis process in the obtaining of the crystalline structures.

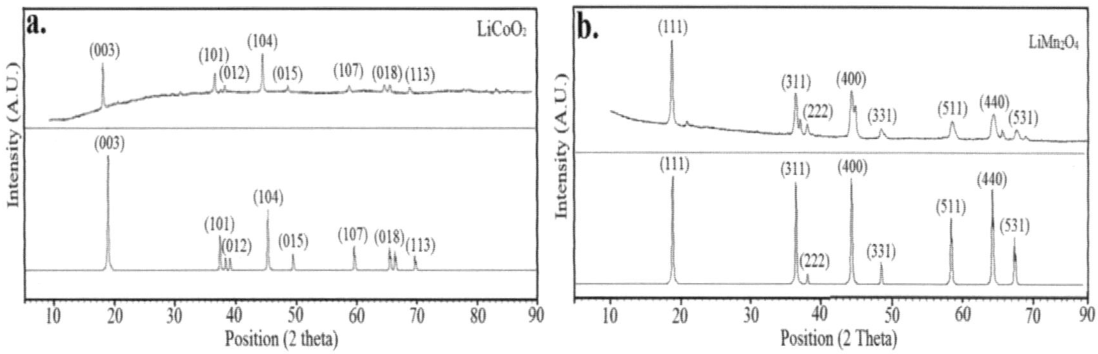

Figure 1. Indexed XRD patterns of $LiCoO_2$ and $LiMn_2O_4$ oxides with its corresponding Rietveld refinement patterns.

The detailed analysis of the diffraction patterns, led to the characterization of the corresponding cell parameters in both systems, which are shown in Table 1, confirming the formation of high crystallized oxides along (0 0 3) and (1 1 1) facets for the $LiMn_2O_4$ and $LiCoO_2$ ceramics respectively.

Table 1. Crystalline characteristics of obtained solids.

Material	Cell parameters (Å)	Crystalline system
$LiCoO_2$	a=2.8166 b=2.8166 c=14.045 α=β=90°	Rhombohedral R-3m (166)
$LiMn_2O_4$	a=8.2400 b=8.2400 c=8.2400 α=β=γ=90°	Cubic Fm-3m (225)

The subsequent refinement of experimental diffraction data using the ELMIX software yielded to find the corresponding unit cells of $LiCoO_2$ and $LiMn_2O_4$ systems, where could be identified the form in which the cell size in both compositions change dramatically, mainly in the case of $LiMn_2O_4$ system [13]. According to Cabrera *et al.*, in the spinel structure with Fm-3m space group the oxygens occupy the 32e positions and the manganese occupies the half of the octahedral positions 16d, leaving the lithium in the tetrahedral 8a position, which gives in the material a superior possibility to diffusion as shown in Figure 2 [14].

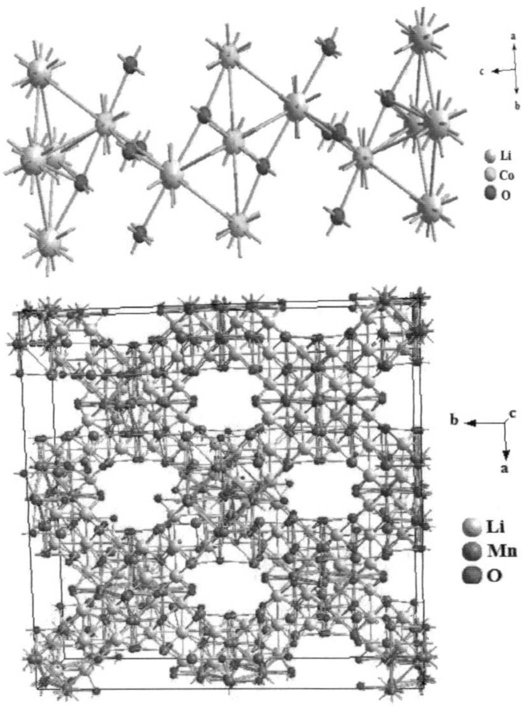

Figure 2. Calculus unit cell for $LiCoO_2$ and $LiMn_2O_4$ systems, obtained by the ELMIX software.

The scanning and transmission electron microscopy analysis, was possible to identify that the solids are composed of homogeneous aggregates with regular morphology with crystallite sizes around 5-8nm with interplanar distances about 0.45nm for the main diffraction signal, in coherence with previous XRD results and works done by Okubo *et. al.*, in similar perovskite compositions [15]. These results demonstrate that synthesis methodology favours the obtaining of pure crystalline phases, attributed to the low temperature consolidation during spontaneous combustion, as shown in Figure 3.

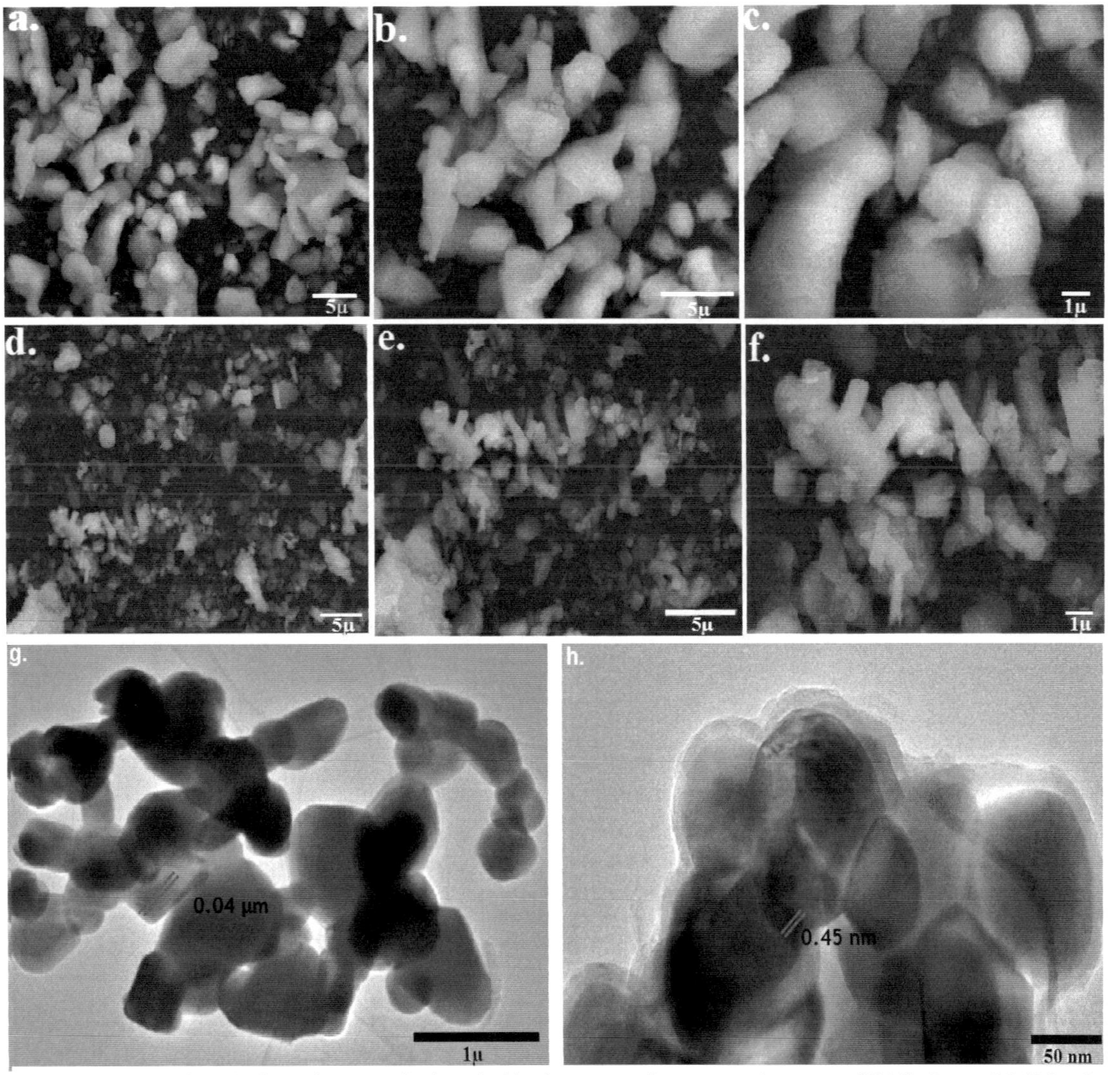

Figure 3. Scanning (a-f) and transmission (g-h) electron microscopy images of $LiCoO_2$ and $LiMn_2O_4$ solids obtained by the combustion technique.

The characterization by XRF and Energy-Dispersive X-Ray Scanning (EDS), allowed evaluating the quantitative composition of samples and assessing the consistency of the proposed and obtained compositions as shown in Table 2.

Table 2. Difference between proposed and obtained compositions of $LiCoO_2$ and $LiMn_2O_4$ oxides obtained by XRF and EDX measurements.

	Proposed composition (%)				Obtained composition (%)			
Material	Li_2O	Co_3O_4	MnO_2	Total	Li_2O	Co_3O_4	MnO_2	Total
$LiCoO_2$	15.25	84.75	0.0	100	15.20	84.79	0.0	99.99
$LiMn_2O_4$	6.26	0.0	93.74	100	6.24	0.0	93.75	99.99

In Figure 4, the signals corresponding with the characteristic K_{α} emission lines are observed, demonstrating in a quantitative form the relative concentrations of each synthesized material, testing the appropriate composition in each case.

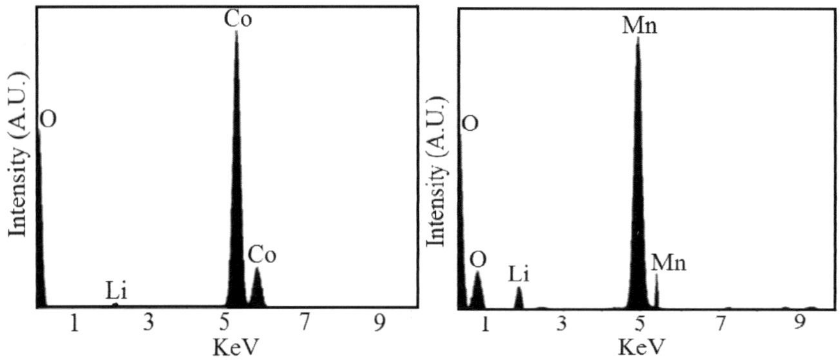

Figure 4. Patterns of energy-dispersive of X-ray (EDX) obtained for $LiCoO_2$ $LiMn_2O_4$ systems.

The analysis by solid-state impedance spectroscopy (IS), allowed to define the behaviour of materials with respect to potential (V), in order to set the values of electrical resistivity and conductivity. The Nyquist plots, shown in Figure 5 display two semicircles one with a smaller diameter related with the $LiCoO_2$ oxide and other with a wide real resistivity related with $LiMn_2O_4$ system, indicating how the charge transfer is more effective in the olivine case, due to the arrangement of atoms in its structure [16-17]. The data expressed in terms of an equivalent circuit, show a capacitive behaviour (C) of the solids, whose resistance is associated with a polarization phenomenon (R_{sol}). While the effect of electron transfer (R) is clear for cobalt composition, the problems related with diffusion phenomenon (W), is evident in manganese composition, in which the main carrier could be attributed to ions movement across the wide crystalline structure according to Bard *et al.* [18].

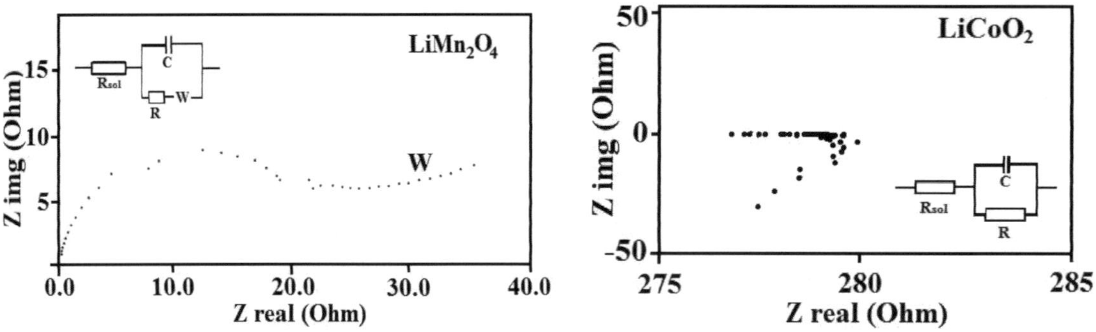

Figure 5. Nyquist plots for $LiCoO_2$ and $LiMn_2O_4$ systems.

The detailed analysis of impedance, confirm a p-type semiconductor behaviour, where the coefficient of chemical diffusion apparently evaluates the resistance values to the transfer of charge in the low frequency region, where the material changes from an electronic conductor to show a phenomenon of resistance to charge transfer, shown in Figure 5. These results, allow evaluating the effect of carriers charge in obtained solids by the effect mobility of oxygen vacancies, which may be formed on the surface of oxides by effect of the applied potential [19-20]. The values derived from the resistivity, allow identify conductivities around 6.2×10^{-3}S and 2.7×10^{-7}S for the $LiCoO_2$ and $LiMn_2O_4$ oxides, in accordance with current state of art for cathodic materials used in design of Li-ion batteries [21].

IOP Conf. Series: Journal of Physics: Conf. Series **935** (2017) 012066 doi:10.1088/1742-6596/935/1/012066

4. Conclusions

The synthesis method based on combustion process permit to obtain high pure crystalline materials in short reaction times at low temperatures with an excellent stoichiometric control. The XRD results, allow establishing the formation of crystalline phases in the nanometric range between 5 and 8nm. The microstructural analysis performed by SEM-TEM analysis corroborate in both cases the obtaining of homogeneous solids with regular morphology, in which is clear that further thermal treatment consolidates the materials and permit to eliminate carbonaceous remnant of previous combustion stage with excellent results. The microcomposition results obtained by means X-ray fluorescence and EDX analysis show a strong consistency between proposed and obtained compositions. Finally, the solid-state impedance analysis, demonstrates that $LiCoO_2$ system has a p-type semiconducting behaviour in contrast with $LiMn_2O_4$ system, in which a mixed behaviour of conductivity was discovered, mainly associated with ions as carrier charge, which is in accordance with the study of these systems for the design of advanced cathodes for lithium-ion batteries.

References

[1] Yazami R, OzawaY 2006 *J. Power Sources* **153** 251
[2] Zhao J, Wang L, He X, Wan C, Jiang C 2010 *J. Electrochem. Sci.* **5** 478–488
[3] Mariño T, Mosqueda Y, Ricardo C, *et al.* 2010 *Revista CENIC* **41** 1-11
[4] Cai Y, Huangn Y, Wang X, Jian D, Tang X 2014 *Ceramics International* **40** 14039–14043
[5] Fergus J 2010 *Journal of Power Sources* **195** 939–954
[6] Guo Z, Konstantinov K, Wang G, Liu H, Dou S 2003 *J. Power Sources* **119** 221-225
[7] Eriksson T 2001 $LiMn_2O_4$ como cátodo de una batería Li-Ion *Acta Universitatis Upsaliensis* **651** 53
[8] Gómez-Cuaspud J A, Valencia-Ríos J 2009 *Rev. Col. Quim.* **38** 289-302
[9] Pacheco C, Gómez Cuaspud J A, *et al.* 2016 *Journal of Physics: Conference Series* **687** 012046
[10] Patarroyo M, Vera E, Pineda Y, Gómez J, *et al.* 2017 *Journal of Physics: Conference Series* **786** 012014
[11] Su C, Yang T, Zhou W, Wang W, Xu X, Shao Z 2016 *J. Mater. Chem. A* **4** 4516
[12] Qu Q, Fu L, Zhan X, Samuelis D, Maier J, Li L, S Tian, Li Z, Wu Y 2011 *Energy Environ. Sci.* **4** 3985
[13] Cabrera S, Benavente F, *et al.* 2001 *Revista Boliviana de Química* **1** 29
[14] Levi M, Gamolsky K, Aurbach D, Heider U, Oesten R 2000 *Electrochemical Acta* **45** 1781-1789
[15] Okubo M, Hosono E, Kim J, Enomoto M, Kojima N, Kudo T, Zhou H, Honma I 2007 *J. Am. Chem. Soc.* 7444-7452
[16] Lou S, Shen B, Zuo P, Yin G, Yang L, Ma Y, Cheng X, Du C, Gao Y 2015 *RSC Adv.***5** 81235
[17] Gómez J 2010 *Síntesis y caracterización de composites nanométricos de silicio para su uso como electrodos en baterías de ión-litio* (Colombia: Universidad de Córdoba) pp 11-23
[18] Bard A J, Faulkner L R 2001 *Electrochemical methods: fundamentals and applications* 2a edition (USA: Wiley) pp 376-380
[19] Andre D, Meiler M, Steiner K, Walz H, Soczka-guth T, Sauer D 2011 *J. of Power Sour.* **196** 5334-5341
[20] Vetter J, Novák P, Wagner M, Veit C, Möller K, Besenhard J, Winter M, Wohlfahrt-mehrens M, Vogler C, Hammouche A 2005 *Journal of Power Sources* **147** 269-281
[21] Gotcu P, Seifert H 2016 *Phys.Chem.Chem.Phys.* **18** 10550

IOP Conf. Series: Journal of Physics: Conf. Series **935** (2017) 012067 doi:10.1088/1742-6596/935/1/012067

Effect of the voltage pulse frequency on the structure of TiO_2 coatings grown by plasma electrolytic oxidation

D A Torres-Cerón[1], F Gordillo-Delgado[1] and S N Moya-Betancourt[1]
[1] Universidad del Quindío, Armenia, Colombia

E-mail: fgordillo@uniquindio.edu.co

Abstract. Plasma Electrolytic Oxidation (PEO) is used to synthetize titanium dioxide (TiO_2) ceramic coatings with the appropriate selection of an electrolyte. The dimension of the micro-cavities and the particle size at the surface can be controlled through the pulse frequency of the voltage that is applied between the electrodes. The change of surface morphology can increase the surface area-to-volume ratio. In this work, PEO of an ASME SB-265 titanium substrate (20×20×1mm) was made in a water solution containing 8g/L Na_3PO_4 and 0.4g/L NaOH. Hence, the coatings were fabricated using voltage pulses of 340V for 10 minutes with a 10% duty cycle and frequencies of 1000, 1500 and 2000Hz. According to the X-ray diffractograms of the obtained samples, the sintering process at 500°C during 1 hour generated Anatase titanium dioxide porous coatings. The grain size decreased approximately from 29nm for 1000 and 1500Hz pulse frequencies until 21nm for 2000Hz. On the other hand, from the micrographs of scanning electron microscopy was possible to see the uniform formation of the micro-cavities with the largest diameter, 900nm, for the lowest frequency value used in PEO.

1. Introduction

Plasma Electrolytic Oxidation (PEO) causes surface modifications of metal substrates and alloys, in order to produce ceramic coatings of high hardness, greater surface area and better tribological properties of the substrate [1]. This technique is environmentally friendly and widely used in the aerospace industry, biomedical implants and in wastewater treatment [2,3]. During the PEO process, the characteristics of the micro-discharges generated in the electrode surface change according to the voltage pulse frequency or the duty cycle; for this reason, these growth parameters are important in defining the structure and morphology of the oxide layer [4], specially the size and shape of the micro-cavities [5-6].

2. Materials and methods

2.1. Preparation of substrates and TiO_2 coatings growth

Titanium sheets, 99% purity grade 2 (ASME SB-265), with dimensions 20×20×1mm was used as electrode-substrate in this work. Before PEO, the plates were polished with 800 and 1000 grit SiC sandpapers successively and subsequently cleaned by ultrasonic bath (BRANSON 1510) in distilled water for 30 minutes [6]. A stainless steel chamber cooled with water was used for containing the electrolyte, an aqueous solution of 8g Na_3PO_4 with addition of 0.4g NaOH in 1L of distilled water. PEO process was carried out using a switching power source, output voltage of 340V. The coatings were made using three voltage pulse frequencies: 1000, 1500 and 2000Hz [6], with a 10% duty cycle for 10 minutes. The sintering treatment of the samples was performed at 500°C for one hour [7].

2.2. Characterization of TiO_2 coatings

The composition of the coatings was studied using grazing-incidence X-ray diffraction in parallel beam geometry with an X-ray diffractometer (XBruker D8-Advance, Colombia), that was operated at 40kV and 40mA, with an X-ray source emitting Cu Kα radiation. The scans were acquired at 2θ range from 20° to 60° [8]. Pictures of the surface of the samples were taken using a Scanning Electron Microscopy (SEM, JEOL JSM-6390LV, México and ZEISS EVO MA 10 SEM, Colombia) and the roughness was measured with a contact-mode Atomic Force Microscope (AFM, CP Hitachi AFM5100N AFM-C).

3. Results and discussion

3.1. Characterization of TiO_2 coatings

A feature of PEO is the formation of specific surface structures as metastable phases under extreme conditions of pressure and temperature; the interaction phenomena that occurs at a first stage of ionization in the discharge area, gives rise to the dissociation of compounds, generating discharges that occur in intervals of time less than 10^{-6}s. At a second stage of condensation there is a temperature drop of about 10^8K/s and the plasma interaction between the electrode-substrate and the electrolyte components yield oxides and other compounds that are melt-quenched at high-temperature on the substrate [9].

The types of spark/arc micro-discharges are observed in the Figures 1 and 2; a vigorous gas evolution after the onset of the treatment system accompanies the whole process. Applying voltages pulses from the switching power supply between the electrodes, in the first moments a conventional anodizing appears [9] with a green-violet pattern on the substrate as shown in the Figure 1. When the voltage is increased until the rupture tension, white sparks/arc with high spatial density are produced as seen in the Figure 2 [8,10].

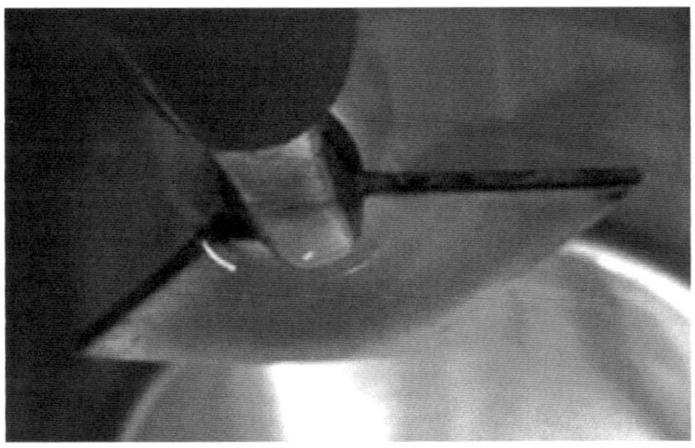

Figure 1. Formation by PEO of green-violet area due to conventional anodizing.

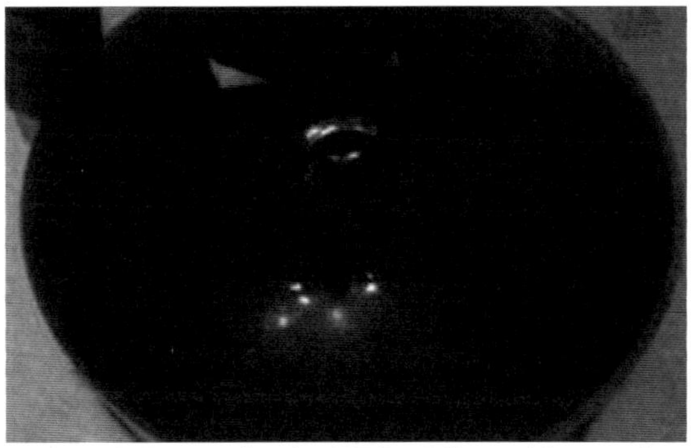

Figure 2. Formation by PEO of fine white sparks during TiO_2 superficial coating.

The X-ray diffraction patterns from the coatings are depicted in the Figures 3, 4 and 5; it can be seen that immediately after PEO all the coatings with the 10% duty cycle are amorphous phase material, possibly containing phosphorus. The peaks in diffractograms related to Anatase TiO_2 nanocrystals are visible for the coatings sintered at 500°C; the corresponding average particle sizes were determined by the Scherrer equation (1):

$$T = \frac{K\,\lambda}{\mathrm{d}\,Cos(\theta)} \tag{1}$$

Where T is the crystallite mean size, K the shape factor usually equal to 0.9, λ the X-ray wavelength (0.154nm), d is the full width at half maximum of the peak at 2θ≈25.4° and θ the relative Bragg angle. In Table 1, the average particle sizes corresponding to Anatase TiO_2 nanocrystals are reported for the used voltage pulses frequencies; the minimum average crystal size, 20.71±0.03nm, was obtained with the highest frequency, while a size of 29nm approximately with the others frequencies.

Table 1. Particle size of nanocrystals in TiO_2 coatings grown by PEO on Ti substrates using voltage pulses frequencies of 1000, 1500, and 2000Hz at 10% duty cycle and sintering temperature of 500°C.

Frequency (Hz)	Duty cycle (%)	Average size (nm)
1000		29.30±0.01
1500	10	29.47±0.02
2000		20.71±0.03

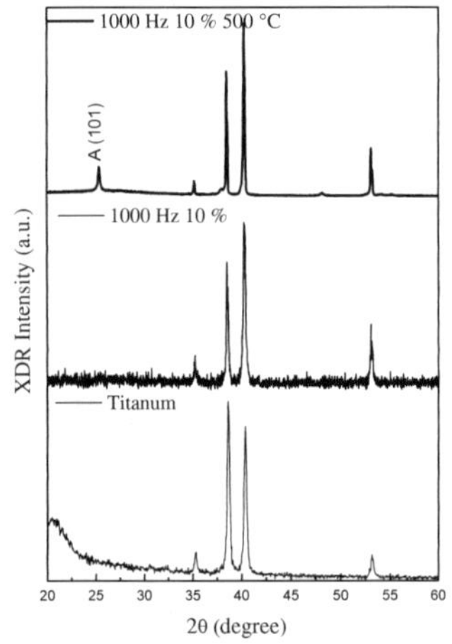

Figure 3. X-ray diffractograms of the TiO_2 coatings grown by PEO with 10% duty cycle, sintering process at 500°C and voltage pulse frequency of 1000Hz.

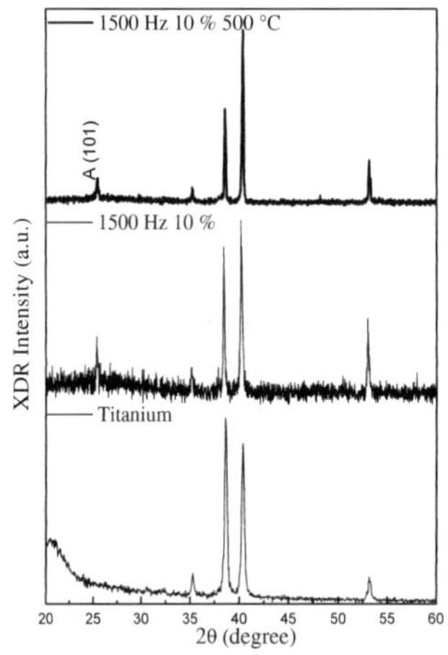

Figure 4. X-ray diffractograms of the TiO_2 coatings grown by PEO with 10% duty cycle, sintering process at 500°C and voltage pulse frequency of 1500Hz.

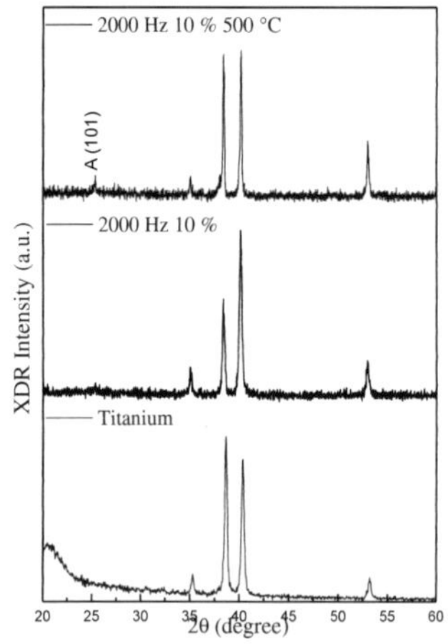

Figure 5. X-ray diffractograms of the TiO_2 coatings grown by PEO with 10% duty cycle, sintering process at 500°C and voltage pulse frequency of 2000Hz.

3.2. Surface morphology of the TiO_2 coatings

The elemental composition of the TiO_2 coatings samples was analyzed by Energy Dispersive Spectroscopy (EDS); Table 2 shows data related to the incorporation of phosphorous (P) and carbon (C) into the titanium oxide coatings, which can be centers of impurities. These species enter into the discharge channels from the electrolyte during PEO process [4]. The amount of P decreases with increasing voltage pulse frequency because the higher energy pulses put more elements of the electrolyte into the coating.

In the Figures 6, 7 and 8, from the SEM micrographs it can be observed that PEO generated micro-cavities on the Ti sheet surface. Uniform microcapsules with average pore diameter of 632nm for 1000Hz, 406nm for 1500Hz and 417nm for 2000Hz [10]; these sizes were estimated over an area of $10\times10\mu m^2$ approximately [10]. The pore surface density, number of micro-cavities per unit area, was $0.15\mu m^{-2}$ for 1000Hz, $0.27\mu m^{-2}$ for 1500Hz and $0.32\mu m^{-2}$ for 2000Hz. The size of the micro-pores varied between 295 and 1184nm with the voltage pulse frequency of 1000Hz in PEO, 60 % of the micro-pores diameter between 295 and 500nm, 20 % between 500 and 800nm and 20 % between 800 and 1184nm; while for 1500Hz frequency this size varied between 202 and 983nm as follows: 85% between 202 and 500nm, 11% between 500 and 800nm and only 4% with a diameter greater than 800nm; Finally, for 2000Hz the size varied between 215 and 681nm as follows: 72% between 202 and 500nm and 28% between 500 and 800nm [11]. In this way, it is possible to deduce that the size of the micro-pores become smaller with increasing voltage pulse frequency in the PEO, thus the frequency controls the micro-discharges power through pulse activation time [6,12].

Table 2. Surface composition by EDS and average micro-cavity diameter of samples grown by PEO (voltage pulse frequency with 10% duty cycle and 500ºC sintering temperature).

Voltage pulse frequency (Hz)	Atoms (%)				Minimum diameter (nm)	Maximal diameter (nm)
	Ti	O	P	C		
1000	21.36	72.29	6.35	--	295	1184
1500	22.82	70.22	4.44	2.51	202	983
2000	16.69	67.95	4.41	7.31	215	743

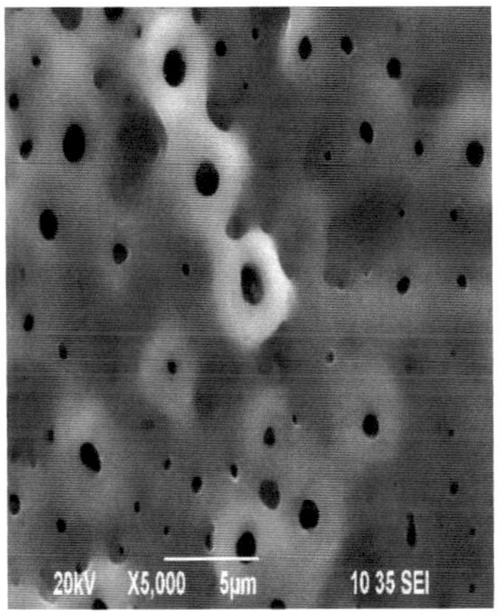

Figure 6. SEM micrographs of TiO_2 coatings grown by PEO with voltage pulse frequency of 1000Hz, 10% duty cycle and 500°C sintering temperature.

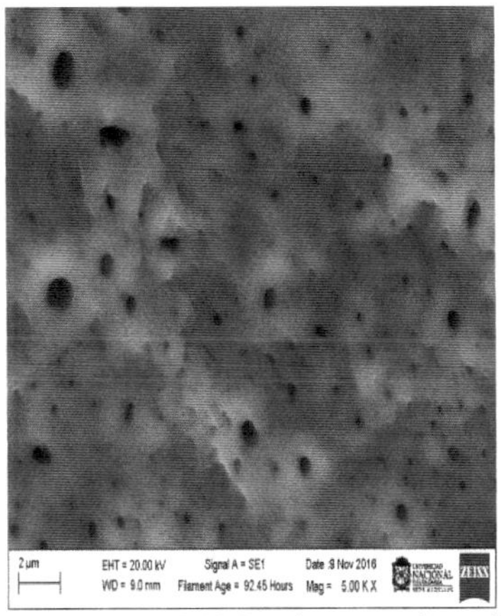

Figure 7. SEM micrographs of TiO_2 coatings grown by PEO with voltage pulse frequency of 1500Hz, 10% duty cycle and 500°C sintering temperature.

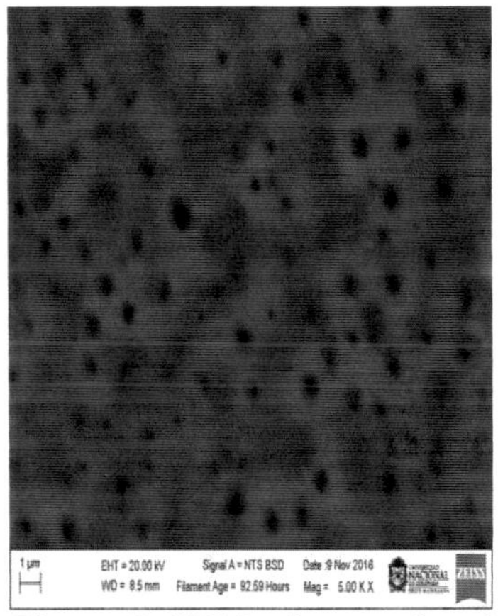

Figure 8. SEM micrographs of TiO_2 coatings grown by PEO with voltage pulse frequency of 2000Hz, 10% duty cycle and 500°C sintering temperature.

The topographic roughness images of 10×10μm^2 of the surface of the coatings are illustrated in the Figures 9, 10 and 11. From the analysis of these micrographs the roughness parameters were registered in the Table 3; where Ra is the arithmetic average of the peak heights and valleys from the mean line, Rms the root mean squared value of these profile height deviations and Rz is the average of the distance from the five highest peaks to the five lowest valleys. It can be seen that the increase of the frequency generated a rise in the micro-cavities depth. In addition, the formation of the oxide layer is observed with perpendicularly oriented pores to the metal surface.

Table 3. Roughness of the TiO_2 coatings from AFM micrographs. Growths by PEO at 10% duty cycle and 500°C sintering temperature.

Voltage pulse frequency in the PEO (Hz)	Ra (nm)	Rz (nm)	Rms (nm)
1000	73	393	89
1500	95	494	235
2000	116	620	144

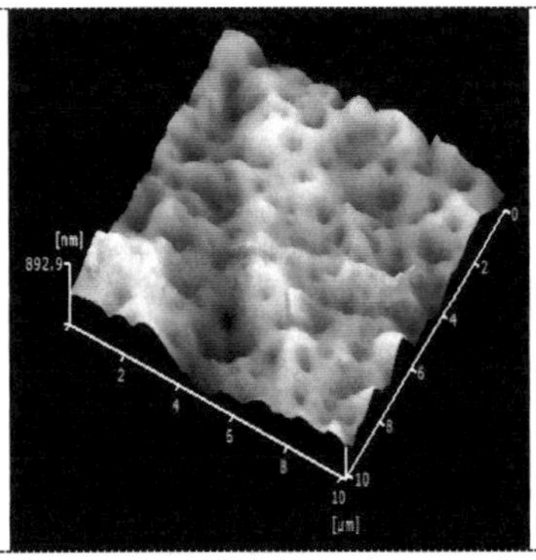

Figure 9. AFM 3D micrographs of samples (10×10μm^2 area) grown by PEO with voltage pulse frequency of 1000Hz, 10% duty cycle and 500°C sintering temperature.

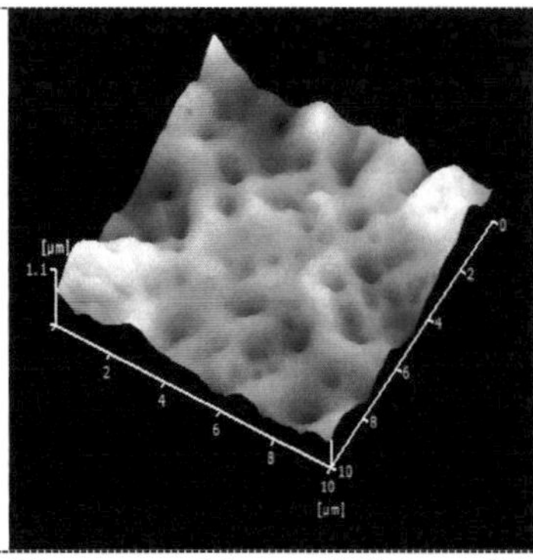

Figure 10. AFM 3D micrographs of samples (10×10μm^2 area) grown by PEO with voltage pulse frequency of 1500Hz, 10% duty cycle and 500°C sintering temperature.

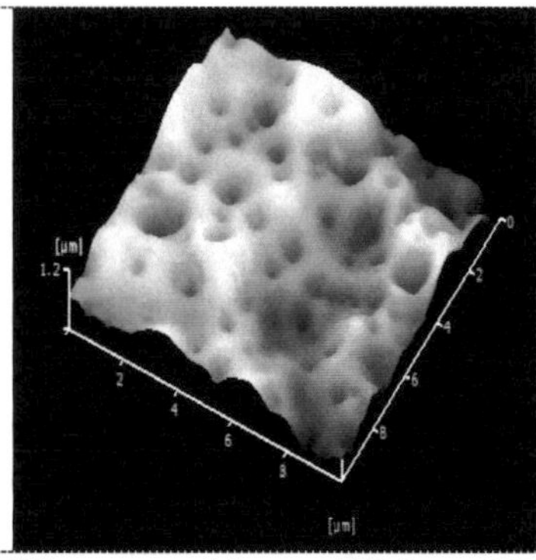

Figure 11. AFM 3D micrographs of samples (10×10μm^2 area) grown by PEO with voltage pulse frequency of 2000Hz, 10 % duty cycle and 500°C sintering temperature.

4. Conclusion

TiO_2 coatings were achieved by the PEO technique using a switching power supply with pulses of 340V, with frequencies of 1000, 1500 and 2000Hz and 10% duty cycle, on ASME SB-265 titanium sheets using an aqueous solution of Na_3PO_4 and NaOH as electrolyte. The X-ray diffractograms showed the formation of Anatase TiO_2 nanocrystals after sintering process at 500°C for 1h. In the oxide layer, the grain size of these nanocrystals decreased approximately from 29nm for 1000 and 1500Hz voltage pulse frequencies in the PEO to 21nm for 2000Hz.

From SEM micrographs, the micro-cavities size diminished when the voltage pulse frequency was increased. The elemental map showed a uniform distribution of the phosphorus and oxygen on the substrate. From AFM micrographs, the roughness of the TiO_2 coatings increased in function of the voltage pulse frequency.

Acknowledgments

The authors thank the financial support of the Chemistry Program and Basic Sciences Faculty of Quindío University through 869 project. Special thanks to the Biomaterials Laboratory of National University of Colombia- Medellín and to CICATA- Mexico City for the SEM measurements. We are very grateful to the Plasma Laboratory of National University of Colombia-Manizales for the AFM images. Acknowledgments to the Interdisciplinary Institute of Sciences of Quindío University for the measurements of XRD.

References

[1] Urbanczyk E, Krzakala A, Kazek-Kesik A, Michalska J, Stolarczyk A, Dercz G and Simka W 2016 Electrochemical modification of Ti-13Nb-13Zr alloy surface in phosphate based solutions *Surf. Coatings Technol.* **291** 79–88

[2] Cheng Y liang, Wu X Q, Xue Z gang, Matykina E, Skeldon P and Thompson G E 2013 Microstructure, corrosion and wear performance of plasma electrolytic oxidation coatings formed on Ti-6Al-4V alloy in silicate-hexametaphosphate electrolyte *Surf. Coatings Technol.* **217** 129–139

[3] Kazek-Kesik A, Dercz G, Kalemba I, Michalska J, Piotrowski J and Simka W 2014 Surface treatment of a Ti6Al7Nb alloy by plasma electrolytic oxidation in a TCP suspension *Arch. Civ. Mech. Eng.* **14** 671–681

[4] Albella J M, Montero I, and Martinez-Duart J M 1987 A Theory of avalanche breakdown oxidation during anodic oxidation *Electrochimica Acta* **32(2)** 255–258

[5] Zhang X, Zhang Y, Chang L, Jiang Z, Yao Z and Liu X 2012 Effects of frequency on growth process of plasma electrolytic oxidation coating *Mater. Chem. Phys.* **132** 909–915

[6] Babaei M, Dehghanian C and Vanaki M 2015 Effect of additive on electrochemical corrosion properties of

IOP Conf. Series: Journal of Physics: Conf. Series **935** (2017) 012067 doi:10.1088/1742-6596/935/1/012067

plasma electrolytic oxidation coatings formed on CP Ti under different processing frequency *Appl. Surf. Sci.* **357** 712–720

[7] Torres D, Gordillo F and Plazas J 2017 Formation of TiO2 nanostructure by plasma electrolytic oxidation for Cr(VI) reduction *J. Phys.: Conf. Ser.* **786** 11001

[8] Gowtham S, Arunnellaiappan T and Rameshbabu N 2016 An investigation on pulsed DC plasma electrolytic oxidation of Cp-Ti and its corrosion behaviour in simulated body fluid *Surf. Coatings Technol.* **301** 63–73

[9] Duan H, Yan C and Wang F 2007 Growth process of plasma electrolytic oxidation films formed on magnesium alloy AZ91D in silicate solution *Electrochim. Acta* **52** 5002–5009

[10] Byeon I-S, Hwang I-J, Choe H-C and Brantley W A 2016 Electrochemically-coated hydroxyapatite films on nanotubular TiNb alloys prepared in solutions containing Ca, P, and Zn ions *Thin Sol. Fil.* **620** 132–138

[11] Song W, Xiaohong W, Wei Q and Zhaohua J 2008 Optoelectronic properties of TiO_2 films prepared by microplasma oxidation method *Mater. Lett.* **62** 1078–1081

[12] Lv G H, Chen H, Gu W C, Li L, Niu E W, Zhang X H and Yang S Z 2008 Effects of current frequency on the structural characteristics and corrosion property of ceramic coatings formed on magnesium alloy by PEO technology *J. Mater. Process. Technol.* **208** 9–13

IMRMPT IOP Publishing
IOP Conf. Series: Journal of Physics: Conf. Series **935** (2017) 012068 doi:10.1088/1742-6596/935/1/012068

Transport spin dependent in nanostructures: Current and geometry effect of quantum dots in presence of spin-orbit interaction

H Paredes-Gutiérrez[1], S T Pérez-Merchancano[2] and C L Beltran-Rios[1]
[1] Universidad Industrial de Santander, Bucaramanga, Colombia
[2] Universidad del Cauca, Popayán, Colombia

E-mail: hparedes@uis.edu.co , sperez@unicauca.edu.co

Abstract. In this work, we study the quantum electron transport through a Quantum Dots Structure (QDs), with different geometries, embedded in a Quantum Well (QW). The behaviour of the current through the nanostructure (dot and well) is studied considering the orbital spin coupling of the electrons and the Rashba effect, by means of the second quantization theory and the standard model of Green's functions. Our results show the behaviour of the current in the quantum system as a function of the electric field, presenting resonant states for specific values of both the external field and the spin polarization. Similarly, the behaviour of the current on the nanostructure changes when the geometry of the QD and the size of the same are modified as a function of the polarization of the electron spin and the potential of quantum confinement.

1. Introduction

In semiconductor nanostructures, spin-orbit coupling produces a splitting of electron states. Here, the spin of the electron generally shows an asymmetric inversion in these structures due to the presence of magnetic fields in the system which are applied in the direction of growth of the quantum mechanism, generating a Zeeman effect that significantly affects the behaviour of the electrons. In this context, the investigation of quantum transport influenced by time independent potentials is of great importance in the study of semiconductor nanostructures, since it is feasible to control the dynamics of electrons using polarization voltages that can lead to practical applications. Experimentally [1-3], different authors have shown how some properties of nanostructures and in particular, the current are dependent on the spin of the carriers: this led to the quantum transport of semiconductor nanostructures information to be controlled by manipulation of the spin using for this purpose, an applied voltage between the contact points of the structure [3]. At present, the study of the effects of particle spin is very important due to the control effects that can be induced in the quantum mechanics, because they produce changes in the coupling of the system and in the structure of the bands of the mechanism. Proof of this are the Rashba and Dresselhaus effects, which simply modify the quantum information in the system [4-10]. Thus, the study of the carriers in QDs with different shape and size embedded in a QW [11-19] have shown the appearance of new effects in the behaviour of the tunnelling current and spin information processes. In these structures, the spin states can be manipulated in a coherent sense. The above contributes to the research in spintronic based on the sensitive manipulation of spin and current of nanostructures and encodes various applications of quantum nanodevices, such as double barrier heterostructure (DBH) and similar [16].

In this paper, we study quantum transport through semiconductor nanostructures involving QDs of different size and embedded geometries that act as a bridge for the information from one contact to another in our quantum mechanism. Here we work with the general quantum theory of quantum transport using the tight binding model and Keldysh's second quantization theory nonequilibrium in Green's formalism of functions to describe the resonant states in the system taking into account the Rashba interaction having a relevant effect on the current intensity [11-19]. We also conduct an exhaustive discussion of physical ideas in which we underlie the concepts that we used, as well as detailed interpretation of our results. We complement our numerical results with analytical models that show the physical importance of our model of semiconductor nanostructure, in which we have embedded different QDs.

2. Model and theoretical calculation

In this work, we study the electron resonant tunnelling through a QD of (InGa)As of different geometries (lens, ring and pyramid) embedded in a DBH of GaAS-(Ga,Al)As, that was grown in the z-direction and is connected to two semiconductor leads, as shown in Figure 1. Our quantum system is subjected to an external electric field applied in the z direction and to an asymmetric confinement potential, facilitating the analysis of the interaction orbital of spin Rashba.

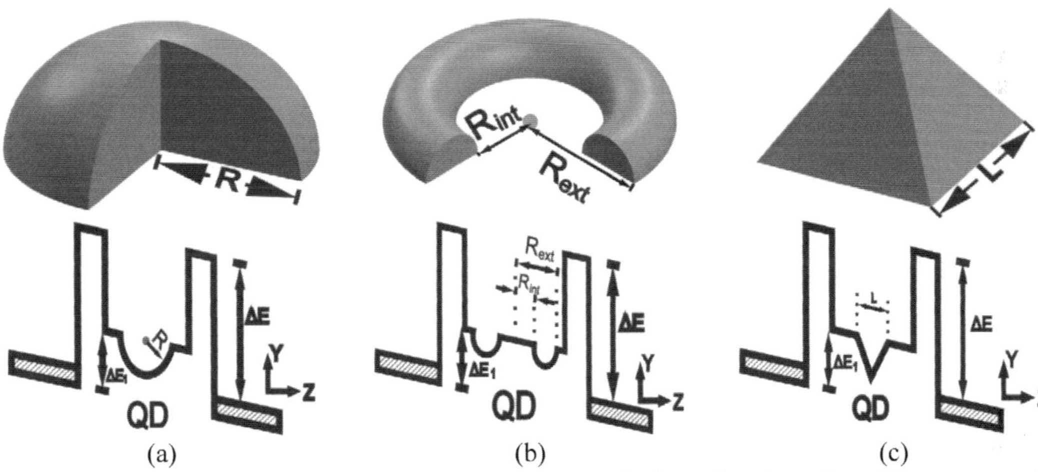

Figure 1. This picture shown the effective potential grown in the z-direction of our DBH system with different QDs embedded: (a) QDs in lens shape of radius R=50Å (b) ring shape of internal radius R_{int}=30Å and external radius R_{ext}=50Å and (c) pyramid shape, whose base is an equilateral triangle of side L=100Å. The barrier width is 80Å, the barrier height $\Delta E = 300meV$, the QW width is 100 Å and the band offset between the BDH and the QD is $\Delta E_1 = 262meV$.

The Hamiltonian $H(r)$ associated to the electron transport through a quantum mechanical system is QD-DBH given by:

$$H(r) = \frac{p^2}{2m^*} + V(r) + H_{SO}(r), \quad (1)$$

where the first term is the kinetic energy, second term is the potential energy and the third term is the Rashba spin orbit (SO) interaction. The $Hso(r)$ term reduces to the form $H_{SO} = \hat{z}/2\hbar \cdot [\alpha(\sigma \times \vec{p}) + (\vec{p} \times \sigma)\alpha]$. The vector $\sigma = (\sigma_x, \sigma_y, \sigma_z)$ represents the Pauli matrices, the vector $\boldsymbol{p}$ of the momentum operator and α Rashba SO interaction strength [4]. The expression H_{SO} can be split in two terms, H_{R1} and H_{R2} [4]. Thus, $H_{SO} = H_{R1} - H_{R2}$, where $H_{R1} = 1/(2\hbar)[\alpha\sigma_x p_y + \sigma_x p_y \alpha]$

. H_{R1} makes a spin precession and the H_{R2} term can cause spin flips between different energy levels. $H_{R2} = (1/\hbar)\alpha\sigma_y$.

The Hamiltonian in equation (1) of the DBH device connected to two leads $(\beta = L, R)$, and considering the Rashba SO interaction (H_{SO} *with* $\alpha \neq 0$) can be written in the standard Anderson model: $H = H_{QD-DBH} + \sum_{\beta=L,R} H_\beta + H_T$, where $H_{QD-BDH} = \sum_{n,s} \varepsilon_n d^\dagger_{ns} d_{ns} + \sum_{m,n} [t^{SO}_{mn} d^\dagger_{m\downarrow} d_{n\uparrow} + H.c]$ is the Hamiltonian from the QD-DBH device, $H_\beta = \sum_{ks} \varepsilon_{k\beta} a^\dagger_{k\beta s} a_{k\beta s}$, for the leads and $H_T = \sum_{k,s} [t_{LDBH} a^\dagger_{Lks} d_s + t_{RDBH} e^{-isK_R L} a^\dagger_{Rks} d_s + H.c]$, is the coupling between the leads and QD-DBH device.

The amount $\hat{n}_{ns} = d^\dagger_{ns} d_{ns}$ is the number operator, $s = \uparrow o \downarrow$ is the spin index, in which it describes the spin state with spin-up or spin down, respectively. n is quantum number for the eigenstates of the single-particle Hamiltonian (1), at the isolated QD region with eigenenergy $\varepsilon_n = \langle n|H|n\rangle$. The second term in the last expression H_{QD-BDH} is the Rashba H_{R2} interaction in the second quantization with $t^{SO}_{mn} = 0.3meV$ intensity [4] are the elements for the non-diagonal matrix that takes into account the transition process among levels when $(n_\uparrow) \rightarrow (m_\downarrow)$ for $n \neq m$. $k\beta$ is the quantum index for β lead with eigenenergy $\varepsilon_{k\beta} = \langle k\beta|H|k\beta\rangle$. H_T is the Rashba interaction in the second quantization that gives place to a spin dependent phase factor $-isK_R L$ with ($K_R = \alpha m^* / \hbar^2$) where m^* the effective mass and L is the dimension of our double-barrier system in the z-direction.

The quantum transport in the DBH device can be solved using the standard Green functions method proposed by Keldysh [15] and the electron current equation with spin up or spin down in our quantum device, which involves two contacts in the semiconductor nanostructure, quantum dots embedded of different geometries and external electric field, can take the form of:

$$I = \frac{2e}{\hbar} \int \frac{d\omega}{2\pi} \mathrm{Re}[t_{LDBH} G^<_{DBHL}(\omega) + t_{RDBH} G^<_{DBHR}(\omega)] \tag{2}$$

where the Green function $G^<(\omega)$ is the transform Fourier of $G^<(t)$ that allows to perform the calculations of our quantum model considering the frequency spectrum or the quantized energies of the system.

3. Results and discussions

For our DBH of GaAS-(Ga,Al)As we consider the effective electronic mass for the GaAs of $m^* = 0.067m_0$, m_0 is the free electron mass, the Fermi level $E_F = 100meV$, the barrier height $\Delta E = 300meV$, the band offset between the BDH and the QD is $\Delta E_1 = 262meV$, the width of the barrier (B) and the well (W) of 80Å and 100Å, $\alpha = 3\times10^{-11} eV\,m$ Rashba SO interaction strength, $t^{SO}_{mn} = 0.3meV$, and $t_{LDBH} = t_{RDBH} = 1$, respectively.

Figure 2(a) shows current vs. voltage for a Rashba effect value of $K_R L = \pi/4$, without QD and with S^- and S^+. We can observe that the current increases in function of the applied voltage in the DBH device presenting certain current peaks voltage values. This happens when an electron in the sea of Fermi in the emitter enters in resonance with a bounded level in the QW. Also, as the applied voltage continues increasing, the electron is no longer in resonance with the bounded level in the QW generating a negative differential current, if the applied voltage continues increasing, this current will continue growing in the system, showing the resonance behaviour for higher voltage values. In addition, we also can observe in Figure 2(a) that the peaks with S⁻ go first into resonance with a greater intensity compared to the peaks with S⁺ and without Rashba effect, showing fundamentally that with this type of polarity (spin down) improves the electronic transportation in the system.

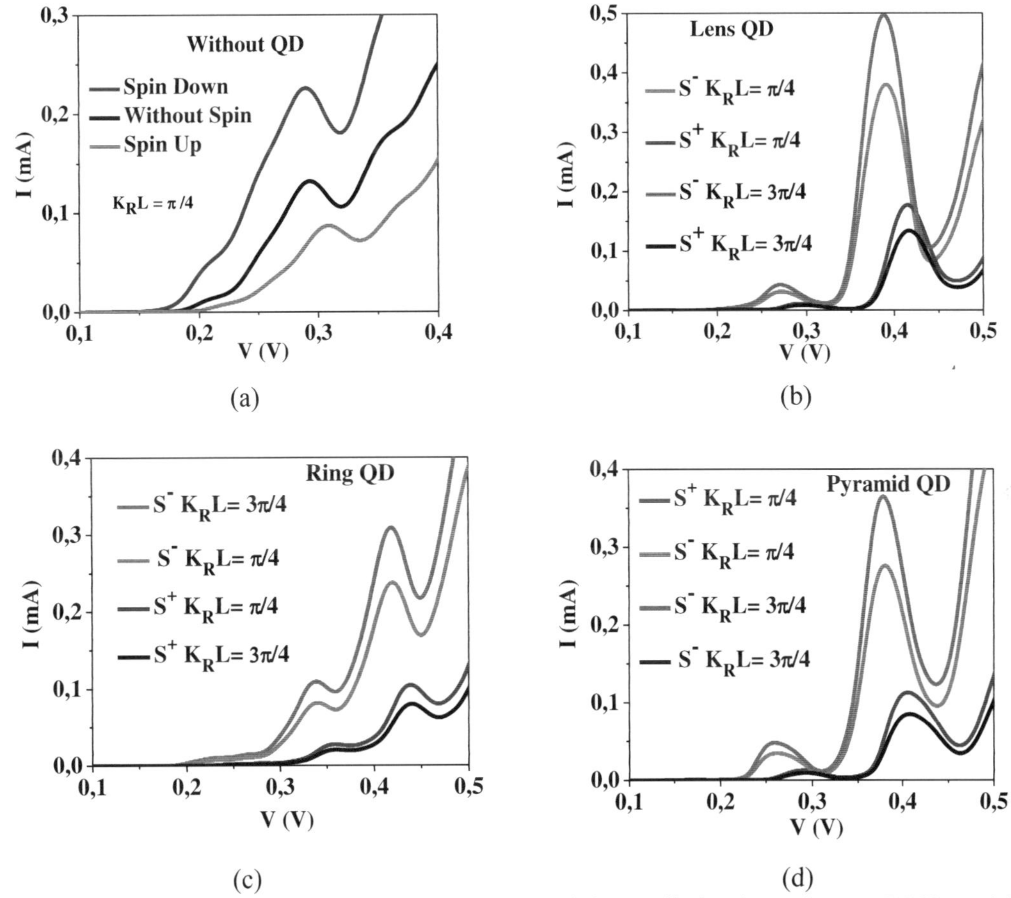

Figure 2. This figure exhibit current as a function of the applied voltage for our DBH model: (a) without QD and three different spin values, (b) lens, (c) ring and (d) pyramid, for two different Rashba effect values $K_R L = \pi / 4$ and $K_R L = 3\pi / 4$ and with spin up (S^+) and spin down (S^-).

Figure 2(b) shows the current vs. voltage for a QD in lens shape of a 50Å radius embedded in a DBH, with S^+, S^- and with two Rashba interaction values $K_R L = \pi / 4$ and $K_R L = 3\pi / 4$. We are shown that the current in function of the applied voltage changes significantly when it is compared to the results obtained in Figure 2(a). Likewise, is shown seem that due to the QD presence in lens shape, there is a higher concentration of resonant states for some voltage values. And, in the case for the spin down, it shows that behaviour the current intensity is lower in the DBH in the phase $K_R L = \pi / 4$ than in $K_R L = 3\pi / 4$. For example, for voltage values close to 0.4V, in the same way, when we analyse the spin up case for the two phases quoted above, it is observed that the current intensity values decrease when the system has embedded lens in the confinement well as expected in the DBH in the presence of QDs. In this context, a higher number of peaks in certain voltage values are also determined, which implies a greater tunnelling of carriers when comparing Figures 2(a) and 2(b).

In Figure 2(b), we observe that the current has a greater intensity for electrons with S^- compared to those of S^+, presenting a shift of the peaks with S^- to lower voltage values. This is because the 2D gas electrons first tunnel into the emitter of the device, when they are in resonance with the states bounded in the QD and then are in resonance with the bounded states in the QW. In the 0.1 - 0.3V voltage range, these tunnelling peaks are associated to the QD linked levels, and between 0.3-0.5V these tunnelling peaks are associated with the confined levels in the QW. In addition, it is observed that the Rashba effect has a significant influence on the tunnelling current, presenting a higher peak for the interaction $K_R L = 3\pi/4$ than for interaction $K_R L = \pi/4$ when the electrons have S^-;and for electrons with S^+ the current intensity behaviour is inversed, being the greater peak with interaction $K_R L = \pi/4$ compared to the curve with interaction $K_R L = 3\pi/4$.

Figure 2(c) shows the current characteristic curve vs voltage for a QD in the shape of a ring with internal radius R_{int}=30Å, and external radius R_{ext}=50Å embedded in a DBH, with spin up and spin down and with two values of Rashba interaction $K_R L = \pi/4$ and $K_R L = 3\pi/4$. In this figure, a higher number of peaks are observed than in previous results for similar voltage values, indicating a higher concentration of resonant states favouring quantum electron tunnelling. This is due to the change of geometry in the QD in the DBH, which indicates that our system is sensitive to the geometric change of dots. It is observed that the current changes its behaviour when the S^+ and S^- cases with phases $K_R L = \pi/4$ and $K_R L = 3\pi/4$ are studied. This behaviour is similar to the one observed in Figure 2(b) only that due to the morphology of the QD, the current behaviour shows higher values for spin down with a phase $K_R L = 3\pi/4$ than for a phase $K_R L = \pi/4$ showing again an improvement of tunnelling when the system is polarized with spin down.

Figure 2(d) shows similar behaviours of the current intensity in function of voltage as in Figure 2(b) and (c), but they show a greater system efficiency when the QD geometry is changed, allowing to conclude in these cases, that the concentration of these resonant peaks is higher when the confinement increases and the QDs morphology changes.

4. Conclusions

In this work, we have studied the behaviour of current as a function of voltage for a double potential barrier system (DBH) that has QDs of different geometry. The study of current intensity was done using a Hamiltonian from the system in the formalism of the second quantization, where we included among others, the spin-orbit coupling. Here was highlighted that the interacting spin orbits Rashba type, is dominant throughout the DBH generating two main effects: it gives rise to an extra spin-dependent phase factor and spin-flip between levels allowing a change in the current intensity behaviour when spin-up or down states are taken into account, indicating that the spin polarization changes. Likewise, we observe that the current intensity in our DBH is susceptible to the presence of QDs and that this is significantly modified when the morphology of the QDs changes.

Acknowledgments

This work has been financed by the "Universidad Industrial de Santander (UIS)" by means of the "Vicerrectoria de Investigación" and by the Vice-rectory for Research (VRI) at "Universidad del Cauca" partial funding for the development of this work, through the research projects: ID 4523 and ID 4524.

References

[1] D Z Y Ting and X Cartoixà 2004 InAs/GaSb/AlSb Resonant Tunneling Spin Device Concepts *Physica E* **20** 350
[2] Thomas Schäpers 2016 *Semiconductor spintronics* (Boston: Walter Gruyter Gmbh)
[3] Peter Michler 2017 *Quantum dots for quantum information technologies* (Switzerland: Springer Nature)
[4] Qin Feng Sun, Jian Wang and Hong Guo 2005 Quantum transport theory for nanostructures with Rashba spin-orbital interaction *Phys. Rev. B* **71** 165310

[5] T Noda, N Koguchi 2006 Current voltage characteristics in double-barrier resonant tunneling diodes with embedded GaAs quantum rings *Physica E* **32** 550

[6] K Gnanasekar, K Navaneethakrishnan 2006 Effects of Rashba spin–orbit interaction on spin-dependent resonant tunneling in ZnSe/$Zn_{1-x}Mn_xSe$ multilayer heterostructures *Physica E* **35** 103

[7] Ernie Pan, Yu Zou, Peter W Chung and Yan Zhang 2009 Interlayer correlation of embedded quantum-dot arrays through their surface strain energy distributions *Phys. Rev. B* **80** 073302

[8] C M Ryu and S Y Cho 1998 Phase evolution of the transmission coefficient in an Aharonov-Bohm ring with Fano resonance *Phys. Rev. B* **58** 3572

[9] S Murakami, N Nagaosa and S C Zhang 2003 Dissipationless quantum spin current at room temperature *Science* **05(301)** 1348

[10] J Sinova, D Culcer, Q Niu, N A Sinitsyn, T Jungwirth and A H MacDonald 2004 Universal intrinsic spin hall effect *Phys. Rev. Lett.* **92** 126603

[11] B K Nikolic, L P Zarbo, and S Souma 2010 *The Oxford Handbook on Nanoscience and Technology: Frontiers and Advances* vol 1, ed A V Narlikar and Y Y Fu (New York: Oxford University Press) chapter 24 pp 814–866

[12] D J Paul 2010 *The Oxford Handbook on Nanoscience and Technology: Frontiers and Advances* vol 3, ed A V Narlikar and Y Y Fu (New York: Oxford University Press) chapter 5 pp 181-204 Dimitry D Vvedensky 2010 The *Oxford Handbook of Nanoscience and Technology: Frontiers and Advances* vol 3, ed. A V Narlikar and Y Y Fu (New York: Oxford University Press) chapter 6 pp 205-243

[13] S M Mirzanian and A A Shokri 2013 Angular dependence of shot noise in the presence of Rashba spin–orbit coupling in semiconductor spintronics junctions *Physica E* **54** 59

[14] N Niketić Vitomir Milanović and Jelena Radovanović 2014 Properties of the resonant tunneling diode in external magnetic field with inclusion of the Rashba effect *Solid State Communications* **189** 52

[15] L V Keldysh 1965 Diagram Technique for Nonequilibrium Processes *Sov. Phys. JETP* **20** 1018

[16] J H Marin, I D Mikhailov and L F García 2007 Charge distribution in quantum dot with trapped exciton *Physica B* **398** 135

[17] D S Smirnov, M M Glazov, E L Ivchenko and L Lanco 2015 Theory of optical spin control in quantum dot microcavities *Phys. Rev. B* **92** 115305

[18] Leonor Chico, A Latge and Luis Brey 2015 Symmetries of quantum transport with Rashba spin–orbit: graphene spintronics *Phys. Chem. Chem* **17** 16469

[19] Ichiro Tanaka, Y Tada, S Nakatani, K Unoa, I Kamiya and H Sakaki 2010 Resonant tunneling of electrons through single self-assembled InAs quantum dot studied by conductive atomic force microscopy *Physica E* **42** 2606

IOP Conf. Series: Journal of Physics: Conf. Series **935** (2017) 012069 doi:10.1088/1742-6596/935/1/012069

2D and 3D virtual interactive laboratories of physics on Unity platform

J D González[1], J H Escobar[1], H Sánchez[1], J De la Hoz[1] and J R Beltrán[1]

[1] Universidad del Magdalena, Santa Marta, Colombia

E-mail: jgonzaleza@unimagdalena.edu.co

Abstract. Using the cross-platform game engine Unity, we develop virtual laboratories for PC, consoles, mobile devices and website as an innovative tool to study physics. There is extensive uptake of ICT in the teaching of science and its impact on the learning, and considering the limited availability of laboratories for physics teaching and the difficulties this causes in the learning of school students, we design the virtual laboratories to enhance studentâĂŹs knowledge of concepts in physics. To achieve this goal, we use Unity due to provide support bump mapping, reflection mapping, parallax mapping, dynamics shadows using shadows maps, full-screen post-processing effects and render-to-texture. Unity can use the best variant for the current video hardware and, if none are compatible, to use an alternative shader that may sacrifice features for performance. The control over delivery to mobile devices, web browsers, consoles and desktops is the main reason Unity is the best option among the same kind cross-platform. Supported platforms include Android, Apple TV, Linux, iOS, Nintendo 3DS line, macOS, PlayStation 4, Windows Phone 8, Wii but also an asset server and Nvidia's PhysX physics engine which is the most relevant tool on Unity for our PhysLab.

1. Introduction

This paper focuses on the development of innovative tool in which the learners could interact with movil virtual laboratories based on the platform Unity that include virtual apparatus and materials and conduct experiments on a movil device as well as on a computer. The knowledge of new Information and Communication Technologies (ICTs) is directing traditional learning environments, especially in theoretical-practical courses, towards new learning environments, which can articulate the direct teaching line, as suggested by Larreamendy-Joerns and Leinhardt (2006) [1]. Those virtual laboratories become a popular alternative because they are more flexible, safe, manageable, clean, cost-efficient and fast than physical experiments [2, 3]. These new innovative and interactive learning scenarios not only allow us to form more competitive citizens within a society that revolves around information and software tools, such as computers, tablets and smart phones, that facilitate communication processes, but also they are advancing at their own pace [4–8]. Several studies have shown the amazing learning effects of virtual enviroments that support students to test hypotheses, explore and analyze data as a experts do [9–11]. Studies conducted by the US Department of Education (2009) show that online instruction, in general, may be more beneficial than the traditional one and shown the advantages of virtual laboratories in concept learning [2, 3, 12]. Given the impact of this new learning environment and the limited availability of laboratories for teaching physics in the institutions of secondary education, the authors developed a series of virtual laboratories with the objective of supporting the work of teachers and the improvement of Compression of students about fundamental concepts of physical phenomena [13–16]. As a result, there is an increasing number of researchers that propose the replacement of physical laboratories by virtual ones. Our virtual laboratories include a guide for each of the experiments, which contributes to reduce the cognitive load and, therefore, facilitate meaningful learning as it was studied [17]. The guide can also be used to prevent random interactions with simulations and to focus on important parameters and concepts. As a result, virtual laboratories can effectively improve student learning concepts. To develop the virtual laboratories on Unity platform we decided to use the video game drive unit, since it is a platform that can be used for developments in Windows, Linux, Android, Apple and for video game consoles, stories such as Xbox-one, Play Station among others and mobile devices. The graphics drive engine uses the OpenGl libraries, for Microsoft Windows, Mac and Linux; Direct3D, for Windows and OpenGl ES, for Android and IOS. In addition to this it supports relief, reflection and parallel mapping, dynamic shadows and rendering of textures that facilitate the design and creation of virtual scenarios in 2D and 3D. Its integrated physics engine provides a set of components that are very useful for the simulation of physical phenomena [18].

The studies through educational website, videoconferencing and different educational platform are not significantly different from regular classroom learning in terms of effectiveness. In light of today's learning applications, which can take advantage of a wide range of the resources, including not only Web-based and multimedia but also using tools as Unity which open the opportunity to develop apps that can improve the way to catch the attention of new generations of young people who grows in a technological world. Our apps are develop to obtain a better understanding of the study of physics in a enviroments that simulate a video game and which can offer a realistic view of the physical phenomenon considered in the virtual laboratories. According to Gordon and Pea (1995) the use of Scientific Visualization (SciVis) through virtual laboratories improves learning because it supports the thought process and the methods followed by the researchers [19]. In (2002) Fisher and Unwin argue that visual interfaces maximize perceptual skills and facilitate understanding of the characteristics of geographic ecosystems [20]. In addition, the simulation of physical phenomena offers students the opportunity to test their hypotheses about them, which in the real world is impossible and test multiple scenarios to get an idea of possible outcomes taking into account these assumptions, for instance, in 2002, Barak and Nater developed a virtual Virtual museum of minerals and molecules: Molecular visualization in a virtual hands-on museum [21].

2. Unity 3D and design of physical laboratories simulation

Unity 3D was developed by Unity Technologies in Denmark and is a game development ecosystem, a powerful rendering engine fully integrated with a complete set of intuitive tools and rapid workflows to create interactive 3D and 2D. One of the key features of Unity 3D is its support for modeling physical properties. For example, one can assemble objects, using a variety of joints. Also, various objects can have mass, compliance, collision detection, etc. One of the main differences from the realXtend platform is that Unity 3D comes from the FPS (First Person Shooting) game arena, whereas realXtend is a multiuser social collaboration environment. The platform of the Unity allows to create new materials through its inspector module in which dialog box of that module is illustrated (see Figure 1). Its rendering power enables a high fidelity of the objects created, which can be integrated into the virtual world or laboratory and its integrated physics engine provides a set of components that are very useful for simulating physical phenomena. For example, the Rigidbody component allows you to recreate object dynamics through the "Is Kinematic" property, changing its value from a script can activate or deactivate the physics of an object as shown in Figure 2 and in Table 1 we lists the functions of the properties of the Rigidbody component. Another of the components of this platform

IOP Conf. Series: Journal of Physics: Conf. Series **935** (2017) 012069 doi:10.1088/1742-6596/935/1/012069

is the Colliders, it is used to determine the shape of the objects when they collide, that is, they simulate the physical properties of the material that supposedly represent. For example, a rubber ball should bounce when dropped. If you want to attach a RigidBody to another or to a fixed point you use the Joints tool, the Hinge Joint command causes the object to rotate around a specific point or axis. A very useful component is the Constant Force, when adding a constant force to a RigidBody, in Figure 3 shows the dialog box of this component. In Figure 4 shows a virtual scene created by this platform, which shows not only its rendering power, but also lighting effects, shadows among others.

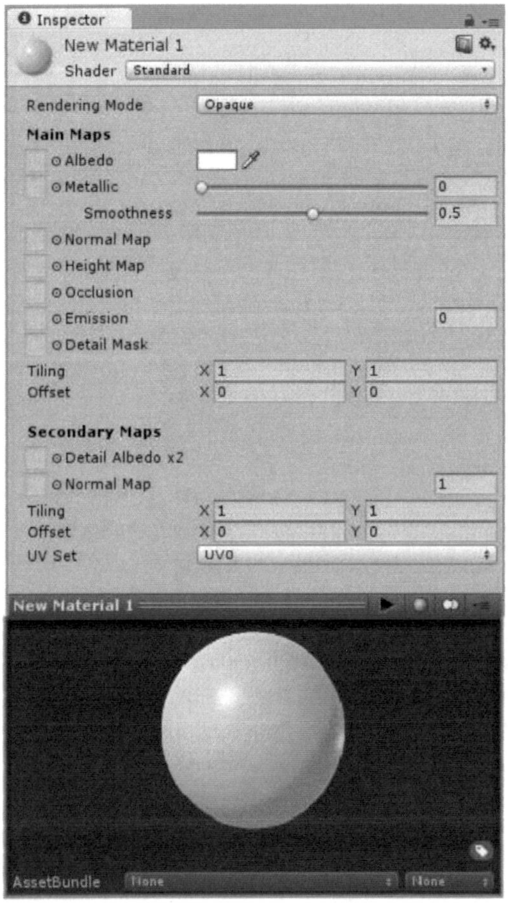

Figure 1. Inspector module dialog box.

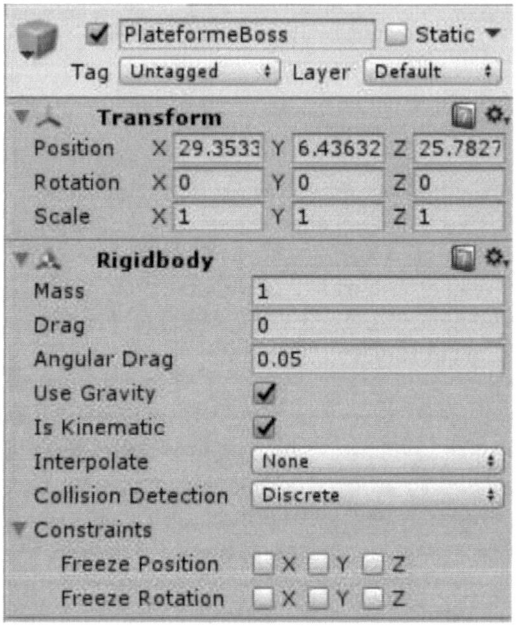

Figure 2. Rigidbody component properties dialog box.

IOP Conf. Series: Journal of Physics: Conf. Series **935** (2017) 012069 doi:10.1088/1742-6596/935/1/012069

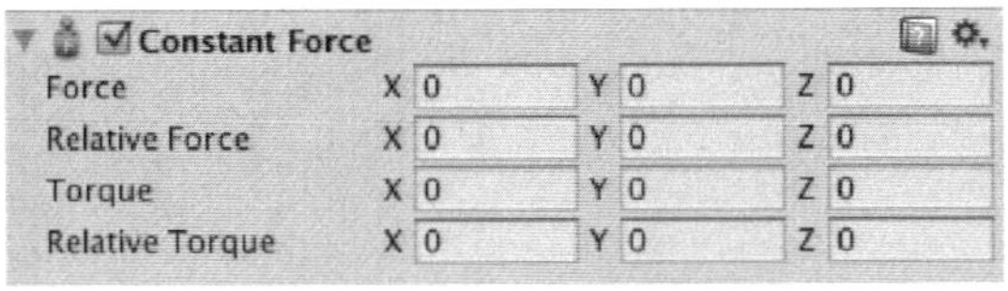

Figure 3. Constant strength component dialog box.

Figure 4. Scenario created using Unity.

Table 1. Rigidbody properties and functions.

Property	Function
Mass	Mass of object
Drag	Resistance to air
Angular Dgag	Related air resistance when rotating
Use gravity	Enables or disables gravitational acceleration
Is Kinematic	Enables or disables the physics engine
Interpolate	Enables or disables interpolation
Collision Detection	Enable or disable collisions
Constraints	Restrictions on the movement of Rigidboby

3. Virtual laboratories

In this paper we present 3 of 22 virtual laboratories developed in an environment in which different devices can be selected to check different laws of physics (see http://simulacionfisica.unimagdalena.edu.co/) in which the area for data entry is modified to obtain results in output area (also see Table 2). We can see that the machines for performing the experiments of free fall, coefficient of friction and parabolic movement (see Figures 5, 6, 7 respectively) shown an enviroment that can attract the attention of the students easily. The virtual laboratories create a powerful versatile tool which is linked to a laboratory guide, which facilitates the learning of each physical phenomena. It is possible to find in this guide questions, tables for recording data to obtain representative graphs of the physical phenomenon, analysis of equations and conclusions area.

We propose a set of laboratories in which the students recognize relevants variables of the simulated phenomenon, the relation exists between them and the physical laws that explain the phenomenon. Each laboratory is accompanied by a structured guide as follows: Name of laboratory, standard, competences to be developed, problem question, curricular area, achievement to develop, indicators of performance, theoretical foundation, simulation, observation of the phenomenon, calculations, results and analysis. This structuring of the guides, evidences that the learning mechanisms to achieve the appropriation of the knowledge on the part of the students are: Learning based on observation, competence-based learning, learning based Scientific Visualization (SciVis)and learning based on abstraction.

IOP Conf. Series: Journal of Physics: Conf. Series **935** (2017) 012069 doi:10.1088/1742-6596/935/1/012069

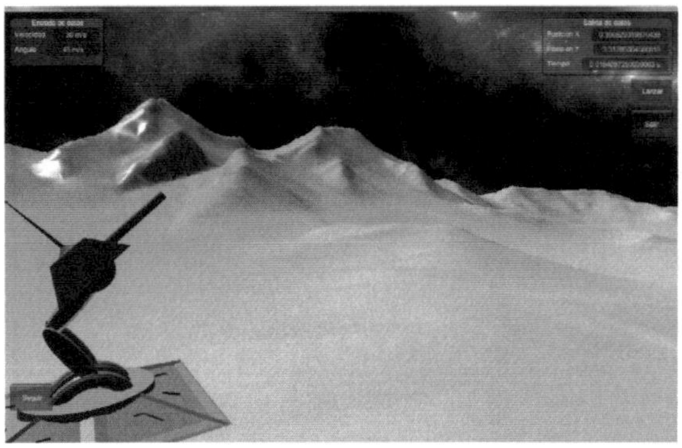

Figure 5. Parabolic movement: Combination of two movements (uniform rectilinear and free fall).

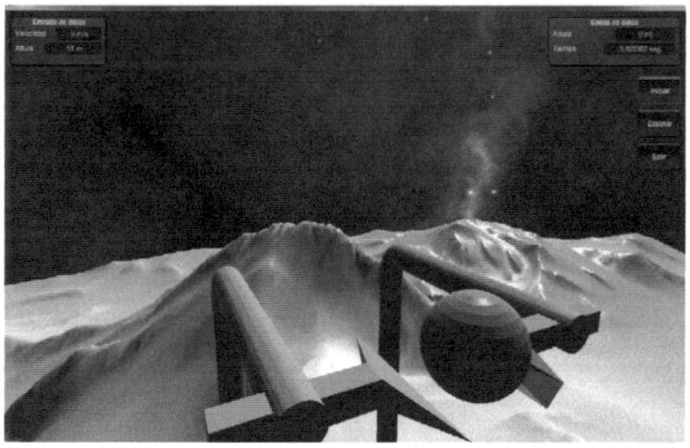

Figure 6. Free fall: Determination of the gravitational acceleration and its relation with the movement of the celestial bodies.

Figure 7. Coefficient of friction: Friction. Determination of the coefficients of static and kinetic friction.

IOP Conf. Series: Journal of Physics: Conf. Series **935** (2017) 012069 doi:10.1088/1742-6596/935/1/012069

Table 2. Virtual laboratories.

Virtual laboratory	Description
Velocity and acceleration	Relationship between the variables distance, time
Simple harmonic movement	Period and frequency in harmonic motion
Simple pendulum	Analysis of period and length of the pendulum
Torsion pendulum	Study of period, inertia and torsion coefficient
Waves	Kinematics of the wave motion
Collisions	Linear momentum and its conservation principle
Newton's second law	Concept of inertial mass. Force and acceleration
Energy	Concept of energy. Kinetic and potential energy
Archimede's principle	Pushing force. Relationship with weight and density
Circuit design	Resistors and capacitors in series and parallel
Kirchhoff Laws	Principle of charge and energy in electrical circuits
Charging and discharge of a capacitor	Loading and unloading a capacitor through a resistor
Lenses 1 and Lenses 2	Notable rays. Formation of images
Snell's Law	Concepts of angles of incidence and refraction
Conducting wire	Study of resistivity. Relationship area and length
Average and instantaneous acceleration	Analysis of ratio of velocity and time
Capacitance	Combination capacitors. Serie and parallel

4. Conclusions

The goal of this paper was to present a novel 2D and 3D dimensional virtual laboratories based on the cross-platform game Unity to open the range to access in the study of physical phenomenons, due to the users can work properly every Web or by download our app in Google Play, App Store and Windows store. In the future our vision is to overcome the barriers that still prevent the wide scale implementation of the distance-learning, through the use of virtual physics experiences. The authors are aware that virtual lab systems and simulators are often used only as an initial step in a student's engineering education and training, followed by more in-depth hands-on experience with real authentic equipment. Nevertheless, need to recognize, based on the current state of art, that the progress in Virtual reality, computer graphics, and world technologies can offer the chance to encrease rapidly the use of virtual laboratory based systems applications, and eventually reduces the need for real world laboratories altogether. Finally, the guides developed in this research are according to the implementation of immersed education and distance learning, due to important relationship between the field of pedagogy and the desing of effective learning.

Acknowledgments

This work was financed by the "Universidad del Magdalena (Fonciencias)" and the Colombian Agency Colciencias through doctoral scholarships 567.

References

[1] Larreamendy-Joerns J and Leinhardt G 2006 *Review of Educational Research* **76** 567
[2] Triona L M and Klahr D 2003 *Cognition and Instruction* **21** 149-173
[3] Zacharia Z C and Constantinou C P 2008 *American Journal of Physics* **76** 425-430
[4] Marshall J A and Young E S 2006 *Journal of Research in Science Teaching* **43** 907-937
[5] van der Meij J and de Jong T 2006 *Learning and Instruction* **16** 199-212
[6] Scheiter K, Gerjets P, Huk T, Imhof B and Kammerer Y 2009 *Learning and Instruction* **19** 481-494
[7] Swaak J, van Joolingen W R and de Jong T 1998 *Learning and Instruction* **8** 235-252
[8] Zacharia Z C 2005 *Journal of Science Education* **27** 1741-1767

[9] Gordon D N and Pea R D 1995 *Journal of Learning Science* **4** 249-279
[10] Sun K T, Lin Y C and Yu C J (2008) *Computers & Education* **50** 1411-1422
[11] Yang K Y and Heh J S 2007 *Journal of Science Education and Technology* **16** 451-461
[12] Klahr D, Triona L M and Williams C 2007 *Journal of Research in Science Teaching* **44** 183-203
[13] Redish E F and Steinberg R N 1999 *Physics Today* **52** 24-30
[14] Reiner M 1999 *Interactive Learning Environments* **7** 31-55
[15] Swaak J and de Jong T 1996 *Studies in Educational Evaluation* **5** 341-362
[16] Sweller J 1994 *Learning and Instruction* 4 295-312
[17] Holzinger A, Kickmeier-Rust M D, Wassertheurer S and Hessinger M 2009 *Computers & Education* **52** 292-301
[18] Unity User Manual 2008 *Unity Technologies.* **5** 3-Q
[19] Gordon D N and Pea R D 1995 *Journal of Learning Science* **4** 249
[20] Fisher P and Unwin D 2002 *Virtual reality in geography* (New York: Taylor & Francis)
[21] Barak P, and E A Nater 2005 The Virtual Museum of Minerals and Molecules: Molecular visualization in a virtual hands-on museum *J. Nat. Resour. Life Sci. Educ.* **34** 67-71

IOP Conf. Series: Journal of Physics: Conf. Series **935** (2017) 012070 doi:10.1088/1742-6596/935/1/012070

Anisotropy included in a nanoscale superconductor: Theoretical development

O J Gaona[1], **J D González**[1] **and J R Beltrán**[1]

[1] Universidad del Magdalena, Santa Marta, Colombia

E-mail: `jgonzaleza@unimagdalena.edu.co`

Abstract. A theoretical procedure to obtain the Ginzburg-Landau (GL) equations with the inclusion of localized anisotropy in the superconducting sample is shown. Using this theory, it is possible to study the vortex structures in two-dimensional mesoscopic superconductors with defects in the presence of a uniform magnetic field. The defects would be included through variation of parameter $\gamma(x, y, z)$ considering different shapes into the sample. This theory would allow to find unconventional vortex states which can show vortex clusters or exhibiting asymmetry. In samples with defects the vortex strings are formed owing to the interactions of vortices with variable Meissner currents through the sample due to spatial change of the anisotropy, but also the thermodynamics parameters of the superconducting sample could be determined as a function of the degree of the anisotropy generated in the defect.

1. Introduction

Whenever a scientific advances is made, the reseachers make a effort to explain it theoretically, in the particular case of superconductivity the experimental discoveries occurs before of the theoretical study in the most of cases. For instance, the London theory [1] was the first theory which explain the results obtained in the laboratory but it took more than two decades, but also in this theory vortices are treated as point-like objects but valuable information regarding to the vortex structure cannot be demonstrated. Over the past 50 years, Landau and Ginzburg [2] developed a phenomenological theory which had great success in explaining macroscopic properties of superconductors but it was also immensely useful for description of mesoscopic superconducting samples [3, 4]. The geometry of the system is important as well as the defects included in the sample, not only to determine equilibrium flux structures, but also to determine the critical parameters of the mesoscopic superconductor [5, 6] and physics becomes richer when a magnetic field is applied to this systems and vortices also show very rich behavior when different kind of defects are included in the superconducting sample [7, 15]. For this reason, this paper hopes to contribute theoretically to the study of superconducting systems that include defects generated by anisotropies in the superconducting sample.

2. Theoretical formalism

The GL theory was implement by solving the equations obtained by minimizing the functional energy that can be expanded in powers of ψ as

$$G_s = G_n + \alpha|\psi|^2 + \frac{\beta}{2}|\psi|^4 + \frac{1}{2m^*}|\left(-i\hbar\nabla - \frac{2e}{c}\vec{A}\right)\psi|^2 + \frac{h^2}{8\pi} - \frac{\vec{h} - \vec{h_0}}{4\pi}\cdot\vec{h_0} \quad (1)$$

Where G_n is the free energy density of a superconductor in the normal state, whereas $\vec{h}$ denotes the applied magnetic field, and $\vec{h_0}$ is the total local field, including the response of the superconductor.

It will be scale used as for the order parameter $|\psi_0|^2 = \frac{-\alpha_0}{\beta}$ which is the order parameter of the Mesissner state.

Now in order to find the equations for ψ and $\vec{A}$ such that their solutions give the minimum value of the functional, we solve the variational problem with respect of ψ^*:

$$\int \left(\alpha\psi\psi^* + \frac{\beta}{2}\psi\psi^*\psi\psi^* + \frac{1}{2m^*}\left(-i\hbar\nabla - \frac{2e}{c}\vec{A}\right)\psi\left(i\hbar\nabla - \frac{2e}{c}\vec{A}\right)\psi^* + \frac{h^2}{8\pi} - \frac{\vec{h}-\vec{h_0}}{4\pi}\vec{h_0}\right) dV = 0 \tag{2}$$

$$\int \left(\alpha\psi\delta\psi^* + 2\frac{\beta}{2}\psi\psi\psi^*\delta\psi^* + \frac{1}{2m^*}\left(-i\hbar\nabla - \frac{2e}{c}\vec{A}\right)\psi\left(i\hbar\nabla - \frac{2e}{c}\vec{A}\right)\delta\psi^*\right) dV = 0 \tag{3}$$

it is obtained:

$$\alpha\psi + \beta\psi|\psi|^2 + \frac{1}{2m^*}\left(-i\hbar\nabla - \frac{2e}{c}\vec{A}\right)^2\psi = 0 \tag{4}$$

and its boundary condition

$$\boldsymbol{n}\cdot\left(-i\hbar\nabla - \frac{2e}{c}\vec{A}\right)^2\psi|_s = 0 \tag{5}$$

where **n** is the unit vector normal to the surface of the superconductor. In order to obtain the equation for $\vec{A}$, we shall minimize the functional with respect to $\vec{A}$:

$$\int \left(\alpha\psi\psi^* + \frac{\beta}{2}\psi\psi^*\psi\psi^* + \frac{1}{2m^*}\left(-i\hbar\nabla - \frac{2e}{c}\vec{A}\right)\psi\left(i\hbar\nabla - \frac{2e}{c}\vec{A}\right)\psi^* + \frac{(\nabla\times\vec{A})^2}{8\pi} - \frac{\nabla\times\vec{A}-\vec{h_0}}{4\pi}\vec{h_0}\right) dV = 0 \tag{6}$$

$$\int \left(\frac{1}{2m^*}\left(-i\hbar\nabla\psi - \frac{2e}{c}\vec{A}\psi\right)\left(-\frac{2e}{c}\delta\vec{A}\psi\right)\left(-\frac{2e}{c}\delta\vec{A}\psi\right)\left(i\hbar\nabla\psi^* - \frac{2e}{c}\vec{A}\psi^*\right)\right) dV + \int \left(\frac{2(\nabla\times\vec{A})(\nabla\times\delta\vec{A})}{8\pi} - \frac{\nabla\times\delta\vec{A}-\vec{h_0}}{4\pi}\vec{h_0}\right) dV = 0 \tag{7}$$

Finally, it is obtained the second equation of GL theory, for the vector potential $\vec{A}$

$$\frac{i\hbar}{2m^*}\left(\psi^*\nabla\psi - \psi\nabla\psi^*\right) - \frac{(e^*)^2}{m^*c}|\psi|^2\vec{A} = \vec{J} = \frac{c}{4\pi}\nabla\times\nabla\times\vec{A} \tag{8}$$

The anisotropy is included in the equations trough in the GL coefficient α, as $\alpha = \alpha_0/\gamma(x,y,z)$. In the presence of such anisotropy, by minimization of equation (1), we obtain the equations for order parameter and vector potential, following the same procedure shown before. Now, normalizing the first GL equation

$$\alpha\psi + \beta\psi|\psi|^2 - \frac{1}{2m^*}\left(-i\hbar\nabla - \frac{2e}{c}\vec{A}\right)^2\psi = 0 \tag{9}$$

$$\alpha\psi + \beta\psi|\psi|^2 = \frac{1}{2m^*}\left(-i\hbar\nabla - \frac{2e}{c}\vec{A}\right)^2\psi \tag{10}$$

IOP Conf. Series: Journal of Physics: Conf. Series **935** (2017) 012070 doi:10.1088/1742-6596/935/1/012070

$$\left(\alpha\psi_0\tilde{\psi} + \beta\psi_0\tilde{\psi}|\psi_0|^2|\tilde{\psi}|^2\right) = \psi_0\frac{1}{2m^*}\left(-i\hbar\nabla - \frac{2e}{c}\vec{A}\right)^2\tilde{\psi} \tag{11}$$

$$\left(\frac{\alpha_0}{\gamma(x,y,z)}\tilde{\psi} - \beta\frac{\alpha_0}{\beta}\tilde{\psi}|\tilde{\psi}|^2\right) = \frac{1}{2m^*}\left(-i\hbar\nabla - \frac{2e}{c}\vec{A}\right)^2\tilde{\psi} \tag{12}$$

Now, the equation is divided by α_0

$$\tilde{\psi}\left(\frac{1}{\gamma(x,y,z)} - |\tilde{\psi}|^2\right) = \frac{\hbar^2}{2m^*\alpha_0}\frac{1}{\xi^2(0)}\left(-i\tilde{\nabla} - \frac{2e}{c\hbar}\xi(0)\vec{A}\right)^2\tilde{\psi} \tag{13}$$

Considering that $\frac{2e}{c\hbar}\xi(0)\vec{A} = \frac{\vec{A}}{H_{c2}\xi(0)} = \tilde{\vec{A}}$ and $\frac{\hbar^2}{2m^*\alpha_0} = \xi^2(0)$, we obtain

$$\left(\frac{1}{\gamma(x,y,z)} - |\tilde{\psi}|^2\right) = \left(-i\tilde{\nabla} - \tilde{\vec{A}}\right)^2\tilde{\psi}. \tag{14}$$

Following the same procedure for the second GL equation:

$$\frac{i\hbar}{2m^*}\left(\psi^*\nabla\psi - \psi\nabla\psi^*\right) - \frac{(e^*)^2}{m^*c}|\psi|^2\vec{A} = \vec{J} = \frac{c}{4\pi}\nabla\times\nabla\times\vec{A} \tag{15}$$

It is normalized the right term as follow:

$$\frac{c}{4\pi}\nabla\times\nabla\times\vec{A} = \frac{c}{4\pi}\frac{H_{c2}}{\xi(0)}\tilde{\nabla}\times\tilde{\nabla}\times\tilde{\vec{A}} \tag{16}$$

Hence

$$\frac{i\hbar e^*}{2m^*}\left(\psi^*\nabla\psi - \psi\nabla\psi^*\right) - \frac{(e^*)^2}{m^*c}|\psi|^2\vec{A} = \frac{c}{4\pi}\frac{H_{c2}}{\xi(0)}\tilde{\nabla}\times\tilde{\nabla}\times\tilde{\vec{A}} \tag{17}$$

$$\frac{4\pi}{c}\frac{\xi(0)}{H_{c2}}\left(\frac{i\hbar e^*}{2m^*}(\psi^*\nabla\psi - \psi\nabla\psi^*) - \frac{(e^*)^2}{m^*c}|\psi|^2\tilde{\vec{A}}H_{c2}\xi(0)\right) \tag{18}$$

$$\left(\frac{i}{2}\frac{\xi^2(0)}{\lambda^2(0)}|\psi_0|^2(\tilde{\psi}^*\tilde{\nabla}\tilde{\psi} - \tilde{\psi}\tilde{\nabla}\tilde{\psi}^*) - \frac{\xi^2(0)}{\lambda^2(0)}|\psi|^2|\tilde{\psi}|^2\tilde{\vec{A}}\right) \tag{19}$$

$$\left(\frac{i}{2}\frac{1}{\kappa}|\psi_0|^2(\tilde{\psi}^*\tilde{\nabla}\tilde{\psi} - \frac{1}{\kappa}|\psi_0|^2|\tilde{\psi}|^2\tilde{\vec{A}}\right) \tag{20}$$

$$\left(\frac{i}{2}|\psi_0|^2(\tilde{\psi}^*\tilde{\nabla}\tilde{\psi} - |\psi_0|^2|\tilde{\psi}|^2\tilde{\vec{A}}\right) = \kappa\tilde{\nabla}\times\tilde{\nabla}\times\tilde{\vec{A}} \tag{21}$$

Finally, the GL equations it is written as follow:

$$\left(\frac{1}{\gamma(x,y,z)} - |\tilde{\psi}|^2\right)\tilde{\psi} = \left(-i\tilde{\nabla} - \tilde{\vec{A}}\right)^2\tilde{\psi}. \tag{22}$$

$$\frac{i}{2}\left((\tilde{\psi}^*\tilde{\nabla}\tilde{\psi} - \tilde{\psi}\tilde{\nabla}\tilde{\psi}^*\right) - |\tilde{\psi}|^2\tilde{\vec{A}} = \kappa\tilde{\nabla}\times\tilde{\nabla}\times\tilde{\vec{A}} \tag{23}$$

3. Conclusions

In summary, we have explained in detail the the theoretical procedure to obtain the Ginzburg-Landau (GL) equations to make numerical experiments to simulate anisotropic samples and is hopes to have presented a brief information which can be of use for researchers in the years to come.

Acknowledgments
This work was partially financed by the Universidad del Magdalena (Fonciencias) and the Colombian agency Colciencias through doctoral scholarships 567.

References
[1] London F, London H 1935 *Proc. Roy. Soc. A* **149** 71
[2] Moshchalkov V V 1999 *Handbook of nanostructured materials and nanotechnology* (New York: Academic Press) p 451
[3] Schweigert V A, Peeters F M 1998 *Phys. Rev. B* **57** 3817
[4] Schweigert V A, Peeters F M 1998 *Phys. Rev. Lett.* **81** 2783
[5] Berdiyorov G R, Yu S H, Xiao Z L, Peeters F M, Hua J, Imre A and Kwok W K 2009 *Phys. Rev. B* **80** 064511
[6] Romaguera A R, Doria M M, Peeters F M 2007 *Phys. Rev. B* **76** R020505
[7] Harada K, Kamimura O, Kasai H, Matsuda T, Tonomura A, Moshchalkov V V 1996 *Science* **274** 1167
[8] Nori F 1996 *Science* **271** 1373
[9] Misko V R, SavelâĂŹev S, Rakhmanov A L, Nori F 2006 *Phys. Rev. Lett.* **96** 127004
[10] González J D, Rincon M, Barba-Ortega J 2014 *Int. Journal of Mod. Phys. B* **28** 1450227
[11] Barba-Ortega J, González J D, Sardella E 2014 *Journal of low Temp. Phys.* **174** 96
[12] Berdiyorov G R, Milošević M V, Peeters F M 2006 *Phys. Rev. Lett.* **96** 207001
[13] Berdiyorov G R, Milošević M V, Peeters F M 2007 *Phys. Rev. B* **76** 134508
[14] Berdiyorov G R, Milošević M V, Peeters F M 2006 *Phys. Rev. B* **74** 174512
[15] Berdiyorov G R, Milošević M V, Peeters F M 2006 *Europhys. Lett.* **74** 493

IOP Conf. Series: Journal of Physics: Conf. Series **935** (2017) 012071 doi:10.1088/1742-6596/935/1/012071

ForistomApp a Web application for scientific and technological information management of Forsitom foundation

L A Saavedra-Duarte[1], A Angarita-Jerardino[1], P A Ruiz[1], H J Dulce-Moreno[1,2], F H Vera-Rivera[1,2] and E D V-Niño[2]
[1] Universidad Francisco de Paula Santander, San José de Cúcuta, Colombia
[2] Foundation of Researchers in Science and Technology of Materials, Colombia

E-mail: fredyhumbertovera@ufps.edu.co , deydannv@gmail.com

Abstract. Information and Communication Technologies (ICT) are essential in the transfer of knowledge, and the Web tools, as part of ICT, are important for institutions seeking greater visibility of the products developed by their researchers. For this reason, we implemented an application that allows the information management of the FORISTOM Foundation (Foundation of Researchers in Science and Technology of Materials). The application shows a detailed description, not only of all its members also of all the scientific production that they carry out, such as technological developments, research projects, articles, presentations, among others. This application can be implemented by other entities committed to the scientific dissemination and transfer of technology and knowledge.

1. Introduction

The Group of Research of Development in Software Engineering (GIDIS) of the "Universidad Francisco de Paula Santander" and the Foundation Of Researchers In Science and Technology Of Materials (FORISTOM) worked together to develop a web application to facilitate the management and divulgation of Information of the research groups, especially FORISTOM information. Among other works between GIDIS and FORISTOM are the development of web applications for the calculation of the dose and the concentration of ions implanted in solid substrates and the development of a 3D simulator of a virtual laboratory where the ionic implantation processes can be studied [1-3]

The FORISTOM foundation aims to create appropriate scenarios where there is the exchange of creative ideas and the transfer of scientific knowledge of research and innovation in the area of science, engineering and technology of materials and the industrial applications, establishing a universal network of cooperation between research groups, universities and industries. The main purpose of the professionals who make up the FORISTOM Foundation, through the network of researchers in science, engineering and materials technology, this work automates their processes and make them known to national and international societies. In this context, it is vital that FORISTOM can have a web space with full autonomy where it can manage the information of the different scientific products, technological developments and innovations (projects, articles, papers, etc.) and enable it to contribute to the Transfer of scientific and technological knowledge. As a result of the above, arises the need to implement a web application that takes control of its processes and that allows to obtain the information of FORISTOM in an efficient way and that in turn manages to make known the impact of this to the community.

As an antecedent to the research, it is worth mentioning the platform of the GIDIS research group of the UFPS, is a web platform of GIDIS that allows to the visitors access and obtain information corresponding to their events, news related to the research group, the lines of research they have, their members and projects [4]. On the other hand, The "Scienti" platform of "Colciencias" is the national level responsible for managing the information corresponding to research groups (category, researchers, articles, papers, research projects, lines of research) and individual researchers [5]. However, despite having all the necessary information, it does not provide a pleasant sight to visitors who are in search of information about a particular group or researcher. Neither allows generate reports or indicators of the production of the group that allow better management. Therefore, ForistomApp is an information system created to record, collect, organize, distribute and improve the visibility of the projects and products created by the Foristom Foundation. Although the initial idea is that ForistomApp is used by the Foristom Foundation this can be parameterized and suitable to any research group.

ForistomApp was implemented using the Sandbox - UFPS platform created by the research group GIDIS, this platform offers a set of tools for the development and deployment of web applications in an easy and automated way in the cloud [6]. ForistomApp is a Rich Internet Application (RIAs) characterized by a common goal of adding new capabilities to the conventional hypertext-based Web. RIAs combine the Web's lightweight distribution architecture with desktop application's interface interactivity and computation power, and the resulting combination improves all the elements of a Web application (data, business logic, communication, and presentation) [7]. It is also a responsive web application that can be used from any device (tablet, cell phone or computer).

2. Development methodology

The FORISTOM Foundation has not a web application that is robust enough to cover all the processes it performs, they have not a reliable page that allows access to reliable information about its projects and research. Knowing this and taking the year 2017 as a reference, it is indispensable that this foundation has its application and can thus automate its processes. It is necessary to have technological support to make the production of the group more efficient and place it in a place in cyberspace so that it can be consulted, thus generating greater visibility and impact towards the national and international community.

For the development of this application, the OpenUp methodology [8, 9] was taken as a reference and the stages of planning, monitoring and control were added. In addition, feedback activities were added at the end of each iteration. The development of the software product is performed through the life cycle known as incremental iterative, which consists of three (3) iterations for the total coverage of product development.

The project began with the compilation of the theoretical foundation that serves as the basis for the beginning of the research process, after which the functional and non-functional requirements of the information system were identified to manage the data corresponding to the FORISTOM foundation (projects, users, events, etc.). This list of prioritized requirements of the product, which acts as a project plan, is provided by the legal representative of FORISTOM foundation, it prioritizes the objectives, balancing the value they bring, being distributed in iterations and deliveries. Once the requirements were obtained and the user's needs were analysed, the team of development proceeded to the design of the graphical user interface, describing the architecture of the information system and the development of the back-end of the system, which is evaluated in the next phase to verify that it really complies with what the final consumer wants.

After each iteration, the monitoring and control process is carried out, resulting in a software version with new functions ready to be used, thus adjusting the prototypes of the application already built, and adds new functions always prioritizing those that bring greater value to the business.

The iterations consist of the following phases:

- Iteration planning: consists of identifying the specific requirements of the iteration, as well as identifying those that are priority.
- Start: in this phase the use cases are made and a sketch of the user interfaces is made.
- Elaboration: The purpose of this phase is to establish the architecture of the software, as well as the definition of class diagrams, conceptual and data.
- Construction: in this phase the use cases are implemented, and the user manual for the iteration is developed.
- Transition: The iteration test plan is executed, debugging found faults and testing implemented corrections.
- Monitoring and control of the iteration: the purpose of this phase is the review of progress in each activity, as well as the reports, their arrears, compliance and feedback.

3. Results

An application was developed that allows the FORISTOM foundation to have control over its members, registering in the application and allowing its projects and products to be published. The application will have an administrator, who will have control over the information of the group on the page, being able to edit the corporate elements such as the mission, vision and research lines, also can manage the news and events of the group. The Figure 1 shows the system map, including all the pages integrated in the application and its dependency.

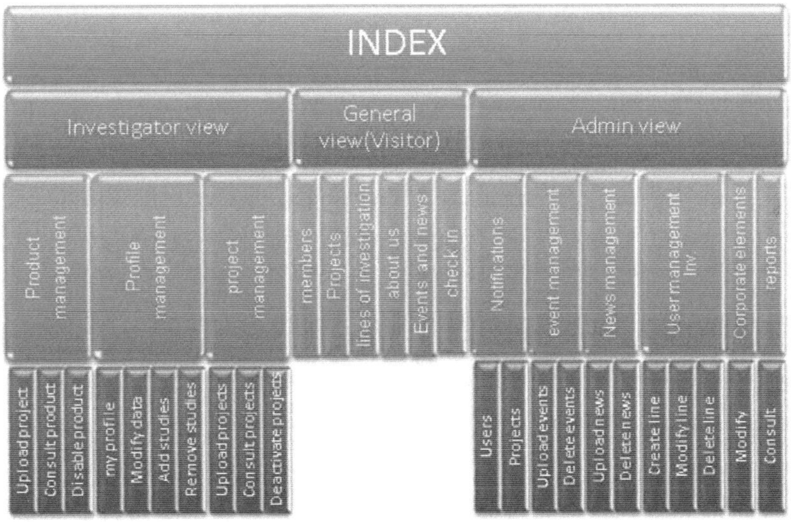

Figure 1. ForistomApp navigation map.

In the proposed solution, the model-view-controller architecture is implemented, together with the façade pattern. The Model View Controller (MVC) is a software architectural style that separates data from an application, user interface, and control logic into three distinct components. This model has proven its validity over the years in all types of applications, and on a multitude of languages and development platforms [10,11].

- The Model: contains a representation of the data that the system handles, its business logic, and its mechanisms of persistence.
- The View, or user interface: composes the information that is sent to the client and the mechanisms of interaction with it.

- The Controller, which acts as an intermediary between the Model and the View, managing the flow of information between them to adapt the data to the needs of each one [10,11].

For its part, the façade pattern is a design pattern that simplifies the communication between two objects of the system [12,13].

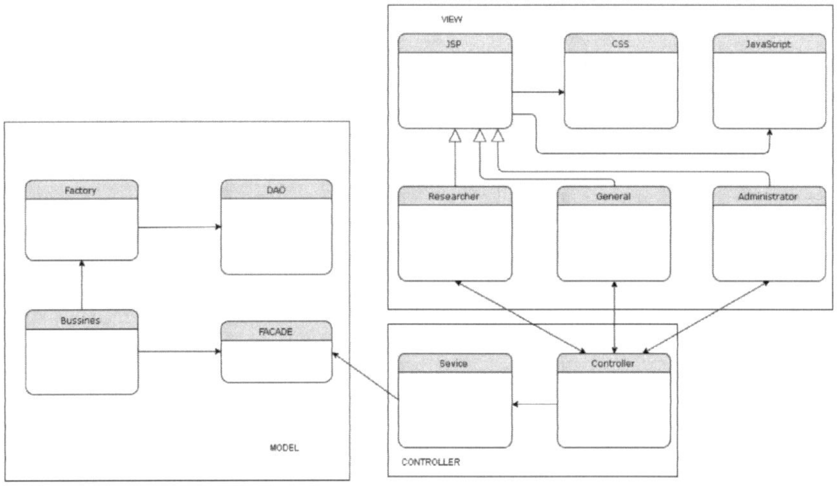

Figure 2. ForistomApp system architecture.

When you enter ForistomApp you will be presented with the overview (see Figure 3), the navigation is done through the menu at the top of the page, where the visitor can access the information shared by the research group. The users of the application can consult projects, news, events, products, research lines and corporate information of the group. All this content will be available in the corresponding top menu items. Also at the top you will find the options Login and Register. For log in, the user must belong to the group and have a user created and have been accepted by the application administrator. Otherwise, you must register by completing a form, entering the relevant data, and waiting for the acceptance by the administrator.

Figure 3. General view ForistomApp.

Once the user has logged on as a researcher (see Figure 4), the user can access his profile and his options to manage the information as a researcher, available through a side menu. From this view the researcher can manage all his personal information and the projects that are advanced within the research group. This one also has the possibility of administering its studies with the title, year of completion and institution.

Figure 4. ForistomApp research view.

The administrator's view (see Figure 4) allows the leader of the group or in charge of the web application, to manage the publications, that is, to accept or reject these according to their criteria and, if these are approved, these will be presented in the overview. It also has the ability to administer the researchers; this includes admitting or rejecting users and deactivating them when one of the members leaves the group. This view also allows the application manager to detail the corporate elements of the group (mission, vision, corporate objectives) and the research lines of the group either active or inactive. This view is the most influential in the application because it depends on the information presented to the public is reliable.

The administrator can generate reports so that the community can know the products developed per month and products of each researcher. In this way, we have the detailed production information of the group that displays the information in real time. Researchers may also have a more orderly handling of their products and they will be able to make known the result of their work. ForistomApp can also be parameterized in corporate colours, logos, images and static information, adapting to different research groups, the administrator can manage the application through a friendly interface. The web application was develop using bootstrap for responsive systems [14].

4. Conclusions

Software as an information management tool of a research group is indispensable, ForistomApp, allows to detail the production of a group, to carry detailed reports of the projects realized and that are advancing. It allows the users to visualize the products developed by the group members with the necessary information. In addition, it is a flexible application that allows the parameterization of corporate colours, logos and static information, making it easily adaptable to any research group.

The construction of the web application ForistomApp, for the Foristom foundation will improve the management of its processes allowing them to take control over their internal tasks and an inventory of people of the group as well as their production.

IOP Conf. Series: Journal of Physics: Conf. Series **935** (2017) 012071 doi:10.1088/1742-6596/935/1/012071

The web application allows the Foristom foundation to be more accessible to the general public, because anyone can see all their production online facilitating that the whole community can accede to it and not only the circle of investigators allowing to download documents elaborated by the working groups and to read in detail the results of the realized investigations.

The Foristom foundation will be known nationally and internationally thanks to the visibility that will be granted by the ForistomApp application, as this is an end user friendly page and with truthful and reliable information provided by the group members and validated by the site administrator.

The web application ForistomApp integrates the collaboration of the research groups in Colombia with information and communication technologies, so that this last one is a constant support to the research processes in the country allowing to detail the products obtained by the investigations carried out, showing monthly reports or in dates ranges established by the administrator, revealing the quantity of products obtained by researchers.

Rererences

[1] F H Vera-Rivera, B R Pérez Gutiérrez, H J Dulce-Moreno, F Duran-Flórez, and E D V-Niño 2016 Estimate of the concentration of implanted ions in solid substrates using a web application *J. of Phys.: Conf. Series* **743** 012016 1-5

[2] F H Vera-Rivera, V D Dugar-Zhabon, H J Dulce-Moreno, and E D V-Niño 2016 Development of a web application for estimate the total charge in an electric discharge *J. Phys.: Conf. Ser.* **687** 012058 1-4

[3] B R P Gutiérrez, F H Vera-Rivera, and E. D V-Niño 2017 Design and implementation of a web application to estimate the surface density of implanted ions in solid substrates *J. Phys.: Conf. Ser.* **786** 012015 1-6

[4] Universidad Francisco de Paula Santander 2014 *Portal web del Grupo de Investigación y Desarrollo de Ingeniería de Software (GIDIS)* (http://gidis.ufps.edu.co/front/#/home)

[5] Departamento Administrativo de Ciencia, Tecnología e Innovación de Colombia 2015 *Plataforma ScienTI, Colombia* (http://www.colciencias.gov.co/scienti)

[6] F. H. Vera-Rivera, B. R. Perez-Gutierrez, and F. J. Torres-Bermudez 2015 Sandbox UFPS - cloud development platform for server management, creation and deployment of web applications of academic use *Res. Comput. Sci.* **101** 65–75

[7] P. Fraternali, G. Rossi, and F. Sánchez-Figueroa 2010 Rich Internet Applications *IEEE Internet Comput.* **14** 9–12

[8] Per Kroll & Bruce MacIsaac 2006 *Agility and Discipline Made Easy: Practices from OpenUP and RUP* 1st edition (USA: Addison-Wesley Professional)

[9] Steve Hansen 2002 Web information systems: the changing landscape of management models and web applications *Proceedings of the 14th international conference on Software engineering and knowledge engineering (Italy)* (USA: ACM)

[10] Staff of the University of Alicante 2015 *Model view controller (MVC) of University of Alicante (Spain)* (http://si.ua.es/es/documentacion/asp- net-mvc-3/1-dia/modelo-vista-controlador-mvc.html)

[11] Cecilio Álvarez 2014 *Diseño con patrones y fachadas, genbetadev* (http://www.genbetadev.com/paradigmas-de- programacion/diseno-con-patrones-y-fachadas)

[12] Krasner G E and Pope S T 1988 A cookbook for using the model-view-controller user interface paradigm in Smalltalk- 80 *Journal of Object-Oriented Programming* **1(3)** 26-48

[13] E Althammer and W Pree 2001 Design and implementation of a MVC-based architecture for ecommerce applications *International Journal of Computers and Applications* **23(3)** 173-185

[14] Bootstrap 2010 *The most popular HTML, CSS, and JS library in the world* (http://getbootstrap.com/)

IOP Conf. Series: Journal of Physics: Conf. Series **935** (2017) 012072 doi:10.1088/1742-6596/935/1/012072

Comparison of different approaches of modelling in a masonry building

M Saba[1] and D Meloni[2]
[1] Universidad de Cartagena, Cartagena, Colombia
[2] University of Cagliari, Piazza d'Armi Cagliari, Italy

E-mail: manuelsaba1987@gmail.com

Abstract. The present work has the objective to model a simple masonry building, through two different modelling methods in order to assess their validity in terms of evaluation of static stresses. Have been chosen two of the most commercial software used to address this kind of problem, which are of S.T.A. Data S.r.l. and *Sismicad12* of Concrete S.r.l. While the *3Muri* software adopts the Frame by Macro Elements Method (FME), which should be more schematic and more efficient, *Sismicad12* software uses the Finite Element Method (FEM), which guarantees accurate results, with greater computational burden. Remarkably differences of the static stresses, for such a simple structure between the two approaches have been found, and an interesting comparison and analysis of the reasons is proposed.

1. Introduction
Modelling a structure is certainly a fundamental issue to tackle very carefully, in fact from that it depends, in large part, the reliability and the interpretation of the results obtained by the software of calculation. Often professionals choose superficially the software for carrying out the structures analysis, and especially in the case of historic structures of masonry, that could lead to underestimate structural problems. The model of the structure should be three-dimensional and adequately represent the spatial distribution of mass, stiffness and strength, in the other hand the discretization of the structure normally carried by the software should be carefully analysed to verify spatial distribution as well in restricted spaces where the discretization is difficult. With the objective to assess the reliability of the results of commercial software, a comparison between different types of approaches has been done, in particular between the methodology Finite Elements Methods (FEM)[1-3], already known and established, implemented in *Sismicad12 software*, and the new modelling technique elaborated from the University of Padua, Italy, called Frame Macro Elements (FME) implemented in *3Muri* software [4-6]. The latter has been developed mainly to carry out analysis of masonry structures, and complies with international regulations as well as *Sismicad12*. A simple case of a masonry structure has been studied, and static analysis with the two softwares mentioned has been carried out.

2. Methodology
Below are exposed the theoretical basis of the two methods, which are then implemented in *Sismicad12* and *3Muri* software to determine stresses and deformations of the structure.

IOP Conf. Series: Journal of Physics: Conf. Series **935** (2017) 012072 doi:10.1088/1742-6596/935/1/012072

2.1. Finite element method

The FEM originally developed for complex problems in structural mechanics, it is the main method used for complex systems [7]. In the FEM, the structural system is discretized by a set of points called nodes interconnected between them. At the nodes can be applied physical properties such as coefficient of thermal expansion, Young's modulus, shear modulus, Poisson's ratio. The fundamental idea is to discretize a given object for obtaining a discrete problem, that is a system of equations with a finite number of unknowns that can be resolved by a solver [8].

2.2. Frame by macro element method

The FME is a strategy for three-dimensional analysis of masonry buildings [9]. As mentioned from the authors of the schematic model [10], structural element modelling strategies are based on the identification of macroscopic structural elements, defined from a geometrical and kinematic point of view through finite elements (solid, shell or frame) and described from a static point of view through their internal generalized forces [11]. Usually two main structural components may be identified: piers and spandrels. This idealization starts from the earthquake damage observation that shows as usually cracks and failure modes are concentrated in such elements (Figure 1) [9,12].

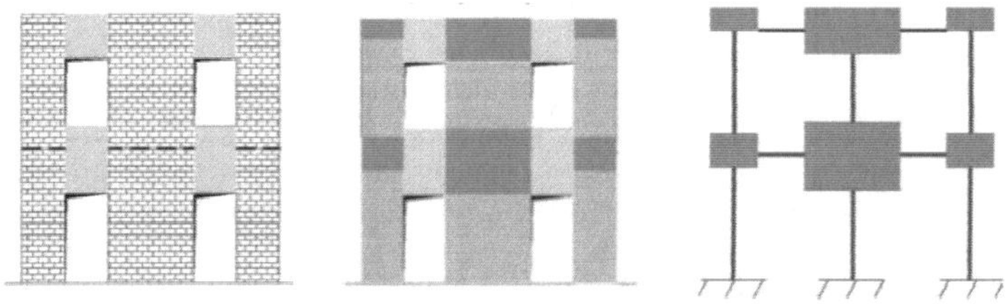

(a) (b) (c)

Figure 1. (a) Identification of spandrels; (b) Identification of nodes; (c) Equivalent frame. Adapted from Lagomarsino et al. [9].

3. Discussion and results

The building object of study is a simple masonry shell size, 4m×7m in plan (*A*), height 3m, and wall thickness 0.3m, with unidirectional rigid floor which is based on the walls P1, P3 and P5 (Figure 2), with masonry blocks of soft stone, tufa and limestone.

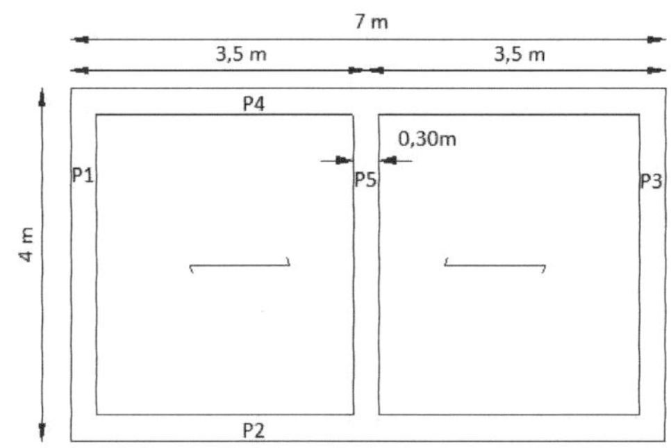

Figure 2. Plan of the masonry box.

Once has been developed the model with both software, and made the static analysis, applying a load q of 11.95kNm^{-1}, it can be clarifying the actual distribution of the loads on the whole structure. The walls P1 and P3, show a significant difference in terms of resistance action, Nu, and applied action N (Table 1), which outlines a totally different approach referring to the perimeter walls, orthogonal to the

IOP Conf. Series: Journal of Physics: Conf. Series **935** (2017) 012072 doi:10.1088/1742-6596/935/1/012072

floor load. Analysing the results relating to the wall P5, *Nu* is basically the same between the two software, contrariwise *N* is significantly different.

Table 1. Result of the Static Analysis done with the two software.

Wall	Quote (m)	N (daN)		Nu (daN)		N/Nu		Verified	
		3Muri	Sism12	3Muri	Sism12	3Muri	Sism12	3Muri	Sism12
P1 and P3	3	5050	8264	66263	32427	0.076	0.254	yes	yes
	1.5	8723	10625	66263	53286	0.132	0.199	yes	yes
	0	12395	13588	66263	32427	0.187	0.419	yes	yes
P5	3	5050	16376	66263	67319	0.076	0.243	yes	yes
	1.5	8723	16154	66263	67319	0.132	0.243	yes	yes
	0	12395	16740	66263	67319	0.187	0.248	yes	yes
P2 and P4	3	8804	278	115960	17807	0.076	0.002	yes	yes
	1.5	15232	10196	115960	117807	0.131	0.008	yes	yes
	0	21659	19108	115960	17807	0.187	0.16	yes	yes

Although the static analysis is satisfied, even in the walls P2 and P4 can be noticed a remarkable difference in terms of applied action *N*. *3Muri* software considered the schematization of the wall and its correspondent loads as shown in Figure 3. Considering the load applied uniformly on the floor, the yellow part falls upon the walls P1 and P3, the blue part upon the P5 and the red part upon P2 and P4.

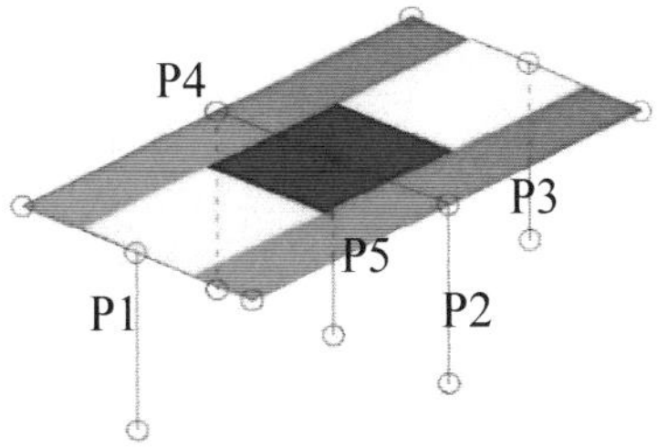

Figure 3. Allocation scheme of loads adopted by *3Muri* software.

Analysing the results, can be said that the walls P1, P3 and P5 at the top are loaded in the same way, instead, the side walls, P2 and P4, also take a share of the load. Figure 4 shows the complex stress distribution inside the building panels with *Sismicad12*, this type of modelling takes into account therefore the enchain between the walls, mutual wind bracing, mutual contrast respect to the instability, and collaboration between the walls to support vertical loads.

In addition, the elastic calculation can be done manually, considering areas of influence, and that can help to clarify and provide a meter of judgment for assessing the consistency of the results presented by the two software. Starting from the given scheme of the floor, the walls parallel to the warping load, P2 and P4, can be consider almost unloaded.

About the walls P1, P3 and P5 perpendicular to the warping load, the loading area can be assumed as in Figure 5, at the top of each bearing wall, thereafter the load on the side walls P1 and P3 can be assumed to be equal to *P*/4 and on the central wall *P*/2, being *P* the total weight on the floor (1).

$$P = qA = 33460\,daN \tag{1}$$

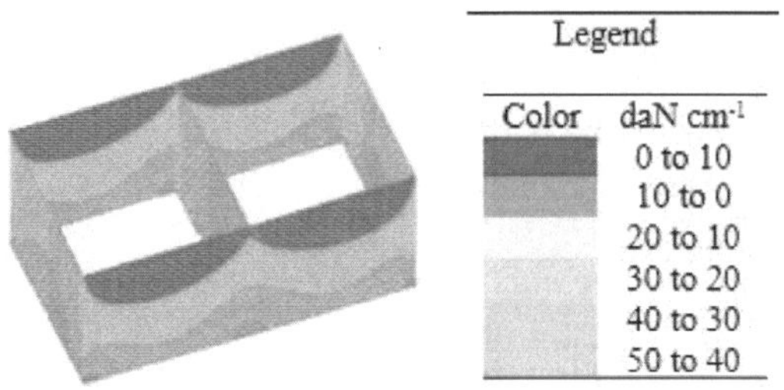

Figure 4. Isometric view of stresses and mapping colors, *Sismicad12*.

It can be immediately noted that the so found values (Figure 5) are quite similar to those resulting from the FEM analysis.

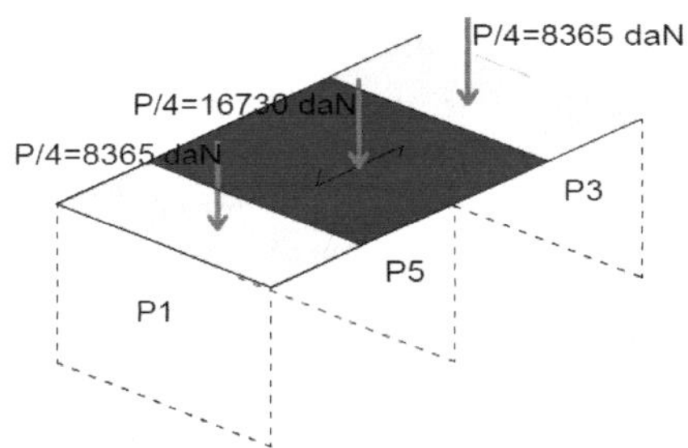

Figure 5. Load distribution on the load bearing walls.

Comparing the results with the FME modelling instead, it is shown a marked difference in the calculated values, in the case of the walls P2 and P4, it is found a value equal to 8800daN, against an almost zero in the FME modelling. Furthermore, the walls P1, P3 and P5 have values equal to 5050daN in the top, with *3Muri*, against 16300daN (wall P5) and 8300daN (walls P1 and P3) of the FEM model.

Looking at the bottom of the bearing walls, where it is necessary to take into account the own weight of the masonry, further assessment can be done.

Table 2. Account of the own weight of the masonry walls P1, P2, P3, P4 and P5.

Walls	t (m)	Length (m)	ho (m)	Limestone $\gamma_{masonry}$ (daN m^{-3})	Safety factor γ_{G1}	Weight of the Masonry (daN)
P1, P3, P5	0.30	4.00	3.00	1600	1.3	7488
P2, P4	0.30	7.00	3.00	1600	1.3	13104

In the FEM modelling, considering the wall P5, it can be seen that the loads in the upper part with the lower part are similar, while in the walls P1 and P3, the upper and lower loads are different, but that difference (5324daN) is not equal to the weight of the wall panel P5 (see Table 2). In the walls P2 and P4 instead, the difference between the upper and lower loads in the FEM analysis, results to be higher (18830daN) than the weight of the wall P2. Instead in the FME modelling the difference between the upper and lower loads on the walls P2 and P4 appears to coincide with the weight of the corresponding masonry piers. Discussing the results of this simplified structure, it seems that the FEM method takes into account the whole box and the redistribution of loads and stresses between the walls in a more complex and detailed way than it does the FME method, consequently leading to results that would seem more reasonable and close to the reality. It is in fact difficult to accept that the walls P2 and P4 in the upper part are subjected to a load equal to 50% of the total load of the floor as in the case of FME modelling (Figure 3). On the other hand, however, in the FEM method, the fact that in the bottom of the wall P5 there is a load similar to the upper part of the same wall, is not convincing in full.

4. Conclusions

Some important conclusions on the modelling of buildings in masonry can be drown, in fact from what has emerged in the case of study, the two approaches used can lead to very different results, both from quality and quantity point of view. In the opinion of the authors such analysis seems to be more in conformity with reality if carried out with the finite element method model, in fact, this model allows to take into account the effects of mutual connection and mutual cooperation between the walls of masonry, with a redistribution of stresses in a more accurate way. The software *3Muri* offers a rigid interpretation of the structure and it is impossible to interact with the creation of nodes and meshes, creating an inaccessible system. Otherwise in the case of the FEM method the integrations made to get from the matrix of the vectors of nodal displacements δ to the actions, are difficult to identify to understand the problem. It can therefore be concluded with some general directions as the fact that these types of modelling in the present state, cannot be generalized to any type of geometry of structure, and may lead to completely different results when applied to complex structures.

Acknowledgments

A special thanks to the companies S.T.A. Data S.r.l. and Concrete S.r.l. This project has been funded with support of the European Commission. This publication reflects the view only of the author, and the Commission cannot be held responsible for any use which may be made of the information contained therein - Elarch Program (project reference number: 552129-em-1-2014-1-it-era mundus-ema21).

References

[1] Sifakis E and Barbič J 2015 *Finite Element Method Simulation of 3D Deformable Solids* (California: Morgan and Claypool Publishers) chapter 3
[2] Dow J O 2015 A Concise Overview of the Finite Element Method (USA: Momentum press Engineering) chapter 1
[3] Morgan K, Hassan O and Weatherill N P 1981 The Finite Element Method *Academy of Engineering Polish Academy of Science Chinese Academy of Sciences National Academy of Science Italy* (Italy: Accademia dei Lincei) **35(4)** 110–114
[4] Lagomarsino S, Galasco A and Penna A 2007 *Non linear macro-element dynamic analysis of masonry buildings* First International Conference on. on Computational Methods in Structural Dynamics and Earthquake Engineering (Greece: ECCOMAS) pp 1-12
[5] S. T. A. DATA S.R.L 2007 *Documento di Validazione Version 1.0* (Italy: S.T.A. DATA) pp 3-78
[6] S. T. A. DATA S.R.L 2013 *3Muri software, Version 5.5* (Italy: S.T.A. DATA) pp 6-88
[7] Zienkiewicz O C, Taylor R L and Fox D 2013 *The Finite Element Method for Solid and Structural Mechanics* 7th Edition (USA: Elsevier Ltd) chapter 1 pp 1-20
[8] Ferreira A J M MATLAB Codes for Finite Element Analysis 2009 *Solids and Structures* vol 157 (Netherlands: Springer)
[9] Lagomarsino S, Penna A, Galasco A and Cattari S 2013 TREMURI program: An equivalent frame model for the nonlinear seismic analysis of masonry buildings *Eng. Struct.* **56** 1787–1799
[10] Tomaževič M, Lutman M and Petković L 1996 Seismic Behavior of Masonry Walls: Experimental Simulation *J. Struct. Eng.* **122(9)** 1040–1047
[11] S. T. A. DATA S.R.L. 2011 *3Muri Manual Version 2011* (Italy: S.T.A. DATA) pp 5-73
[12] Penna A, Lagomarsino S and Galasco A 2014 A nonlinear macroelement model for the seismic analysis of masonry buildings *Earthq. Eng. Struct. Dyn.* **43(2)** 159–179

IOP Conf. Series: Journal of Physics: Conf. Series **935** (2017) 012073 doi:10.1088/1742-6596/935/1/012073

Simulation of the electromagnetic field in a cylindrical cavity of an ECR ions source

A Estupiñán[1], E A Orozco[1], V D Dugar-Zhabon[1] and M T Murillo Acevedo[3]

[1] Universidad Industrial de Santander, Bucaramanga, Colombia
[2] Universidad Manuela Beltrán, Bucaramanga, Colombia
[3] Universidad Santo Tomás, Bucaramanga, Colombia

E-mail: alex.estupinan@saber.uis.edu.co

Abstract. Now there are numerous sources for multicharged ions production, each being designed for certain science or technological objectives. Electron cyclotron resonance ion sources (ECRIS) are best suited for designing heavy ion accelerators of very high energies, because they can generate multicharged ion beams at relatively great intensities. In these sources, plasma heating and its confinement are effected predominantly in minimum-B magnetic traps, this type of magnetic trap consist of two current coils used for the longitudinal magnetic confinement and a hexapole system around the cavity to generate a transversal confinement of the plasma. In an ECRIS, the electron cyclotron frequency and the microwave frequency are maintained equal on a quasi-ellipsoidal surface localized in the trap volume. It is crucial to heat electrons to energies sufficient to ionize K- and L- levels of heavy atoms. In this work, we present the preliminary numerical results concerning the space distribution of TE_{111} microwave field in a cylindrical cavity. The 3D microwave field is calculated by solving the Maxwell equations through the Yee's method. The magnetic field of minimum-B configuration is determined using the Biot-Savart law. The parameters of the magnetic system are that which guarantee the ECR surface location in a zone of a reasonably high microwave tension. Additionally, the accuracy of electric and magnetic fields calculations are checked.

1. Introduction

Initially, the Electron Cyclotron Resonance Ions Sources (ECRIS) were elaborated and fabricated as a heavy particle injection part for cyclotron systems [1]. Harnessing of the low-emittance multicharged ion beams in cyclotron accelerators permits to achieve high energies that are quadratic in the ion charge. Thereafter the ECRIS found its application in the ion implantation technology, mass-spectrometry, technology of Nano layer production, low-impulse propulsion engines, etc. [2–5]. The conditions and processes that lead to multicharged ion generation in ECRIS is discussed in [1, 6–9]. Of all the parameters that determine the ion yield and the ion charge states, two of them are of importance: the magnetostatic field profile and microwave field distribution in the trap volume, especially the ECR zone location in the magnetostatic trap field. In this work, we present some results on the numerical simulation of the TE_{111} microwave field in a cylindrical resonant cavity of 10 cm long and 9.1 cm in diameter, which is typical of some real ECRIS. Calculating the magnetic field it is important to keep in mind that the ECR zone should be as large as possible but should not touch the cavity walls. This trap field is determined using the Biot-Savart law [10].

IOP Conf. Series: Journal of Physics: Conf. Series **935** (2017) 012073 doi:10.1088/1742-6596/935/1/012073

2. Theoretical formalism

A physical scheme of a typical ECRIS is shown in Figure 1(a).The ECR plasma is confined by the minimum-B magnetic field formed by two current coils (1) and the hexapole permanent magnet system (2). The discharge chamber (3) made of copper is placed inside the magnetic system. The discharge chamber is fed by microwaves from a magnetron generator along the chamber axis (4),(5) y (6). In our case, a microwave fed line transversal to the chamber axis is used, being presented in Figure 1(b). In this scheme, two current coils (7) are disposed bilaterally symmetrical with the cavity (1). The hexapole bars (6) are set on the cavity cylinder surface parallel to each other in such a way that the fed line passes between the bars. The magnetic system is shown in Figure 1(b).

The fed line consists of a 2.45 GHz magnetron mounted on a waveguide that relates to the chamber through a rectangular window of $7.65 \times 1.0\ cm^2$ in size.

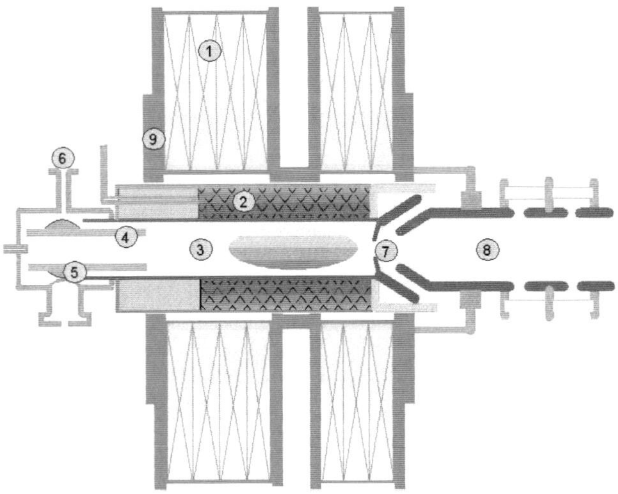

Figure 1. (*a*) Schematic diagram of the ECR ion source: 1-coils, 2-hexapole, 3-plasma chamber, 4-coaxial line, 5-tuner, 6-RF injection, 7-exit hole, 8-Einzel lens, 9-yoke. (Figure taken from [11]). (*b*) Three-dimensional physical scheme of the simulate system: 1-cavity, 2-window, 3-ferrite isolator, 4-waveguide, 5-magnetron, 6-hexapole, 7-coils.

The electromagnetic field evolution in the cavity is calculated by using the UPML method, where the fields components evolve according to the next set of equations [12]:

$$\frac{\partial H_z}{\partial y} - \frac{\partial H_y}{\partial z} = \frac{\partial}{\partial t}(K_y D_x) + \frac{\sigma_y}{\epsilon_0} D_x \tag{1}$$

$$\frac{\partial H_x}{\partial z} - \frac{\partial H_z}{\partial x} = \frac{\partial}{\partial t}(K_z D_y) + \frac{\sigma_z}{\epsilon_0} D_y \tag{2}$$

$$\frac{\partial H_y}{\partial x} - \frac{\partial H_x}{\partial y} = \frac{\partial}{\partial t}(K_x D_z) + \frac{\sigma_x}{\epsilon_0} D_z \tag{3}$$

and

IOP Conf. Series: Journal of Physics: Conf. Series **935** (2017) 012073 doi:10.1088/1742-6596/935/1/012073

$$\frac{\partial E_z}{\partial y} - \frac{\partial E_y}{\partial z} = -\frac{\partial}{\partial t}(K_y B_x) - \frac{\sigma_y}{\epsilon_0} B_x \tag{4}$$

$$\frac{\partial E_x}{\partial z} - \frac{\partial E_z}{\partial x} = -\frac{\partial}{\partial t}(K_z B_y) - \frac{\sigma_z}{\epsilon_0} B_y \tag{5}$$

$$\frac{\partial E_y}{\partial x} - \frac{\partial E_x}{\partial y} = -\frac{\partial}{\partial t}(K_x B_z) - \frac{\sigma_x}{\epsilon_0} B_z \tag{6}$$

where:

$$\breve{D}_x = \epsilon_0 \frac{S_z}{S_x} \breve{E}_x \ , \ \breve{D}_y = \epsilon_0 \frac{S_x}{S_y} \breve{E}_y \ , \ \breve{D}_z = \epsilon_0 \frac{S_y}{S_z} \breve{E}_z \tag{7}$$

and

$$\breve{B}_x = \mu \frac{S_z}{S_x} \breve{H}_x \ , \ \breve{B}_y = \mu \frac{S_x}{S_y} \breve{H}_y \ , \ \breve{B}_z = \mu \frac{S_y}{S_z} \breve{H}_z \tag{8}$$

where

$$S_x = K_x + \frac{\sigma_x}{j\omega\epsilon_0} \quad ; \quad S_y = K_y + \frac{\sigma_y}{j\omega\epsilon_0} \quad ; \quad S_z = K_z + \frac{\sigma_z}{j\omega\epsilon_0}. \tag{9}$$

In our simulations K_x, K_y y K_z are chosen equal to the unity and the electrical conductivities $\sigma_x, \sigma_y, \sigma_z$ are chosen equal to zero in all space except the microwave port where $\sigma_x = (x/d)^3 \sigma_x^{max}$, $d = 10\Delta x$ is the width of the UPML with with the spatial step Δx in x direction, x is the distance from the border towards the interior of the UPML and $\sigma_x^{max} = 0.8(m+1)/(\Delta x \sqrt{\mu_0/\varepsilon_0})$ with $m = 3$ when 10 cells are used in the UPML zone as in this case. The UPML zone is used to avoid non physical reflections.

This set of equations (1)-(6) is written in a finite difference scheme using the Yee method and solved in the traditional way [13].

The static magnetic field of this trap is obtained by the superposition of the field produced by two systems. The first one is the coil magnetic field, where each coil is simulated as a superposition of circular turns with stationary current. The second one is the hexapole magnetic field. This field is calculated using the magnetic potential method because all hexapole magnets are out of the confinement region. The field produced for one turn in the point (x, y, z) in Gaussian units is:

$$\vec{B}(x,y,z) = \frac{RI}{c} \int_0^{2\pi} \frac{(cos\theta(z-z_s)\hat{\imath} + sin\theta(z-z_s)\hat{\jmath} - [ysin\theta + xcos\theta - R]\hat{k})}{[(x^2+y^2-2R(xcos\theta+ysin\theta)+R^2+(z-z_s)^2]^{3/2}} d\theta. \tag{10}$$

Where R is radius of turn, which center is along z axis at point $(0, 0, z_s)$, I is electric current and c is the light velocity [10].

The second one is the magnetic field from magnets which conform the sixtupole system. This magnetic field can be calculated through the method of gradient of scalar magnetic potential since, none magnet exists in confinement region [14]:

$$\phi_m(\vec{r}) = \int_{s_{(v)}} \frac{\vec{M}(\vec{r}')}{|\vec{r}-\vec{r}'|} \cdot d\vec{s}' - \int_v \frac{\nabla' \cdot \vec{M}(\vec{r}')}{|\vec{r}-\vec{r}'|} dv' \tag{11}$$

where ϕ_m is the scalar magnetic potential, $s_{(v)}$ is the magnet boundary surface, v is the magnet volume.

IOP Conf. Series: Journal of Physics: Conf. Series **935** (2017) 012073 doi:10.1088/1742-6596/935/1/012073

3. Results and discussion

The numerical simulation of the cavity microwave field is realized under the above-mentioned physical parameters. To excite a microwave field of 14 kV/cm tension, an input power of 569 kW is injected into the cavity through a TE_{10} waveguide. Such a high level of the microwave power are related with the use of an not optimal injection system for our simulations. For this system the quality factor of the cavity is too low (about 50). However, for an standard experimental system this quality factor may be of the order of 10000, therefore the same electric field strength inside the cavity could be obtained with a injection power of the order of kW. In the simulations, a perfect electric conductor (PEC) approximation is taken for describing the electrical characteristics of both the waveguide and cavity walls. Uniaxial perfectly matched layer (UPML) method [12] is applied for calculating the microwave power injected through the cavity window. The simulations are fulfilled on a rectangular 3D mesh at the spatial steps $\Delta x = \Delta y = 0.07\ cm, \Delta z = 0.2\ cm$ at a time step $\Delta t = 2.07\ ps$, that are chosen in accordance with the Courant stability condition [15].

In Figure 2(a) one can see the distribution of the electric field strength in the cavity middle cross section ($z = L_c/2$, L_c is the cavity length), and its vector pattern Figure 3 shows the electric field strength in longitudinal plane $y = 0$. There is an obviously good agreement between this distribution and the TE_{111} pattern that can be obtained from the well-known analytical expressions. The calculation accuracy of the electric and magnetic fields in the cavity obtained by solutions of $\nabla \cdot \vec{B} = 0$ and $\nabla \cdot \vec{E} = 0$, equations in a finite difference form is found no worse than to 10^{-7}.

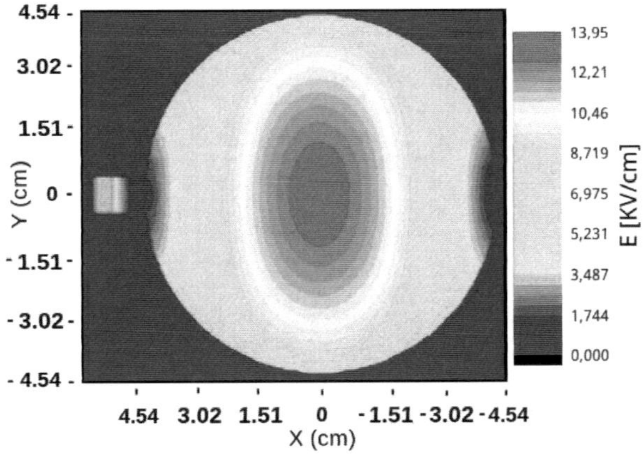

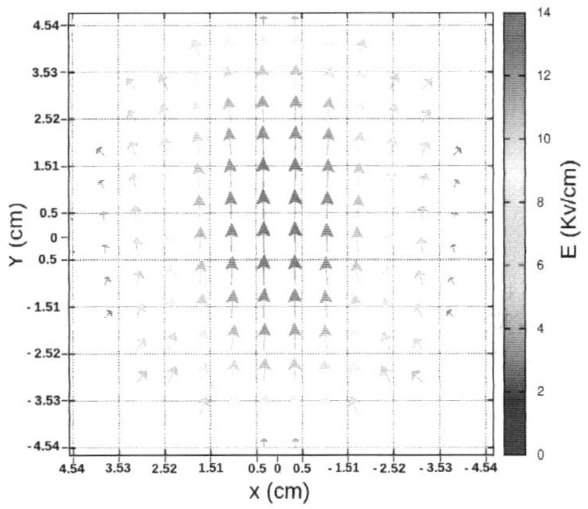

Figure 2. (*a*) The electric field strength in the cross section $z = L_c/2$. (*b*) The corresponding electric vector field.

From electric field distribution presented in Figure 2 and Figure 3, it follows that the yellow zone corresponding to the electric field of order of 10 kV/cm is appropriate for resonance energizing of electrons. From this standpoint, it could be profitable if the ECR zone coincided with the yellow microwave zone. Two current coils and six permanent magnetic bars form the magnetic field of minimum-B type. The geometry of the coils and bars, the degree of bar magnetization and coil current are chosen so that the ECR zone $B(x, y, z) = (m\omega)/(-e) = 87.53\ mT$, coincides accurately with the yellow microwave zone (see Figure 2(a)). Here m and $-e$ are the electron mass and the magnitude of electric charge of electron, ω is the microwave angular frequency. The space shape of the ECR zone calculated in accordance with the above-mentioned method is shown in Figures 4. One can see from Figures 2 and Figures 4 that the

ECR and the microwave yellow zones do not coincide perfectly but the fact that the ECR zone oscillates around the yellow zone provides some expectations that no-coincidence of these zones has not an adverse effect on the ECR heating efficiency.

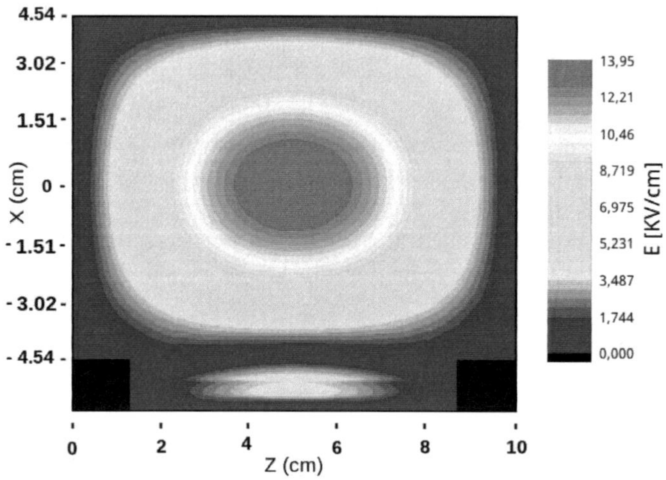

Figure 3. Electric field strength in the longitudinal plane $y = 0$.

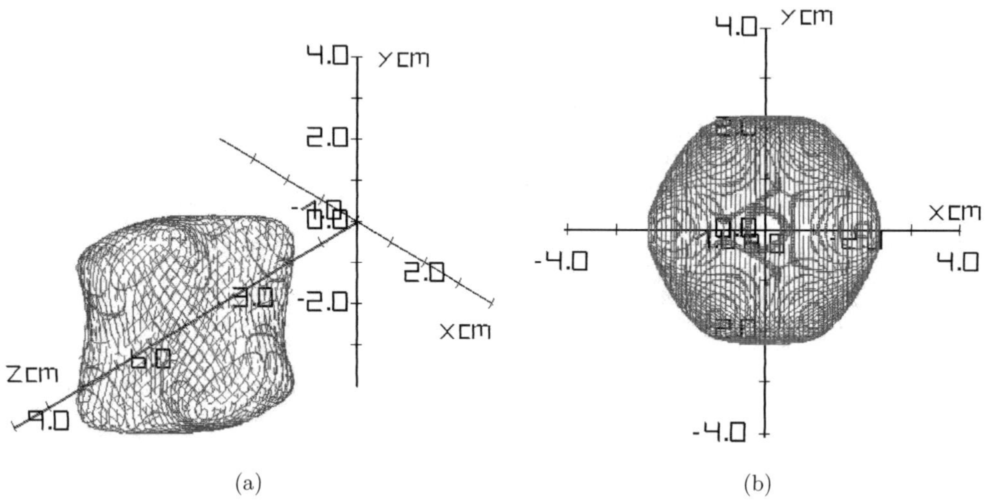

Figure 4. (*a*) 3D view of the ECR Zone. (*b*) The ECR zone in the cross section $z = L_c/2$.

4. Conclusions

By choosing the trap magnetic field shape of ECR ion source, we proceed from the microwave field distribution in the TE_{111} cavity because it is much easier to manipulate the magnetic field shape than the microwave distribution in a cavity. It is shown that in the chosen minimum-B trap the ECR zone must be of an ellipsoidal type shape. In the near future, the numerical simulation of plasma heating in a minimum-B trap by the particle-in-cell method will be realized.

IOP Conf. Series: Journal of Physics: Conf. Series **935** (2017) 012073 doi:10.1088/1742-6596/935/1/012073

Acknowledgments
This work is supported by "Facultad de Ciencias and Escuela de Física, Universidad Industrial de Santander (UIS), Colombia".

References

[1] Geller R 1996 *Electron cyclotron resonance ion sources and ECR plasmas* (Bristol: Institute of Physics Publishing)
[2] Chen F F 2012 *Introduction to plasma physics* (New York: Springer International Publishing AG)
[3] Brown I G 2004 *The physics and technology of ion sources* (Germany: John Wiley & Sons)
[4] Kawai Y, Ikegami H, Sato N, Matsuda A and Uchino K 2010 *Industrial plasma technology: Applications from environmental to energy technologies* (Germany: John Wiley & Sons)
[5] Nakamura T and Riedmüller W 1974 *AIAA Journal* **12** 661–668
[6] Xie Z Q 1998 *Review of Scientific Instruments* **69** 625–630
[7] Dougar-Jabon V D, Umnov A M and Suescun Diaz D 2002 *Review of Scientific Instruments* **73** 629–631
[8] Dougar Jabon V D, Vivas F, Chacon Velasco A J and Umnov A M 1998 *Review of Scientific Instruments* **69** 671–673
[9] Dougar-Jabon V D, Umnov A M and B K V 1996 *Review of Scientific Instruments* **67** 1152–1154
[10] Murillo M T and Otero O 2016 vol 687 p 012022
[11] Burau H, Widera R, Honig W, Juckeland G, Debus A, Kluge T, Schramm U, Cowan T E, Sauerbrey R and Bussmann M 2010 *IEEE Transactions on Plasma Science* **38(10)** 2831–2839
[12] Taflove A and Hagness S C 2005 *Computational electrodynamics: The finite-difference time-domain method* (USA: Artech house Inc.)
[13] Yee K 1966 *IEEE Transactions on antennas and propagation* **14** 302–307
[14] Acevedo Murillo M T, Dugar-Zhabon V D and Otero O 2016 *Journal of Physics: Conference Series* vol 687 (IOP Publishing) p 012023
[15] Dugar-Zhabon V D, González J D and A O E 2016 vol 687 p 012077

IOP Conf. Series: Journal of Physics: Conf. Series **935** (2017) 012074 doi:10.1088/1742-6596/935/1/012074

Adsorption of arsenate on Fe-(hydr)oxide

N Y Acelas[1] and E Flórez[1]
[1] Universidad de Medellín, Medellín, Colombia

E-mail: nyacelas@udem.edu.co

Abstract. Adsorption using metal oxide materials has been demonstrated to be an effective technique to remove hazardous materials from water, due to its easy operation, low cost, and high efficiency. The high number of oxyanions in aquatic ecosystems causes serious pollution problems. Removal of arsenate ($H_2AsO_4^-$), is one of the major concerns, since it is a highly toxic anion for life. Within the metal oxides, the iron oxide is considered as a suitable material for the elimination of oxyanions. The adsorption of $H_2AsO_4^-$ on Fe-(hydr)oxide is through the formation of inner or outer sphere complexes. In this work, through computational methods, a complete characterization of the adsorbed surface complexes was performed. Three different pH conditions were simulated (acidic, intermediate and basic), and it was found that, the thermodynamic favourability of the different adsorbed complexes was directly related to the pH. Monodentate complex (MM_1) was the most thermodynamically favourable complex with an adsorption energy of -96.0kJ/mol under intermediate pH conditions.

1. Introduction

Arsenic (As) is mobilized to the environment through natural processes (weathering, biological activities, volcanic emissions), and anthropogenic (mining, fossil fuels, pesticides, etc.) and, even at low concentrations is toxic to humans [1]. The presence of arsenic in surface (rivers, lakes, reservoirs) and underground waters (aquifers) constitutes a major threat to health. Therefore, the World Health Organization (WHO), the European Union (Directive 98/83) and the US Environmental Protection Agency (USEPA) have established a reduction of the permitted limit of arsenic in drinking water from 50 a 10μg/l [2].

Arsenic is mainly found as the oxyanion, $H_2AsO_4^-$, when the pH value in water is <6.9. An efficient method to control the contamination of aquatic ecosystem caused by arsenate is to binding it at water/mineral interfaces. Infrared (IR) spectra is particular useful to explore protonation and adsorption mechanism at solid/liquid interfaces [3]. Previous studies have indicated that the IR vibrational spectra of arsenate on several minerals (such as goethite, ferrihydrite and anatase) contain abundant coordination information [4,5]. It has been reported that the predominate adsorption species when arsenate interacts with hydrated iron oxides are inner sphere complexes [6,7]. However, the exact geometry of the surface complex (monodentate or bidentate) is unknown.

There is still some discussion on this topic and the effect of factors such as, pH and thermodynamic favourability of each surface complex is not clear. Therefore, computational methods are an alternative to study adsorption problems [8-10]. The goals of this work are: (i) to help in the understanding of the factors driving thermodynamic preferences for the adsorption of arsenate over Fe-(hydr)oxide and (ii) to understand the modes of adsorption in inner and outer sphere complexes as a function of pH.

IOP Conf. Series: Journal of Physics: Conf. Series **935** (2017) 012074 doi:10.1088/1742-6596/935/1/012074

2. Computational details and models

Following models previously worked by some authors [8,9] the iron oxide was simulated using two iron atoms in octahedral coordination with 10 atoms of oxygen. In order to simulate various pH conditions, the -OH^-/-H_2O ratios was adjusted in the initial clusters as follows: cluster with 4/6 OH^- to H_2O with net charge of +2 correspond to acidic pH; 1/1 ratio have net +1 charges corresponding to intermediate pH conditions, and basic conditions correspond to those with the 6/4 ratio and a net charge of 0. The Gibbs free energies of adsorption (equation 1) for the different surface complexes under the three simulated pH conditions were estimated using a series of stoichiometric ratios.

$$\Delta G_{ads} = G_{complex} - [G_{H_2AsO_4^-} + G_{cluster}] \quad (1)$$

The specie of Arsenate used for the adsorption process was the oxyanion, $H_2AsO_4^-$. To simulate its molecular properties in aqueous phase it was microsolvated with six water molecules. The other species involved in the adsorption processes such as: $[OH(H_2O)_7]^-$, $[(OH)_2(H_2O)_6]^{2-}$, $[OH(H_2O)_6]^-$, $(H_2O)_8$, $(H_2O)_7$, and $(H_2O)_6$ were obtained by the use of the ASCEC algorithm, a stochastic-like search strategy [11]. The optimization of each complex under different pH conditions was carried out using the PBE0 functional and the DEF2TZVP basis set on O, H, and as atoms. The LANL2DZ relativistic electron core potential (RECP) was used for Fe. To verify that the optimized structures were minimal in the potential energy surface harmonic vibrational frequencies were calculated (i.e., no imaginary frequencies). Taking into account that the adsorption process occurs in aqueous medium, the short-range solvent effect was simulated by the inclusion of six molecules of water during the optimization process of each specie. We have shown before [8] that long–range solvent effects heavily influence adsorption free energies. To account long-range solvent effect polarized continuum model (IEFPCM) was used. All calculations were performed with the Gaussian 09 program [12]. In our previous work [9] it was shown that for clusters containing Fe^{3+} atoms, their stability is favoured in high spin states and therefore, all the calculations carried out here were made in the cluster of iron in multiplicity of 11.

3. Results and discussion

Figure 1 shows the clusters used to simulate the different adsorption processes of arsenate anion under the three simulated pH conditions. All the geometric parameters in our calculations for the Fe-(hydr) oxide clusters are in very good agreement with the experimental geometric parameters reported for goethite [13]. It indicates that this is a suitable model to do the simulations.

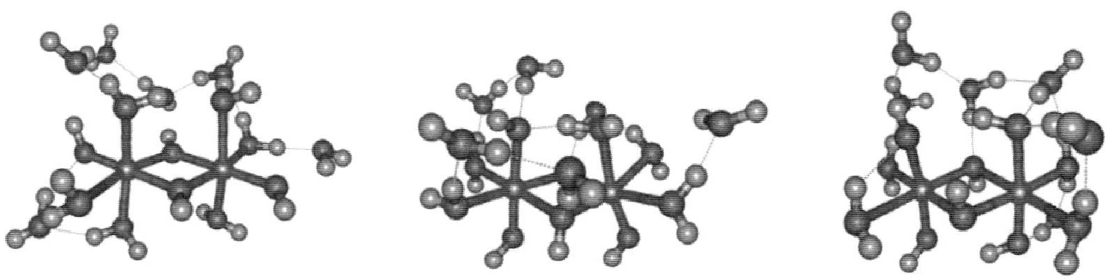

Acidic: $[Fe_2(OH)_4(H_2O)_6 \cdot (H_2O)_6]^{2+}$ **Intermediate:** $[Fe_2(OH)_5(H_2O)_5 \cdot (H_2O)_6]^{+}$ **Basic:** $[Fe_2(OH)_6(H_2O)_4 \cdot (H_2O)_6]^{0}$

Figure 1. DFT-calculated clusters of Fe-(hydr) oxide under acidic (charge+2), intermediate (charge +1), and basic (charge 0) pH conditions. Red, blue, and grey spheres denote O, Fe, and H atoms, respectively. The total spin multiplicity of 11 was considered for all cases.

3.1. Formation of surface complexes

In the literature there are several reports related to the surface models for arsenate complexation on oxides [1,14]. Inner sphere complexes, Monodentate Mononuclear (MM) and Bidentate Binuclear (BB) have been reported as the most likely adsorption modes of arsenate species when they are coordinated

IOP Conf. Series: Journal of Physics: Conf. Series **935** (2017) 012074 doi:10.1088/1742-6596/935/1/012074

with metal oxides [14]. Figure 2 shows the adsorption reaction to form the inner sphere complex, BB, under acidic pH conditions.

The arsenate adsorption on Fe-(hydr)oxide as inner sphere complexes (BB or MM) can be explained as a ligand exchange process between the oxide surface functional groups and the oxygen atoms of the anion. BB complexes are formed by exchanging two functional groups present on the oxide surface while MM complexes are formed by the exchange of one functional group of the oxide surface.

The functional group present on the oxide surface depends on the pH conditions, as follows: under acidic pH conditions $-H_2O$ functional groups will be exchanged; at basic pH conditions the exchanged will be with only $–OH^-$ functional groups and at intermediate pH conditions, both groups $-H_2O$ and $-OH^-$ are exchanged, which leads to the possibility to form two types of monodentate complexes, MM_1 complexes if the group $-H_2O$ is exchanged and MM_2 with the exchange of $-OH^-$ group.

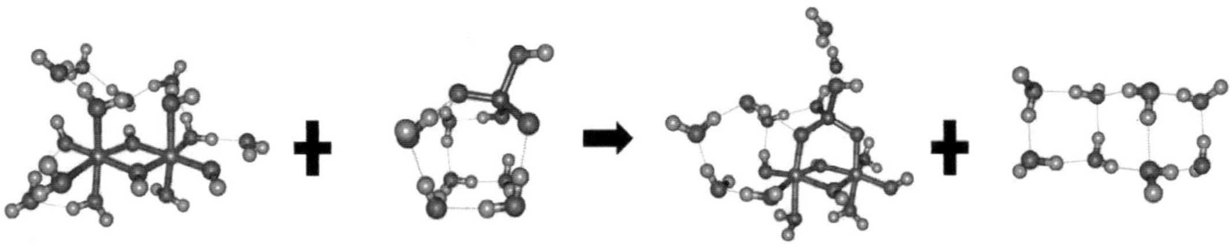

$[Fe_2(OH)_4(H_2O)_6.(H_2O)_6]^{2+}$ $[AsO_2(OH)_2(H_2O)_6]^-$ $[Fe_2(OH)_4(H_2O)_4.AsO_2(OH)_2(H_2O)_6]^{+1}$ $(H_2O)_8$

Figure 2. Adsorption of arsenate as bidentate complex on Fe-(hydr)oxide under acidic pH conditions. Red, blue, green, and grey denote O, Fe, As and H atoms, respectively.

The adsorption of arsenate as outer-sphere complex (H-bonded), it is suggested that occurs by intermolecular interaction (hydrogen bonds) of the reactive oxygen atoms of arsenate and the surface functional groups of the oxide, without ligand exchange. Similar pathways have been reported in the literature for the adsorption of oxoanions on metal oxides [1,8-10]. Figure 3 shows optimized structures of all inner and outer sphere complexes.

The prediction of the thermodynamic favourability of the different adsorption modes by experimental tools is difficult and through IR spectroscopy is possible to obtain a mix of all adsorption modes [15]. Therefore, these experimental models need to be complemented by computational results. Surface complexation modelling [1,8-10,14,16] has the capability to predict the behaviour and the nature of adsorbed arsenate species as a functions of environmental parameters, such as pH.

3.2. Gibbs free energies of adsorption

It is important to determinate the Gibbs free energy to identify the most favourable adsorption mode under certain experimental conditions. Gibbs free energy was calculated according to equation (1), by subtracting the total energy of the reactant species from the total energy of the product species in aqueous phases (Figure 3).

Figure 4 shows the energy curves of arsenate adsorption on Fe-(hydr)oxide. It can be seen that the thermodynamic favourability for adsorption as an inner and outer sphere complexes is directly related to pH. Under acidic pH conditions both types of adsorption are energetically favourable from -87.01kJ/mol to -37.03kJ/mol, whereas at basic pH conditions the thermodynamic favourability for the different adsorption modes decreases. For BB complex an adsorption energy of +36.10kJ/mol was found.

From Figure 4, it can be concluded that although at intermediate pH conditions the most thermodynamically favourable compound (MM_1=-96.00kJ/mol) is found, under acidic pH conditions there are high variety of favourable complexes. BB complex formed under acidic pH condition is only +9kJ/mol more energetic that the MM_1 complex generated under intermediate pH conditions. Our results are in agreement with those reported by some authors as follows: (i) Ladeira *et. al.* 2001. [17] demonstrated that during the adsorption of arsenate on aluminium oxide clusters, the BB specie was more stable than the MM complexes. (ii) Sherman and Randall, 2003. [18] using DFT and EXAFS data,

IOP Conf. Series: Journal of Physics: Conf. Series **935** (2017) 012074 doi:10.1088/1742-6596/935/1/012074

suggested that the adsorption of arsenate on iron hydroxide clusters was as a BB complex. And, (iii) Kubicki, 2005 [19] showed by using computational methods that the formation of BB complexes during the adsorption of arsenate on iron hydroxide clusters resulted in good agreement with spectroscopic data.

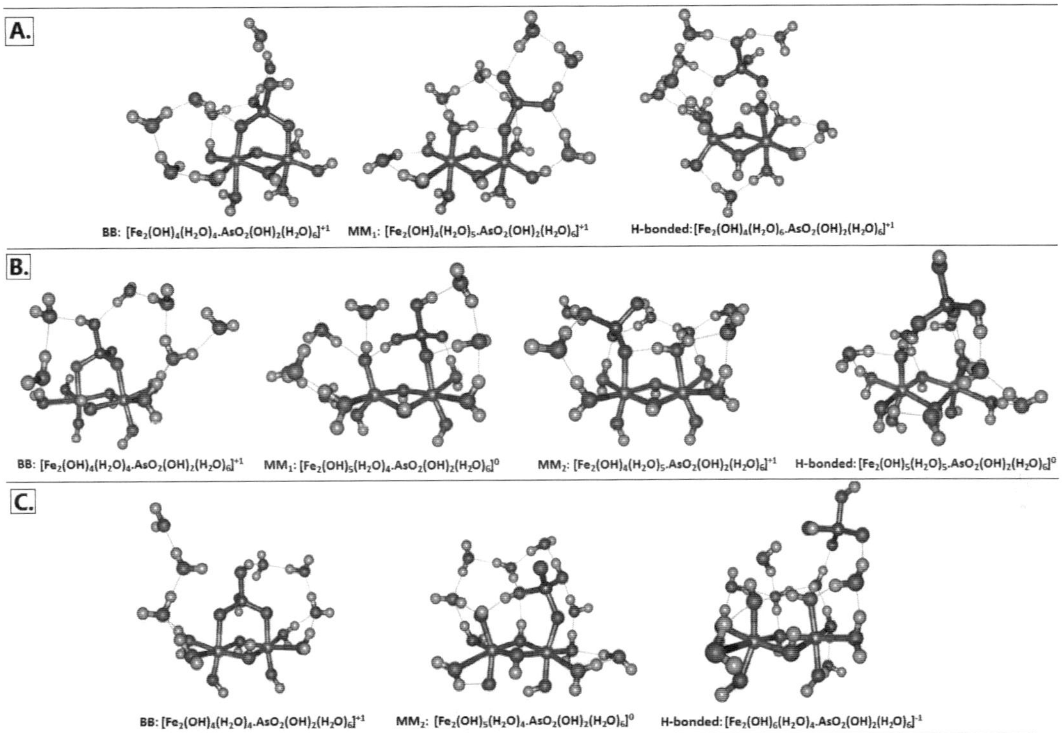

Figure 3. DFT-calculated structures of inner and outer sphere adsorption products of arsenate on Fe-(hydr)oxide (a) acidic; (b) intermediate; and (c) basic pH conditions. Note: Red, blue, green, and grey denote O, Fe, As and H atoms, respectively. MM_1: Monodentate mononuclear complex exchanged to one -H_2O surface functional group and MM_2: Monodentate mononuclear complex exchanged to one –OH^- surface functional group.

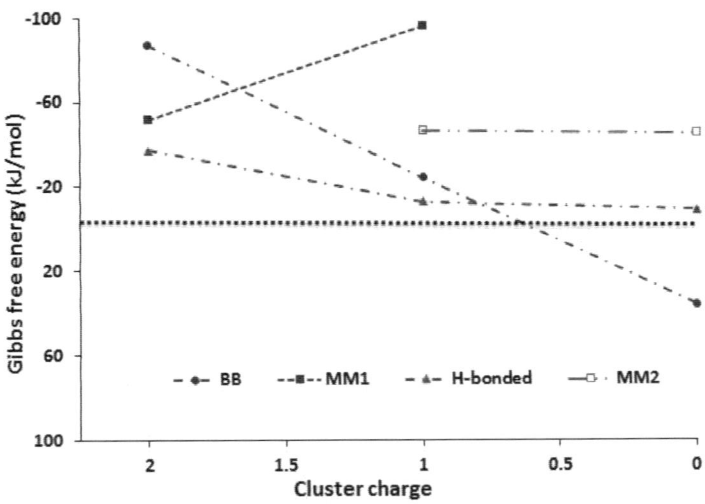

Figure 4. Gibbs free energies of arsenate adsorption on various protonated Fe-(hydr)oxide surfaces at different pH conditions. MM_1: Monodentate mononuclear complex exchanged to one -H_2O surface functional group and MM_2: Monodentate mononuclear complex exchanged to one –OH^- surface functional group.

IOP Conf. Series: Journal of Physics: Conf. Series **935** (2017) 012074 doi:10.1088/1742-6596/935/1/012074

According to above, it is possible to conclude that during the adsorption process of arsenate on Fe-(hydr)oxide a mixture of thermodynamically favourable surface complexes is generated and computational tools allow to characterize them separately. BB configuration is most consistent with spectroscopic data, but Gibbs free energies of adsorption suggested that the MM_1 configuration is energetically more stable.

4. Conclusion

In this work, a characterization of the potential energy surface for the adsorption of arsenate on Fe-(hydr)oxide surfaces under aqueous environments at three different pH conditions (acidic, intermediate and basic) was made. Our results show that monodentate (MM_1) complex under intermediate pH conditions and bidentate complex (BB) under acidic pH conditions are preferred, and we support this claim by the calculated adsorption energies of -96.00kJ/mol and -87.01kJ/mol for MM_1 and BB, respectively.

Acknowledgments

The authors are grateful to "Universidad de Medellín", for financing the project. N Y Acelas thanks "Colciencias" for the PhD scholar-ship.

References

[1] Otte K, Schmahl W and Pentcheva R 2013 *J. Phys. Chem. C* **117(30)** 15571
[2] Grossl P et al 1997 *Environ. Sci. Technol.* **31(2)** 321
[3] Pena M et al 2006 *Environ. Sci. Technol.* **40(4)**1257
[4] Jia Y et al 2007 *Geochim. Cosmochim. Acta.* **71(7)** 1643
[5] Arai Y and Sparks D 2001 *J. Colloid Interface Sci.* **241(2)** 317
[6] Waychunas G, Davis J and Fuller C 1995 *Geochim. Cosmochim. Acta* **59(17)** 3655
[7] Waychunas G et al 1993 *Geochim. Cosmochim. Acta.* **57(10)** 2251
[8] Acelas N et al 2013 *Comp. Theor. Chem.***1005** 16
[9] Acelas N et al 2017 *Inorg Chem.* **56(9)** 5455
[10] Acelas N and Flórez E 2017 *Desalination Water Treat.* **60** 88
[11] Pérez J 2008 *ASCEC V-02* Theoretical Chemical Physics Group (Medellín: Universidad de Antioquia)
[12] Frisch M et al 2009 *Gaussian 09* I W Revision D 01
[13] Bargar J et al 2000 *Geochim. Cosmochim. Acta.* **64(16)** 2737
[14] Fukushi K and Sverjensky D 2007 *Geochim. Cosmochim. Acta.* **71(15)** 3717.
[15] Adamescu A et al 2009 *Can. J. Chem.* **88(1)** 65
[16] Adamescu A et al 2011 *Environ. Sci. Technol.* **45(24)** 10438
[17] Ladeira A et al 2001 *Geochim. Cosmochim. Acta.* **65(8)** 1211
[18] Sherman D and Randall S 2003 *Geochim. Cosmochim. Acta.* **67**(22) 4223
[19] Kubicki J 2005 *J. Am. Chem. Soc.* **915** 104

IOP Conf. Series: Journal of Physics: Conf. Series **935** (2017) 012075 doi:10.1088/1742-6596/935/1/012075

Effect of conduction band nonparabolicity on Aharonov-Bohm oscillations in n-type InAs/GaAs quantum ring

L F García[1], **I D Mikhailov**[1] **and H Paredes**[1]
[1] Universidad Industrial de Santander, Bucaramanga, Colombia

E-mail: lgarciar@uis.edu.co

Abstract. We study the effect of nonparabolicity of the conduction band on the energy spectrum of the off-axis donors in InAs crater-like quantum rings in framework of the Kane Hamiltonian. By using ta simple numerical procedure we find the low-lying energies dependencies on the radius, height and width of the ring and on the distance of the donor location from the axis. The comparison of our results with those obtained in one-band effective mass approximation reveals an essential influence of the non-parabolicity of the conduction band on the off-axis energy spectra in InAs/GaAs quantum rings. In addition, we analyse the effect of the external magnetic field applied along the symmetry axis on the Aharonov-Bohm oscillations of lower energy levels.

1. Introduction

A carefully controlled introduction of impurities into a semiconductor crystal allows us modify their properties in order to use them in semiconductor science and technology. The theory of electronic states of a hydrogenic donor impurity in the semiconductor bulk developed earlier in framework the Effective Mass Approximation (EMA) gives an excellent accordance with the experimental results [1]. Later, one-band EMA theory has been widely used for calculating the electronic states of a hydrogenic donor impurity in quantum wells [2-4], quantum well wires [5-6] and quantum dots [7-9]. In this paper, we analyse the applicability of one-band EMA for hydrogenic impurity placed at self-assembled quantum dots and quantum rings (QRs), where the excessive electron released from the donor to the conduction band is strongly confined within a narrow region and therefore can has a large kinetic energy. Under these conditions, it is important to take into account non-parabolicity of the conduction band, which is essential especially for semiconductors with a small gap between the conduction and the valence bands, such as the *InAs* nanostructures. A simplest way to analyse the effect of the non-parabolicity related to the mixing of the valence and the conduction bands consists in application of the Kane model, in which it is considered the Kp-mixing of the states in the conduction band Γ_6 and in the valence bands Γ_8, Γ_7 but it is neglected the impact of the faraway valence bands [10]. Below, we by using this model analyse the energies of the donor impurity placed in an axially symmetrical volcano-shaped QR as functions of the donor position, the geometrical parameters of the nanostructure and the external magnetic field applied along the symmetry axis.

2. Theoretical model

Although the method used below is applicable for quantum dots of arbitrary geometry, we consider here a simple model of an embedded in *GaAs* matrix axially symmetrical crater-like *InAs* nanostructure, schematically presented in Figure 1, whose radial dependence of the thickness $h(\rho)$ is:

IOP Conf. Series: Journal of Physics: Conf. Series **935** (2017) 012075 doi:10.1088/1742-6596/935/1/012075

$$h(\rho) = d_0 \Big/ \left[1 + (\rho - R_0)^2 / w^2\right] \tag{1}$$

Here d_0, R_0 are the thickness and the radius of the crater's rim, R_h, d_h are the radius and the thickness of the central hole, and, w is a parameter that characterizes the width of the crater. In the model, B is the homogeneous magnetic fields directed along the symmetry axis Z. We simulate the *InAs*/*GaAs* structure by using the material parameters for *InAs* inside the layer are $E_g^{(1)} = 0.42\ eV$, $\Delta_1 = 0.38\ eV$ and $m_1^* = 0.024 m_0$. The parameters for *GaAs* surrounded matrix are $E_g^{(2)} = 1.52\ eV$, $\Delta_2 = 0.34\ eV$ and $m_2^* = 0.067 m_0$, whereas the mismatch of the band gap at the junction is equal to $V_0 = 0.7\left(E_g^{(2)} - E_g^{(1)}\right) = 0.77 eV$ [11-12]. In what follows, we use as dimensionless units of the length and the energy, the effective Bohr radius $a_0^* = 4\pi\varepsilon_0\varepsilon\hbar^2 / m_1^* e^2 \approx 28.65$ nm and the effective Rydberg $R_y^* = e^2 / 8\pi\varepsilon_0\varepsilon a_0^* \approx 1.93 \cdot 10^{-3} eV$, respectively, and the magnetic field strengths $\gamma = e\hbar B / 2m_1^* R_y^* \approx 1.14 B(T)$, being $\varepsilon = 12.5$ the dielectric constant of *InAs* material.

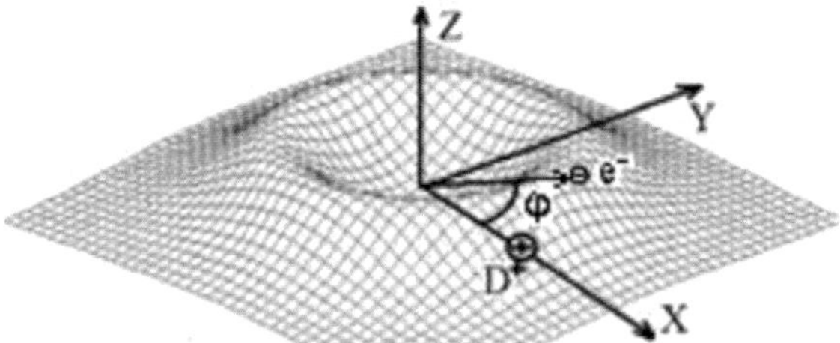

Figure 1. Model of the uniform QR.

Earlier, we show that for actual non-homogeneous thin *InAs*/*GaAs* semiconductor layers with $d_0 \ll R_0$ the wave functions of carrier confined satisfy a two-dimensional Schrödinger equation with the adiabatic potential, given in the dimensionless units by the relation [6]:

$$V_a(\rho) \approx 9.5 / h(\rho) = 9.5 \cdot d_0^{-1} + 9.5 \cdot (\rho - R_0)^2 / d_0 \cdot w^2 \tag{2}$$

The corresponding differential equation in polar coordinates for the excessive electron released from the donor placed at the point separated from the symmetry axis at the distance R_i then in the framework of the Kane model we represent in the following form [6]:

$$\begin{gathered} H_{fast}\psi(\rho,\varphi) + H_{slow}\psi(\rho,\varphi) = \eta(E) E\psi(\rho,\varphi);\ \eta(E) = \frac{\left(E/E_g^{(1)} + 1\right)\left[E\Big/\left(E_g^{(1)} + \Delta^{(1)}\right) + 1\right]}{E\Big/\left(E_g^{(1)} + 2\Delta^{(1)}/3\right) + 1}; \\ H_{fast} = -\frac{\partial^2}{\partial\rho^2} - \frac{1}{\rho}\frac{\partial}{\partial\rho} + \frac{\gamma^2\rho^2}{4} + V_a(\rho);\ H_{slow} = -\frac{1}{\rho^2}\frac{\partial^2}{\partial\varphi^2} - i\gamma\frac{\partial}{\partial\varphi} - \frac{2}{\sqrt{\rho^2 + R_i^2 - 2\rho \cdot R_i \cos\varphi}} \end{gathered} \tag{3}$$

The first term in the Hamiltonian (3) describes the fast oscillation of the electron in the radial direction about the crater's rim with the frequency $\omega_{fast} \approx \sqrt{\gamma^2/4 + 9.5/d_0 \cdot w^2}$, while the second term corresponds to the slow non-uniform rotation around the symmetry axis with the frequency $\omega_{slow} \approx 1/R_0$. For typical QRs dimensions, $R_0 \geq 1a_0*$, $d_0 \leq 0.1a_0*$; $w \leq 0.4a_0*$ the slow frequency $\omega_{slow} \leq 1$ is essentially smaller than fast frequency $\omega_{fast} \geq 25$ and therefore one can use the following two-step adiabatic procedure, for the wave function, presented as $\psi(\rho,\varphi) = f_0(\rho)\Phi(\varphi)$:

$$H_{fast}(\rho) f_0(\rho) = E_0 f_0(\rho) \Rightarrow \overline{H}_{slow}(\varphi) = \langle f_0(\rho) | H_{slow}(\rho,\varphi) | f_0(\rho) \rangle + E_0 \tag{4a}$$

$$\overline{H}_{slow}(\varphi)\Phi(\varphi) = \eta(E_D) E_D \Phi(\varphi) \tag{4b}$$

At the first step of this procedure, one should find the exact energy E_0 and the wave function $f_0(\rho)$ of the ground state of two-dimensional harmonic oscillator with the Hamiltonian H_{fast}. The function $f_0(\rho)$ one can use then for finding the averaged Hamiltonian $\overline{H}_{slow}(\varphi)$ whose eigenvalues give low-lying energies E_D of the donor corresponding to different rotational states. In our numerical procedure, we solve equation (4b) by using the Fourier series expansion method.

3. Results and discussions

In order to analyse the effect of the non-parabolicity on the spectral properties of the electron of the excessive electron released from the donor to the conduction band and of the QR we have performed numerical calculations for low-lying energy levels as functions of the external magnetic field. Below we present results only for QRs with two different rim's radii radius $R_0 = 20nm$ and $R_0 = 40nm$, the rim height $d_0 = 4nm$ and the rim's width and $R_e = 8nm$. In Figure 2, we display the lower energies of the electron, released from the donor to the *InAs* conduction band as functions of the external magnetic field for QR for three different distances from the donor location up to the symmetry axis, $R_i = 0$ $R_i = 12.5nm$, and $R_i = 20nm$, respectively. In the upper part of Figures 2(a, b, c) we show energies calculated in the one-band parabolic approximation, assuming $\eta(E_D) = 1$ in equation (4b), whereas in the lower part in Figures 2(d), (e) and (f) there are the energies obtained for the Kane model.

Is to observe various essential differences between energies dependencies calculated in EMA and obtained for the Kane model. First, the energies obtained for the Kane model are significantly lower than those calculated in EMA. The difference is about $300Ry*$ or $0.6eV$. This large difference is due to the effect of the conduction band nonparabolicity. While the in the parabolic model the confinement may increase the kinetic energy without any limits, in framework of the Kane model such increase is limited. The larger the kinetic energy, the larger becomes the effective mass, resulted in suppressing of the effect of the confinement. That is, the effect of the non- parabolicity of the conduction band provides an essential lowering of the electron energies in nanostructures and diminishes significantly any quantum size effect related to the quantum confinement.

Second, for on-axis donor $(R_i = 0)$ in Figure 2(a, d) the period of the Aharonov-Bohm oscillations is the same for both models and satisfies well-known relation $\Delta\gamma = 2/R_0^2$, but the amplitudes of the oscillations and gaps between sublevels in the Kane model are notably lower. Due to the presence of the axial symmetry, in this case, the confinement potential along the crater's rim is constant, and all solutions of the corresponding wave equation are periodic and all states are rotational. The periodic splitting and the degeneration of rotational states in opposite direction in the presence of the magnetic field conduces to the Aharonov-Bohm oscillations of the energy levels. As the donor is displaced from the axis, breaking the axial symmetry in the confinement potential along the crater's rim appears a quantum well provided by the electrostatic attraction to the donor, and some lower states becomes localized, i.e. are transformed from a rotational state to a vibrational. In Figures 2(b) and (c) one can observe that the lowest rotational state splits from other sublevels and the gap grows substantially as the donor approaches to the crater's rim making the quantum well more and more profound. A similar process also occurs in Figures 2(e)-(f) obtained for Kane model, but the splitting in this case is notably smaller.

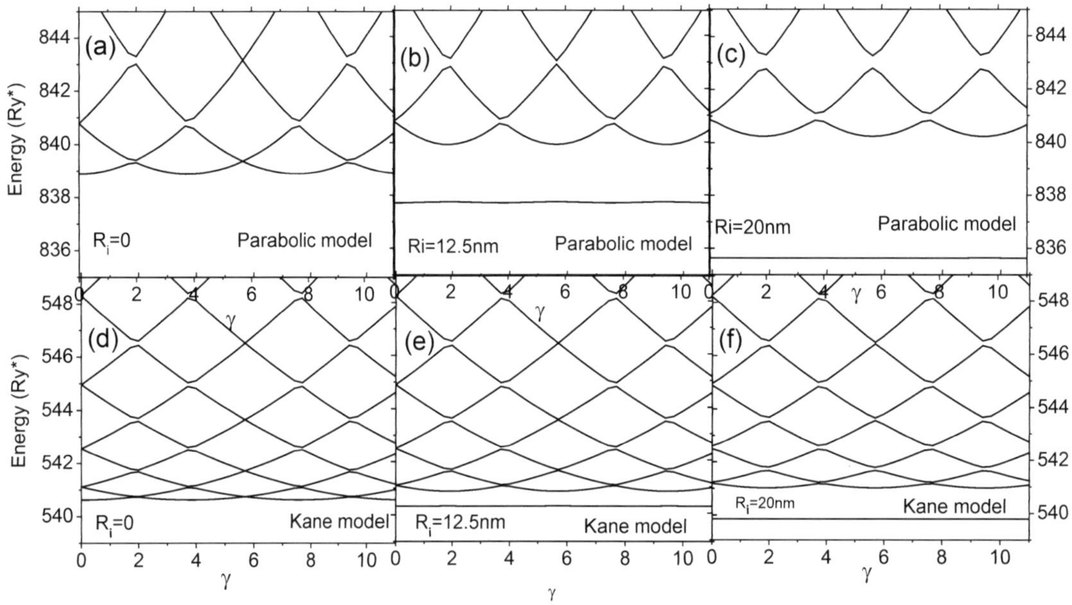

Figure 2. Donor energies, counted from the top of the valence band, as functions of the external magnetic field in InAs/GaAs QRs with the crater's rim radius $R_0 = 20nm$ and three different distances R_i from the donor location up to the symmetry axis, calculated for one-band parabolic model (a), (b), (c) and for the Kane model (d), (e), (f).

One of principal factors that defines spectral of the system is the interplay between the electrostatic attraction, which holds the electron and donor together, and the structural confinement that keeps electron in a rotational state along the crater's rim within a narrow circular region. The contribution of the electrostatic potential energy is inversely proportional tor rim's radius $V_c \propto 1/R_0$ while the contribution of the rotational kinetic energy $K \propto 1/R_0^2$. As the radius R_0 increases, the contribution of the electron-donor attraction becomes dominant, furthering the localization of the rotational states. Therefore, one could expect a modification of the Aharonov-Bohm oscillations with an increase of the dimension of the QR.

In Figure 3, we present donor's energies dependencies on the magnetic field for QR with double crater's rim radius $R_0 = 40nm$. It can be seen that a double increase in the radius leads to a fourfold decrease of periods and the amplitudes of the Aharonov-Bohm oscillation in all cases independently of the distance from the donor location from the symmetry axis. It is due to the decrease of rotational kinetic energy. Nevertheless, the gaps between the vibrational states (levels, splitting downwards) and the lowest rotational state has not almost changed for all off-axis donor positions. With the rise of the QR dimension, the relative value of gaps for localized states increases, that agrees with the growing contribution of the electrostatic attraction. In addition, one can observe in Figures 3(c)-(f) unlike of Figures 2(c)-(f) the appearance of the second localized level.

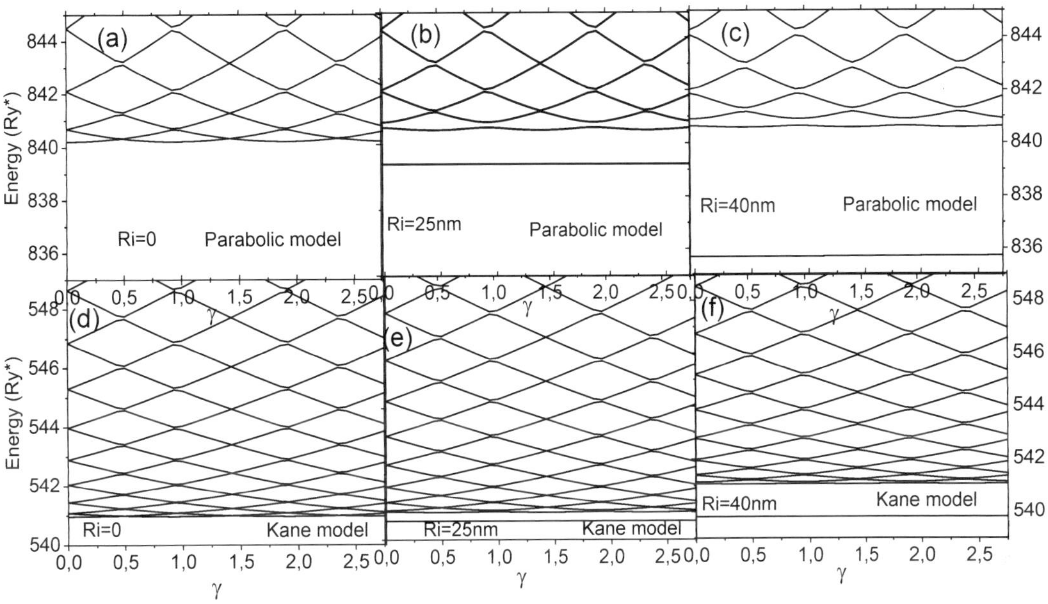

Figure 3. The same as in Figure 2, but for QRs with the crater's rim radius $R_0 = 40nm$.

4. Conclusion

we propose a simple method in framework of the Kane model for theoretical analysis of the effect of non-parabolicity on the energies of off-axis donor impurities introduced in InAs/GaAs crater-shaped quantum rings. Our analysis shows essential changes that introduce the non-parabolicity of the conduction band in donor impurity energies, obtained by means the effective mass approximation, lowering significantly energy levels, diminishing the amplitudes of the Aharonov-Bohm oscillations and gaps between sublevels, and suppressing the effect of the localization of the electron rotation along the crater's rim. Our method can be widely applied in the calculation of the electronic states of hydrogenic donor impurities in nanostructures of other different shapes.

Acknowledgements

This work was financed by the «Universidad Industrial de Santander (UIS) » through the «Vicerrectoría de Investigación y Extensión (VIE), DIEF de Ciencias (Cod. 5124)» and the «Patrimonio autónomo del Fondo Nacional de financiamiento para la ciencia, la tecnología y la innovación Francisco José De Caldas-Contrato RC– No. 275-2011» and the Cod. No. 1102-05-16923 subscribed with «Colciencias».

References

[1] W Kohn and J M Luttinger 1955 *Phys. Rev.* **98** 915
[2] G Bastard 1981 *Phys. Rev.* B **24** 4714
[3] C Mailhiot, Y C Chang, T C McGill 1982 *Phys. Rev.* B **26** 4449
[4] R L Greene, K K Bajaj 1985 *Phys. Rev.* B **31** 913
[5] G W Bryant 1985 *Phys. Rev.* B **31** 7812
[6] A Latge, N Porras-Montenegro, L E Oliveira 1992 *Phys. Rev.* B **45** 6742
[7] C Bose, K Sarkarc 1998 Physica B **253** 238
[8] N Porras-Montenegro, S T Perez-Merchancano, A Latge 1993 *J. Appl. Phys.* **74** 7624
[9] I D Mikhailov, L F García, J H Marín 2006 *J. Phys.: Condens. Matter.* **18** 9493
[10] E Kane 1957 *J. Phys. Chem. Solids* **1(4)** 249

IOP Conf. Series: Journal of Physics: Conf. Series **935** (2017) 012075 doi:10.1088/1742-6596/935/1/012075

[11] L F García, S Yu Revinova, I D Mikhailov 2015 *Physica E* **71** 101
[12] L F García, S Yu Revinova, I D Mikhailov 2017 *Superlattices and Microstructures* **103** 151

IOP Conf. Series: Journal of Physics: Conf. Series **935** (2017) 012076 doi:10.1088/1742-6596/935/1/012076

Electrons acceleration in a TE_{113} cylindrical cavity affected by a static inhomogeneous magnetic field

V E Vergara[1], **J D González**[1], **J R Beltrán**[1] **and E A Orozco**[2]

[1] Universidad del Magdalena, Santa Marta, Colombia
[2] Universidad Industrial de Santander, Bucaramanga, Colombia

E-mail: jgonzaleza@unimagdalena.edu.co

Abstract. The relativistic dynamics of an electron accelerated in a cylindrical cavity mode TE_{113} in the presence of a static inhomogeneous magnetic field is studied. This type of acceleration is known as Spatial AutoResonance Acceleration (SARA). The magnetic field profile is such that it keeps the phase difference between the electric microwave field and the electron velocity vector within the acceleration phase band. We study the dynamic of the electron through simulations of the relativistic Newton-Lorentz equation. Numerical experiments with TE_{113} cylindrical microwave fields of a frequency of $2.45GHz$ and an amplitude of the order of $7kV/cm$ show that it is possible accelerate the electrons up to energies of the order of $300keV$. This energy is about of 30% higher than those obtained in previous studies by using the TE_{112} mode.

1. Introduction

The study of cyclotron resonance has been extensively studied and has led to several mechanisms to accelerate charged particles [1–13]. The spatial autoresonance acceleration SARA [7–10,12,13] has emerged as an excellent option to accelerate electric charged particles by using cylindrical TE_{11p} or rectangular TE_{10p} modes ($p = 1, 2, ..$). For example, particles that reach energies of the order of 230 keV by using cylindrical TE_{112} mode or rectangular TE_{102} mode have alrady been reported [7, 12, 13]. An X ray source based on in the SARA concept has been certificated recently [14].

In the SARA mechanisms the axial motion is restricted by the diamagnetic force limiting the energy which can be achieved. Such force arise from the longitudinal inhomogeneity of the magnetostatic field. One way to reach higher value energies is to use higher order TE_{11p} or TE_{10p} modes. In this work we show that it is possible to accelerate an electrons beam up to energies of the order of $300keV$ by using a cylindrical TE_{113} mode of a frequency of $2.45GHz$ and an amplitude of the order of the order $7kV/cm$.

The manuscript is organized as follows. Section 2 represents our theoretical approach. Section 3 is about the results for the conditions selected for the autoresonant system. Our findings are summarized in section 4.

2. Theoretical formalism

The physical scheme of the autoresonant accelerator is shown in Figure 1. A magnetron emits $2.45GHz$ microwaves into the cylindrical cavity (2) through a waveguide (4).The magnetostatic field is generated by a set of four axisymmetric coils (3). An electron gun (1) inject a electons beam along the cavity axis, which describe a helical trajectory (5) which is afected by both the electric and the magnetic field inside the cavity.

Figure 1. Physical scheme of the autoresonant system: (1)-electron gun, (2)-cavity, (3)-magnetic coils system, (4)-magnetrón, (5)-Electron trajectory.

The electric and magnetic field components of a cylindrical TE_{113} mode circular polarized excited in the cavity are described by the following expresions:

$$E_r^{hf} = -2E_0 \frac{J_1(K_\perp r)}{K_\perp r} sen(K_\parallel z) cos(\theta - \omega t), \tag{1}$$

$$E_\theta^{hf} = 2E_0 J_1'(K_\perp r) sen(K_\parallel z) sen(\theta - \omega t), \tag{2}$$

$$B_r^{hf} = 2E_0 \frac{K_\parallel}{\omega} J_1'(K_\perp r) cos(K_\parallel z) cos(\theta - \omega t), \tag{3}$$

$$B_\theta^{hf} = -2E_0 \frac{K_\parallel}{\omega} \frac{J_1(K_\perp r)}{K_\perp r} cos(K_\parallel z) sen(\theta - \omega t), \tag{4}$$

$$B_z^{hf} = 2E_0 \frac{K_\perp}{\omega} J_1(K_\perp r) sen(K_\parallel z) cos(\theta - \omega t), \tag{5}$$

where $K_\perp = S_{11}/a$, $S_{11} = 1.8411$, $K_\parallel = 3\pi/d$. d and a are the length and radius of the cavity, respectively. E_0 is the electric field strength and

$$\omega = c(K_\parallel^2 + K_\perp^2)^{1/2} \tag{6}$$

is the angular frequency of the microwave field.

For our simulations we consider a cylindrical cavity whose length and diameter are $30cm$ and $9.1cm$, respectively. An appropiate magnetostatic field profile is obtained using the parameters showed in Table 1. where R_i, R_e, L_b , and z are the internal radius, the external radius, the width of each coil, and the positions of the coils, respectively, and J is the coil current density.

Table 1. Magnetic coil system parameters.

Bobina	$R_i(cm)$	$R_e(cm)$	$L_b(cm)$	$J(A/mm^2)$	$z(cm)$
1	6.0	20.0	6.0	1.39	−5.75
2	6.0	20.0	7.5	1.08	8.25
3	6.0	20.0	6.9	1.18	19.50
4	6.0	20.0	6.1	2.07	32.00

The two-dimensional profile of the magnetic field in units of $B_0 = m_e\omega/e = 0,0875T$ is shown in the Figure 2.

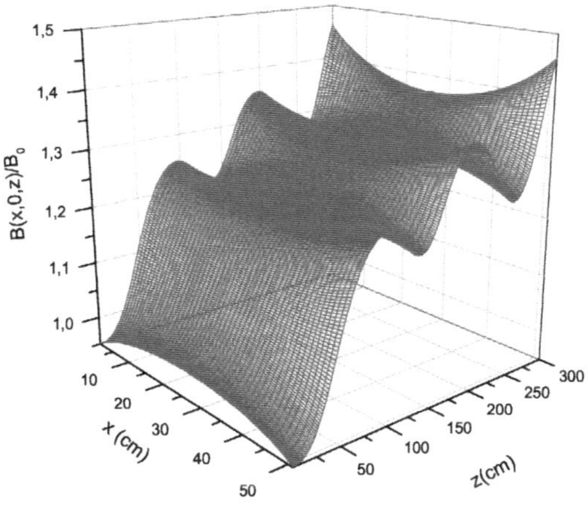

Figure 2. The profile of the magnetostatic field in the $y = 0$ plane.

The single particle dynamics is described with the following motion equations:

$$\frac{d\vec{r}}{dt} = \vec{v}, \qquad \frac{d(\gamma m_e \vec{v})}{dt} = -e(\vec{E} + \vec{v} \times \vec{B}) \tag{7}$$

Where $\gamma = [1 - (v/c)^2]^{-1/2}$ is the relativistic factor, $-e$ and m_e are the charge and mass of an electron, $\vec{E} = \vec{E}^{hf}$ and $\vec{B} = \vec{B}^{hf} + \vec{B}^{coils}$ are the total electric and magnetic fields at the electron position, respectively. The equation (7) is solved numerically in a finite difference scheme by using the Boris method, which let us calculate the electron position, particle energy and longitudinal velocity at any time step.

In the simulations all the quantities are taken in dimensionless form: $\vec{E}^* = \vec{E}/(-B_0 c)$, $\vec{B}^* = \vec{B}/(-B_0)$, $\vec{r}^* = \vec{r}/r_{l0}$ and $t^* = \omega t$; where: $B_0 = m_e\omega/e$ is the magnetic field corresponding to the classic cyclotron resonance for the electrons and $r_{l0} = c/\omega$ is the relativistic Larmor radius.

3. Results and discussion

The numerical simulations is run under the above-mentioned physical parameters. The Figure 3(a) shows the spatial evolution of the energy of a single electron injected with an energy of $20keV$ into the cavity for three different electric field strength of the TE_{113} mode: $7.2kV/cm$ (red line), $7.4kV/cm$ (green line) and $7.6kV/cm$ (blue line). It is shown that the electron can be accelerated up to energies of the order of $300keV$ for the 3 cases.

The Figure 3(b) shows the evolution of the phase-shift between the electric field and the transversal velocity of the electron when the electron advances along the z-direction. In the injection position $z = 0$, where the electric field has a node, the phase-shift φ acquire the value $\pi/2$ due to the deviation produced by the magnetic field component of the microwave field. Then the phasing is produced by the electric field up to the exact resonance $\varphi = \pi$ in the position $z = 6cm$. Despite of the electron acceleration is not produced in exact resonance, the phase-shift is maintained in the range $\pi/2 < \varphi < 3\pi/2$. In such range the microwave electric field can transfer power efectively to the electron, in this sense this interval is known as *acceleration band* [7].

There is some intervals in the Figure 3(b) where the phase-shift is out of the *acceleration band*, e.g. for the blue line case these intervals are $10 - 12.5cm$, $19 - 20cm$ and $25 - 29cm$. It is easy to check that the electron energy decrease in these intervals (See Figure 3(a)). We can note that the phase-shift changes a π value ($\varphi \rightarrow \varphi + \pi$) in the positions $z = 10cm$ and $z = 20cm$ because the electric field of TE_{113} has nodes in such points.

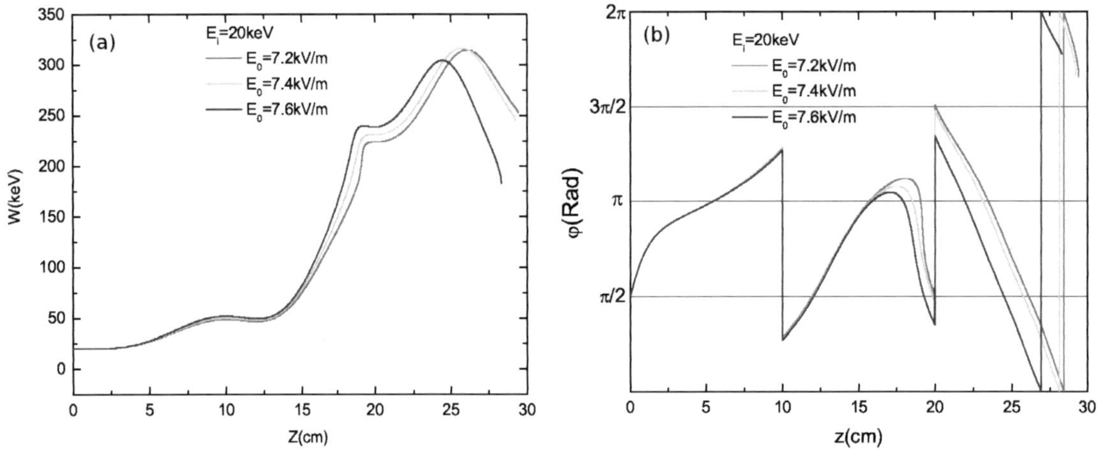

Figure 3. (a) Spatial evolution of the energy of a single electron for different electric field strength of the TE_{113} mode. (b) The corresponding phase-shift between the electric field and the transversal velocity of the electron.

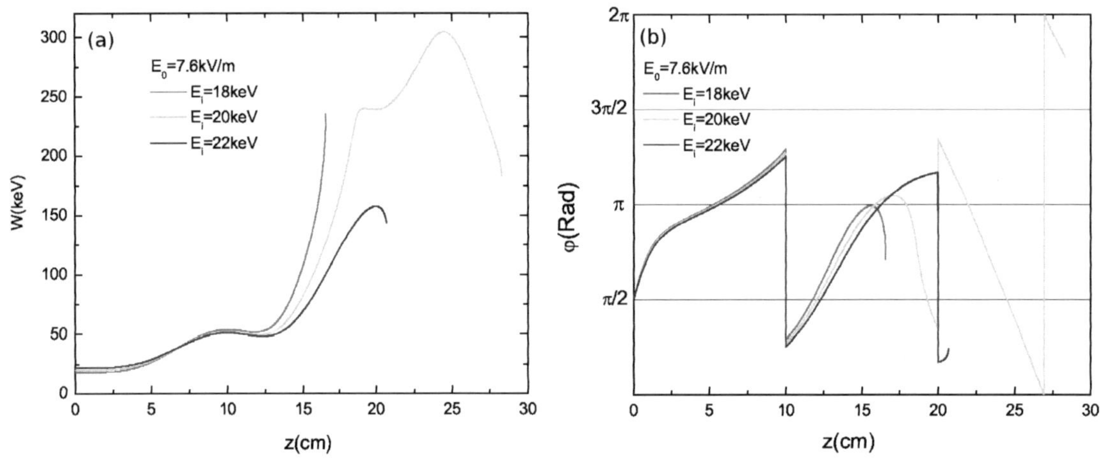

Figure 4. (a) Spatial evolution of the energy of a single electron for different injection energies. (b) The corresponding phase shift between the electric field and the transversal velocity of the electron.

The Figure 4(a) shows the spatial evolution of the energy of a single electron for different injection energies while Figure 4(b) the corresponding phase shift between the electric field and the transversal velocity of the electron by using a electric field strength $E_0 = 7.6kV/cm$ for all cases. The relation between these graphics can be understood using a similar analysis as was previously described for the Figure 3. We can see that the electron injected with an energy of $20keV$ advances a greater longitudinal distance, thereby prolonging the resonant interaction with the microwave field and increasing its energy. To understand this behaviour let us check the Figure 5, which shows the longitudinal velocity for the case considered in the Figure 4. For all energies injections appears a diamagnetic force which acts in the direction opposite to the magnetic field gradient and impedes the advance of the electrons into a higher magnetic field. However for the green line case the electron accelerates longitudinally in the reange 18 − 22cm (see Figure 5). This effect is attributed to the magnetic component of the microwave field whic depend of the phase-shift φ and z coordinate of each electron.

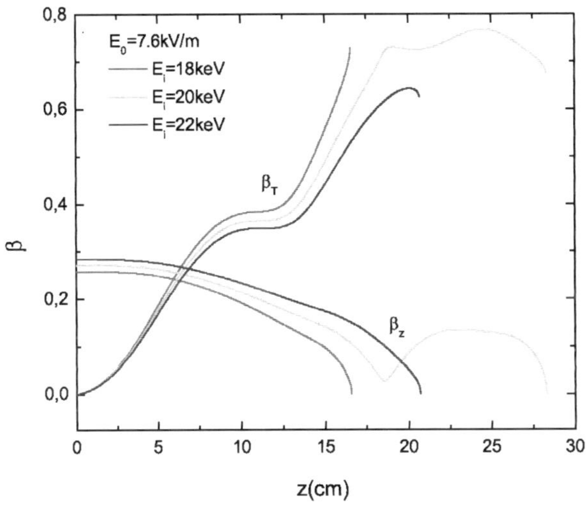

Figure 5. Transversal ($\beta_T = v_T/c$) and longitudinal ($\beta_Z = v_z/c$) velocities for the case presented in Figure 4.

4. Conclusions

The realized numerical experiment shows that electrons can be accelerated up to energies of the order of $300keV$ in spatial autoresonance acceleration conditions by using a TE_{113} mode. This results could be usefull to design a hard X-ray device based on de SARA concept which have many applications, e.g. in airport security.

Acknowledgments

This work was partially financed by the "Universidad del Magdalena (Fonciencias)" and the Colombian agency Colciencias through doctoral scholarships 567.

References

[1] A Kolomenskii and A Levedev 1962 *Dokl Akad Nauk USSR* **145** 1251
[2] V Davydovskii 1962 *Zh Eksp. Teor. Fiz.* **43** 886
[3] V Milantiev 2013 *Physics Uspekhi* **56** 823
[4] K S Golovanivsky 1983 *IEEE Trans. PLasma Sci.* **11** 28
[5] Gal 1989 *IEEE Trans. PLasma Sci.* **17** 622
[6] R Geller and K S Golovanivsky 1999 *Nucl. Instrum. Methods Phys. Res. Sect. B* **68** 479
[7] V Dugar-Zhabon, E A Orozco and A Umnov 2008 *Phys. Rev. Spec. Top. Acceler. and Beams* **11** 041302 1-5
[8] V Dugar-Zhabon and E A Orozco 2009 *Phys. Rev. Spec. Top. Acceler. and Beams* **12** 041301 1-8
[9] V Dugar-Zhabon and E A Orozco 2010 *IEEE Transaction on Plasma Science* **38(10)** 2980
[10] V Dugar-Zhabon and E A Orozco 2009 *Journal of Plasma Fusion Research Series* **8** 473-476
[11] H Gutierrez, E A Orozco and V Dugar-Zhabon 2011 *Rev. Colombiana de Física* **44(3)** 220-224
[12] V Dugar-Zhabon, E A Orozco and A M Herrera 2016 *Journal of Physics: Conference Series* **687** 012076
[13] V Dugar-Zhabon, J D González and E A Orozco 2016 *Journal of Physics: Conference Series* **687** 012077
[14] V D Dugar-Zhabon and E A Orozco Ospino 2015 *Compact self-resonant X-ray source* (USA: Patent office)

Institute of Physics
Dirac House, Temple Back
Bristol BS1 6BE UK

ISSN: 1742-6588
ISBN 978-1-5108-5516-8

9 781510 855168

CURRAN ASSOCIATES INC.
proceedings
.com